REPRESENTATIONS OF REAL AND p-ADIC GROUPS

LECTURE NOTES SERIES
Institute for Mathematical Sciences, National University of Singapore

Series Editors: Louis H. Y. Chen and Yeneng Sun
Institute for Mathematical Sciences
National University of Singapore

Published

Vol. 1 Coding Theory and Cryptology
edited by Harald Niederreiter

Vol. 2 Representations of Real and p-Adic Groups
edited by Eng-Chye Tan & Chen-Bo Zhu

Lecture Notes Series, Institute for Mathematical Sciences,
National University of Singapore
Vol. 2

REPRESENTATIONS OF REAL AND p-ADIC GROUPS

Editors

Eng-Chye Tan
Chen-Bo Zhu
National University of Singapore

SINGAPORE UNIVERSITY PRESS
NATIONAL UNIVERSITY OF SINGAPORE

World Scientific

Published by

Singapore University Press
Yusof Ishak House, National University of Singapore
31 Lower Kent Ridge Road, Singapore 119078

and

World Scientific Publishing Co. Pte. Ltd.
5 Toh Tuck Link, Singapore 596224
USA office: Suite 202, 1060 Main Street, River Edge, NJ 07661
UK office: 57 Shelton Street, Covent Garden, London WC2H 9HE

British Library Cataloguing-in-Publication Data
A catalogue record for this book is available from the British Library.

REPRESENTATIONS OF REAL AND *p*-ADIC GROUPS

ISBN 981-238-779-X

Printed in Singapore by World Scientific Printers (S) Pte Ltd

Contents

Foreword vii

Preface ix

Three Uncertainty Principles for an Abelian Locally Compact Group
Tomasz Przebinda 1

Lectures on Representations of p-adic Groups
Gordan Savin 19

Lectures on Harmonic Analysis for Reductive p-adic Groups
Stephen DeBacker 47

On Classification of Some Classes of Irreducible Representations of Classical Groups
Marko Tadić 95

Dirac Operators in Representation Theory
Jing-Song Huang and Pavle Pandžić 163

On Multiplicity Free Actions
Chal Benson and Gail Ratcliff 221

Multiplicity-Free Spaces and Schur–Weyl–Howe Duality
Roe Goodman 305

Foreword

The Institute for Mathematical Sciences at the National University of Singapore was established on 1 July 2000 with funding from the Ministry of Education and the University. Its mission is to provide an international center of excellence in mathematical research and, in particular, to promote within Singapore and the region active research in the mathematical sciences and their applications. It seeks to serve as a focal point for scientists of diverse backgrounds to interact and collaborate in research through tutorials, workshops, seminars and informal discussions.

The Institute organizes thematic programs of duration ranging from one to six months. The theme or themes of each program will be in accordance with the developing trends of the mathematical sciences and the needs and interests of the local scientific community. Generally, for each program there will be tutorial lectures on background material followed by workshops at the research level.

As the tutorial lectures form a core component of a program, the lecture notes are usually made available to the participants for their immediate benefit during the period of the tutorial. The main objective of the Institute's Lecture Notes Series is to bring these lectures to a wider audience. Occasionally, the Series may also include the proceedings of workshops and expository lectures organized by the Institute. The World Scientific Publishing Company and the Singapore University Press have kindly agreed to publish jointly the Lecture Notes Series. This volume on "Representations of Real and p-Adic Groups" is the second of this Series. We hope that through regular publication of lecture notes the Institute will achieve, in part, its objective of promoting research in the mathematical sciences and their applications.

February 2004

Louis H. Y. Chen
Yeneng Sun
Series Editors

Preface

A semester-long program on Representation Theory of Lie Groups was held at the Institute for Mathematical Sciences (IMS) at the National University of Singapore (NUS) from July 2002 to January 2003. This is the third research program of IMS since it started operations in July 2001. The goal of the program was to explore recent advances in representation theory of both real and p-adic groups. The program had three sub-themes: (i) representation of p-adic groups; (ii) unitary representations of real reductive groups; (iii) multiplicity-free actions and representations.

As part of the program, tutorials related to the sub-themes were conducted by leading experts in the fields. These tutorials covered the fundamentals of representation theory as well as some of its recent developments and were meant for graduate students and researchers who would like to prepare themselves for original research in the fields. The current volume collects the expanded lecture notes of these tutorials. In the following, we give a brief indication of the ranges of topics represented in this volume.

The first five articles are on the sub-themes (i) and (ii). The article by Tomasz Przebinda directs us to some recent developments in harmonic analysis on locally compact abelian groups, specifically three Uncertainty Principles. In Gordan Savin's notes, he gives an elementary and elegant introduction, with exercises, to representations of p-adic reductive groups. The material represents real classics from the past quarter century. Stephen DeBacker's article examines some distinguished classes of distributions on reductive p-adic groups and their Lie algebras and discusses one of the deepest connections between them: the Harish-Chandra-Howe local character expansion (and its ramifications). The lecture notes of Marko Tadić deal with the problem of classification of some important classes of irreducible representations of both p-adic and real classical groups (with more emphasis on p-adic groups), in particular the classification of the square-integrable representations modulo cuspidal data. Jing-Song Huang and Pavle Pandžić's article examines the role Dirac operators play in representation theory and discusses various topics surrounding their proof of the Vogan's conjecture on Dirac cohomology, an important development in the algebraic aspects of unitary representations.

The last two articles in this volume are on the sub-theme of multiplicity-free actions. Here Chal Benson and Gail Ratcliff present a comprehensive treatment of these actions, covering topics such as the classification of linear multiplicity-free actions and eigenvalues of invariant differential operators. Finally Roe Goodman's lecture notes explore the fundamental duality between the irreducible representations occuring in a linear group action and irreducible representations of the commuting algebra, which of course, is a recurrent theme in many parts of representation theory.

Besides the local organizers (us), the other members of the Organizing Committee are Jian-Shu Li (Hong Kong University of Science and Technology and Co-Chairman), Jeffrey Adams (University of Maryland), Kyo Nishiyama (Kyoto University), Dipendra Prasad (Mehta Research Institute) and Gordan Savin (University of Utah). We are very much grateful to their invaluable services. We would also like to express our deep appreciation to Roger Howe (Yale University and Chairman, Scientific Advisory Board of the IMS) who helped to conceptualize scientific aspects of the program and was equally valuable in many other aspects throughout this program. Thanks also to all the participants of this program for their support and stimulating interactions during those few months!

We would like to take this opportunity to thank Louis Chen, Director of IMS, for his leadership in creating an exciting environment for mathematical research in IMS and for his guidance throughout our program. The expertise and dedication of all IMS staff contributed essentially to the success of this program. Financial support to the program was generously provided by the IMS and by a grant from the Faculty of Science, NUS. Last but not least, we would like to record our appreciation to the Department of Mathematics, NUS for freeing some of our obligations during this period, and more importantly for providing all kinds of support to our work ever since we joined as faculty members more than a decade ago.

Eng-Chye Tan and Chen-Bo Zhu
National University of Singapore
Singapore

Three Uncertainty Principles for an Abelian Locally Compact Group

Tomasz Przebinda
Department of Mathematics
University of Oklahoma
Norman, OK 73019, USA
E-mail: tprzebinda@math.ou.edu

0. Introduction

The purpose of this article is to direct reader's attention to some recent developments concerning three Uncertainty Principles: the classical Heisenberg-Weyl Uncertainty Principle, The Hirschman-Beckner Uncertainty Principle based on the notion of the entropy, and the Donoho-Stark Uncertainty Principle. We state all three in the first section with some further explanations and proofs in the following sections.

1. The main results

Let $S(\mathbb{R})$ denote the Schwartz space on $\mathbb{R}$, [10], and let $f \in S(\mathbb{R})$ be a real valued function. Then, by the Cauchy inequality,

$$\int_{\mathbb{R}} |xf(x)|^2\,dx \cdot \int_{\mathbb{R}} |f'(x)|^2\,dx \geq \left(\int_{\mathbb{R}} xf(x)f'(x)\,dx\right)^2.$$

Moreover,

$$\int_{\mathbb{R}} xf(x)f'(x)\,dx = \int_{\mathbb{R}} x\frac{1}{2}\left(f(x)^2\right)'\,dx$$
$$- x\frac{1}{2}f(x)^2\Big|_{-\infty}^{\infty} - \int_{\mathbb{R}} \frac{1}{2}f(x)^2\,dx = -\frac{1}{2}\parallel f \parallel_2^2.$$

Thus

$$\int_{\mathbb{R}} |xf(x)|^2\,dx \cdot \int_{\mathbb{R}} |f'(x)|^2\,dx \geq \frac{\parallel f \parallel_2^4}{4}. \tag{1}$$

If the equality holds in (1), then there is a real number a such that

$$f'(x) = -2axf(x),$$

so that

$$f(x) = \pm e^{-ax^2-c}, \tag{2}$$

where $c \in \mathbb{R}$. Since $\| f \|_2 < \infty$, the number a must be positive. Let us write f as the inverse Fourier transform of the Fourier transform $\hat{f}$ of f:

$$f(x) = \int_{\mathbb{R}} e^{2\pi i x\xi} \hat{f}(\xi)\, d\xi.$$

Then

$$f'(x) = \int_{\mathbb{R}} 2\pi i \xi\, e^{2\pi i x\xi} \hat{f}(\xi)\, d\xi$$

and therefore

$$\int_{\mathbb{R}} |f'(x)|^2\, dx = \int_{\mathbb{R}} |2\pi i \xi \hat{f}(\xi)|^2\, d\xi = 4\pi^2 \int_{\mathbb{R}} |\xi \hat{f}(\xi)|^2\, d\xi.$$

Thus for $\| f \|_2 = 1$,

$$\int_{\mathbb{R}} |xf(x)|^2\, dx \cdot \int_{\mathbb{R}} |\xi \hat{f}(x)|^2\, dx \geq \frac{1}{16\pi^2}. \tag{3}$$

A few more easy steps lead to the following theorem of Herman Weyl, (see [20]).

Theorem 1.1: (H. Weyl, 1931) *Let $f \in S(\mathbb{R})$, with $\| f \|_2 = 1$. Set*

$$\mu = \int_{\mathbb{R}} x|f(x)|^2\, dx, \quad \sigma^2 = \int_{\mathbb{R}} (x-\mu)^2 |f(x)|^2\, dx,$$
$$\tilde{\mu} = \int_{\mathbb{R}} \xi |\hat{f}(\xi)|^2\, d\xi, \quad \tilde{\sigma}^2 = \int_{\mathbb{R}} (\xi - \tilde{\mu})^2 |\hat{f}(\xi)|^2\, d\xi.$$

Then

$$\sigma \cdot \tilde{\sigma} \geq \frac{1}{4\pi},$$

and the equality occurs if and only if

$$f(x) = e^{-ax^2-bx-c},$$

where $a > 0$, and $b, c \in \mathbb{C}$. In other words f is a constant multiple of a translation

$$f(x) \to f(x + x_0)$$

and a modulation

$$f(x) \to e^{iy_0 x} f(x)$$

of a Gaussian

$$e^{-ax^2}.$$

The above theorem states that a function and its Fourier transform cannot both be arbitrarily concentrated. In computer applications one deals often with finite cyclic groups, rather than with the real line. In this context there is no obvious or straightforward generalization of the theorem 1.1, because the quantities involved $(\mu, \sigma, ...)$ do not seem to make sense. In order to circumvent this difficulty, Donoho and Stark, [7], have introduced a different, elementary, measure of uncertainty, which we explain below.

Let $A = \{0, 1, 2, 3, ..., N-1\}$ be a finite cyclic group of order $|A| = N$. Let $\hat{A}$ denote the dual group of all the characters (i.e. group homomorphisms $\hat{a} : A \to \mathbb{C}^\times$). It is customary, and sometimes convenient, to identify A with $\hat{A}$ by the formula

$$\hat{a}(b) = e^{\frac{2\pi i}{N} ab} \qquad (a, b \in A).$$

For a function $f : A \to \mathbb{C}$ define a Fourier transform of f by

$$\hat{f}(\hat{b}) = \sum_{a \in A} f(a)\hat{b}(-a) \qquad (\hat{b} \in \hat{A}).$$

Theorem 1.2: (Donoho-Stark, 1989, [7]) *For any non-zero function* $f : A \to \mathbb{C}$,

$$|supp\, f| \cdot |supp\, \hat{f}| \geq |A|.$$

The equality occurs if and only if f is a constant multiple of a translation

$$f(a) \to f(a + c)$$

and a modulation

$$f(a) \to \hat{b}(a) f(a)$$

of the indicator function of a subgroup of A.

In applications one deals often with multi-dimensional signals, i.e. with functions defined on a finite product of finite cyclic groups. Hence it is natural to ask for a generalization of tho theorem 1.2 to this more general context. This has been done by K. Smith.

Theorem 1.3: (K. Smith, 1990, [19]) *The above theorem 1.2 holds for any finite Abelian group A (a finite direct product of finite cyclic groups).*

The theorem 1.3 also follows from theorem 1.10 below. In addition, a proof which uses no more than basic concepts from finite dimensional linear algebra over complex numbers and the structure of finite abelian groups is available in [12].

Clearly, the cardinality of the support is not the most precise measure of the concentration of a function. One possible improvement is based on the notion of entropy.

Definition 1.4: (based on Shannon, 1948, [18]) Let μ be a non-negative measure on a measure space M. Let $\phi : M \to [0, \infty)$ be a probability density function, i.e.

$$\int_M \phi(x)\, d\mu(x) = 1.$$

The entropy of ϕ is defined as

$$H(\phi) = -\int_M \phi(x)\, log(\phi(x))\, d\mu(x),$$

where the *log* stands for the natural logarithm, whenever the integral converges.

In his 1957 paper, [9], Hirschman has proven the following theorem and stated the following conjecture.

Theorem 1.5: *Let $f \in S(\mathbb{R})$, with $\| f \|_2 = 1$. Then*

$$H(|f|^2) + H(|\hat{f}|^2) \geq 0.$$

Conjecture 1.6: *Let $f \in S(\mathbb{R})$, with $\| f \|_2 = 1$. Then*

(a) $$H(|f|^2) + H(|\hat{f}|^2) \geq log(\tfrac{e}{2}).$$

(Notice that since $\frac{e}{2} > 1$, $log(\frac{e}{2}) > 0$.)

(b) *The equality holds in (a) if and only if f is a constant multiple of a translation and a modulation of a Gaussian e^{-ax^2}, $a > 0$.*

As an application of his L^p, L^q estimates for the Fourier transform, Beckner proved part (a) of Hirschman's conjecture.

Theorem 1.7: (Beckner, 1975, [2], [3]) *Part (a) of Hirschman's Conjecture is true.*

Theorem 1.8: (Özaydın-Przebinda, 2000, [16]) *Part (b) of Hirschman's Conjecture is true.*

Let A be a finite Abelian group and let α be a Haar measure on A. Thus α is a positive constant multiple of the counting measure on A. Let $\hat{\alpha}$ be the dual Haar measure on the dual group $\hat{A}$, in the sense that for a function $f : A \to \mathbb{C}$, the Fourier transform and the inverse Fourier transform are given by

$$\begin{aligned} \hat{f}(\hat{b}) &= \int_A f(a)\hat{b}(-a)\, d\alpha(a), \\ f(a) &= \int_{\hat{A}} \hat{f}(\hat{b})\hat{b}(a)\, d\hat{\alpha}(\hat{b}). \end{aligned} \tag{4}$$

The following theorem is known. For the idea of a proof, various particular cases and generalizations see [9], [14], [13] and [6].

Theorem 1.9: *Let $f \in L^2(A, \alpha)$, with $\| f \|_2 = 1$. Then*

$$H(|f|^2) + H(|\hat{f}|^2) \geq 0.$$

Theorem 1.10: (Özaydın-Przebinda, 2000, [16]) *The equality holds in the above theorem 1.9 if and only if f is a constant multiple of a translation and a modulation of the indicator function of a subgroup of A.*

Corollary 1.11: (DeBrunner-Özaydın-Przebinda, 2001, [17]) *The discretization of the minimizers for the entropy inequality on $\mathbb{R}$ does not give minimizers for the entropy inequality on any finite cyclic group A.*

In other words, it is certain that no discretization of a signal defined on the real line and best concentrated in the time-frequency plane, will lead to a signal defined on a finite cyclic group which is also best concentrated in the corresponding finite time-frequency plane.

In a mathematically natural search for the most general theorem, which would generalize all the cases considered above, we arrive at the notion of a locally compact Abelian group, (see [8]). The integration and the notion of the Fourier transform are both well established for such groups.

Let A be a locally compact Abelian group. As was explained to the author by Michael Cowling, a result of Ahern and Jeweet [1], together with [8], 9.8, imply that A is isomorphic to the direct product of a finite number of copies of $\mathbb{R}$ and an Abelian locally compact group B, which contains an open compact subgroup:

$$A = \mathbb{R}^n \times B. \tag{5}$$

Let $\hat{A}$ be the Pontryagin dual of A. Then $\hat{A} = \mathbb{R}^n \times \hat{B}$, where $\hat{B}$ also contains an open compact subgroup. Let α be a Haar measure on A and let $\hat{\alpha}$ be the

Haar measure on $\hat{A}$, dual to α, so that the Fourier transform and the inverse Fourier transform are given by the formulas (4). Let $U(L^2(A,\alpha))$ be the group of unitary (norm preserving) operators on the Hilbert space $L^2(A,\alpha)$, and let $G \subseteq U(L^2(A,\alpha))$ be the group generated by all the translations, all modulations and by the multiplications by complex numbers of absolute value 1. This is the Heisenberg group attached to the Abelian group A.

Theorem 1.12: (Özaydın-Przebinda, 2000, [16]) *For any function* $f \in L^2(A,\alpha)$, *with* $\| f \|_2 = 1$, *such that*

$$(*) \qquad \| f \|_1 < \infty \quad \text{and} \quad \| \hat{f} \|_1 < \infty,$$

the following inequality holds

$$(a) \qquad H(|f|^2) + H(|\hat{f}|^2) \geq n \, log(\tfrac{e}{2}).$$

The set of functions for which equality occurs in (a) coincides with the union of orbits

$$(b) \qquad G \cdot f,$$

where $f = g \otimes h$, g *is a Gaussian on* $\mathbb{R}^n$, *and* h *an appropriate constant multiple of the indicator function of a (open-compact) subgroup of* B.

2. The support inequality for a finite Abelian group

In this section A stands for a finite Abelian group and α for the Haar measure given by $\alpha(a) = \frac{1}{\sqrt{|A|}}$, $a \in A$. Denote the translations and modulations by

$$\begin{aligned} T_c &: f(a) \to f(a+c), \\ M_{\hat{b}} &: f(a) \to \hat{b}(a) f(a). \end{aligned} \tag{6}$$

Notice that

$$T_b M_{\hat{b}} T_c M_{\hat{c}} = z M_{\hat{b}\hat{c}} T_{b+c} \qquad (b, c \in A, \; \hat{b}, \hat{c} \in \hat{A}), \tag{7}$$

where z is a complex number of absolute value 1. Indeed, the action of the operator on the left hand side on a function $f(a)$ may be represented as

$$f(a) \xrightarrow{M_{\hat{c}}} \hat{c}(a) f(a) \xrightarrow{T_c} \hat{c}(a+c) f(a+c) \xrightarrow{M_{\hat{b}}} \hat{b}(a) \hat{c}(a+c) f(a+c)$$

$$\xrightarrow{T_b} \hat{b}(a+b) \hat{c}(a+b+c) f(a+b+c) = (\hat{b}(b)\hat{c}(b+c))(\hat{b}(a)\hat{c}(a)) f(a+(b+c)).$$

In particular the Heisenberg group attached to A may be written as

$$G = \{ z M_{\hat{b}} T_c; \; \hat{b} \in \hat{A}, \; c \in A \}. \tag{8}$$

It is esthetically pleasing, and convenient for applications, to know that an orthonormal basis of a space of functions may be indexed by a group. This is the case for an orthonormal wavelet basis for functions defined on the real line, [5], [15]. The lemma below states that such bases may be chosen from among our minimizers on a finite Abelian group. In contrast, the translations and/or modulations of Gaussians defined on the reals, will not be orthogonal.

Lemma 2.1: (see [7], [17] for the cyclic case) *Let B be a subgroup of A and let $\mathbb{I}_B$ denote the indicator function of B. Let $\beta \in \mathbb{C}$ be such that $\| \beta\mathbb{I}_B \|_2 = 1$. Then, up to constant multiples of absolute value 1, the orbit $G \cdot \beta\mathbb{I}_B$ is an orthonormal basis of the Hilbert space $L^2(A, \alpha)$.*

Proof: For $c \in A$ and $\hat{c} \in \hat{A}$ we have

$$M_{\hat{c}}T_c\beta\mathbb{I}_B(a) = \hat{c}(a)\beta\mathbb{I}_B(a+c) = \hat{c}(a)\beta\mathbb{I}_{B-c}(a).$$

Hence, in order to parametrize the orbit, we may choose the $\hat{c}$ modulo $B^{\perp}(= \{\hat{a} \in \hat{A};\ \hat{a}|_B = 1\})$, and the c modulo B. Thus the cardinality of the orbit, modulo the multiplications by complex numbers of absolute value 1, is equal to

$$|\hat{A}/B^{\perp}| \cdot |A/B| = |B| \cdot |A/B| = |A| = dim\, L^2(A, \alpha).$$

Let $d \in A$ and let $\hat{d} \in \hat{A}$. Then,

$$\int_A M_{\hat{c}}T_c\beta\mathbb{I}_B(a) \cdot \overline{M_{\hat{d}}T_d\beta\mathbb{I}_B(a)}\, d\alpha(a) = |\beta|^2 \int_A \hat{c}(a)\overline{\hat{d}(a)}\mathbb{I}_{B-c}(a)\mathbb{I}_{B-d}(a)\, d\alpha(a).$$

For this quantity to be non-zero, we must have $B - c = B - d$. Then it is equal to

$$|\beta|^2 \int_{B-c} \hat{c}(a)\overline{\hat{d}(a)}\, d\alpha(a) = |\beta|^2 \int_B \hat{c}(a+c)\overline{\hat{d}(a+c)}\, d\alpha(a)$$
$$= |\beta|^2 \int_B \hat{c}(a)\hat{d}(a)^{-1}\, d\alpha(a) \cdot \hat{c}(c)\hat{d}(c)^{-1},$$

which is non-zero if and only if $\hat{c}\hat{d}^{-1} \in B^{\perp}$. Then, modulo multiplication by the constant $\hat{c}(c)\hat{d}(c)^{-1}$, the result is

$$|\beta|^2 \int_B d\alpha(a) \| \beta\mathbb{I}_B \|_2^2 = 1.$$ □

Corollary 2.2: (see [17] for the cyclic case) *In terms of Lemma 2.1, the orbit $G \cdot \beta\mathbb{I}_B$ consists of functions such that*

(a) $$H(|f|^2) = H(|\hat{f}|^2)$$

or equivalently, such that

(b) $$|supp\, f| = |supp\, \hat{f}|,$$

if and only if

(c) $$|B| = \sqrt{|A|}.$$

Proof: Since the entropy and the cardinality of the support for f and for $\hat{f}$ are invariant under the action of the Heisenberg group G, it suffices to consider $f = \beta \mathbb{I}_B$. Then, by a straightforward computation,

$$\hat{f} = \alpha(B)\beta\mathbb{I}_{B^\perp}.$$

Hence, $|supp\, \hat{f}| = |B^\perp| = |A|/|B|$, and since $|supp\, f| = |B|$, the equivalence of (b) and (c) follows.

Furthermore

$$H(|\beta\mathbb{I}_B|^2) = -\int_A |\beta|^2\mathbb{I}_B(a)^2\, log(|\beta|^2\mathbb{I}_B(a)^2)\, d\alpha(a) = -|\beta|^2 log(|\beta|^2)\,\alpha(B),$$

and similarly,

$$H(|\beta\alpha(B)\mathbb{I}_{B^\perp}|^2) = -|\beta|^2|\alpha(B)|^2 log(|\beta|^2|\alpha(B)|^2)\,\hat{\alpha}(B^\perp).$$

Thus (a) is equivalent to

$$log(|\beta|^2) = log(|\beta|^2|\alpha(B)|^2)\alpha(B)\hat{\alpha}(B^\perp).$$

Since, by a simple computation,

$$\alpha(B)\hat{\alpha}(B^\perp) = 1,$$

we see that (a) is equivalent to $\alpha(B)^2 = 1$. But, by our assumption on α, $\alpha(B) = |B|/\sqrt{|A|}$, so (a) is equivalent to (c). □

For reader's convenience we reproduce here a lemma which is the key to the proof of theorem 1.2, leaving the complete proof of the theorem as an exercise. (See [17] for an exposition.)

Lemma 2.3: [7] *Let $f : A \to \mathbb{C}$ be a non-zero function. Then under the identification $\hat{A} = A$, $\hat{a}(b) = e^{\frac{2\pi i}{|A|}ab}$, the function $\hat{f}$ cannot have more than $|supp\, f|$ consecutive zeros.*

Proof: Let $supp\, f = \{a_1, a_2, a_3, ..., a_m\}$. Let $\omega = e^{-\frac{2\pi i}{|A|}}$. Then

$$\begin{bmatrix} \hat{f}(0) \\ \hat{f}(1) \\ . \\ . \\ . \\ \hat{f}(m-1) \end{bmatrix} = \begin{bmatrix} 1 & 1 & .. & 1 \\ \omega^{a_1} & \omega^{a_2} & .. & \omega^{a_m} \\ . & . & .. & . \\ . & . & .. & . \\ . & . & .. & . \\ \omega^{a_1(m-1)} & \omega^{a_2(m-1)} & .. & \omega^{a_m(m-1)} \end{bmatrix} \cdot \begin{bmatrix} \hat{f}(a_1) \\ \hat{f}(a_2) \\ . \\ . \\ . \\ \hat{f}(a_m) \end{bmatrix}.$$

By Vandermonde, the above matrix is invertible. Thus the vector on the left hand side is non-zero. Since a translation of $\hat{f}$ is equivalent to a modulation of f, which does not change the support of f, the lemma follows. □

In the remainder of this section we provide an elementary proof of the part of theorem 1.3, which describes the minimizers for the support inequality (which we assume), if A is a finite direct product of groups of order 2: $A = (\mathbb{Z}/2\mathbb{Z})^N$. (See theorem 2.5 below.) This proof is a simplified version of the argument developed in [12].

Lemma 2.4: *Let B and C be subgroups of A, such that*

$$A = B \oplus C.$$

For a function $f : A \to \mathbb{C}$ let

$$f_c(b) = f(b + c) \qquad (b \in B,\ c \in C).$$

Suppose

$$|supp\, f| \cdot |supp\, \hat{f}| = |A| \quad and \quad f_c \neq 0 \quad for\ all \quad c \in C.$$

Then

(a) $$|supp\, f_c| \cdot |supp\, \hat{f}_c| = |B| \qquad (c \in C),$$

(b) $$|supp\, f_c| = \tfrac{1}{|C|}|supp\, f| \qquad (c \in C),$$

(c) $$|supp\, \hat{f}_c| = |supp\, \hat{f}| \qquad (c \in C),$$

(d) $$supp\, \hat{f}_c = P(supp\, \hat{f}) \qquad (c \in C),$$

where $P : \hat{A} = \hat{B} \oplus \hat{C} \ni \hat{b} + \hat{c} \to \hat{b} \in \hat{B}$.

Proof: Clearly,

$$supp\, f = \bigcup_{c \in C} ((supp\, f_c) + c). \tag{9}$$

Hence,

$$|supp\, f| = \sum_{c\in C} |supp\, f_c|. \tag{10}$$

By the Fourier inversion formula on C we have

$$\hat{f}_c(\hat{b}) = \frac{1}{\sqrt{|C|}} \sum_{\hat{c}\in\hat{C}} \hat{f}(\hat{b}+\hat{c})\hat{c}(c) \qquad (\hat{b}\in\hat{B},\ c\in C). \tag{11}$$

Hence,

$$supp\, \hat{f}_c \subseteq P(supp\, \hat{f}) \qquad (c\in C), \tag{12}$$

and therefore

$$|supp\, \hat{f}_c| \le |supp\, \hat{f}| \qquad (c\in C). \tag{13}$$

Suppose $c\in C$ is such that

$$|C|\cdot|supp\, f_c| < |supp\, f|.$$

Then, by (13),

$$|C|\cdot|supp\, f_c|\cdot|supp\, \hat{f}_c| < |supp\, f|\cdot|supp\, \hat{f}| = |A|.$$

Hence,

$$|supp\, f_c|\cdot|supp\, \hat{f}_c| < |B|,$$

which contradicts the fact that $f_c \neq 0$. (Here we use the assumption that we have the support inequality for the finite cyclic groups of the form $(\mathbb{Z}/2\mathbb{Z})^N$.) Thus

$$|C|\cdot|supp\, f_c| \ge |supp\, f| \qquad (c\in C). \tag{14}$$

Clearly, (10) and (14) imply (b). Further (b) and (13) imply that for all $c\in C$,

$$\begin{aligned}|supp\, f_c|\cdot|supp\, \hat{f}_c| &= \frac{1}{|C|}|supp\, f|\cdot|supp\, \hat{f}_c| \\ &\le \frac{1}{|C|}|supp\, f|\cdot|supp\, \hat{f}| = |A|/|C| = |B|.\end{aligned}$$

Thus (a) follows. Part (c) follows from (a) and (b). Part (d) follows from (12) and (c). □

Theorem 2.5: *Let $A = (\mathbb{Z}/2\mathbb{Z})^N$. Suppose $f : A \to \mathbb{C}$ is a minimizer, i.e.*

$$|supp\, f|\cdot|supp\, \hat{f}| = |A|.$$

Then, up to a constant multiple, a modulation and a translation, f is the indicator function of a subgroup of A.

Proof: We proceed via the induction on N. It is easy to check that the statement holds for $N = 2$.

Let $f : A \to \mathbb{C}$ be a minimizer. Suppose there is a proper subgroup $B \subseteq A$, such that $supp\, f \subseteq B$. Then there is a subgroup $C \subseteq A$ such that $A = B \oplus C$. (In order to have some standard linear algebra at the disposal, it might be easier here to view A a vector space of dimension N over the field $\mathbb{Z}/2\mathbb{Z}$ of two elements.) In these terms

$$f = (f|_B) \otimes \delta,$$

where δ is the Dirac delta at zero on C. Hence

$$\hat{f} = (\widehat{f|_B}) \otimes \hat{\delta},$$

where $\hat{\delta}$ is a constant multiple of the function $\mathbb{I}_C$. We see from the above two equations that $f|_B$ is a minimizer on B. Thus, by the inductive assumption $f|_B$, and hence f has the desired form.

Suppose there is a proper subgroup $B \subseteq A$ and an element $c \in A$ such that $supp\, f \subseteq B + c$. Let $g(a) = f(a + c)$, $a \in A$. Then g is a minimizer on A and $supp\, g \subseteq B$. Hence, by the previous argument, g and hence f has the desired form.

Let $A = B \oplus C$ as in Lemma 2.4, with $|C| = 2$. By the previous two cases we may assume that

$$f_c \neq 0 \qquad (c \in C).$$

By Lemma 2.4 (a), each f_c is a minimizer on B. Therefore, by the inductive assumption, for each $c \in C$, $supp\, \hat{f}_c$ is a coset of a subgroup of $\hat{B}$. By Lemma 2.4 (d) these cosets do not depend on $c \in C$. Thus there is a subgroup $D \subseteq \hat{B}$ and an element $\hat{b} \in \hat{B}$ such that

$$sup\, \hat{f}_c = D + \hat{b}_0 \qquad (c \in C).$$

Replacing f by an appropriate translation of f we may assume that each $\hat{f}_c$ is a constant on its support. Thus there is a function $h : C \to \mathbb{C}$ such that

$$\hat{f}_c = \mathbb{I}_{D+\hat{b}_0} \otimes h(c) \qquad (c \in C).$$

Therefore

$$\hat{f} = \mathbb{I}_{D+\hat{b}_0} \otimes \hat{h}. \tag{15}$$

We see from (15) that h is a minimizer for C. But

$$f = \hat{\mathbb{I}}_{D+\hat{b}_0} \otimes h$$

and our assumption ($f_c \neq 0$ for all $c \in C$) implies that $supp\, h = C$. Hence, $|supp\, \hat{h}| = 1$. Thus $supp\, \hat{h} = \{\hat{c}_0\}$, for some $\hat{c}_0 \in \hat{C}$. Therefore, by (15),

$$supp\, \hat{f} = (D + \hat{b}_0) + \hat{c}_0 = D + (\hat{b}_0 + \hat{c}_0).$$

Thus $supp\, \hat{f}$ is a coset of a proper subgroup of $\hat{A}$. Since $\hat{f}$ is a minimizer, our previous argument implies that $\hat{f}$, and hence f, has the desired property.

□

3. A few words on the notion of the entropy

In this section we recall a few basic facts, which indicate that the notion of the entropy is quite natural and useful. For more details we refer the reader to [4] and [18].

Let X be a discrete random variable with the probability distribution function $p(x) = P(X = x)$. Set

$$H(X) = H(p) = -\sum_x p(x) log_2(p(x)). \tag{16}$$

Example 3.1: Suppose X has a uniform distribution over $2^3 = 8$ outcomes. Then

$$H(X) = -\sum_{x=1}^{8} \frac{1}{8} log_2(\frac{1}{8}) = 3.$$

This agrees with the number of bits needed to describe X in binary:

$$c(0) = 000,\ c(1) = 001,\ c(2) = 010,\ c(3) = 011,\ c(4) = 100,$$
$$c(5) = 101,\ c(6) = 110,\ c(7) = 111.$$

Example 3.2: Consider a horse race, with eight horses taking part. Suppose the probability of winning the race is distributed as follows:

$$p(0) = 1/2,\ p(1) = 1/4,\ p(2) = 1/8,\ p(3) = 1/16,$$
$$p(4) = p(5) = p(6) = p(7) = 1/64.$$

Then $H(X) = 2$. We may encode the horses in binary as follows:

$$c(0) = 0,\ c(1) = 10,\ c(2) = 110,\ c(3) = 1110,\ c(4) = 111100,$$
$$c(5) = 111101,\ c(6) = 111110,\ c(7) = 111111.$$

Let $l(c(x))$ be the length of the code word $c(x)$. Then

$$l(c(0)) = 1,\ l(c(1)) = 2,\ l(c(2)) = 3,\ l(c(3)) = 4,\ l(c(5)) = ... = l(c(7)) = 6.$$

Hence, the expected value of $l(c(X))$ is

$$E(l(c(X)) = 1 \cdot \frac{1}{2} + 2 \cdot \frac{1}{4} + 3 \cdot \frac{1}{8} + 4 \cdot \frac{1}{16} + 6 \cdot \frac{1}{64} = 2.$$

Thus

$$H(X) = E(l(c(X)).$$

This last equality is not a coincidence, as we shall explain below. For details see [4].

A source code for a discrete random variable X is a mapping c from the range of X to

$$\{0,1\} \cup \{0,1\}^2 \cup \{0,1\}^3 \cup ... .$$

If a codeword $c(x)$ belongs to $\{0,1\}^k$, let $l(c(x)) = k$ denote the length of $c(x)$. The code is called instantenous if no codeword is a prefix to any other code word. For such codes one can recognize the separate words $c(x_1)$, $c(x_2)$, ... by looking at the string $c(x_1)c(x_2)$.... For instance in our second example, the string 010110111111 is made of the codewords 0, 10, 110, 111111.

Theorem 3.3: ([18], see also [4]) *For an instantenous code c we have,*

$$E(l(c(X)) \geq H(X).$$

Moreover, there is an instantenous code c such that

$$E(l(c(X)) < H(X) + 1.$$

A theorem of Shannon, which we quote below, gives a characterization of the entropy as a measure of uncertainty of the outcome of an experiment.

Let $P = \{p_1, p_2, p_3, ..., p_n\}$ be a finite probability sequence, i.e. $p_j \geq 0$, $\sum_j p_j = 1$.

Theorem 3.4: (Shannon, 1948, [18]) *Suppose H is a function defined on finite probability sequences such that*

(a) $H(p_1, p_2, p_3, ..., p_n)$ *is continuous,*

(b) $H(\frac{1}{n}, \frac{1}{n}, \frac{1}{n}, ..., \frac{1}{n})$ *is monotonically increasing as a function of* n,

(c) *if* $Q_j = \{q_{j1}, q_{j2}, q_{j3}, ...\}$, $j = 1, 2, 3, ..., n$, *are probability sequences, then* $H(\bigcup_{j=1}^n p_j Q_j) = H(P) + \sum_{j=1}^n p_j H(Q_j)$.

Then, up to a constant multiple,

$$H(P) = -\sum_{j=1}^{n} p_j log(p_j).$$

Suppose M is a finite set and μ is a positive multiple of the counting measure on M. For a function $f : M \to \mathbb{C}$ and for $1 \leq p < \infty$ we have the L^p norm of f defined by

$$\| f \|_p = \left(\int_M |f(x)|^p \, d\mu(x) \right)^{1/p}.$$

There is a simple and explicit connection between the entropy (see Definition 1.4) and the L^p norms, expressed in the lemma below, which may be verified by a straightforward computation (left to the reader). See [20] for more explanations.

Lemma 3.5: *Let* $p = p(t) = \frac{1}{t}$, $0 < t \leq 1$. *Then for any function* $f : M \to \mathbb{C}$,

(a) $$\frac{d}{dt} log(\| f \|_p) = -\int_M \frac{|f(x)|^p}{\|f\|_p^p} log(\frac{|f(x)|^p}{\|f\|_p^p}) \, d\mu(x).$$

In particular, for f *with* $\| f \|_1 = 1$,

(b) $$\frac{d}{dt} log(\| f \|_p)|_{p=1} = H(|f|),$$

and for f *with* $\| f \|_2 = 1$,

(b) $$\frac{d}{dt} log(\| f \|_p)|_{p=2} = H(|f|^2).$$

4. The entropy inequality for a finite Abelian group

In this section we give a proof of theorem 1.10 for a finite Abelian group A. For the general case we refer the reader to [16].

The proof is based on the following four basic theorems.

Hölder's Inequality [11]:

$$\int_M |\phi(x)\psi(x)| \, dx \leq \| \phi \|_p \| \psi \|_q, \quad \frac{1}{p} + \frac{1}{q} = 1.$$

Plancherel's Formula [8]:

$$\| \hat{f} \|_2 = \| f \|_2 .$$

Riesz-Thorin-Young Inequality ([21], Chapter 9, (1.11)):

$$\| \hat{f} \|_{1/(1-t)} \leq \| f \|_{1/t} \quad (\frac{1}{2} \leq t \leq 1).$$

Hopf's Maximum Principle ([11], Theorem 3.1.6'): *Let $D \subseteq \mathbb{C}$ be an open unit disc, and let $u : D \to \mathbb{R}$ be a harmonic function which extends to a continuous function on the closure $\overline{D}$ of D, $u : \overline{D} \to \mathbb{R}$. Suppose z is a point on the boundary of D such that $u(z) \geq u(z')$ for all $z' \in \overline{D}$, and the directional derivative of u at z along the radius which ends at z, is zero. Then*

$$u(z) = u(z') \qquad \text{for all } z' \in \overline{D}.$$

Let $f : A \to \mathbb{C}$ be a minimizer. Consider the following function

$$F(z) = \int_{\hat{A}} (|f|^{2z}\frac{f}{|f|})\hat{\ }(\hat{a})\, |\hat{f}(\hat{a})|^{2z}\, \overline{\frac{\hat{f}(\hat{a})}{|\hat{f}(\hat{a})|}}\, d\hat{\alpha}(\hat{a}) \qquad (z \in \mathbb{C},\ \frac{1}{2} \leq Re(z) \leq 1), \tag{17}$$

where $\frac{f}{|f|} = 0$ outside the support of f, and similarly for $\frac{\hat{f}}{|\hat{f}|}$. A straightforward application of Hölder's inequality and the Riesz-Thorin Theorem shows that for $\frac{1}{2} \leq x \leq 1$, $y \in \mathbb{R}$, $p = \frac{1}{x}$ and q defined by the equation $\frac{1}{p} + \frac{1}{q} = 1$ (with $q = \infty$ if $p = 1$), we have

$$\begin{aligned} |F(x+iy)| &\leq \| (|f|^{2x+i2y}\frac{f}{|f|})\hat{\ } \|_q \cdot \| |\hat{f}|^{2x+i2y} \|_p \\ &\leq \| |f|^{2x+i2y} \|_p \cdot \| |\hat{f}|^{2x+i2y} \|_p = \| f \|_2 \cdot \| \hat{f} \|_2 = 1. \end{aligned} \tag{18}$$

The function F is analytic in the open strip (17) and continuous in the closed strip. A straightforward calculation shows that

$$\begin{aligned} F'(z) = &\int_{\hat{A}} (|f|^{2z}\frac{f}{|f|} log(|f|^2))\hat{\ }(\hat{a})\, |\hat{f}(\hat{a})|^{2z}\, \overline{\frac{\hat{f}(\hat{a})}{|\hat{f}(\hat{a})|}}\, d\hat{\alpha}(\hat{a}) \\ + &\int_{\hat{A}} (|f|^{2z}\frac{f}{|f|})\hat{\ }(\hat{a})\, |\hat{f}(\hat{a})|^{2z}\, \overline{\frac{\hat{f}(\hat{a})}{|\hat{f}(\hat{a})|}} log(|\hat{f}(\hat{a})|^2)\, d\hat{\alpha}(\hat{a}). \end{aligned}$$

Hence, by the Plancherel formula,

$$F'(\frac{1}{2}) = -H(|f|^2) - H(|\hat{f}|^2). \tag{19}$$

Since f is a minimizer, the right hand side of the equation (19) is zero. In particular $Re\, F(z)$ is a real valued harmonic function on the interior of the disc of radius $\frac{1}{4}$ centered at $z = \frac{3}{4}$, which achieves the maximum at $z = \frac{1}{2}$ and has derivative equal to zero at this point. Hence the Hopf's Maximum

Principle implies that $Re\,F(z) = 1$ on the disc. Hence, $F(z) = 1$ on the disc. In particular,

$$1 = F(1) = \int_{\hat{A}} (|f|\hat{f})(\hat{a})|\hat{f}(\hat{a})|\overline{f(\hat{a})}\, d\hat{\alpha}(\hat{a}). \tag{20}$$

The formula (20) may be rewritten as

$$1 = \int_{\hat{A}}\int_{A} |f(a)|^2|\hat{f}(\hat{a})|^2\hat{a}(-a)\frac{f(a)}{|f(a)|}\frac{\overline{\hat{f}(\hat{a})}}{|\hat{f}(\hat{a})|}\, d\alpha(a)\, d\hat{\alpha}(\hat{a}). \tag{21}$$

Since,

$$1 = \int_{\hat{A}}\int_{A} |f(a)|^2|\hat{f}(\hat{a})|^2\, d\alpha(a)\, d\hat{\alpha}(\hat{a})$$

the equation (21) implies that

$$1 = \hat{a}(-a)\frac{f(a)}{|f(a)|}\frac{\overline{\hat{f}(\hat{a})}}{|\hat{f}(\hat{a})|} \qquad (a \in supp\, f, \hat{a} \in supp\, \hat{f}). \tag{22}$$

Hence,

$$\hat{a}(-a) = \frac{\overline{f(a)}}{|f(a)|}\frac{\hat{f}(\hat{a})}{|\hat{f}(\hat{a})|}$$

Thus for $\hat{a} \in supp\, \hat{f}$

$$\hat{f}(\hat{a}) = \int_{A} f(a)\hat{a}(-a)\, d\alpha(a) = \int_{A} f(a)\frac{\overline{f(a)}}{|f(a)|}\, d\alpha(a)\frac{\hat{f}(\hat{a})}{|\hat{f}(\hat{a})|} = \| f \|_1 \frac{\hat{f}(\hat{a})}{|\hat{f}(\hat{a})|}.$$

Therefore

$$|\hat{f}(\hat{a})| = \| f \|_1 \qquad (\hat{a} \in supp\, \hat{f}), \tag{23}$$

and similarly

$$|f(a)| = \| \hat{f} \|_1 \qquad (a \in supp\, f). \tag{24}$$

The statement (23) implies that the function $|\hat{f}|$ is constant on its support. Since $\| \hat{f} \|_2 = 1$, the constant is equal to $\hat{\alpha}(supp\, \hat{f})^{-1/2}$. Hence,

$$H(|\hat{f}|^2) = log(\hat{\alpha}(supp\, \hat{f})).$$

Similarly

$$H(|f|^2) = log(\alpha(supp\, f)).$$

Since f is a minimizer,

$$log(\alpha(supp\, f)\cdot \hat{\alpha}(supp\, \hat{f})) = 2\left(H(|f|^2) + H(|\hat{f}|^2)\right) = 0.$$

Therefore

$$\alpha(supp\, f) \cdot \hat{\alpha}(supp\, \hat{f}) = 1. \tag{25}$$

We may assume that $0 \in supp\, \hat{f}$ and $0 \in supp\, f$. Then (23) implies

$$\left| \int_A f(a)\, d\alpha(a) \right| = \int_A |f(a)|\, d\alpha(a).$$

Therefore there is $\lambda \in \mathbb{C}$ such that $f = \lambda |f|$. Hence (23) may be rewritten as

$$\left| \int_A \lambda |f(a)| \hat{a}(-a)\, d\alpha(a) \right| = \int_A |f(a)|\, d\alpha(a) \quad (\hat{a} \in supp\, \hat{f}). \tag{26}$$

Therefore

$$supp\, \hat{f} \subseteq (-supp\, f)^{\perp}, \tag{27}$$

where for a subset $S \subseteq A$, $S^{\perp} = \{\hat{a} \in \hat{A};\ \hat{a}|_S = 1\}$. Similarly (24) implies

$$supp\, f \subseteq (supp\, \hat{f})^{\perp}. \tag{28}$$

By dualizing (27) and (28) we deduce

$$\begin{aligned} -\, supp\, f &\subseteq (-supp\, f)^{\perp\perp} \subseteq (supp\, \hat{f})^{\perp}, \text{ and} \\ supp\, \hat{f} &\subseteq (supp\, \hat{f})^{\perp\perp} \subseteq (supp\, f)^{\perp}. \end{aligned} \tag{29}$$

But, as is well known and easy to check,

$$\hat{\alpha}((supp\, \hat{f})^{\perp}) \cdot \alpha((supp\, \hat{f})^{\perp\perp}) = 1. \tag{30}$$

By combining (25), (29) and (30) we see that the inclusions (29) are equalities. In particular $supp\, f$ is a subgroup of A and f is invariant under the translations by this subgroup. Thus f is a constant multiple of the indicator function of a subgroup of A, as claimed.

Acknowledgement

Part of these notes was written during the author's visit to the Institute for Mathematical Sciences (IMS) at the National University of Singapore. The author thanks IMS for its support and hospitality.

References

1. P. Ahern and R. Jewett, "Factorization of locally compact Abelian groups", *Illinois Jour. Math.*, **9** (1965), 230–235.
2. W. Beckner, "Inequalities in Fourier Analysis", *Annals of Math.*, **102** (1975), 159–182.
3. W. Beckner, "Pitt's Inequality and the Uncertainty Principle", *Proceedings of the AMS*, **123** (1995), 1897–1905.
4. T. Cover and J. Thomas, *Elements of information theory*, John Wiley & Sons, Inc., New York, 1991.
5. I. Daubechies, *Ten Lectures on Wavelets*, Society for Industrial and Applied Mathematics (SIAM), Philadelphia, 1992.
6. A. Dembo, T. M. Cover and J. A. Thomas, "Information Theoretic Inequalities", *EEE Transactions on Information Theory*, **37** (1991), 1501–1518.
7. D. L. Donoho and P. B. Stark, "Uncertainty Principles and Signal Recovery", *SIAM Journal of Applied Mathematics*, **49** (1989), 906–931.
8. E. Hewitt and K. A. Ross, *Abstract Harmonic Analysis*, Springer Verlag, 1963.
9. I. I. Hirschman, Jr., "A Note on Entropy", *Amer. Jour. Math.*, **79** (1957), 152–156.
10. L. Hörmander, *The Analysis of Linear Partial Differential Operators, I*, Springer Verlag, 1983.
11. L. Hörmander, *Notions of Convexity*, Birkhäuser, 1994.
12. E. Matusiak, M. Özaydın and T. Przebinda, "The Donoho - Stark Uncertainty Principle for a Finite Abelian Group", *preprint*, available at http://crystal.ou.edu/~tprzebin/papers.html.
13. H. Maassen, "A discrete entropic uncertainty relation" *Quantum Probability and Applications*, **5** (1988), 263–266.
14. H. Maassen and J. Uffink, "Generalized Entropic Uncertainty Relations", *Phys. Rev. Lett.* , **60** (1988), 1103–1106.
15. M. Özaydın and T. Przebinda, "Platonic Orthonormal Wavelets", *Applied and Computational Harmonic Analysis*, **4** (1997), 351–365.
16. M. Özaydın and T. Przebinda, "An Entropy-based Uncertainty Principle for a Locally Compact Abelian Group", *preprint*, available at http://crystal.ou.edu/~tprzebin/papers.html.
17. T. Przebinda, V. DeBrunner and M.Özaydın, "The Optimal Transform for the Discrete Hirschman Uncertainty Principle", *IEEE Transactions on Information Theory*, **47** (2001), 2086–2090.
18. C. E. Shannon, "A Mathematical Theory of Communication", *The Bell System Technical Journal"*, **27** (1948), 379–656.
19. K. T. Smith, "The Uncertainty Principle on Groups", *SIAM Journal of Applied Mathematics*, **50** (1990), 876–882.
20. M. Wickerhauser, *Adapted wavelet analysis from theory to software*, Cambridge University Press, 1990.
21. A. Zygmund, *Trigonometric series, second edition*, volumes I and II combined, A K Peters, Ltd., Wellesley, MA, 1994.

Lectures on Representations of p-adic Groups

Gordan Savin
Department of Mathematics
University of Utah
Salt Lake City, UT 84112, USA
E-mail: savin@math.utah.edu

1. Introduction

This text contains an elementary introduction, with exercises, to representations of p-adic reductive groups. The text is intended to be more accessible then the standard references, such as the paper of Bernstein and Zelevinsky [1], and (unpublished) notes of Casselman. The material presented here has been known for the past quarter century, and has undergone many modifications in that period. Thus, it would be hard (if not impossible) to acknowledge properly contributions of many people who have worked on the subject. However, it is safe to say that Bernstein, Casselman, Harish-Chandra, Howe and Jacquet (in alphabetical order) are some of the most prominent contributors to the subject.

The text is organized as follows. Sections 2-4 contain definitions and preliminary results on p-adic fields, structure of $GL_n(F)$ over a p-adic field F, and smooth representations. Then, in order to keep the exposition as simple as possible, we restrict ourselves to $GL_2(F)$. However, the topics and their proofs are chosen so that they easily generalize to $GL_n(F)$ and other reductive groups.

In Section 5 and 6 we introduce induced and cuspidal representations, and prove that irreducible smooth representations are admissible. Sections 7 and 8 are devoted to describing the composition factors of induced representations. In section 9 we discuss unitarizable representations, and construct the complementary series for $GL_2(F)$. Sections 10 and 11 are devoted to two examples. In section 10 we show that the Steinberg representation is square integrable, and in section 11 we construct one cuspidal representation.

Finally, in Sections 12 and 13, we go back to $GL_n(F)$ and describe the composition factors of (regular) induced representations. This result, due to Rodier [4], is based on the combinatorics of the root system, and the reduction to the special case of $GL_2(F)$ (obtained in Section 7). As such, it gives a good introduction to further, more advanced topics.

Acknowledgments: This material was presented during a week long tutorial in the Institute for Mathematical Sciences at the National University of Singapore. The author would like to thank the people associated with the institute, especially Eng-Chye Tan and Chen-Bo Zhu, for their hospitality and support. Thanks are also due to Jian Shu Li, Allen Moy, and Marko Tadić for useful comments and suggestions at various stages.

2. The field $\mathbb{Q}_p$

This is a crash course on the field of p-adic numbers. Let $\mathbb{Z}$ denote, as usual, the ring of integers. Define the p-adic absolute value on $\mathbb{Z}$ as follows. Let $x \neq 0$ be an integer. Write $x = p^a m$ where m is relatively prime to p. Then

$$|x|_p = \frac{1}{p^a}.$$

We also put $|0|_p = 0$. This absolute value satisfies the usual properties: the multiplicativity, $|x|_p \cdot |y|_p = |xy|_p$, and the triangle inequality. In fact we have an even stronger property:

$$|x+y|_p \leq \max\{|x|_p, |y|_p\}.$$

It follows that $d(x,y) = |x-y|_p$ defines a metric on $\mathbb{Z}$. The ring of p-adic integers $\mathbb{Z}_p$ is the completion of $\mathbb{Z}$ with respect to this metric.

To understand this ring we can proceed as follows. Let $B(x,r)$ be the (closed) ball of radius $r \geq 0$, centered at x. Note that $\mathbb{Z}$ can be written as a union of p disjoint balls of radius $1/p$, centered at $0, 1, \ldots, p-1$:

$$\mathbb{Z} = B(0, \frac{1}{p}) \cup B(1, \frac{1}{p}) \cup \ldots \cup B(p-1, \frac{1}{p}).$$

Note that $B(i, \frac{1}{p})$ is simply $i + p\mathbb{Z}$, a coset of the maximal ideal (p). Similarly, $\mathbb{Z}$ can be written as a union of p^n disjoint balls of radius $1/p^n$, centered at all reminders modulo p^n. There are two consequences of this: First, $\mathbb{Z}$ is totally bounded, so its completion $\mathbb{Z}_p$ is compact. Second, the completion of $\mathbb{Z}$ is the union of completion of individual balls of radius $1/p^n$, which implies that

$$\mathbb{Z}/p^n\mathbb{Z} = \mathbb{Z}_p/p^n\mathbb{Z}_p.$$

Exercise. Let x be an integer relatively prime to p. Show that the multiplicative inverse of x exists in $\mathbb{Z}_p$. Hint: for every n, there exist integers y_n and z_n such that $xy_n + p^n z_n = 1$.

The ring $\mathbb{Z}_p$ has (p) as unique maximal ideal. The field of fractions $\mathbb{Q}_p$ is obtained by adjoining $1/p$. It follows that

$$\mathbb{Q}_p = \cup_{-\infty<i<\infty} p^i\mathbb{Z}_p.$$

Next, we shall describe smooth additive characters of $\mathbb{Q}_p$. A character ψ is smooth if it is trivial on $p^i\mathbb{Z}_p$ for some some i. The minimal such i is called the conductor of ψ. We shall now construct a canonical character of conductor 0. Note that $\mathbb{Q}_p/\mathbb{Z}_p$ is naturally isomorphic to the p-torsion part of $\mathbb{Q}/\mathbb{Z}$. The group $\mathbb{Q}/\mathbb{Z}$ has a natural additive character given by

$$\psi(x) = e^{2\pi i x}.$$

The restriction of ψ to $\mathbb{Q}_p/\mathbb{Z}_p$ is the additive character of conductor 0. Any other character is of the form $\psi_a(x) = \psi(ax)$ for an element a in $\mathbb{Q}_p$.

We shall now say a couple of words about characters of $\mathbb{Q}_p^\times$. Note that $\mathbb{Q}_p^\times = \langle p \rangle \times \mathbb{Z}_p^\times$, where $\langle p \rangle \cong \mathbb{Z}$ is the free group generated by p. The group $\mathbb{Z}_p^\times$ has a filtration by subgroups $1 + p^i\mathbb{Z}_p$ with $i \geq 1$. A character of $\mathbb{Q}_p^\times$ is smooth if it is trivial on some $1 + p^i\mathbb{Z}_p$. Moreover, it is called unramified, if it is trivial on $\mathbb{Z}_p^\times$. Every unramified character χ is completely determined by its value $\chi(p)$. In particular, the group of unramified characters is isomorphic to $\mathbb{C}^\times$.

Any other p-adic field F is simply a finite extension of $\mathbb{Q}_p$ (see [3]). The integral closure of $\mathbb{Z}_p$ will be denoted by $\mathcal{O}$. It has a unique maximal ideal (ϖ), which is the radical of the ideal (p). Let q be the order of the residue field $\mathcal{O}/\varpi\mathcal{O}$. We shall normalize the absolute value on F so that

$$|\varpi| = \frac{1}{q}.$$

3. Structure of $GL_n(F)$

Let F be a p-adic field, and $G = GL_n(F)$. Let $K = GL_n(\mathcal{O})$. Then K is a (maximal) open, compact subgroup of G. It has a filtration by the principal congruence subgroups K_i of G defined by

$$K_i = \{g \in G \mid g \equiv 1 \mod \varpi^i\}.$$

These groups are normal in K, and provide a fundamental system of neighborhoods of 1 in G, defining a topology of G. We have the following important decompositions in G:

3.1. *Cartan decomposition*

Let $\Lambda \cong \mathbb{Z}^n$ be the group of diagonal matrices

$$\begin{pmatrix} \varpi^{m_1} & & \\ & \ddots & \\ & & \varpi^{m_n} \end{pmatrix}$$

where $m_1, \ldots, m_n$ are integers. Let Λ^+ be the subset of Λ consisting of diagonal matrices such that $m_1 \geq \ldots \geq m_n$. Then

$$G = K\Lambda^+K.$$

3.2. *Bruhat-Tits decomposition*

Let B denote the group of upper triangular matrices in G, and N the subgroup of B consisting of matrices with 1 on the diagonal. Let W be the group of permutation matrices, which is isomorphic to the permutation group S_n. The group W is called the Weyl group of G. Then

$$G = NWB.$$

3.3. *Iwasawa decomposition*

$$G = BK.$$

Each of the three decompositions given here can be proved by a variant of row-column reduction. This is deceptively simple and somewhat misleading. An entirely different approach is needed for general reductive groups. Thus, we give only a hint here, in form of an exercise.

Exercise. Prove the Cartan decomposition. Hint: Let g be in $GL_m(F)$. Let g_{ij} be the entry of g with the largest absolute value. Multiplying g by elementary matrices with coefficients in $\mathcal{O}$ (so they are in K) turn all entries *in the i-th row and j-th column (except g_{ij}) into 0. Proceed by induction on n.*

A critical difference between real and p-adic groups is that N is a union of open compact subgroups. For example, if $G = GL_2(F)$, then N is a union of

$$N_i = \{\begin{pmatrix} 1 & x \\ 0 & 1 \end{pmatrix} \mid x \in \varpi^i \mathcal{O}\}.$$

As we shall see later, one consequence of this fact is that $SL_n(F)$ does not have non-trivial finite dimensional representations.

3.4. *Haar Measure*

Recall that the congruence subgroups K_i give a system of open neighborhoods of 1 which defines the topology of G. Let $C_0^\infty(G)$ be the set of locally constant, compactly supported functions on G. In particular, for every function f there exist a sufficiently small congruence subgroup K_i, such that f can be written as a linear combination

$$f = \sum_j c_j \cdot Char(g_j K_i)$$

where $Char(g_j K_i)$ is the characteristic function of the right coset $g_j K_i$. Now put

$$\mu(f) = \int_G f(g)\, dg = \frac{1}{[K : K_i]} \sum_j c_j.$$

This defines a (unique) left G-invariant measure on G such that $vol(K) = 1$.

Exercise. Show that the measure is right G-invariant. Hint: Let g be in G, and define a left G-invariant measure μ_g *by*

$$\mu_g(f) = \mu(R_g(f))$$

where $R_g(f)$ *is the right translate of f by g. By uniqueness of the left G-invariant measure,* $\mu_g = c(g)\mu$ *for some positive constant* $c(g)$. *Show that* $c(g)$ *is a character, and then use that* $[G,G] \cdot Z$ *(Z is the center of G) is of finite index in G.*

4. Smooth representations

Let V be a vector space over the field of complex numbers $\mathbb{C}$. A *smooth* representation of G on V is a homomorphism

$$\pi : G \to GL(V)$$

such that every vector v in V is fixed by a (sufficiently small) congruence subgroup. We shall refer to V as a (smooth) G-module. For every congruence subgroup K_i, let V^{K_i} be the subspace of all vectors fixed by K_i. A representation is *admissible* if V^{K_i} is finite dimensional for every i. It turns out that any irreducible smooth module is in fact admissible. We shall give a proof of this fact for $GL_2(F)$.

Proposition 4.1: (Schur's Lemma) *Let V be an irreducible smooth representation of G, and $A : V \to V$ be a non-trivial homomorphism. Then $A = \mu \cdot I$ for some μ in $\mathbb{C}$. (Here I is the identity operator on V).*

Proof: First of all note that V has countable dimension. Indeed, V is spanned by $\pi(G)v$ for any non-zero vector v in V. Since v is fixed by some K_i, the statement follows, as G/K_i is countable (from the following exercise, for example).

Exercise. Show that for every λ in Λ^+, the index of $K_\lambda = \lambda K \lambda^{-1} \cap K$ in K is sandwiched by

$$\delta(\lambda)^{-1} \leq [K : K_\lambda] \leq [K : K_1] \cdot \delta(\lambda)^{-1}$$

where $\delta(\lambda) = \prod_{i<j} q^{m_i - m_j}$.

Assume that $A \neq \mu \cdot I$ for every μ in $\mathbb{C}$. Then irreducibility of V implies that $A - \mu \cdot I$ is bijective. In particular, $R_\mu = 1/(A - \mu \cdot I)$ is well defined for every μ in $\mathbb{C}$. We now need the following:

Exercise. Let v be a non-zero vector in V. Assuming that $A - \mu \cdot I$ are bijective, show that $R_\mu v$ are linearly independent vectors in V.

But we know that V has countable dimension. This is a contradiction, and the proposition is proved. □

Proposition 4.2: *Let V be a smooth representation. Then:*

- *If the representation V is finitely generated, then there exists an irreducible quotient.*
- *Otherwise, there exits two submodules $V'' \subset V'$ such that V'/V'' is irreducible.*

Proof: The second statement clearly follows from the first. To prove the first statement, let E be a finite dimensional subspace generating V. If U is a proper submodule, then $E \cap U$ is a proper subspace of E. Since E is finite dimensional, there exists a maximal subspace $E_{\max}$ of E which is an intersection with a proper invariant subspace. Let $\mathcal{U}$ be the family of all invariant subspaces U such that $E \cap U = E_{\max}$. By the Zorn's lemma, the family $\mathcal{U}$ has a maximal element $U_{\max}$. Clearly, $V/U_{\max}$ is irreducible. The proposition is proved. □

Proposition 4.3: *Let V be a finitely generated smooth G-module. Then V is also finitely generated over B.*

Proof: Fix a finite set of generators. Since V is smooth, there exists a congruence subgroup K_i fixing all generators. Since K_i is of finite index in K, the proposition easily follows from the Iwasawa decomposition. □

An interesting feature of smooth representations it that the trivial representation is the unique finite-dimensional representations of $SL_n(F)$.

Proposition 4.4: *Let (π, V) be a finite dimensional irreducible smooth representation of $GL_n(F)$. Then $V \cong \mathbb{C}$, and $\pi(g) = \chi(det)$ for some character χ of $F^\times$.*

Proof: Since V is smooth and finite dimensional, there exists a sufficiently small open compact subgroup $N_c \subset N$ acting trivially on V.

Exercise. Let n be any element in N. Then there exits an element t in T such that $tnt^{-1} = n_c$ is in N_c.

Since $\pi(n)v = \pi(t^{-1})\pi(n_c)\pi(t)v = v$, it follows that N acts trivially on V. The same argument shows that $\bar{N}$ acts trivially on on V, and the proposition follows from the following exercise.

Exercise. $SL_n(F)$ is generated by N and $\bar{N}$. Hint: Use row-column reduction. □

For any smooth G representation V, let V^* be the space of all linear functionals on V. Let $\langle v, v^* \rangle$ be the usual pairing between V and V^*. Then G acts on V^* by

$$\langle v, \pi^*(g)v^* \rangle = \langle \pi(g^{-1})v, v^* \rangle.$$

Let $\tilde{V}$ be the set of smooth vectors in V^*, and let $\tilde{\pi}$ denote the restriction of π^* to $\tilde{V}$. The representation $\tilde{V}$ is called the contragredient dual of V. A priori, it may not be clear that $\tilde{V}$ is non-trivial. However, we have the following simple proposition:

Proposition 4.5: *Let V be a smooth representation of G. Then $\tilde{V}^{K_i}$ is isomorphic to the set of linear functionals on V^{K_i}. In particular, $\tilde{V}$ is admissible, if and only if V is so.*

Proof: Let v be any vector in V. Then v is fixed by K_j for some sufficiently small j which we can assume to be less then i. Define

$$P(v) = \frac{1}{[K_i : K_j]} \sum_{k \in K_i/K_j} \pi(k)v.$$

Clearly, $P(v)$ does not depend on the choice of j, and P is a K_i-invariant projection from V onto V^{K_i}. Now any functional on V^{K_i} composed with P defines an element in $\tilde{V}^{K_i}$. Conversely, any functional in $\tilde{V}^{K_i}$ clearly factors through P. The proposition is proved. □

5. Induced representations, case of $GL_2(F)$

Let χ_1 and χ_2 be two characters of the multiplicative group $F^\times$. Let χ be the character of T, the group of diagonal matrices in G, defined by

$$\chi\begin{pmatrix} a_1 & 0 \\ 0 & a_2 \end{pmatrix} = \chi_1(a_1)\chi_2(a_2).$$

We shall now describe the so-called parabolic induction, which produces a smooth representation of G from a character of T. Since $B/N = T$, the character χ can be pulled back to B, and we can consider the smooth induced representation

$$\mathrm{Ind}_B^G(\chi) = \{f : G \to \mathbb{C} \mid f(ntg) = |a_1/a_2|^{1/2}\chi(t)f(g)\}^\infty,$$

where here, of course, we have considered only the smooth functions on G. The meaning (and necessity) of the term $\delta^{1/2} = |a_1/a_2|^{1/2}$ will become more clear as we develop more theory. Basically, this factor compensates for the choice of the Borel subgroup B containing T, as we could have defined the induction using $\bar{B}$, the group of lower triangular matrices.

The induced representations are admissible. Indeed, the Iwasawa decomposition implies that

$$dim\ Ind_B^G(\chi)^{K_i} \le [K : K_i].$$

Our goal is to understand the composition factors of the induced representations. To do so, we need to introduce an important functor which is dual, in a certain sense, to the parabolic induction.

5.1. *Jacquet Functor*

We first define a functor from the category of smooth N-modules to the category of vector spaces. Let V be a smooth N-module, and put

$$V_N = V/V(N)$$

where $V(N)$ is the linear span of $\pi(n)v - v$ for all $n \in N$ and $v \in V$. The functor $V \rightsquigarrow V_N$ is exact, which will be of fundamental importance:

Theorem 5.1: *Let* $0 \to V' \to V \to V'' \to 0$ *be an exact sequence of* N*-modules. Then*

$$0 \to V'_N \to V_N \to V''_N \to 0$$

is exact.

Proof: The right exactness is clear. We only need to check that $V'_N \to V_N$ is injective. This is equivalent to $V' \cap V(N) = V'(N)$, which will follow from the following important characterization of $V(N)$. □

Lemma 5.2: *Let* V *be a smooth* N*-module. A vector* v *is in* $V(N)$ *if and only if*

$$\int_{N_j} \pi(n)v \; dn = 0$$

for a sufficiently large open compact subgroup N_j*. In particular,* $V(N) = \cup_{j\in\mathbb{Z}} V(N_j)$*, where* $V(N_j)$ *is the subspace of all* v *in* V *such that the above integral vanishes.*

Proof: If v is in $V(N)$, then v is a finite linear combination of terms $\pi(n)u - u$. Let N_j be sufficiently large so that it contains all n appearing in that linear combination. Then the integral over N_j will be 0. □

Exercise. Prove the other direction of the Lemma. Hint: The vector v *is invariant under some* N_i*.*

Now assume that V is a smooth G-module. Then V_N is naturally a smooth T-module, and $V \rightsquigarrow V_N$ is called the Jacquet functor. The Jacquet functor and the parabolic induction are related by the Frobenius reciprocity:

$$Hom_G(V, Ind_B^G(\chi)) = Hom_T(V_N, \delta^{1/2}\chi)$$

Exercise. The Frobenius reciprocity is a tautology. Prove it!

Proposition 5.3: *Let V be an irreducible smooth G-module such that $V_N \neq 0$. Then V is a submodule of $Ind(\chi)$ for some character χ of T.*

Proof: By Proposition 4.3 we know that V is finitely generated over B. Thus V_N is also finitely generated over B, and it has an irreducible quotient by Proposition 4.2. The proposition follows from the Frobenius reciprocity. □

6. Cuspidal representations, case of $GL_2(F)$

We have just seen that a smooth irreducible representation V such that $V_N \neq 0$, embeds into a principal series representation. Our concern in this section will be the representations such that $V_N = 0$; so called cuspidal representations. Since the center Z of G is not compact, so it will be more convenient to work with a smaller group

$$G^\circ = \{g \in GL_2(F) \mid det(g) \in \mathcal{O}^\times\}.$$

Note that ZG° is of index 2 in G, so representations of G° can be easily related to those of G.

Exercise. Let (π, V) be an irreducible smooth G-module. Then the restriction of V to G° is either irreducible, or a direct sum of two non-isomorphic G°-modules. Moreover, the restriction reduces if and only if $\pi \cong \pi \otimes \gamma$ where γ is the unique non-trivial character of G trivial on ZG°.

Definition 6.1: A smooth representation (π, V) of G° is called cuspidal if $V_N = 0$.

We shall now show that the matrix coefficients of cuspidal representations are compactly supported. A matrix coefficient of V is a (smooth) function

$$\varphi_{v,\tilde{v}}(g) = \langle \pi(g)v, \tilde{v} \rangle$$

where v is in V, and $\tilde{v}$ in $\tilde{V}$, the contragredient of V.

Proposition 6.2: *Let (π, V) be a cuspidal (not necessarily irreducible) representation of G°. Then the matrix coefficients of π are compactly supported.*

Proof: Let U and $\tilde{U}$ be the linear span of $\pi(K)v$ and $\tilde{\pi}(K)\tilde{v}$, respectively. Both spaces are finite dimensional, since the representations are smooth and

K is compact. Since U is cuspidal, by Lemma 5.2, we can find a sufficiently large i such that such that

$$\int_{N_{-i}} \pi(n)u \, dn = 0$$

for all $u \in U$, and $\tilde{U}$ is contained in $\tilde{V}^{K_i}$. Let

$$\lambda = \begin{pmatrix} \varpi^a & 0 \\ 0 & \varpi^{-a} \end{pmatrix}.$$

Then, for any u in U, and any $\tilde{u}$ in $\tilde{U}$,

$$\int_{N_i} \langle \pi(\lambda)u, \tilde{\pi}(n)\tilde{u} \rangle \, dn = \int_{N_k} \langle \pi(\lambda)\pi(n)u, \tilde{u} \rangle \, dn,$$

where $k = i - 2a$. Thus, if $a \geq i$ then the right hand side is 0. Since $\tilde{u}$ is in $\tilde{V}^{K_i}$, the left hand side is equal to $\langle \pi(\lambda)u, \tilde{u} \rangle$, up to a non-zero factor (the measure of N_i). Therefore, $\varphi_{v,\tilde{v}}(k\lambda k') = 0$ for any two elements k and k' in K. The proposition follows form the Cartan decomposition. □

Proposition 6.3: *Let V be an irreducible cuspidal representation of G°, and $P : U \to V$ a non-trivial homomorphism of smooth G°-modules. Then the above sequence splits, that is there exists a homomorphism $s : V \to U$ such that $P \circ s(v) = v$ for all v in V.*

Proof: Let $\tilde{V}$ be the contragredient dual of V. Pick v any vector in V. We first define a skew-linear (with respect to the complex conjugation) map $i : \tilde{V} \to V$ by

$$i(\tilde{v}) = \int_{G^\circ} \overline{\varphi}_{v,\tilde{v}}(g)\pi(g)v \, dg.$$

Since $\varphi_{v,\tilde{\pi}(h)\tilde{v}} = \varphi_{v,\tilde{v}}(h^{-1}g)$, the map $\tilde{i}$ is a skew-linear homomorphism. Since V is irreducible, i must be onto. Finally, since

$$\langle i(\tilde{v}), \tilde{v} \rangle = \int_{G^\circ} |\varphi_{v,\tilde{v}}(g)|^2 \, dg > 0,$$

we see that i is injective, as well. Now pick any u in U such that $P(u) = v$. Define, analogously to i, a skew-linear map from $\tilde{V}$ to U by

$$j(\tilde{v}) = \int_{G^\circ} \overline{\varphi}_{v,\tilde{v}}(g)\pi(g)u \, dg.$$

(Here π denotes the action of G on U). Now put $s = j \circ i^{-1}$. Clearly, s is a linear homomorphism from V to U, and by Schur's Lemma $P \circ s$ has to be a multiple of the identity operator. Since $P(s(v)) = v$, for our fixed vector v, the proposition follows. □

Exercise. The proof of the above proposition shows that smooth irreducible cuspidal representations are admissible. Hint: The map i defines a skew-linear isomorphism of $\tilde{V}$ and V. Now use the fact that $\tilde{V}^{K_i}$ is equal to the linear dual of V^{K_i} (Proposition 4.5).

Corollary 6.4: *Every irreducible smooth representation V of $GL_2(F)$ is admissible.*

Proof: We have just seen that cuspidal representations are admissible. Otherwise, the representation is a submodule of a principal series representation by Proposition 5.3, so it is again admissible. □

7. Decomposing principal series

In this section we shall describe the composition series of $Ind_B^G(\chi)$. Some critical calculations will be performed in Section 8.

Proposition 7.1: *Let V be an irreducible subquotient of $Ind_B^G(\chi)$. Then $V_N \neq 0$.*

Proof: If $V_N = 0$, then V is cuspidal, so it has to be a submodule as well. But then by Frobenius reciprocity $V_N \neq 0$. Contradiction. □

Since $V_N \neq 0$ for any subquotient, the following proposition shows that the length of a composition series of $Ind_B^G(\chi)$ is at most two. The proof is based on the Bruhat-Tits decomposition $B \cup BwN$ of G, where

$$w = \begin{pmatrix} 0 & 1 \\ 1 & 0 \end{pmatrix}.$$

Proposition 7.2: *Let χ be a character of T, and χ^w be the character obtained by conjugating χ by w, which simply means permuting χ_1 and χ_2. Then*

$$0 \to \delta^{1/2}\chi^w \to Ind_B^G(\chi)_N \to \delta^{1/2}\chi \to 0.$$

Proof: Let $Ind_B^G(\chi)_w$ be the B-submodule of functions in $Ind_B^G(\chi)$ supported on the big cell BwN. We shall use the filtration

$$Ind_B^G(\chi)_w \subseteq Ind_B^G(\chi)$$

and the exactness of $V \rightsquigarrow V_N$ to calculate $Ind_B^G(\chi)$.

Define a functional α on $Ind_B^G(\chi)$ by

$$\alpha(f) = f(1).$$

Since

$$\alpha(\pi(nt)f) = \delta^{1/2}\chi(t) \cdot \alpha(f)$$

we have shown that $\delta^{1/2}\chi$ is a quotient of $Ind_B^G(\chi)_N$. Clearly, $Ind_B^G(\chi)_w$ is the kernel of α. Define a functional α_w on $Ind_B^G(\chi)_w$ by

$$\alpha_w(f) = \int_N f(wn)\, dn.$$

Exercise. Show that the kernel of α_w is precisely $Ind_B^G(\chi)_w(N)$. Hint: Note that $Ind_B^G(\chi)_w$ is isomorphic to $C_0^\infty(N)$, as an N-module, and apply Lemma 5.2.

An easy calculation shows that

$$\alpha_w(\pi(nt)f) = \delta^{1/2}\chi^w(t) \cdot \alpha_w(f).$$

The proposition is proved. □

We now note a very important consequence of the above result. If $\chi \neq \chi^w$ (such χ is called regular) then the above exact sequence splits. This implies that α_w extends to $Ind_B^G(\chi)$, and thus by the Frobenius reciprocity defines a non-zero intertwining operator

$$A_w(\chi) : Ind_B^G(\chi) \to Ind_B^G(\chi^w)$$

Proposition 7.3: *The following gives a complete answer to decomposing principal series representations of $GL_2(F)$.*

- *The induced representation $Ind_B^G(\chi)$ is irreducible if $\chi_1/\chi_2 \neq |\cdot|^{\pm 1}$.*
- *Otherwise, the composition series is of length two, with one of the subfactors a one-dimensional representation.*

Proof: The singular case $\chi = \chi^w$ will be dealt with later. Assume that $\chi \neq \chi^w$. Then the Frobenius reciprocity implies that

$$Hom(Ind_B^G(\chi), Ind_B^G(\chi))$$

is one-dimensional, so every intertwining operator is a multiple of the identity I. In particular, it follows that

$$A_w(\chi^w) \circ A_w(\chi) = c(\chi) \cdot I$$

for some scalar $c(\chi)$ (called c-function).

Lemma 7.4: *If $\chi \neq \chi^w$, then $Ind_B^G(\chi)$ reduces if and only if $c(\chi) = 0$.*

To verify this lemma, recall that $Ind_B^G(\chi)$ can have at most two subquotients. Let V be a submodule. Then, V is a proper submodule if and only if $V_N = \delta^{1/2}\chi$. In that case $Ind_B^G(\chi)$ has a proper quotient U and $U_N = \delta^{1/2}\chi^w$. The Frobenius reciprocity implies that $A_w(\chi)$ annihilates V. Likewise, $A_w(\chi^w)$ annihilates U, which is a submodule of $Ind_B^G(\chi^w)$. It follows that $A_w(\chi^w) \circ A_w(\chi) = 0$. The lemma is proved.

The first part of the proposition follows as soon as we calculate $c(\chi)$. This will be done in the next section. Thus, assume that $\chi_1/\chi_2 = |\cdot|^{\pm 1}$. In fact, since subquotients of $Ind_B^G(\chi)$ are the same as subquotients of $Ind_B^G(\chi^w)$, it suffices to consider $\chi_1/\chi_2 = |\cdot|^{-1}$. In this case $\gamma(det)$ appears as a submodule, where

$$\gamma = \chi_1|\cdot|^{1/2} = \chi_2|\cdot|^{-1/2}.$$

If γ is trivial, then the other (infinite-dimensional) subquotient is called the Steinberg representation. Obviously, for other γ, the infinite dimensional subquotient is just the twist of the Steinberg representation by $\gamma(det)$. □

8. c-function

The purpose of this section is to obtain an explicit formula for the c-function $c(\chi)$. We also obtain a result which plays a crucial role in proving that $Ind_B^G(\chi)$ is irreducible for singular χ. In order to keep notation simple, we shall restrict ourselves to the family $I(s) = Ind_B^G(\chi)$ where

$$\chi\begin{pmatrix} a_1 & 0 \\ 0 & a_2 \end{pmatrix} = |a_1/a_2|^{s/2}.$$

If $s \neq 0$ let

$$A_w(s) : I(s) \to I(-s)$$

be the intertwining operator which, via the Frobenius reciprocity, corresponds to the functional α_w. Recall the definition of α_w. On the subspace of functions f such that $f(1) = 0$, it is defined by

$$\alpha_w(f) = \int_N f(wn)\, dn,$$

and, if $s \neq 0$, then α_w is the unique extension to $I(s)$. In order to calculate $A_w(-s) \circ A_w(s)$, we need to obtain an explicit formula for α_w on the whole $I(s)$. In other words, our task is to *regularize* this integral for any f in $I(s)$.

To do so, we shall now present some key technical ingredients. Let dx be the Haar measure on F, and for every s in $\mathbb{C}$, let λ_s be the functional on $C_0^\infty(F^\times)$ defined by

$$\lambda_s(f) = \int_{F^\times} f(x)|x|^s \frac{dx}{|x|}.$$

Let π denote the (right) regular representation of $F^\times$ on $C_0^\infty(F^\times)$. The functional λ_s is equivariant with respect to π in the sense that

$$\lambda_s(\pi(a)f) = |a|^{-s}\lambda_s(f).$$

Proposition 8.1: *The functional λ_s extends to an equivariant functional on $C_0^\infty(F)$ if and only if $s \neq 0$. The extension is unique if $s \neq 0$.*

Proof: Let Λ_s be an extension, if any. Let f be in $C_0^\infty(F)$. Then the function $f - \pi(\omega)f$ is in $C_0^\infty(F^\times)$. In particular,

$$\Lambda_s(f - \pi(\varpi)f) = \int_{F^\times} [f(x) - f(\varpi x)]|x|^s \frac{dx}{|x|}.$$

On the other hand, since Λ_s is equivariant, we have $\Lambda_s(f - \pi(\varpi)f) = (1 - |\varpi|^{-s})\Lambda_s(f)$. Putting things together,

$$(1 - q^s)\Lambda_s(f) = \int_{F^\times} [f(x) - f(\varpi x)]|x|^s \frac{dx}{|x|}.$$

This formula shows that the extension exists and is unique if $1 - q^s$ is not zero. Otherwise, which happens precisely when $s = 0$, the left hand side is always zero, while the right hand side can be easily arranged to be non-zero (take, for example, the characteristic function of $\mathcal{O}$). In particular, the functional λ_0 does not extend. The proposition is proved. □

We shall now apply this idea to our situation. For any f in $I(s)$ define

$$f'(g) = |\varpi|^{-(s+1)/2} \cdot \pi \begin{pmatrix} \varpi & 0 \\ 0 & 1 \end{pmatrix} f(g).$$

Next note that $f - f'$ vanishes at 1, so it is supported on the big double coset NwB. Arguing exactly as in the proof of the previous proposition we see that an extension of α_w to $I(s)$ exists if and only if $s \neq 0$, in which case it is given by

$$\alpha_w(f) = (1-q^{-s})^{-1}\int_N (f-f')(wn)\, dn.$$

Corollary 8.2: $Hom_G(I(0), I(0)) = \mathbb{C}$.

This corollary follows from the Frobenius reciprocity, and the fact that α_w does not extend to $I(0)$. Combined with the fact that $I(0)$ is completely decomposable, which will be shown in the next section, it will imply that $I(0)$ is irreducible.

Proposition 8.3: *Let* $L(s) = 1/(1-q^{-s})$. *Then*

$$c(s) = \frac{L(-s)L(s)}{L(1-s)L(1+s)}.$$

In particular, $c(s) = 0$ *if and only if* $s = \pm 1$.

Proof: In order to calculate $A_w(-s) \circ A_w(s)$, we need to make the formula for $\alpha_w(f)$ more explicit. Clearly, f is K_i-invariant for some i, and we shall show that $f(1)$ determines $f(wn)$ for all n in $N \setminus N_{-i+1}$. Indeed, note the following relation in $SL_2(F)$:

$$\begin{pmatrix} 1 & x \\ 0 & 1 \end{pmatrix} = \begin{pmatrix} 1 & 0 \\ x^{-1} & 1 \end{pmatrix}\begin{pmatrix} 0 & x \\ -x^{-1} & 0 \end{pmatrix}\begin{pmatrix} 1 & 0 \\ x^{-1} & 1 \end{pmatrix}.$$

Let $j \leq -i$, and assume that x is in $\varpi^j\mathcal{O} \setminus \varpi^{j+1}\mathcal{O}$. Since f is right K_i-invariant, and left N-invariant, the relation implies that

$$f(w\begin{pmatrix} 1 & x \\ 0 & 1 \end{pmatrix}) = |x|^{-s-1}f(1).$$

In particular, it follows that $f - f'$ is supported on N_{-i}.

Next, note that the Iwasawa decomposition implies that $I(s)$ has a unique element f_s such that $f_s(gk) = f_s(g)$ for all k in K, and $f_s(1) = 1$. The above calculation can be made very precise for f_s:

Exercise. Show that

$$\alpha_w(f_s) = L(s)/L(1+s).$$

Since $A_w(f_s)$ is K-invariant, it must be a multiple of f_{-s}. The above exercise shows that the multiple is $L(s)/L(1+s)$. The proposition follows. □

9. Unitary representations

A smooth representation of G is called unitarizable if it admits a positive definite G-invariant hermitian form. In this section we shall first show that $Ind_B^G(\chi)$ is unitarizable if χ is a unitary character ($\chi \cdot \bar{\chi} = 1$). Combined with Corollary 8.2 this will (finally) show that $I(0)$ is irreducible. We finish this section by showing that the representations $I(s)$ with $0 < s < 1$ (the so called complementary series) are also unitarizable.

The Iwasawa decomposition implies that each representation $I(s)$ has a unique (up to a non-zero constant) K-invariant functional given by

$$f \mapsto \int_K f(k)\, dk.$$

Since the trivial representation of G is a quotient of $I(1)$, the K-invariant functional is also G-invariant if $s = 1$. Now we can easily construct a positive definite G-invariant hermitian form on $Ind_B^G(\chi)$. Let f and g be any two functions in $Ind_B^G(\chi)$. If χ is unitary, then $f \cdot \bar{g}$ is in $I(1)$, so the hermitian form

$$\langle f, g \rangle = \int_K f(k)\bar{g}(k)\, dk$$

is clearly positive definite and G-invariant.

Proposition 9.1: *The representation $I(0)$ is irreducible.*

Proof: Since $I(0)$ is unitarizable, and of finite length, it will decompose as a direct sum of irreducible constituents. In particular, $Hom(I(0), I(0))$ is isomorphic to a direct sum of matrix algebras $M_n(\mathbb{C})$, one for each irreducible constituent of multiplicity n. On the other hand, we know that $Hom(I(0), I(0)) = \mathbb{C}$. The proposition follows. □

Our last topic for $GL_2(F)$ will be a construction of the unitary structure on $I(s)$ for $-1 < s < 1$. We first need to renormalize the intertwining operators $A_w(s)$, so that they are defined at the singular point $s = 0$ as well. Recall that $A_w(s)$ corresponds, via the Frobenius reciprocity, to the functional

$$\alpha_w(f) = L(s) \int_N (f - f')(wn)\, dn.$$

This is well defined only if $s \neq 0$ since $L(s)$ has a pole at $s = 0$. Let $\mathbf{A}_w(s)$ be the operator which corresponds to the functional $\alpha_w L(1+s)/L(s)$. Then $\mathbf{A}_w(s)$ is well defined for every $-1 < s < 1$, and

$$\mathbf{A}_w(s) f_s = f_{-s}.$$

We can now easily define a G-invariant hermitian form on $I(s)$ as follows. Let f and g be two functions in $I(s)$. Then $\mathbf{A}_w(s)f$ is in $I(-s)$, and as above,

$$\langle f, g \rangle = \int_K (\mathbf{A}_w(s)f)(k)\bar{g}(k)\, dk$$

defines a G-invariant sesquilinear form on $I(s)$. Apriori, it is not clear that this form is hermitian. However, Schur's Lemma implies that an irreducible smooth G-module can have only one sesquilinear G-invariant form, up to a non-zero scalar. Thus the form $\overline{\langle g, f \rangle}$ must be proportional to the form $\langle f, g \rangle$. Since $\langle f_s, f_s \rangle = 1$, it follows that they are equal, so our form is hermitian, after all.

Next, note that $\mathbf{A}_w(0)$ is the identity operator on $I(0)$. Indeed, we know that it is a multiple of the identity operator, but the formula for the spherical vector shows that it is in fact equal to the identity operator. It follows that the form is positive definite on $I(0)$. We shall use this observation to prove the following proposition.

Proposition 9.2: *If $0 < s < 1$, then the G-invariant hermitian form $\langle \cdot, \cdot \rangle$ on $I(s)$ is positive definite.*

Proof: Since $I(s)$ is irreducible for $-1 < s < 1$ the G-invariant form must be non-degenerate (otherwise the kernel of the form would be an invariant subspace). Moreover, as the following exercise shows, the restriction of this form to every finite-dimensional subspace of K_i-invariants is also non-degenerate.

Exercise. Let V be a smooth G-module with a non-degenerate G-invariant hermitian form. Then the restriction of the form to V^{K_i} is also non-degenerate. Hint: Use the K_i-invariant projection operator P from V onto V^{K_i}.

Every element in $I(s)$ is completely determined by it restriction to K. In this way, we can identify, in a K-invariant fashion, $I(s)^{K_i}$ with the (finite-dimensional) space E_i of functions on $B \cap K \backslash K / K_i$. By the above exercise, the invariant form on $I(s)$ induces a non-degenerate form $\langle \cdot, \cdot \rangle_s$ on E_i, which is positive definite for $s = 0$. The proposition follows from the following exercise.

Exercise. Let E be a finite dimensional vector space, and $\langle \cdot, \cdot \rangle_s$ a family of non degenerate hermitian symmetric forms, depending continuously on the parameter s in a connected open set in $\mathbb{R}$. If $\langle \cdot, \cdot \rangle_{s_0}$ is positive definite for some s_0 in the open set, then it is positive definite for all s. □

10. Steinberg representation

By Proposition 7.3 we now know that the principal series $I(s)$ decomposes if and only if $s = \pm 1$, in which case it has a composition series of length 2. One subquotient is the trivial representation, and the other is the Steinberg representation V. Recall that

$$V_N = \delta.$$

We shall show that the Steinberg representation is *square integrable*, which means that the matrix coefficients of V restricted to G^0 are square integrable. The results of this section are, in large part, due to Casselman.

Proposition 10.1: *Let V be the Steinberg representation of G. Then the matrix coefficients of V are square integrable, when restricted to G^0.*

Proof: To prove this proposition, we need a result on the asymptotics of matrix coefficients of V. Recall that

$$\lambda_a = \begin{pmatrix} \varpi^a & 0 \\ 0 & \varpi^{-a} \end{pmatrix}.$$

Lemma 10.2: *Let V be the Steinberg representation of G. Then, for every v in V, and $\tilde{v}$ in $\tilde{V}$, there exists a positive integer i such that*

$$\langle \pi(k\lambda_{a+i}k')v, \tilde{v} \rangle = \delta(\lambda_a)\langle \pi(k\lambda_i k')v, \tilde{v} \rangle.$$

for every non-negative integer a, and any two elements k and k' in K.

Proof: The proof of this lemma is analogous to the proof of Proposition 6.2. Let U and $\tilde{U}$ be the finite dimensional subspaces generated by v and $\tilde{v}$ over K, respectively. Let u be in U, and for every positive integer a, define

$$u_a = \pi(\lambda_a)u - \delta(\lambda_a)u.$$

Since $V_N = \delta$ it follows that u_1 is in $V(N)$. Recall that $V(N)$ is a union of $V(N_j)$, where $V(N_j)$ is the kernel of the operator

$$P_j(u) = \int_{N_j} \pi(n)u\, dn.$$

Take i large enough so that every u_1 is contained in $V(N_{-i})$ and $\tilde{U}$ is contained in $\tilde{V}^{K_i}$. This is our choice of i.

Exercise: Show that $\pi(\lambda_1)V(N_{-i}) \subset V(N_{-i+2})$. *In particular,* $\pi(\lambda_1)V(N_{-i})$ *is again contained in* $V(N_{-i})$.

Since

$$u_a = \pi(\lambda_1)u_{a-1} - \delta(\lambda_1)u_{a-1},$$

an induction argument based on the previous exercise implies that u_a is in $V(N_{-i})$ for all positive integers a. Thus, for every u in U,

$$P_{-i}(u_a) = 0$$

and if $\tilde{u}$ is in $\tilde{U}$,

$$\int_{N_i} \langle \pi(\lambda_i)u_a, \tilde{\pi}(n)\tilde{u}\rangle\, dn = \int_{N_{-i}} \langle \pi(\lambda_i)\pi(n)u, \tilde{u}\rangle\, dn = 0.$$

It follows that $\langle \pi(\lambda_i)u_a, \tilde{u}\rangle = 0$, as desired. The lemma is proved. □

We can now finish the proof of the proposition. Recall the Cartan decomposition

$$G^0 = \cup_{a=0}^{\infty} K\lambda_a K.$$

The measure of $K\lambda_a K$ is equal to the index of $K \cap K\lambda_a K$ in K, which is essentially (see the first exercise in Section 4) $\delta(\lambda_a)^{-1} = q^{2a}$. This and Lemma 10.2 imply that

$$\int_{G^0} |\varphi_{v,\tilde{v}}|^2\, dg \le C \sum_{a=0}^{\infty} q^{-2a},$$

for some positive constant C. This series clearly converges, and the proposition is proved.

11. A cuspidal representation of G^0

In this section we present an example of a cuspidal representation. To make the example as simple as possible, the trick (which the author learned from Allen Moy) is to restrict to the case $p = 2$. Thus, in this section only, our groups are defined over the dyadic field $\mathbb{Q}_2$.

Let $\mathbb{F}_2 = \{0, 1\}$ be the residual field of $\mathbb{Q}_2$. Consider the projective line

$$\mathbb{P}(2) = \{(1,0), (1,1), (0,1)\}$$

over $\mathbb{F}_2$. The action of $K/K_1 = GL_2(2)$ on $\mathbb{P}(2)$ gives an isomorphism of $GL_2(2)$ with S_3, the symmetric group on 3 letters. Let ϵ be the usual sign character on S_3. Then one easily checks that

$$\epsilon \begin{pmatrix} 1 & 1 \\ 0 & 1 \end{pmatrix} = -1.$$

Pull ϵ back to K, and define $V = ind_K^{G^0}(\epsilon)$. The symbol *ind* stands for the induction with compact support. In particular, V consists of compactly supported functions on G^0 such that $f(kg) = \epsilon(k)f(g)$ for all k in K and g in G^0.

Proposition 11.1: *The representation V is an irreducible cuspidal representation of G^0.*

Proof: We shall first prove that $V_N = 0$. First of all, note that

$$\int_{N \cap K} \epsilon(n)\, dn = 0.$$

Let e_K in V be the function supported on K, such that $e_K(k) = \epsilon(k)$ for all k in K. The Iwasawa decomposition implies that any element in V is a linear combination of $\pi(b)e_K$ where b is in B. Let $N_b = b^{-1}(N \cap K)b$. The vanishing of the integral above implies that

$$\int_{N_b} \pi(n)\pi(b)e_K\, dn = 0.$$

It follows that V is cuspidal by Lemma 5.2. It remains to prove that it is irreducible.

Let $H_\epsilon(G^0//K)$ be the Hecke algebra of compactly supported functions on G^0 such that

$$f(kgk') = \epsilon(k)f(g)\epsilon(k').$$

for all k and k' in K. As usual, the multiplication in $H_\epsilon(G^0//K)$ is defined by convolution of functions

$$(f_1 * f_2)(g) = \int_{G^0} f_1(h) f_2(h^{-1}g)\, dg.$$

Note that $H_\epsilon(G^0//K)$ is contained in V, and equal to $V^{K,\epsilon}$, the subspace consisting of all functions f in V such that $\pi(k)f = \epsilon(k)f$ for all k in K. The element e_K defined above is the identity element in $H_\epsilon(G^0//K)$.

Lemma 11.2: *There is a natural isomorphism* $Hom_{G^0}(V,V) \cong H_\epsilon(G^0//K)$.

Proof: Let A be in $Hom_{G^0}(V,V)$. Since V is generated by e_K, the operator A is completely determined by its value $A(e_K)$. Since $A(e_K)$ lies in $V^{K,\epsilon}$, which is the same as $H_\epsilon(G^0//K)$, the map

$$A \mapsto A(e_K)$$

furnishes a canonical injection from $Hom_{G^0}(V,V)$ to $H_\epsilon(G^0//K)$. Conversely, an element T in $H_\epsilon(G^0//K)$ defines an element A in $Hom_{G^0}(V,V)$ by

$$A(f) = T * f.$$

Since $A(e_K) = T * e_K = T$, we see that $A \mapsto A(e_K)$ is surjective as well. The lemma is proved. □

Lemma 11.3: *The algebra* $H_\epsilon(G^0//K)$ *is one-dimensional.*

Proof: Recall the Cartan decomposition

$$G^0 = \cup_{a=0}^{\infty} K\lambda_a K$$

where

$$\lambda_a = \begin{pmatrix} 2^a & 0 \\ 0 & 2^{-a} \end{pmatrix}.$$

Let f be in $H_\epsilon(G^0//K)$. We shall show that f is a multiple of e_K. Clearly, f is determined by its values at all λ_a. Let

$$n = \begin{pmatrix} 1 & 2^a \\ 0 & 1 \end{pmatrix} \text{ and } n' = \begin{pmatrix} 1 & 1 \\ 0 & 1 \end{pmatrix}.$$

If $a \geq 1$, then n is in K_1 and $\epsilon(n) = 1$. On the other hand, recall that $\epsilon(n') = -1$. It follows that

$$f(\lambda_a) = f(n\lambda_a) = f(\lambda_a n') = -f(\lambda_a).$$

Thus, $f(\lambda_a) = 0$ if $a \geq 1$, and f must be a multiple of e_K. The lemma is proved. □

We can now finish the proof of proposition. Since V is generated by e_K, Proposition 4.2 implies that there exists an irreducible quotient V' of V. Let P be the projection from V onto V'. Since V' is cuspidal, by Proposition 6.3 there is a splitting $s : V' \to V$ such that $P \circ s = Id_{V'}$. But

$$s \circ P \in Hom_{G^0}(V, V) = \mathbb{C} \cdot Id_V$$

so $s \circ P$ must be equal to Id_V. It follows that $V = V'$, and the proposition is proved.

12. The root system for $GL_n(F)$

The purpose of this section is to introduce the root system for GL_n. It is a combinatorial object which provides us with a language to study representations of $GL_n(F)$. (For more information on root systems see [2]).

Let $\mathbb{R}$ be the field of real numbers, and consider the space $\mathbb{R}^n$ with the standard basis $e_1, \ldots, e_n$. With respect to this basis, we shall identify every element x in $\mathbb{R}^n$ with an n-tuple $(x_1, \ldots, x_n)$. The group S_n acts on $\mathbb{R}^n$ by permuting the entries of $(x_1, \ldots, x_n)$. This action preserves the inner product

$$\langle x, y \rangle = \sum_{i=1}^{n} x_i y_i.$$

Let

$$\Omega = \{x \in \mathbb{R}^n \mid \sum_i x_i = 0\}.$$

Note that the group S_n preserves Ω. In fact, it is an irreducible representation. We shall now describe the action of S_n by means of Euclidean reflections. Let

$$\Psi = \{e_i - e_j \mid i \neq j\},$$

be a finite set in Ω. Its elements are called *roots.* Any root α defines a reflection w by

$$w(x) = x - \langle \alpha, x \rangle \cdot \alpha.$$

Exercise. Let $\alpha = e_i - e_j$. *Show that the corresponding* w *is simply the permutation of* i *and* j.

It follows from the exercise that the reflections w define the representation of S_n on Ω. Clearly, the fixed points of the reflection corresponding to the root $\alpha = e_i - e_j$ is the hyperplane $\{x_i = x_j\}$, defined by the equation $x_i = x_j$. Consider the open subset obtained by removing all hyperplanes $\{x_i = x_j\}$

$$\Omega^\circ = \Omega \setminus (\cup\{x_i = x_j\}).$$

The connected components of Ω° are called *chambers.* We shall point out a particular chamber given by

$$C^+ = \{x \in \Omega \mid x_1 > \ldots > x_n\}.$$

Next, note that the entries of any x in Ω° are pair-wise different, so there exists a unique permutation which puts them in a decreasing order. This means that Ω° is a disjoint union of $w(C^+)$ as w runs thru S_n. In particular, every chamber C is equal to

$$C = \{x \in \Omega \mid x_{i_1} > \ldots > x_{i_n}\}$$

for some permutation $(i_1, \ldots, i_n)$.

The choice of C^+ gives us a set of generators of S_n as follows. Note that the closure of C^+ is obtained by adding parts (called *walls*) of hyperplanes $\{x_i = x_{i+1}\}$. The corresponding roots

$$\{e_1 - e_2, \ldots, e_{n-1} - e_n\}$$

are called simple. Note that the simple roots form a basis of Ω. Let w_i be the (simple) reflection defined by $e_i - e_{i+1}$. We claim that the reflections w_i generate S_n. Of course, since w_i is nothing but the permutation $(i, i+1)$; this is a well known fact. We shall give a proof based on the fact that S_n acts simply transitively on the set of chambers. Let w be an element in S_n. Pick x^+ in C^+ and x in $w(C^+)$ such that the segment between them avoids the singularities of $\Omega \setminus \Omega^\circ$. Since those singularities form a codimension 2 subvariety, a generic choice of x^+ and x will do. The number of hyperplanes intersected by the segment is clearly independent of the particular choices, is the *length* $\ell(w)$ of w. The chambers containing the segment form a *gallery*

$$C^+, C_1, \dots C_{\ell(w)} = w(C^+).$$

Any two consecutive chambers in this gallery share a wall. In particular, there exists a simple reflection w_{i_1} such that $C_1 = w_{i_1}(C^+)$. Next, note that the walls of C_1 correspond to reflections w_{i_1} (the shared wall) and $w_{i_1} w_j w_{i_1}^{-1}$ for all $j \neq i_1$. In particular, there exists i_2 such that $C_2 = w_{i_1} w_{i_2} w_{i_1}^{-1}(C_1)$, which implies $C_2 = w_{i_2} w_{i_1}(C^+)$. Continuing in this fashion, we can obtain $w_{i_1}, \dots, w_{i_{\ell(w)}}$ such that

$$w_{i_{\ell(w)}} \cdot \ldots \cdot w_{i_1}(C^+) = w(C^+).$$

Exercise. Make a picture of the root system for $GL_3(F)$. The roots form a regular hexagon. Let w be the element in W such that $w(C^+) = C^-$. Express w as a product of simple reflections by picking a gallery between C^+ and C^-.

13. Induced representations for $GL_n(F)$

Let $e_1, e_2, \dots, e_n$ be the standard basis of F^n. Let V_r be the span of $e_1, \dots, e_r$. Consider a partial flag

$$V_{i_1} \subset V_{i_2} \subset \ldots \subset V_{i_m}$$

and let P be its stabilizer in G. Then P is called a parabolic subgroup of G, and it admits the Levi decomposition $P = MN$, where M is isomorphic to the product of $GL(V_i/V_{i-1})$, and U is the unipotent radical of P. Note that the minimal parabolic subgroup B corresponds to the full flag.

If (τ, E) is a smooth representation of M, then we can induce it to G by

$$Ind_P^G(\tau) = \{f : G \to \tau \mid f(umg) = \delta_P^{1/2}\tau(m)f(g)\}^\infty.$$

The character δ_P of M is defined so that for every locally constant, compactly supported function h on U we have

$$\int_U h(mum^{-1})\, du = \delta_P(m) \int_U h(u)\, du.$$

As a particular case of interest, consider the minimal parabolic subgroup $B = TU$. Note that any smooth character χ of T is given by

$$\chi\begin{pmatrix} a_1 & & \\ & \ddots & \\ & & a_n \end{pmatrix} = \chi_1(a_1)\cdot\ldots\cdot\chi_n(a_n)$$

for some smooth characters $\chi_1 \ldots \chi_n$ of $F^\times$. Assume that χ is regular, which means that $\chi \neq \chi^w$ for any w in W. In the case of $GL_n(F)$ this simply means that $\chi_i \neq \chi_j$ if $i \neq j$. As in the case of $GL_2(F)$, the Bruhat-Tits decomposition can be used to show that

$$Ind_B^G(\chi)_N = \bigoplus_{w\in W} \delta^{1/2}\chi^w.$$

It turns out that $V_N \neq 0$ for any irreducible subquotient, just as in the case of $GL_2(F)$. In particular, each irreducible subquotient is completely determined by V_N. Let S be the set of all $\{i,j\}$ such that $\chi_i/\chi_j = |\cdot|^{\pm 1}$. Let

$$\Omega_\chi = \Omega \setminus (\cup_S\{x_i = x_j\})$$

where $\{x_i = x_j\}$ is the hyperplane in Ω defined, of course, by the given equation. A result of Rodier states that the irreducible subquotients corresponds to the connected components of Ω_χ. We shall here give a special case of that result, describing V_N for the unique irreducible submodule V. (Since any irreducible subquotient of $Ind_B^G(\chi)$ is a submodule of $Ind_B^G(\chi^w)$ for some w in W, the general case easily follows.)

Let Ω_χ^+ be the connected component containing the positive chamber C^+. Define W_χ to be the set of all Weyl group elements w such that

$$w(C^+) \subset \Omega_\chi^+.$$

Proposition 13.1: *Let χ be a regular character. Let V be the unique irreducible submodule of $Ind_B^G(\chi)$ (normalized induction). Then*

$$V_N = \bigoplus_{w\in W_\chi} \delta^{1/2}\chi^w.$$

Proof: The proof of this statement is combinatorial, and based on the special case of $GL_2(F)$, proved in Proposition 7.3. Let P_i be the parabolic subgroup corresponding to the flag with V_i omitted. Note that its Levi factor M_i has exactly one factor isomorphic to GL_2, and let w_i $(= w)$ be the permutation matrix in that factor. Assume that $\chi_i/\chi_{i+1} \neq |\cdot|^{\pm 1}$, so w_i is in W_χ. Then, by Proposition 7.3

$$Ind_B^{P_i}(\chi) \cong Ind_B^{P_i}(\chi^{w_i}).$$

Inducing both representations further up to G, this gives $Ind_B^G(\chi) \cong Ind_B^G(\chi^{w_i})$. In particular, χ^{w_i} is a summand of V_N. For a general element w of length m in W_χ we have to consider a gallery

$$C^+, C_1, \ldots, C_m = w(C^+)$$

which is contained in Ω_χ^+ (since Ω_χ^+ is convex), and repeat the above argument m-times. This shows that

$$V_N \subseteq \bigoplus_{w \in W_\chi} \delta^{1/2}\chi^w.$$

The opposite inclusion will follow from the following lemma:

Lemma 13.2: *Assume that $\chi_i/\chi_j = |\cdot|^{\pm 1}$ for some $i < j$. If χ^w is a summand of V_N then C^+ and $w(C^+)$ are on the same side of the hyperplane $\{x_i = x_j\}$.*

To be specific, assume that $\chi_i/\chi_j = |\cdot|$. Let w' be the permutation $(i+1, j)$, and $\chi' = \chi^{w'}$ be the character of T obtained by permuting χ_j and χ_{i+1}. Then $\chi_i'/\chi_{i+1}' = |\cdot|$. Inducing in stages (first from B to P_i), and using the second part of Proposition 7.3 ($GL_2(F)$-reducibility) we get an inclusion

$$Ind_{P_i}^G(\eta) \subset Ind_B^G(\chi'),$$

where η is a character of M_i, which on the $GL_2(F)$-factor is given by composing the determinant with the character

$$\chi_i|\cdot|^{1/2} = \chi_{i+1}|\cdot|^{-1/2}.$$

A more general form of the Bruhat-Tits decomposition implies that

$$Ind_{M_i}^G(\eta)_N = \bigoplus_{w \in W_i} \delta^{1/2}(\chi')^w$$

where W_i is the set of all w such that $w(i) < w(i+1)$. This, however, is equivalent to $ww'(i) < ww'(j)$, which happens if and only if C^+ and $ww'(C^+)$ are on the same side of the hyperplane $\{x_i = x_j\}$. Summarizing, we have

$$Ind_{M_i}^G(\eta)_N = \bigoplus_{w(C^+) \subseteq \{x_i > x_j\}} \delta^{1/2}\chi^w.$$

□

Corollary 13.3: *Let χ be a regular character, and let $\Omega_1, \ldots \Omega_m$ be the connected components of Ω_χ. Then $Ind_B^G(\chi)$ has m irreducible subquotients $V_1, \ldots, V_m$ so that*

$$(V_i)_N = \bigoplus_{w(C+)\subseteq\Omega_i} \delta^{1/2}\chi^w.$$

Exercise. Show that $Ind_B^G(\delta^{1/2})$ has 2^{n-1} subquotients. Hint: Try first the special case of $GL_3(F)$.

References

1. J. Bernstein and A. Zelevinsky, *Representations of the group $GL_n(F)$, where F is a non-archimedean local field*, Russian Math. Surveys 31 (1976), 1-68.
2. J. Humphreys, *Introduction to Lie Algebras and representation theory*, Graduate Texts in Mathematics 9, Springer-Verlag, 1978.
3. N. Koblitz, *p-adic Numbers, p-adic Analysis, and Zeta Functions*, Second Edition, Graduate Texts in Mathematics 58, Springer-Verlag, 1984.
4. F. Rodier, *Décomposition de la série principale des groupes réductifs p-adiques*, 408-424, Lecture Notes in Mathematics 880, Springer-Verlag, 1981.
5. M. Tadić, *Representations of classical p-adic groups*, 129-204, Pitman Res. Notes Math. Ser. 311, Longman, Harlow 1994.

Lectures on Harmonic Analysis for Reductive p-adic Groups

Stephen DeBacker
Department of Mathematics
Harvard University
Cambridge, MA 02138, USA
E-mail: debacker@math.harvard.edu

1. Introduction

In his paper *The characters of reductive p-adic groups* [14] Harish-Chandra outlines his philosophy about harmonic analysis on reductive p-adic groups. According to this philosophy, there are two distinguished classes of distributions on the group: orbital integrals and characters. Similarly, there are two classes of distributions on the Lie algebra which are interesting: orbital integrals and their Fourier transforms. The real meat of his philosophy states that we ought to treat orbital integrals on both the group and its Lie algebra in the "same way" and similarly, we should think of characters and the Fourier transform of orbital integrals in the same way. This philosophy has many manifestations (see, for example, Robert Kottwitz's excellent article [12]). In this series of lectures, we will examine the various distributions discussed above and discuss one of the deepest connections between them: the Harish-Chandra–Howe local character expansion. Because of requests on the part of participants, I will spend much time reviewing the basics of p-adic fields and discussing some of the uses of Moy-Prasad filtrations in the representation theory of reductive p-adic groups. As such, at a small cost in terms of generality, I will concentrate on those techniques which use this theory. These lectures and the notes are meant as an informal and elementary introduction to the material. For complete, rigorous proofs, please see the references.

Very little of the material in this set of lectures is original. I have borrowed heavily from the work and lectures of Harish-Chandra, Roger Howe,

Robert Kottwitz, Allen Moy, Gopal Prasad, Paul Sally, Jr., and J.-L. Waldspurger, among others. I thank the organizers of this conference, in particular, Eng-Chye Tan and Chen-Bo Zhu, for inviting me and allowing me to present this series of tutorials. I thank Jeff Adler, Amritanshu Prasad, Joe Rabinoff, Loren Spice, and Chen-Bo Zhu for their helpful comments on earlier drafts of these notes.

2. Basics

2.1. *An introduction to the p-adics*

The usual introductory mathematical analysis course proceeds roughly as follows: The class agrees to agree that the set of natural numbers

$$\{1, 2, 3, \ldots\}$$

is very natural and therefore a good place to begin the course. In order to form a group with respect to addition, the additive identity and additive inverses are tossed in to the mix to give us the integers

$$\mathbb{Z} := \{\ldots, -4, -3, -2, -1, 0, 1, 2, 3, \ldots\}.$$

This set does not form a group with respect to multiplication; it is therefore enlarged to form $\mathbb{Q}$, the field of rational numbers. Everything so far has been very natural.

At this point, the incompleteness of the rationals is demonstrated by proving that the square root of two is not rational. To compensate for this, the fact that the rationals are ordered (that is, there is a notion of nonpositive and nonnegative) is invoked to define the absolute value, $|\cdot|$, of any rational number q:

$$|q| = \begin{cases} q & \text{if } q \geq 0 \\ -q & \text{if } q < 0. \end{cases}$$

As usual, the absolute value gives you a metric with respect to which you complete the rationals. We recall that two Cauchy sequences $\{q_n\}$ and $\{q'_n\}$ of rational numbers are said to be equivalent with respect to $|\cdot|$ provided that $\{q_n - q'_n\}$ is a Cauchy sequence which converges to zero. The set of real numbers is then defined to be the set of equivalence classes of Cauchy sequences with respect to $|\cdot|$, that is, the completion of $\mathbb{Q}$ with respect to $|\cdot|$.

It turns out, however, that the normal absolute value provides just one of an infinite number of incompatible ways to complete the rational numbers[a], and, from the proper perspective, this completion is a somewhat unnatural object.

Let p be any prime. If q is a nonzero rational number, then there is a unique integer k such that $q = p^k \cdot a/b$ with $p \nmid a$ and $p \nmid b$. We can then define the *p-adic absolute value*, $|\cdot|_p$, on $\mathbb{Q}$ by setting $|q|_p = 0$ if $q = 0$ and $|q|_p = p^{-k}$ otherwise. The p-adic absolute value has the following properties.

Exercise 2.1.1: If r_1 and r_2 are rational numbers, then

(1) $|r_1|_p \geq 0$, and $|r_1|_p = 0$ if and only if $r_1 = 0$,
(2) $|r_1 \cdot r_2|_p = |r_1|_p \cdot |r_2|_p$, and
(3) $|r_1 + r_2|_p \leq \max(|r_1|_p, |r_2|_p)$.

Exercise 2.1.2: In the last item of the previous exercise, show that if $|r_1|_p \neq |r_2|_p$ then the inequality is an equality. Is the converse of this statement true?

From Exercise 2.1.1, it follows that we can define a metric on $\mathbb{Q}$ with respect to the p-adic absolute value. We define $\mathbb{Q}_p$ to be the completion of $\mathbb{Q}$ with respect to $|\cdot|_p$. The p-adic absolute value on $\mathbb{Q}$ extends continuously (and uniquely) to a p-adic absolute value $|\cdot|_p : \mathbb{Q}_p \to \{0, p^k \mid k \in \mathbb{Z}\}$. We define the *valuation* ν_p on $\mathbb{Q}_p$ by $|x|_p = p^{-\nu_p(x)}$ for $x \in \mathbb{Q}_p^\times$ and $\nu_p(0) = \infty$. Thus $\nu_p(p^m) = m$ for $m \in \mathbb{Z}$.

We remark that Ostrowski's theorem tells us that the only nondiscrete, locally compact fields of characteristic zero are the field of real numbers, the field of complex numbers, and the finite extensions of $\mathbb{Q}_p$. That is, these are the only fields of characteristic zero on which integration naturally makes sense.

2.2. *The structure of $\mathbb{Q}_p$ and additive characters*

If we define $\mathbb{Z}_p$ to be the set of $x \in \mathbb{Q}_p$ such that $|x|_p \leq 1$, then $\mathbb{Z}_p$ is a maximal compact open subring in $\mathbb{Q}_p$. As in [33], it can be identified with the completion of $\mathbb{Z}$ with respect to $|\cdot|_p$. We call $\mathbb{Z}_p$ the *ring of integers* in $\mathbb{Q}_p$. The subset of $\mathbb{Z}_p$ consisting of those elements of $\mathbb{Z}_p$ with p-adic absolute value less than one forms a maximal ideal $(p) = p\mathbb{Z}_p$ of $\mathbb{Z}_p$. The quotient

[a]The field of real numbers is, up to isomorphism, the unique totally ordered completion of the rationals.

$\mathbb{Z}_p/(p)$ is isomorphic to $\mathbb{F}_p$, the field with p elements. It follows that we can write

$$\mathbb{Z}_p = \coprod_{0 \leq i \leq (p-1)} (i + p\mathbb{Z}_p).$$

As a slight tangent, we remark that Hensel's lemma (see nearly any book on local fields) tells us that we can lift a generator of $\mathbb{F}_p^\times$ to $\mathbb{Z}_p^\times$, the group of units in $\mathbb{Z}_p$. Thus, the square root of two is an element of, for example, $\mathbb{Q}_7$.

The set of compact open subrings $p^n\mathbb{Z}_p$ for $n \geq 0$ forms a neighborhood basis of zero. The sets $p^n\mathbb{Z}_p$ also make sense for $n < 0$, and we note that

$$\mathbb{Q}_p = \bigcup_{n \in \mathbb{Z}} p^n \mathbb{Z}_p.$$

Since $\mathbb{Z}_p^\times = \mathbb{Z}_p \setminus p\mathbb{Z}_p$, it follows immediately that

$$\mathbb{Q}_p = \{0\} \cup \coprod_{n \in \mathbb{Z}} p^n \mathbb{Z}_p^\times . \tag{1}$$

From this one can show that for each $x \in \mathbb{Q}_p^\times$ if $|x|_p = p^{-m}$, then there exist unique coefficients $a_n \in \{0, 1, 2, \ldots, (p-1)\}$ such that

$$x = \sum_{n=m}^{\infty} a_n \cdot p^n.$$

This is called the *p-adic expansion* of x. For example, the 5-adic expansion of $1/3$ is

$$2 + 3 \cdot (5) + 1 \cdot (5^2) + 3 \cdot (5^3) + 1 \cdot (5^4) + 3 \cdot (5^5) + \cdots .$$

If $x \in \mathbb{Q}_p$ and $|x|_p = p^{-m}$, then we define the *tail* of x to be the integer

$$t(x) = \begin{cases} \sum_{n=m}^{0} a_n \cdot p^n & \text{if } |x|_p = p^{-m} \text{ and } m \leq 0, \text{ and} \\ 0 & \text{if } x \in p\mathbb{Z}_p \ . \end{cases}$$

We then define

$$\Lambda(x) = e^{\frac{2 \cdot \pi \cdot i \cdot t(x)}{p}} .$$

Exercise 2.2.1: Show that Λ defines an additive character of $\mathbb{Q}_p$ and that the restriction to $p\mathbb{Z}_p$ of Λ is trivial, yet the restriction of Λ to $\mathbb{Z}_p$ is not trivial.

It is a fact that $\mathbb{Q}_p$ is isomorphic to the Pontrjagin dual $\widehat{\mathbb{Q}_p}$ of $\mathbb{Q}_p$ via the map $a \mapsto \big(b \mapsto \Lambda(ba)\big)$.

2.3. *General notation*

From now on, let k denote a finite extension of $\mathbb{Q}_p$, let R denote the ring of integers of k, and let ϖ denote a *uniformizer* so that $\wp = \varpi R$ where $\wp$ is the prime ideal. We define $\wp^m := \varpi^m \cdot R$ for $m \in \mathbb{Z}$. We let $\mathfrak{f} := R/\wp$ denote the *residue field*, and we let $q = |\mathfrak{f}|$. As before, we fix a valuation ν such that $\nu(k) = \mathbb{Z}$. We fix an additive character Λ of k that is trivial on $\wp$ and not trivial on R.

We let G denote the general linear group realized as the set of $n \times n$ matrices with entries in k having nonzero determinant. We let $\mathfrak{g}$ denote the Lie algebra of G. We realize $\mathfrak{g}$ as $\mathrm{M}_n(k)$, the set of $n \times n$ matrices with entries in the field k, having the usual bracket operation. Let dX denote a Haar measure on $\mathfrak{g}$ and let dg denote a Haar measure on G.

We let B denote the Borel subgroup of G consisting of upper triangular matrices, and we let T denote the maximal split torus consisting of diagonal matrices.

The map $(X, Y) \mapsto \mathrm{tr}(X \cdot Y)$ from $\mathfrak{g} \times \mathfrak{g}$ to k defines a G-invariant, non-degenerate, symmetric, bilinear form on $\mathfrak{g}$. As such, it allows us to identify $\mathfrak{g}$ with the Pontrjagin dual $\widehat{\mathfrak{g}}$ of $\mathfrak{g}$ via the map $X \mapsto \big(Y \mapsto \Lambda(\mathrm{tr}(X \cdot Y))\big)$.

A compact, open, R-submodule of $\mathfrak{g}$ is called a *lattice*. For example, for each integer i, we can define the *standard filtration lattice*

$$\mathfrak{k}_i := \varpi^i \cdot \mathrm{M}_n(R)$$

of $\mathfrak{g}$. As another example, consider the *Iwahori filtration lattices* $\mathfrak{b}_{i/n}$ defined by

$$\mathfrak{b}_{i/n} := \{Y \in \mathfrak{g} \,|\, Y_{jk} \in \varpi^{\lceil \frac{j-k+i}{n} \rceil} \cdot R\}.$$

Note that for all integers i, j, we have $\varpi^j \cdot \mathfrak{k}_i = \mathfrak{k}_{i+j}$ and $\varpi^j \cdot \mathfrak{b}_{i/n} = \mathfrak{b}_{j+\frac{i}{n}}$.

More concretely, for $n = 2$ we have

$$\mathfrak{k}_1 = \begin{pmatrix} \wp & \wp \\ \wp & \wp \end{pmatrix}$$

and

$$\mathfrak{b}_0 = \begin{pmatrix} R & R \\ \wp & R \end{pmatrix} \supset \mathfrak{b}_{1/2} = \begin{pmatrix} \wp & R \\ \wp & \wp \end{pmatrix} \supset \mathfrak{b}_1 = \varpi \cdot \mathfrak{b}_0.$$

For $n = 3$ we have

$$\mathfrak{b}_0 = \begin{pmatrix} R & R & R \\ \wp & R & R \\ \wp & \wp & R \end{pmatrix} \supset \mathfrak{b}_{1/3} = \begin{pmatrix} \wp & R & R \\ \wp & \wp & R \\ \wp & \wp & \wp \end{pmatrix} \supset \mathfrak{b}_{2/3} = \begin{pmatrix} \wp & \wp & R \\ \wp & \wp & \wp \\ \wp^2 & \wp & \wp \end{pmatrix} \supset \mathfrak{b}_1 = \varpi \cdot \mathfrak{b}_0.$$

From these examples, it is easy to see that the image of $\mathfrak{b}_0$ in $\mathfrak{k}_0/\mathfrak{k}_1 = \mathrm{M}_n(\mathfrak{f})$ is a Borel subalgebra with nilradical equal to the image of $\mathfrak{b}_{1/n}$ in $\mathfrak{k}_0/\mathfrak{k}_1 =$

$\mathrm{M}_n(\mathfrak{f})$. (We remark that in all of these examples, the lattice in $\mathfrak{g}$ is a direct sum of copies of lattices of k.)

We have similar filtrations of G by compact open subgroups. The *standard filtration subgroups* are defined by $K_0 = \mathrm{GL}_n(R)$ and $K_i = 1 + \mathfrak{k}_i$ for positive integers i. We note that, up to conjugacy, K_0 is the unique maximal compact open subgroup of G. We also define the Iwahori subgroup $B_0 = \mathfrak{b}_0^\times$ and, for positive integers i, we define the *Iwahori filtration subgroups* $B_{i/n} = 1 + \mathfrak{b}_{i/n}$. It is easy to check that if $0 \leq k \leq j$, then K_j is a normal subgroup of K_k and the lattices $\mathfrak{k}_m$ ($m \in \mathbb{Z}$) are invariant under the action of K_j. Similar statements apply to the Iwahori filtration lattices and subgroups.

If $0 < j \leq k \leq 2j$, then $\mathfrak{k}_j/\mathfrak{k}_k$ is an abelian group which is isomorphic via the map induced by $X \mapsto (1+X)$ to K_j/K_k. Similarly, $\mathfrak{b}_{j/n}/\mathfrak{b}_{k/n}$ is isomorphic to $B_{j/n}/B_{k/n}$.

Given a lattice L in $\mathfrak{g}$ we define the dual lattice L^* of L by

$$L^* = \{Y \in \mathfrak{g} \mid \mathrm{tr}(X \cdot Y) \in \wp \text{ for all } X \in L\}.$$

For example, for any integer i, the dual lattice of $\mathfrak{k}_i$ is $\mathfrak{k}_{(1-i)}$, and the dual lattice of $\mathfrak{b}_{i/n}$ is $\mathfrak{b}_{(1-i)/n}$. We let $C_c^\infty(\mathfrak{g})$ denote the set of functions on $\mathfrak{g}$ which are compactly supported, complex valued, and locally constant. We recall that a function f on $\mathfrak{g}$ is said to be *locally constant* if for each $X \in \mathfrak{g}$ there exists a lattice L_X such that $f(X+\ell) = f(X)$ for all ℓ in L_X. If we assume that our f is compactly supported, then this lattice can be chosen uniformly. We let $C_c^\infty(G)$ denote the functions on G which are compactly supported, complex-valued, and locally constant (that is, translation invariant with respect to some compact open subgroup of G).

For $f \in C_c^\infty(\mathfrak{g})$ we define $\hat{f}$, the *Fourier transform* of f, by the formula

$$\hat{f}(X) = \int_{\mathfrak{g}} f(Y) \cdot \Lambda(\mathrm{tr}(X \cdot Y))\, dY$$

for X in $\mathfrak{g}$. For example, if $[L]$ denotes the characteristic function of a lattice L in $\mathfrak{g}$, then $\widehat{[L]}$ is $[L^*]$ up to a constant. We normalize the measure dX on $\mathfrak{g}$ so that $\hat{\hat{f}}(X) = f(-X)$ for all $f \in C_c^\infty(\mathfrak{g})$ and $X \in \mathfrak{g}$.

Exercise 2.3.1: Show that this normalization of measures implies that for L sufficiently small, the measure of the lattice L with respect to dX is given by one over the square root of the index of L in L^*, that is,

$$\mathrm{meas}_{dX}(L) = [L^* : L]^{-1/2}.$$

Exercise 2.3.2: Show that the map $f \mapsto \hat{f}$ from $C_c^\infty(\mathfrak{g})$ to itself is a bijection.

Finally, we normalize dg so that the map $X \mapsto (1 + X)$ takes dX into dg. So, for example, we have that for all $j \in \mathbb{Z}_{>0}$,

$$\text{meas}_{dX}(\mathfrak{k}_j) = \text{meas}_{dg}(K_j).$$

3. Moy-Prasad filtrations

In this section we describe the Moy-Prasad filtration lattices of $\mathfrak{g}$ and subgroups of G.

3.1. *The apartment of T*

Recall that T is the group consisting of diagonal matrices in G. We write $t \in T$ as $t = (t_1, t_2, \ldots, t_n)$ with $t_j \in k^\times$ for $1 \leq j \leq n$. Let $\Phi \subset \mathbf{X}^*(T) = \text{Hom}(T, k^\times)$ denote the set of roots of G with respect to T (that is, the nontrivial eigencharacters for the action of T on $\mathfrak{g}$). More explicitly

$$\Phi = \{\alpha_{ij} \mid 1 \leq i \neq j \leq n \text{ and } \alpha_{ij}(t) = t_i/t_j\}.$$

With respect to our Borel subgroup B we let Δ denote the set of simple roots; that is, $\Delta := \{\alpha_{i(i+1)} \mid 1 \leq i \leq (n-1)\}$. We let Φ^+ denote the set of positive roots in Φ. We fix a Chevalley basis

$$\{Z, H_\beta, X_\gamma \mid \beta \in \Delta \text{ and } \gamma \in \Phi\}.$$

Here

$$(Z)_{kl} = \delta_{kl}$$

and

$$(H_{\alpha_{ij}})_{kl} = \begin{cases} 1 & \text{if } k = l = i \\ -1 & \text{if } k = l = (i+1) \\ 0 & \text{otherwise,} \end{cases}$$

and

$$(X_{\alpha_{ij}})_{kl} = \delta_{ik}\delta_{jl}.$$

We see that the center of $\mathfrak{g}$ is $k \cdot Z$ and the H_β together with Z form a basis for $\mathfrak{t}$, the Lie algebra of T.

For example, for $\mathrm{GL}_2(k)$ we have

$$Z = \begin{pmatrix} 1 & 0 \\ 0 & 1 \end{pmatrix}, \; H_{\alpha_{12}} = \begin{pmatrix} 1 & 0 \\ 0 & -1 \end{pmatrix}, \; X_{\alpha_{12}} = \begin{pmatrix} 0 & 1 \\ 0 & 0 \end{pmatrix}, \text{ and } X_{\alpha_{21}} = \begin{pmatrix} 0 & 0 \\ 1 & 0 \end{pmatrix}.$$

Let $Z(G) \leq T$ denote the center of G. For $1 \leq k \leq n$ define $\lambda_k \in \mathbf{X}_*(T) = \mathrm{Hom}(k^\times, T)$ by setting

$$(\lambda_k(s))_{ij} = \begin{cases} s & \text{if } i = j = k \\ 1 & \text{if } i = j \neq k \\ 0 & \text{otherwise,} \end{cases}$$

for $s \in k^\times$.

With respect to our choice of a Chevalley basis, we can identify $\mathcal{A} = \mathcal{A}(T)$, the *apartment*[b] corresponding to T, with the real vector space

$$(\mathbf{X}_*(T) \otimes \mathbb{R})/(\mathbf{X}_*(Z(G)) \otimes \mathbb{R}).$$

Let x_0 denote the origin in $\mathcal{A}$. The apartment of T is an $(n-1)$-dimensional Euclidean space spanned by the set $\{x_0 + \bar{\lambda}_k \mid 1 \leq k \leq n\}$. Note that these spanning vectors satisfy the relation

$$0 = \sum_k \bar{\lambda}_k.$$

3.2. *A simplicial structure for the apartment of T*

We let

$$\Psi := \{\delta + n \mid \delta \in \Phi \text{ and } n \in \mathbb{Z}\}$$

denote the set of affine roots of G with respect to T and ν. For $\psi = \delta + n \in \Psi$ we let $\dot{\psi} = \delta \in \Phi$.

If $\psi = \delta + n \in \Psi$ and $\sum \lambda_i \otimes r_i \in \mathbf{X}_*(T) \otimes \mathbb{R}$, then we define

$$\psi(\sum \lambda_i \otimes r_i) = n + \sum r_i \langle \delta, \lambda_i \rangle.$$

Here $\langle \, , \rangle$ denotes the usual pairing between characters and cocharacters. Note that if $v_T \in \mathbf{X}_*(T)$ and $v_Z \in \mathbf{X}_*(Z(G))$, then $\psi(v_T + v_Z) = \psi(v_T)$. Consequently, we can and shall think of an affine root as a function on $\mathcal{A}(T)$. For $\psi \in \Psi$, we define

$$H_\psi = \{x \in \mathcal{A}(T) \mid \psi(x) = 0\}.$$

[b] In these notes, we are always looking at the reduced building and apartments.

The H_ψ are hyperplanes in $\mathcal{A}(T)$ and they provide us with a simplicial decomposition of $\mathcal{A}(T)$.

For example, for $\mathrm{GL}_2(k)$ the apartment $\mathcal{A}(T)$ is one-dimensional and it is spanned by the vectors $x_0 + \bar{\lambda}_1 = x_0 - \bar{\lambda}_2$. See Figure 1.

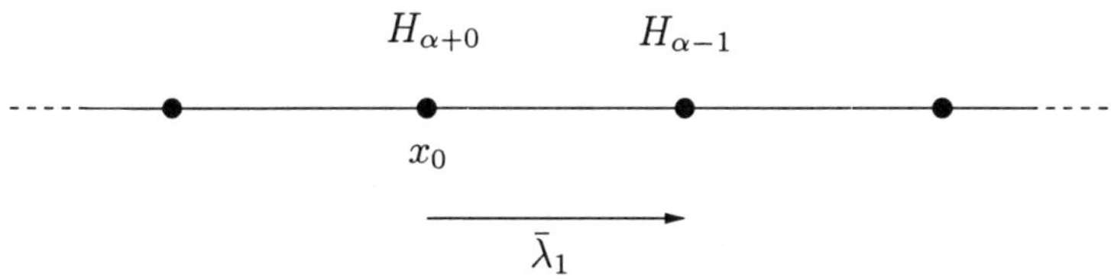

Fig. 1. The standard apartment for $\mathrm{GL}_2(k)$.

For $\mathrm{GL}_3(k)$ the apartment $\mathcal{A}(T)$ is two dimensional, and it is spanned by the vectors $x_0 + \bar{\lambda}_1$, $x_0 + \bar{\lambda}_2$, and $x_0 + \bar{\lambda}_3 = x_0 - (\bar{\lambda}_1 + \bar{\lambda}_2)$. See, for example Figure 2.

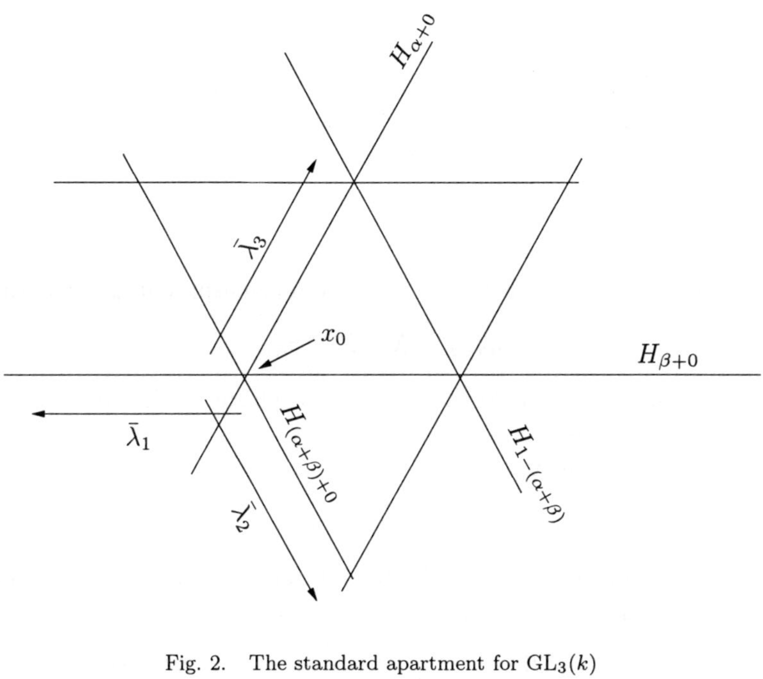

Fig. 2. The standard apartment for $\mathrm{GL}_3(k)$

The maximal facets in this simplicial decomposition of $\mathcal{A}(T)$ are called *alcoves* or *chambers* of $\mathcal{A}(T)$. For $\mathrm{GL}_2(k)$, the alcoves are open line segments. For $\mathrm{GL}_3(k)$, an alcove is the interior of an equilateral triangle. For $\mathrm{GL}_4(k)$, an alcove is the interior of a regular tetrahedron.

3.3. *Some subgroups of G*

Note that for all $\alpha \in \Phi$, the root group

$$U_\alpha := 1 + k \cdot X_\alpha$$

is naturally isomorphic to k as an additive group. From (1), k, regarded as an additive group, has a natural filtration indexed by $\mathbb{Z} \cup \{\pm\infty\}$. Namely,

$$\varpi^\infty \cdot R := \{0\} \subset \cdots \subset \varpi^2 \cdot R \subset \varpi \cdot R \subset R = \varpi^0 \cdot R \subset \varpi^{-1} \cdot R \subset \cdots \subset k.$$

Similarly, U_α has a natural filtration indexed by $\mathbb{Z} \cup \{\pm\infty\} \cong \{\psi \in \Psi \mid \dot{\psi} = \alpha\} \cup \{\pm\infty\}$. The problem is: how do we decide which $\psi \in \{\psi \in \Psi \mid \dot{\psi} = \alpha\}$ corresponds to which subgroup of U_α? The solution to our problem lies with our choice of a Chevalley basis. The choice of this basis determines the subgroup $K_0 = \mathrm{GL}_n(R)$. We define

$$U_{\alpha+0} := U_\alpha \cap K_0.$$

This, along with the requirement $U_{\alpha+1} \subset U_{\alpha+0}$, completely determines a natural indexing of subgroups in U_α by the set $\{\psi \in \Psi \mid \dot{\psi} = \alpha\}$.

For example, for $\mathrm{GL}_3(k)$, we have

$$U_{\alpha_{12}+j} = \begin{pmatrix} 1 & \varpi^j \cdot R & 0 \\ 0 & 1 & 0 \\ 0 & 0 & 1 \end{pmatrix}.$$

Similarly, we have a filtration of $\mathfrak{g}_\alpha$, the α-eigenspace in $\mathfrak{g}$, by setting

$$\mathfrak{g}_{\alpha+0} := \mathrm{M}_n(R) \cap \mathfrak{g}_\alpha.$$

Note that for all $\psi \in \Psi$ we have an isomorphism of the corresponding additive subgroups $\mathfrak{g}_\psi$ and U_ψ via the map $X \mapsto (1 + X)$.

The maximal compact open subgroup

$$T_0 = \{(t_1, t_2, \ldots, t_n) \in T \mid t_i \in R^\times\}$$

of T also has a filtration indexed by the integers. Because of future considerations, we define, for $r \in \mathbb{R}_{\geq 0}$,

$$T_r := \{t \in T_0 \mid \nu(1 - \chi(t)) \geq r \text{ for all } \chi \in \mathbf{X}^*(T)\}.$$

We also define

$$T_{r+} := \{t \in T_0 \,|\, \nu\big(1 - \chi(t)\big) > r \text{ for all } \chi \in \mathbf{X}^*(T)\}.$$

It is a matter of definition to see that $T_r = T_{\lceil r \rceil}$, and, for $i \in \mathbb{Z}_{\geq 0}$, if $i < r \leq (i+1)$, then

$$T_r = \{(t_1, t_2, \ldots, t_n) \in T_0 \,|\, t_j \in 1 + \varpi^{(i+1)} \cdot R \text{ for } 1 \leq j \leq n\}.$$

We can define filtrations of $\mathfrak{t} = \mathrm{Lie}(T)$ in a similar fashion. For example, for $\mathrm{GL}_3(k)$, we have

$$\mathfrak{t}_{5.2} = \begin{pmatrix} \varpi^6 \cdot R & 0 & 0 \\ 0 & \varpi^6 \cdot R & 0 \\ 0 & 0 & \varpi^6 \cdot R \end{pmatrix}.$$

3.4. *The Moy-Prasad filtrations*

Suppose $r \in \mathbb{R}$ and $x \in \mathcal{A}(T)$. In [27, 28], Moy and Prasad define filtration lattices of $\mathfrak{g}$ according to the formulae

$$\mathfrak{g}_{x,r} := \mathfrak{t}_r \oplus \sum_{\{\psi \in \Psi \,|\, \psi(x) \geq r\}} \mathfrak{g}_\psi$$

and

$$\mathfrak{g}_{x,r^+} := \mathfrak{t}_{r^+} \oplus \sum_{\{\psi \in \Psi \,|\, \psi(x) > r\}} \mathfrak{g}_\psi.$$

For some people, it is easier to process these definitions when they are presented in the following form.

$$\mathfrak{g}_{x,r} := \mathfrak{t}_r \oplus \sum_{\alpha \in \Phi} \sum_{\substack{m \in \mathbb{Z} \\ (\alpha+m)(x) \geq r}} \mathfrak{g}_{\alpha+m}$$

and

$$\mathfrak{g}_{x,r^+} := \mathfrak{t}_{r^+} \oplus \sum_{\alpha \in \Phi} \sum_{\substack{m \in \mathbb{Z} \\ (\alpha+m)(x) > r}} \mathfrak{g}_{\alpha+m}.$$

Similarly, if $r \geq 0$, then they define

$$G_{x,r} := \langle T_r, U_\psi \rangle_{\{\psi \in \Psi \,|\, \psi(x) \geq r\}}$$

and

$$G_{x,r^+} := \langle T_{r^+}, U_\psi \rangle_{\{\psi \in \Psi \,|\, \psi(x) > r\}}.$$

These lattices (resp., subgroups) are referred to as the Moy-Prasad filtration lattices of $\mathfrak{g}$ (resp., subgroups of G).

It follows immediately from the definitions that if $X \in \mathfrak{g}_{x,r}$ and $Y \in \mathfrak{g}_{x,s}$, then $X \cdot Y \in \mathfrak{g}_{x,(r+s)}$. The following exercise is highly recommended; we have previously considered these statements for the congruence and Iwahori filtrations.

Exercise 3.4.1: Suppose $s \in \mathbb{R}$ and $r \in \mathbb{R}_{\geq 0}$.

(1) Up to a constant we have $\widehat{[\mathfrak{g}_{x,s}]} = [\mathfrak{g}_{x,(-s)+}]$.
(2) If $g \in G_{x,r}$, then $g\mathfrak{g}_{x,s}g^{-1} = \mathfrak{g}_{x,s}$ and $g\mathfrak{g}_{x,s^+}g^{-1} = \mathfrak{g}_{x,s^+}$.
(3) If $0 < r \leq s \leq 2r$, then $X \mapsto (1 + X)$ induces an abelian group isomorphism of $\mathfrak{g}_{x,r}/\mathfrak{g}_{x,s}$ and $G_{x,r}/G_{x,s}$.

3.4.1. *The special case $r = 0$*

We first attempt to understand these definitions when $r = 0$. In this case, if x is a point in $\mathcal{A}(T)$, then $G_{x,0}$ is called the *parahoric subgroup* attached to x and $G_{x,0^+}$ is called the *pro-unipotent radical* of $G_{x,0}$. Let F be a facet in $\mathcal{A}$ and let $x, y \in F$. It follows from the definitions of both the simplicial structure of $\mathcal{A}$ and the Moy-Prasad filtrations that $G_{x,0} = G_{y,0}$, $\mathfrak{g}_{x,0} = \mathfrak{g}_{y,0}$, $G_{x,0^+} = G_{y,0^+}$, and $\mathfrak{g}_{x,0^+} = \mathfrak{g}_{y,0^+}$. Thus, it is enough to understand these filtrations on a facet by facet basis, and it is natural to label the filtrations using facets rather than points. For $\mathrm{GL}_2(k)$, Figures 3 and 4 describe the subgroups $G_{F,0}$ and $G_{F,0^+}$ for F a facet in the apartment of T.

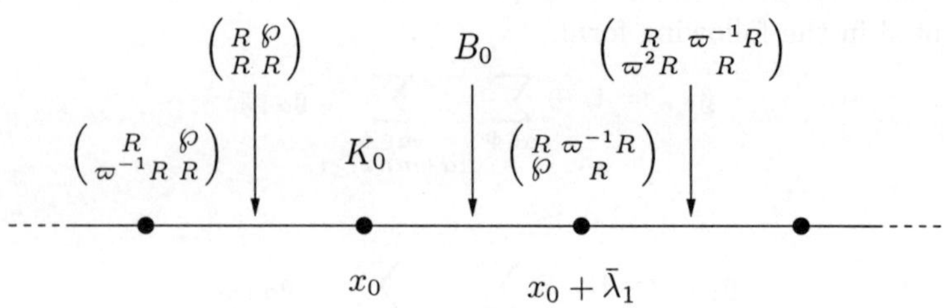

Fig. 3. Parahoric subgroups $G_{F,0}$ for $\mathrm{GL}_2(k)$

Note that as we move to the right, that is, in the positive direction as defined by the spherical Weyl chamber corresponding to B, the positive root spaces "expand" and the negative root spaces "shrink". This simple observation will play a significant role in what is to come.

It is highly recommended that the reader complete the following exercise.

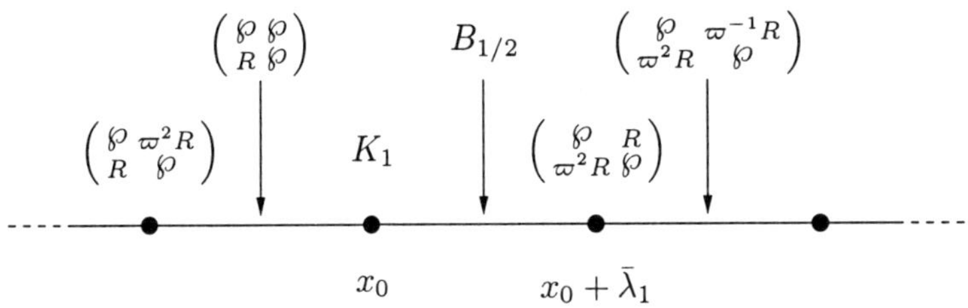

Fig. 4. The subgroups $G_{F,0+}$ for $\mathrm{GL}_2(k)$

Exercise 3.4.2: Make diagrams similar to those above, but for the filtration lattices $\mathfrak{g}_{F,0}$ and $\mathfrak{g}_{F,0+}$. Now, do the same for $\mathrm{GL}_3(k)$.

The origin x_0 is the facet in $\mathcal{A}$ defined by the intersection of the hyperplanes $H_{\alpha+0}$ for $\alpha \in \Phi$. We see from the above examples (and definitions) that $G_{x_0,0} = K_0$ and $G_{x_0,0^+} = K_1$.

For $\mathrm{GL}_n(k)$, up to conjugation, the facets are indexed by

$$\mathcal{P}(n) := \{(\mu_1, \mu_2, \dots, \mu_\ell) \mid \mu_1 \geq \mu_2 \geq \cdots \geq \mu_\ell \geq 1 \text{ and } \sum_{i=1}^{\ell} \mu_i = n\},$$

the set of ordered partitions of n. We briefly describe how this works. We let C_0 be the alcove in $\mathcal{A}$ with vertices $\{v_1 = x_0 = x_0 + \sum_{i=1}^{n} \bar{\lambda}_i, v_2 = v_1 - \bar{\lambda}_1, v_3 = v_2 - \bar{\lambda}_2, \dots, v_n = \bar{\lambda}_n\}$. To $\mu \in \mathcal{P}(n)$ we attach the facet F_μ of dimension ℓ with vertices $v_1, v_{(\mu_1+1)}, v_{(\mu_1+\mu_2+1)}, \dots, v_{(\mu_1+\mu_2+\cdots+\mu_{\ell-1}+1)}$. The parahoric subgroup $G_{F_\mu,0}$ can be described as $K_1 \cdot Q_\mu(R)$ where Q_μ is the "standard" parabolic subgroup of G containing B which corresponds to the partition $\mu = (\mu_1, \mu_2, \dots, \mu_\ell)$. The pro-unipotent radical $G_{F_\mu,0^+}$ is then the inverse image in $G_{F_\mu,0}$ of the unipotent radical of the image of $G_{F_\mu,0}$ in $\mathrm{GL}_n(\mathfrak{f}) \cong K_0/K_1$. Note that if $\mu, \nu \in \mathcal{P}(n)$ with $\mu < \nu$ in the usual partial ordering of ordered partitions, then $Q_\mu \supset Q_\nu$ and similarly for the associated parahoric subgroups.

More generally, note that if F_1 and F_2 are facets in $\mathcal{A}(T)$ with $F_1 \subset \overline{F_2}$ where $\overline{F_2}$ denotes the closure of F_2, then it follows that

$$G_{F_1,0^+} \subset G_{F_2,0^+} \subset G_{F_2,0} \subset G_{F_1,0}.$$

Moreover, as the examples and exercises above show, $G_{F_2,0}/G_{F_1,0^+}$ is a parabolic subgroup of the connected reductive group $G_{F_1,0}/G_{F_1,0^+}$. The unipotent radical of $G_{F_2,0}/G_{F_1,0^+}$ is $G_{F_2,0^+}/G_{F_1,0^+}$ and $G_{F_2,0}/G_{F_1,0^+}$ has Levi component isomorphic to $G_{F_2,0}/G_{F_2,0^+}$.

As a tangent, we also observe that if $W = N_G(T)/T \cong (N_G(T) \cap K_0)/T_0$ denotes the Weyl group, then there exist $|W|$ alcoves in $\mathcal{A}(T)$ which contain x_0 in their closure. This is because these alcoves correspond to the Borel subgroups in $\mathrm{GL}_n(\mathfrak{f}) = K_0/K_1 = G_{x_0,0}/G_{x_0,0+}$ which contain "$T(\mathfrak{f})$".

3.4.2. *The Moy-Prasad filtrations for arbitrary r*

For arbitrary r, there does not exist such a nice description of what is happening. We first consider $\mathrm{GL}_2(k)$. In Figure 5 we have identified the

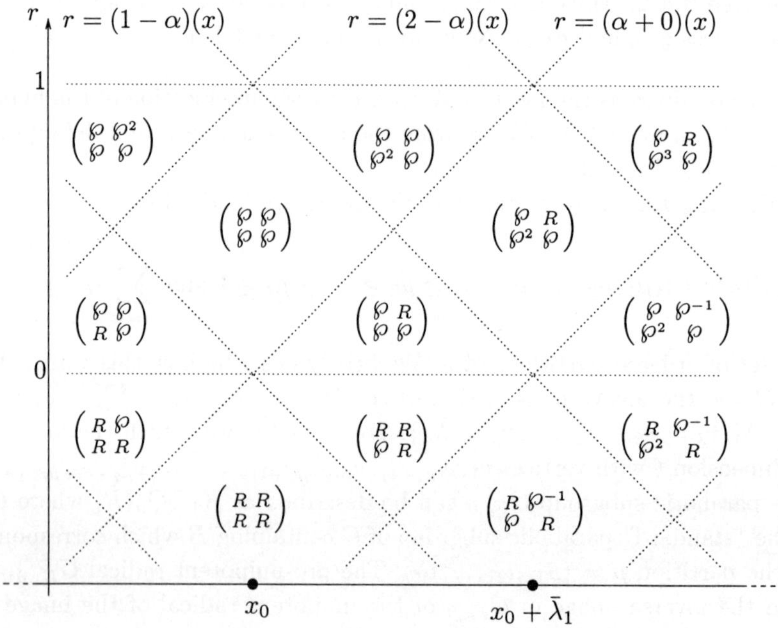

Fig. 5. $(\mathfrak{gl}_2)_{x,r}$

apartment of T with the horizontal axis. The vertical axis measures r. Given a pair $(x,r) \in \mathcal{A}(T) \times \mathbb{R}$, we wish to describe the lattice $\mathfrak{g}_{x,r}$. Note that the plane has been divided into open convex polygons. The diagonal dotted lines are the graphs of the equations $r = \psi(x)$ where $\psi \in \Psi$. These measure where the root subgroup filtrations "change". The horizontal dotted lines are the graphs of the equations $r = n$ where $n \in \mathbb{Z}$. These measure where the toral subgroup filtrations "change". Each of the convex polygons is labeled by a lattice: if (x,r) belongs to a convex polygon, then $\mathfrak{g}_{x,r}$ is the lattice so identified. For purposes of assigning lattices to every point in the figure, the convex polygons are "closed at the top". We note that $\mathfrak{g}_{x,r} = \mathfrak{g}_{x,r^+}$ unless the point (x, r) lies on a dotted line. Finally, note that if we fix r and move to the right, then the positive root space "expands" while the negative root space "shrinks".

Fortunately, even though life is not so nice for arbitrary r, there exists a wonderful result of Moy and Prasad[c]. Define

$$\mathcal{O} := \{x \in \mathcal{A}(T) \mid x \text{ is the barycenter of a facet}\}.$$

An element of $\mathcal{O}$ is called an *optimal point*[d].

Lemma 3.4.3: (Moy and Prasad) *Suppose $z \in \mathcal{A}(T)$ and $r \in \mathbb{R}$. There exist points $x, y \in \mathcal{O}$ such that*

$$\mathfrak{g}_{x,r} \subset \mathfrak{g}_{z,r} \subset \mathfrak{g}_{y,r}$$

and there exist points $x', y' \in \mathcal{O}$ such that

$$\mathfrak{g}_{x',r^+} \subset \mathfrak{g}_{z,r^+} \subset \mathfrak{g}_{y',r^+}.$$

A similar pair of statements can be made for the Moy-Prasad filtration subgroups.

For example, for $\mathrm{GL}_3(k)$, the optimal points, up to the action of G (see below), are the points $\bar{\lambda}_1 + \bar{\lambda}_2 + \bar{\lambda}_3$, $\frac{\bar{\lambda}_3}{2}$ and $\frac{3\bar{\lambda}_3 + 2\bar{\lambda}_2 + \bar{\lambda}_1}{3}$.

Exercise 3.4.4: Check that, up to conjugacy, the points listed above are the optimal points for $\mathrm{GL}_3(k)$. Describe their associated filtration lattices $\mathfrak{g}_{x,r}$ and $\mathfrak{g}_{x,r^+}$.

4. *G*-domains

Recall that the ultimate goal of these notes is to relate certain invariant distributions on G with certain invariant distributions on $\mathfrak{g}$. Since the distributions in question are invariant, we will need sets in G and $\mathfrak{g}$ which are invariant. Such a set can never be compact, so the most we can hope for is to have sets which are invariant, open, and closed. A set with these properties is called a *G-domain.*

As we show in this section, it is possible to use the Moy-Prasad filtrations to define G-domains which have very nice properties. In particular, we will

---

[c]You will not find the result stated exactly as follows, but it is easy to derive this formulation.

[d]Beware, $\mathcal{O}$ plays many roles in these notes. However, there is no possibility of confusion.

define a family of G-domains which form a "neighborhood basis" of $\mathcal{N}$, the set of nilpotent elements.

By using the Cartan decomposition of G (see [33]), in [18] (Lemma 2.4) Howe demonstrates that, for all integers i, the $\mathrm{GL}_n(k)$-orbit of $\mathfrak{k}_i$ is contained in $\mathfrak{k}_i + \mathcal{N}$. From, for example, [13] (Lemma 12.2), it is also true that for any compact set $\omega \subset \mathfrak{g}$, there exists a lattice $\mathcal{L} \subset \mathfrak{g}$ such that ${}^G\omega \subset \mathcal{L} + \mathcal{N}$. We analyze these statements from the perspective of Moy-Prasad filtration lattices. After doing this, we look at the situation on the group.

4.1. *Some comments on the Bruhat-Tits building of G*

Before we can continue, we must recall some facts about $\mathcal{B}$, the *reduced Bruhat-Tits building of G*. To each maximal split torus T' in G we can attach an apartment $\mathcal{A}(T')$ exactly as we did above. The building of G can then be thought of as the "gluing" together of all these various apartments. In Figure 6 we present a picture of $\mathcal{B}$ for $\mathrm{GL}_2(k)$. An apartment in $\mathcal{B}$ is

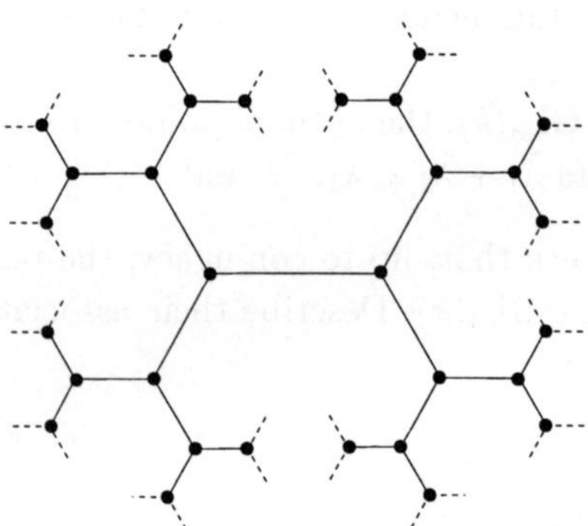

Fig. 6. A picture of the building of $\mathrm{GL}_2(k)$

the image of any continuous injective map from $\mathbb{R}$ which maps integers to vertices.

Just as any two maximal split tori in G are conjugate, there is a natural action of G on $\mathcal{B}(G)$ with respect to which any apartment can be carried into any other. Moreover, a version of the two body problem can be solved: given any two points in the building, there is an apartment which contains both of them. Combining these two facts we have: for any two points x, y in $\mathcal{B}$, there exists a $g \in G$ such that $gx, gy \in \mathcal{A}(T)$. (Here gx denotes the image of x under the action of g on $\mathcal{B}(G)$.) Finally, the action of G on $\mathcal{B}$ is semisimple; that is, for $h \in G$, either there is a point in $\mathcal{B}$ which h fixes, or there is a line (i.e., a one-dimensional subspace of some apartment) on which h acts by nontrivial translation.

Suppose $x \in \mathcal{B}$ and $r \in \mathbb{R}$. We define the Moy-Prasad filtration lattices $\mathfrak{g}_{x,r}$ and $\mathfrak{g}_{x,r^+}$ as follows. Choose $g \in G$ so that $gx \in \mathcal{A}(T)$.

$$\mathfrak{g}_{x,r} := g^{-1} \cdot \mathfrak{g}_{gx,r} \cdot g \text{ and } \mathfrak{g}_{x,r^+} := g^{-1} \cdot \mathfrak{g}_{gx,r^+} \cdot g.$$

One can check that these definitions make sense. For $r \geq 0$ we define the subgroups $G_{x,r}$ and G_{x,r^+} in a similar fashion.

Finally, we note that $\mathcal{B}$ is endowed with a nontrivial invariant metric, denoted dist. Moreover, with respect to dist, $\mathcal{B}$ has nonpositive sectional curvature.

4.2. *G-domains for the Lie algebra*

We begin with a generalization of the result of Howe discussed above; the proof of this result (from [1]) nicely illustrates the benefit of working with the Moy-Prasad filtration lattices.

Lemma 4.2.1: *Let $x, y \in \mathcal{B}$, and let $r \in \mathbb{R}$. Then $\mathfrak{g}_{x,r} \subset \mathfrak{g}_{y,r} + \mathcal{N}$.*

Proof: Since $\mathfrak{g}_{x,r} \subset \mathfrak{g}_{y,r} + \mathcal{N}$ if and only if $\mathfrak{g}_{gx,r} \subset \mathfrak{g}_{gy,r} + \mathcal{N}$ for $g \in G$, we may assume that x and y are elements of $\mathcal{A}(T)$.

Choose $\vec{v} \in \mathbf{X}_*(T) \otimes \mathbb{R}$ such that $x = y + \vec{v}$ (working modulo $\mathbf{X}_*(Z(G)) \otimes \mathbb{R}$). Let B' be a Borel subgroup determined by $\vec{v}$. That is, B' has a Levi decomposition $B' = TN'$ such that for all roots $\alpha \in \Phi^+_{N'}$, the set of roots which are positive with respect to N', we have $\langle \alpha, \vec{v} \rangle \geq 0$.

Let $\bar{B}' = T\bar{N}'$ denote the parabolic opposite $B = TN'$ and let $\mathfrak{g} = \bar{\mathfrak{n}}' + \mathfrak{t} + \mathfrak{n}'$ denote the associated Lie algebras. We have

$$\begin{aligned}
\mathfrak{g}_{x,r} &= \mathfrak{t}_r \oplus \sum_{\{\psi \in \Psi \,|\, \psi(x) \geq r\}} \mathfrak{g}_\psi \\
&= \mathfrak{t}_r \oplus \sum_{\{\psi \in \Psi \,|\, \dot{\psi} \in \Phi \smallsetminus \Phi^+_{N'} \text{ and } \psi(x) \geq r\}} \mathfrak{g}_\psi \oplus \sum_{\{\psi \in \Psi \,|\, \dot{\psi} \in \Phi^+_{N'} \text{ and } \psi(x) \geq r\}} \mathfrak{g}_\psi \\
&= \mathfrak{t}_r \oplus (\mathfrak{g}_{x,r} \cap \bar{\mathfrak{n}}') \oplus (\mathfrak{g}_{x,r} \cap \mathfrak{n}') \\
&\subset \mathfrak{g}_{y,r} + \mathfrak{n}' \subset \mathfrak{g}_{y,r} + \mathcal{N}.
\end{aligned}$$

□

In fact, a kind of converse to the above lemma is true. Namely, if $X \in \mathfrak{g}$ belongs to $\mathfrak{g}_{x,r} + \mathcal{N}$ for all $x \in \mathcal{B}$, then there is a point $y \in \mathcal{B}$ such that

$X \in \mathfrak{g}_{y,r}$. This result along with the above lemma gives us the following theorem.

Theorem 4.2.2:

$$\bigcup_{x\in\mathcal{B}} \mathfrak{g}_{x,r} = \bigcap_{x\in\mathcal{B}} (\mathfrak{g}_{x,r} + \mathcal{N}).$$

In order to simplify our notation, we define

$$\mathfrak{g}_r := \bigcup_{x\in\mathcal{B}} \mathfrak{g}_{x,r}.$$

From its definition, $\mathfrak{g}_r$ is open and invariant. From the above theorem we have that $\mathfrak{g}_r$ is closed. Consequently, $\mathfrak{g}_r$ is a G-domain.

We now consider some examples. The set $\mathfrak{g}_0$ is usually referred to as the set of compact or integral elements of $\mathfrak{g}$. The set $\mathfrak{g}_{0+}$ is usually called the set of topologically nilpotent elements in $\mathfrak{g}$; it consists of those $X \in \mathfrak{g}$ such that powers of X tend to zero in the p-adic topology. These G-domains have very natural interpretations; namely,

$$\mathfrak{g}_r = \{X \in \mathfrak{g} \,|\, \nu(e) \geq r \text{ for all eigenvalues } e \text{ of } X\ \}.$$

(If E is a finite extension of k, then there exists a unique extension of ν to E.) For $\mathrm{GL}_n(k)$ and $X \in \mathfrak{g}$, the value of $\nu(e_X)$ must lie in the set $\{k/n \,|\, k \in \mathbb{Z}\}$.

Of course, just as we have Moy-Prasad lattices of the form $\mathfrak{g}_{x,r^+}$, we also can define $\mathfrak{g}_{r^+} := \bigcup_{x\in\mathcal{B}} \mathfrak{g}_{x,r^+}$. These G-domains satisfy the obvious analogues of the results discussed above. From the previous paragraph, we have that $\mathfrak{g}_r \neq \mathfrak{g}_{r+}$ implies that $r = k/n$ for some $k \in \mathbb{Z}$.

Exercise 4.2.3: Define the subspace D_r of $C_c^\infty(\mathfrak{g})$ by

$$D_r := \sum_{x\in\mathcal{B}} C(\mathfrak{g}/\mathfrak{g}_{x,r})$$

where the sum is interpreted as follows. A function f belongs to D_r if and only if f can be written as a finite sum $f = \sum_i f_i$ with $f_i \in C_c(\mathfrak{g}/\mathfrak{g}_{x_i,r})$ for some $x_i \in \mathcal{B}$. We can define D_{r+} in a similar way. Show that the Fourier transform gives us bijective maps from D_{r+} to $C_c^\infty(\mathfrak{g}_{-r})$ and from D_r to $C_c^\infty(\mathfrak{g}_{(-r)+})$.

4.3. *G-domains in G.*

We now turn our attention to the group-side of things. Unlike the Lie algebra, an element of the group can only belong to a Moy-Prasad filtration subgroup if it first belongs to a parahoric. It turns out that the interesting part of the story here rests in proving the equality

$$\bigcup_{x\in\mathcal{B}} G_{x,0} = \bigcap_{x\in\mathcal{B}} G_{x,0}\cdot\mathcal{U}$$

where $\mathcal{U}$ denotes the set of unipotent elements in G. Once this result is known to be true, it is very easy to establish the equality

$$G_r = \bigcap_{x\in\mathcal{B}} G_{x,r}\cdot\mathcal{U}$$

where $r \geq 0$ and $G_r = \bigcup_{x\in\mathcal{B}} G_{x,r}$. As above, it follows immediately that G_r is a G-domain.

We show how to prove the equality when $r = 0$.

Lemma 4.3.1:

$$G_0 = \bigcap_{x\in\mathcal{B}} G_x\cdot\mathcal{U}.$$

Proof: We first show that the right-hand side is a subset of the left-hand side. We will argue by contradiction. Suppose that $g \in \bigcap_{x\in\mathcal{B}} G_{x,0}\cdot\mathcal{U}$ does not belong to G_0.

Since the action of G on $\mathcal{B}$ is semisimple, we either have that there is a point x in $\mathcal{B}$ which g fixes or a line ℓ in an apartment $\mathcal{A}'$ of $\mathcal{B}$ on which g acts by nontrivial translation.

In the first case, we can write $g = h\cdot u$ with $h \in G_{x,0}$ and $u \in \mathcal{U}$. Since u is unipotent, it must live in $G_{y,0}$ for some $y \in \mathcal{B}$. But, from a result of Eugene Kushnirsky this implies that $u \in G_{x,0}$ [10] (Lemma 4.5.1).

In the latter case, there exists a facet F' in $\mathcal{A}'$ such that $F'\cap\ell$ is open in ℓ. For all $x, y \in F'$ we have $G_{y,0} = G_{x,0}$. By hypothesis, there exist elements $h \in G_{x,0}$ and $u \in \mathcal{U}$ such that $g = uh$. Since u is unipotent, there exists $w \in \mathcal{B}$ which is fixed by u. We have that for all $y \in F'\cap\ell$

$$\operatorname{dist}(w, y) = \operatorname{dist}(w, uy) = \operatorname{dist}(w, uhy) = \operatorname{dist}(w, gy).$$

Thus, we have , for $x, y \in F'\cap\ell$ a picture something like that described in Figure 7.

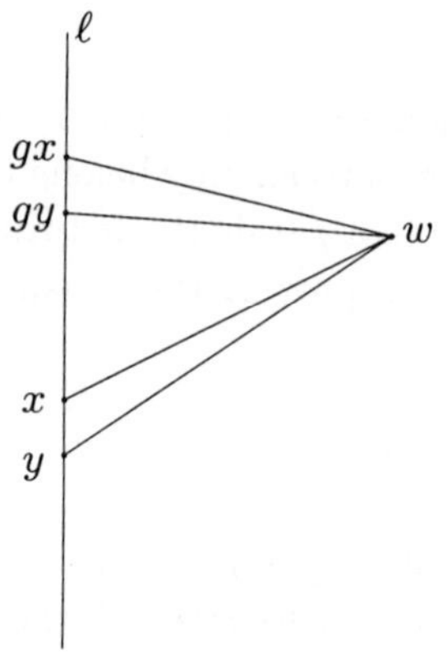

Fig. 7.

However, since $\mathcal{B}$ has nonpositive sectional curvature, the line segment from x to w must be shorter than the line segments from y and gy to w. Similarly, the segments from gy to w must be shorter than the segments from x and gx to w. Consequently, we have

$$\text{dist}(x, w) < \text{dist}(gy, w) < \text{dist}(x, w),$$

a contradiction.

We now show that the left-hand side is a subset of the right-hand side.

We need to show that for $x, y \in \mathcal{B}$, we have $G_x \subset \mathcal{U} \cdot G_y$. As in the proof for the Lie algebra, we may assume that x and y both belong to $\mathcal{A}(T)$. Let $B' = TN'$ be a Borel subgroup of G so that the (spherical) chamber in $\mathcal{A}(T)$ determined by N' is invariant under translation by the vector $(y-x)$.

Let $\bar{N}'$ be the unipotent radical of the parabolic opposite $B = TN'$. From [3] we can write $G = N' \cdot \bar{N}' \cdot N' \cdot T$. With some work, it follows that we can write

$$G_{x,0} = N'_x \cdot \bar{N}'_x \cdot N'_x \cdot T_0$$

where $N'_x = N' \cap G_{x,0}$ and $\bar{N}'_x = \bar{N}' \cap G_{x,0}$. Because of the way in which N' was chosen, we have $N'_x \subset G_{y,0}$. Thus, if $g \in G_x$, then there exist $n_1, n_2 \in N_x$, $\overline{n} \in \overline{N}_x$, and $t \in T_0$ such that

$$\begin{aligned} g &= n_1 \cdot \overline{n} \cdot n_2 \cdot t \\ &= (n_1 \cdot \overline{n} \cdot n_1^{-1}) \cdot (n_1 \cdot n_2 \cdot t) \\ &\in \mathcal{U} \cdot G_y. \end{aligned}$$

□

4.4. *A neighborhood basis for the nilpotent cone*

We remark that

$$\mathcal{N} = \bigcap_r \mathfrak{g}_r$$

and

$$\mathcal{U} = \bigcap_r G_r.$$

It follows that the G-domains we have defined provide us with a neighborhood basis of the nilpotent cone (resp., unipotent variety) consisting of open, invariant, closed neighborhoods.

5. Nilpotent orbital integrals as distributions

A distribution on $\mathfrak{g}$ is any element of the linear dual of $C_c^\infty(\mathfrak{g})$. The aim of this section is to convince ourselves that integrating against a nilpotent orbit defines a distribution on $\mathfrak{g}$. We follow the argument of Ranga-Rao [29].

5.1. *Orbital integrals*

Fix an element $Y \in \mathfrak{g}$. Let $\mathcal{O}_Y = {}^G Y := \{gYg^{-1} \mid g \in G\}$ denote the G-orbit of Y. We identify the orbit of Y with the homogeneous space $G/C_G(Y)$. Thus, the tangent space to the orbit of Y at the point Y is identified with $\mathfrak{g}/C_{\mathfrak{g}}(Y)$.

We define an alternating bilinear form $\langle\, ,\, \rangle_Y$ on $\mathfrak{g}$ by $\langle A, B\rangle = \mathrm{tr}(Y \cdot [A,B])$ for $A, B \in \mathfrak{g}$. Fix an element $B \in \mathfrak{g}$. A calculation shows that $\langle A, B\rangle = 0$ for all $A \in \mathfrak{g}$ if and only if $B \in C_{\mathfrak{g}}(Y)$. Since we have a similar statement when we switch the roles of A and B, it follows that $\langle\, ,\, \rangle_Y$ induces a nondegenerate alternating form on $\mathfrak{g}/C_{\mathfrak{g}}(Y)$. Thus, the dimension of the orbit is even, say $2m$.

Similarly, for each ${}^gY = gYg^{-1} \in {}^GY$, we have a nondegenerate alternating symmetric form on $\mathrm{Tan}_{{}^gY}(\mathcal{O}_Y)$. Consequently, there exists a nondegenerate, invariant two-form ω for $\mathcal{O}_Y$ and from this we can form a nonzero, invariant volume form $\omega \wedge \omega \wedge \cdots \wedge \omega$ (m times) on $\mathcal{O}_Y$. Let $X_1, X_2, \ldots, X_{2m}$ be coordinates for $\mathfrak{g}/C_{\mathfrak{g}}(Y)$. For each $1 \le i \le 2m$, fix a one-form dX_i and associated measure $|dX_i|$ normalized so that for all $f \in C_c^\infty(\mathfrak{g}/C_{\mathfrak{g}}(Y))$, we have $\hat{\hat{f}}(X) = f(-X)$, where the Fourier transform is taken with respect to the measure $\prod_{i=1}^{2m} |dX_i|$. There exists a locally convergent power series $f(X_1, X_2, \ldots, X_{2m})$ so that (locally)

$$\omega = f(X_1, X_2, \ldots, X_{2m})\, dX_1 \wedge dX_2 \wedge \cdots \wedge dX_{2m}.$$

Define

$$|\omega| := |f(X_1, X_2, \ldots, X_{2m})| \prod_{i=1}^{2m} |dX_i|\,.$$

Thus, we have an invariant measure on $\mathcal{O}_Y$. (Since $\mathrm{GL}_n(k)$ is unimodular, it follows from standard measure theory results that $C_G(Y)$ is unimodular as well.)

Exercise 5.1.1: Check that this definition is invariant of the various choices we have made.

5.2. *A framing of the problem*

If Y is a semisimple element of $\mathfrak{g}$, then $\mathcal{O}_Y$ is closed in $\mathfrak{g}$. Consequently, if $f \in C_c^\infty(\mathfrak{g})$, then the restriction of f to $\mathcal{O}_Y$ is an element of $C_c^\infty(\mathcal{O}_Y)$. So it makes sense to define the distribution $\mathcal{O}_Y \colon C_c^\infty(\mathfrak{g}) \to \mathbb{C}$ by setting

$$\mathcal{O}_Y(f) := \int_{G/C_G(Y)} f({}^gY)\, dg^*.$$

for $f \in C_c^\infty(\mathfrak{g})$. Here dg^* denotes the invariant measure on $G/C_G(Y)$ defined above.

What if Y is not semisimple? In particular, what if Y is nilpotent? In this case, we need to do some work to show that for arbitrary $f \in C_c^\infty(\mathfrak{g})$ the integral

$$\int_{G/C_G(Y)} f({}^gY)\, dg^*$$

makes sense.

5.3. *The basic notation associated to nilpotent orbital integrals*

We begin by recalling the familiar parameterization of nilpotent orbits in $\mathfrak{g}$. We then discuss various properties of nilpotent orbits which will be important in the sequel.

From basic linear algebra we know that the nilpotent orbits in $\mathfrak{g}$ can be parameterized by $\mathcal{P}(n)$, the set of ordered partitions of n. To $\mu \in \mathcal{P}(n)$ we associate the nilpotent element $X_\mu \in \mathcal{N} \cap \mathfrak{g}_{x_0,0}$ having Jordan canonical form corresponding to μ. That is, if $\mu = (\mu_1, \mu_2, \ldots, \mu_k)$, then in the ith

block of size $\mu_i \times \mu_i$ we put the matrix with ones on the superdiagonal and zeroes elsewhere. For example, for $\mu = (2,2,1) \in \mathcal{P}(5)$ we have

$$X_\mu = \begin{pmatrix} 0 & 1 & 0 & 0 & 0 \\ 0 & 0 & 0 & 0 & 0 \\ 0 & 0 & 0 & 1 & 0 \\ 0 & 0 & 0 & 0 & 0 \\ 0 & 0 & 0 & 0 & 0 \end{pmatrix}.$$

Note that the nilpotent orbits in $\mathfrak{g}_{x_0,0}/\mathfrak{g}_{x_0,0^+} \cong \mathrm{M}_n(\mathfrak{f})$ are also indexed by $\mathcal{P}(n)$ and the map taking X_μ to the image of X_μ in $\mathrm{M}_n(\mathfrak{f})$ gives a bijective correspondence between $\mathcal{O}(0)$, the set of nilpotent orbits in $\mathfrak{g}$, and the set of nilpotent orbits in $\mathrm{M}_n(\mathfrak{f})$.

We let $\mathcal{O}_\mu$ denote the G-orbit of X_μ. Note that if $\mu < \mu'$ in the usual partial order on $\mathcal{P}(n)$, then $\mathcal{O}_\mu \subset \bar{\mathcal{O}}_{\mu'}$.

Exercise 5.3.1: Show that if $\mathcal{O}$ is a nilpotent orbit such that $\mathcal{O} \cap (X_\mu + \mathfrak{g}_{x_0,0^+}) \neq \emptyset$, then $\mathcal{O}_\mu$ is contained in the p-adic closure of $\mathcal{O}$. Moreover, $\mathcal{O}_\mu \cap (X_\mu + \mathfrak{g}_{x_0,r^+}) = {}^{G_{x_0,r^+}} X_\mu$ for all non-negative r.

Note that if $X \in \mathcal{N}$, and $t \in k^\times$, then $tX \in \mathcal{N}$ and in fact, tX and X are $\mathrm{GL}_n(k)$ conjugate. (In general, it follows from Jacobson-Morosov that t^2X and X are conjugate, but in GL_n we can do better.) So, $\mathcal{N}$ and each nilpotent orbit are closed with respect to scaling.

We can associate a parabolic subgroup $P(\mu)$ of G to μ as follows: For all positive integers j, the matrix X_μ^j acts on $V = k^n$, and we define $V_j = \ker(X_\mu^j) \subset V$. Define $P(\mu) = \{g \in G \,|\, g \cdot V_j \subset V_j \text{ for all positive } j\ \}$. Then X_μ lies in the nilradical $\mathfrak{n}(\mu)$ of the Lie algebra of $P(\mu)$, and the $P(\mu)$-orbit of X_μ is dense in $\mathfrak{n}(\mu)$ and equal to $\mathcal{O}_\mu \cap \mathfrak{n}(\mu)$.

We need most of the basic facts about $\mathfrak{sl}_2(k)$-triples. For a good reference, see [4]. We can complete X_μ to an $\mathfrak{sl}_2(k)$-triple (Y_μ, H_μ, X_μ). Explicitly, the ith block of H_μ is given by the element $\mathrm{diag}((\mu_i - 1), (\mu_i - 3), \ldots, (1-\mu_i))$ and the ith block of Y_μ is given by certain entries on the super-subdiagonal. We let $\lambda_\mu \in \mathbf{X}_*(G)$ denote the associated one-parameter subgroup, that is, the ith block of $\lambda_\mu(t)$ looks like $\mathrm{diag}(t^{(\mu_i-1)}, t^{(\mu_i-3)}, \ldots, t^{(1-\mu_i)})$. For $i \in \mathbb{Z}$, we define

$$\mathfrak{g}(i) = \{X \in \mathfrak{g} \,|\, {}^{\lambda_\mu(t)}X = t^i X\}.$$

We have $\mathfrak{g} = \sum_i \mathfrak{g}(i)$. For $j \in \mathbb{Z}$, define $\mathfrak{g}(\geq j) := \sum_{i \geq j} \mathfrak{g}(i)$. We let $\mathfrak{p}_\mu$ denote the parabolic subalgebra $\mathfrak{g}(\geq 0)$ with Levi subalgebra $\mathfrak{m}_\mu = \mathfrak{g}(0)$ and nilradical $\mathfrak{n}_\mu$ $(= \mathfrak{g}(\geq 1))$. We let P_μ denote the corresponding parabolic subgroup with Levi decomposition $M_\mu N_\mu$. The P_μ-orbit of X_μ is equal to ${}^{M_\mu}X_\mu + \mathfrak{g}(\geq 3)$ and ${}^{M_\mu}X_\mu$ is an open and dense subset of $\mathfrak{g}(2)$. Finally, it is true that $C_G(X_\mu) \subset P_\mu$.

5.4. *A sketch of the proof that nilpotent orbital integrals define distributions*

Fix $\mu \in \mathcal{P}(n)$. Let $f \in C_c^\infty(\mathfrak{g})$.

The Iwasawa decomposition (see [33]) tells us that we can write $G = G_{x_0,0}P_\mu$. Fix a Haar measure dk on K and a left Haar measure $d_\ell p$ on P_μ so that

$$\int_G h(g)\,dg = \int_{P_\mu} d_\ell p \int_{G_{x_0,0}} h(kp)\,dk$$

for all $h \in C_c^\infty(G)$, the space of complex-valued, compactly supported, locally constant functions on G. Because it will be important, we recall that for all $p_0 = m_0 n_0 \in P_\mu = M_\mu N_\mu$ we have $d_\ell(p_0 p p_0^{-1}) = \left|\det\left(m_0|_{\mathfrak{n}_\mu}\right)\right|^{-1} d_\ell p$.

Define $\bar{f} \in C_c^\infty(\mathfrak{g})$ by $\bar{f}(X) := \int_{G_{x_0,0}} f({}^kX)\,dk$. Let us assume for the moment that

$$\int_{P/C_G(X_\mu)} \bar{f}({}^pX)\,d_\ell p^* \tag{*}$$

converges. (Here $d_\ell p^*$ is the quotient measure.) In this case, we have

$$(*) = \int_{P/C_G(X_\mu)} \int_K f({}^{kp}X)\,dk\,d_\ell p^* = \int_{G/C_G(X_\mu)} f({}^gX_\mu)\,dg^*,$$

and so we can define the invariant distribution $\mathcal{O}_\mu$ by

$$\mathcal{O}_\mu(f) = \int_{G/C_G(X_\mu)} f({}^gX_\mu)\,dg^*.$$

for $f \in C_c^\infty(\mathfrak{g})$.

We now sketch the proof of why (*) makes sense. Let dZ be a Haar measure on $\mathfrak{g}(\geq 2)$. We'd like to define a function F_0 on $\mathfrak{g}(\geq 2)$ so that

$$(*) = \int_{{}^{M_\mu}X_\mu + \mathfrak{g}(\geq 3)} \bar{f}(Z)\,|F_0(Z)|\,dZ.$$

For $p_0 = m_0 n_0$ as above, we have $d(p_0 Z p_0^{-1}) = \left|\det\left(m_0|_{\mathfrak{g}(\geq 2)}\right)\right|^{-1} dZ$. Consequently, it is sufficient to produce a nonzero function F_0 with the property $F_0({}^{p_0}Z) = \det\left(p_0|_{\mathfrak{g}(1)}\right)$.

The map $X \mapsto (Y \mapsto \mathrm{tr}(Y \cdot X))$ induces an isomorphism of $\mathfrak{g}(1)$ with $\mathfrak{g}(-1)^*$. Just as at the beginning of this section, for $Z \in \mathfrak{g}(2)$, we can define an alternating bilinear form $\langle\, , \rangle_Z$ on $\mathfrak{g}(-1)$. When $Z = X_\mu$, the form is also nondegenerate. Consequently, both $\mathfrak{g}(1)$ and $\mathfrak{g}(-1)$ are even dimensional (of dimension $2n'$). Let a_Z be the matrix which represents $\langle\, , \rangle_Z$ with respect to some fixed basis of $\mathfrak{g}(-1)$. From [22] (Theorem 6.4) there is a polynomial (Pfaffian) Pf of degree n' on $\mathfrak{g}(2)$ so that $\det(a_Z) = (\mathrm{Pf}(Z))^2$. Moreover, since $\det(a_{({}^m Z)}) = \det\left(m^{-1}|_{\mathfrak{g}(-1)}\right)^2 \cdot \det(a_Z)$, the polynomial Pf transforms in the manner we desire. For $Z \in \mathfrak{g}(\geq 2)$, set $F_0(Z) = \mathrm{Pf}(Z_2)$ where Z_2 denotes the image of Z under the projection map from $\mathfrak{g}(\geq 2)$ to $\mathfrak{g}(2)$.

Example 5.4.1: For the partition $(3,2) \in \mathcal{P}(5)$, we have that the dimension of $\mathfrak{g}(-1)$ is four. The space $\mathfrak{g}(2)$ is spanned by the Chevalley basis elements $X_{\alpha_{12}}$, $X_{\alpha_{23}}$ and $X_{\alpha_{45}}$. If $Z = xX_{\alpha_{12}} + yX_{\alpha_{23}} + zX_{\alpha_{45}}$, then $F_0(Z) = xy$.

5.5. *An important calculation*

For $j > 0$ we calculate $\mathcal{O}_\mu([X_\mu + \mathfrak{k}_j])$.

Proposition 5.5.1:

$$\mathcal{O}_\mu([X_\mu + \mathfrak{k}_j]) = q^{((1-2j)\cdot\dim(\mathcal{O}_\mu)/2)}$$

Proof: Suppose that $m > j$ is very large. Let $\mathcal{O}_\mu^m := {}^{K_m}X_\mu = \{{}^k X_\mu \mid k \in K_m\}$. It follows from Exercise 5.3.1 that

$$\mathcal{O}_\mu([X_\mu + \mathfrak{k}_j]) = [K_j : K_m \cdot C_{K_j}(X_\mu)] \cdot \mathcal{O}_\mu([\mathcal{O}_\mu^m]).$$

Now, by definition, we have

$$\begin{aligned}\mathcal{O}_\mu([\mathcal{O}_\mu^m]) &= \mathrm{meas}_{dg^*}(\mathcal{O}_\mu^m)\\ &= \mathrm{meas}_{|dX_1|\cdot|dX_2|\cdots|dX_{2m}|}\left((\mathfrak{k}_m + C_{\mathfrak{g}}(X_\mu))/C_{\mathfrak{g}}(X_\mu)\right)\\ &= [\mathfrak{k}_{1-m} + C_{\mathfrak{g}}(X_\mu) : \mathfrak{k}_m + C_{\mathfrak{g}}(X_\mu)]^{-1/2}\\ &= \frac{[C_{\mathfrak{k}_{(1-m)}}(X_\mu) : C_{\mathfrak{k}_m}(X_\mu)]^{1/2}}{[\mathfrak{k}_{(1-m)} : \mathfrak{k}_m]^{1/2}}.\end{aligned}$$

On the other hand,

$$[K_j : K_m \cdot C_{K_j}(X_\mu)] = \frac{[K_j : K_m]}{[C_{K_j}(X_\mu) : C_{K_m}(X_\mu)]}.$$

After putting the two pieces together and doing a bit of calculation, we arrive at

$$\mathcal{O}_\mu([X_\mu + \mathfrak{k}_j]) = \frac{[C_{\mathfrak{k}_0}(X_\mu) : C_{\mathfrak{k}_j}(X_\mu)]}{[\mathfrak{k}_0 : \mathfrak{k}_j]} \cdot \frac{[\mathfrak{k}_0 : \mathfrak{k}_1]^{1/2}}{[C_{\mathfrak{k}_0}(X_\mu) : C_{\mathfrak{k}_1}(X_\mu)]^{1/2}}.$$

The proposition follows immediately. □

5.6. *Nilpotent orbital integrals are homogeneous*

Suppose $t \in k^\times$. For $f \in C_c^\infty(\mathfrak{g})$ we define the *dilation* $f_t \in C_c^\infty(\mathfrak{g})$ of f by t via $f_t(X) = f(tX)$. For $\mu \in \mathcal{P}(n)$ we have

$$\mathcal{O}_\mu(f_t) = \int_{G/C_G(X_\mu)} f({}^g(tX_\mu))\, dg^*.$$

Consequently, since nilpotent orbits are closed with respect to scaling and the invariant measure on a nilpotent orbit is unique up to a constant, $\mathcal{O}_\mu(f_t)$ and $\mathcal{O}_\mu(f)$ must differ by a constant. A calculation shows that

$$\mathcal{O}_\mu(f_t) = |t|^{\frac{\dim(\mathcal{O}_\mu)}{2}} \mathcal{O}_\mu(f).$$

For example, from Proposition 5.5.1 we have

$$\mathcal{O}_\mu([\varpi^{(-2j+1)} X_\mu + \mathfrak{k}_{1-j}]) = 1. \tag{2}$$

6. The Fourier transforms of invariant distributions

6.1. *Basics*

A *distribution* on $\mathfrak{g}$ is any element of the linear dual of $C_c^\infty(\mathfrak{g})$ (no topological restrictions). We denote the subspace of invariant distributions on $\mathfrak{g}$ by $J(\mathfrak{g})$[e].

Suppose $T \in J(\mathfrak{g})$. We define the Fourier transform $\hat{T} \in J(\mathfrak{g})$ of T by

$$\hat{T}(f) := T(\hat{f})$$

for $f \in C_c^\infty(\mathfrak{g})$.

It is a remarkable fact that $\hat{T}$ is represented by a locally integrable function; that is, there is a function $\hat{T} \in L^1_{\mathrm{loc}}(\mathfrak{g})$ such that for all $f \in C_c^\infty(\mathfrak{g})$,

$$\hat{T}(f) = \int_{\mathfrak{g}} \hat{T}(X) \cdot f(X)\, dX.$$

Unfortunately, describing this function is beyond our abilities in all but the simplest situations.

Example 6.1.1: Consider the trivial nilpotent orbital integral $\mathcal{O}_{(1,1,\ldots,1)} \in J(\mathfrak{g})$. For $f \in C_c^\infty(\mathfrak{g})$ we have

$$\hat{\mathcal{O}}_{(1,1,\ldots,1)}(f) = \hat{f}(0) = \int_{\mathfrak{g}} f(X) \cdot \Lambda(\mathrm{tr}(0 \cdot X))\, dX = \int_{\mathfrak{g}} f(X)\, dX$$

[e]According to Howe, he chose the letter J because I (for invariant) was already taken, and J follows the letter I in the alphabet.

and we also have

$$\hat{\mathcal{O}}_{(1,1,\ldots,1)}(f) = \int_{\mathfrak{g}} f(X) \cdot \hat{\mathcal{O}}_{(1,1,\ldots,1)}(X)\, dX.$$

We therefore conclude that $\hat{\mathcal{O}}_{(1,1,\ldots,1)} = 1$.

We let $\mathfrak{g}^{\text{r.s.s.}}$ denote the set of regular semisimple elements in $\mathfrak{g}$; this is a dense open subset of $\mathfrak{g}$. In our situation, $\mathfrak{g}^{\text{r.s.s.}}$ consists of those elements of $\mathfrak{g}$ having distinct eigenvalues.

The main idea of this section is to present a very elegant description of $\hat{T}(X)$ when $T \in J(\mathfrak{g})$ and $X \in \mathfrak{g}^{\text{r.s.s.}}$. (From this description, it will follow that $\hat{T}$ is represented by a locally constant function on $\mathfrak{g}^{\text{r.s.s.}}$.) The existence (of a form) of this expression was conjectured by Paul Sally, Jr. and proved by Reid Huntsinger [21]. The proof follows an argument of Harish-Chandra [13].

Originally, Huntsinger and Sally were only interested in studying the behavior of the Fourier transform of a nilpotent orbital integral. We shall temporarily restrict our attention to this situation. Suppose $\mu \in \mathcal{P}(n)$. For $f \in C_c^\infty(\mathfrak{g})$ we have

$$\begin{aligned}
\int_{\mathfrak{g}} f(X) \cdot \hat{\mathcal{O}}_\mu(X)\, dX &= \hat{\mathcal{O}}_\mu(f) = \mathcal{O}_\mu(\hat{f}) \\
&= \int_{G/C_G(X_\mu)} \hat{f}({}^g X_\mu)\, dg^* \\
&= \int_{G/C_G(X_\mu)} \int_{\mathfrak{g}} f(X) \cdot \Lambda\big(\operatorname{tr}({}^g X_\mu \cdot X)\big)\, dX\, dg^* \\
\text{“}=\text{”}& \int_{\mathfrak{g}} f(X) \int_{G/C_G(X_\mu)} \Lambda\big(\operatorname{tr}({}^g X_\mu \cdot X)\big)\, dg^*\, dX.
\end{aligned}$$

So, if we could justify the equality in quotation marks, we'd have

$$\begin{aligned}
\hat{\mathcal{O}}_\mu(X) &= \int_{G/C_G(X_\mu)} \Lambda\big(\operatorname{tr}({}^g X_\mu \cdot X)\big)\, dg^* \\
&= \mathcal{O}_\mu\big(Y \mapsto \Lambda(\operatorname{tr}(Y \cdot X))\big).
\end{aligned}$$

This is nearly correct; we now describe what is true.

Theorem 6.1.2: (Reid Huntsinger) *Let K be any compact open subgroup of G and let dk denote the normalized Haar measure on K. For all $X \in \mathfrak{g}^{r.s.s.}$ we have*

$$\hat{\mathcal{O}}_\mu(X) = \mathcal{O}_\mu\big(Y \mapsto \int_K \Lambda(\operatorname{tr}(Y \cdot {}^k X))\, dk\big).$$

Remark 6.1.3: For reasons which will become clear later, we remark that the function $\hat{\mathcal{O}}_\mu$ is canonical. On the face of things, it depends on two choices: the additive character Λ and the choice of a measure on $\mathcal{O}_\mu$. However, our choice of the measure on $\mathcal{O}_\mu$ was not arbitrary; it depended on Λ. As one can verify, this dependence makes $\hat{\mathcal{O}}_\mu$ canonical.

Remark 6.1.4: In our situation, since every nilpotent orbit is Richardson, there is a very nice description, due to Howe, of the function $\hat{\mathcal{O}}_\mu$. Namely, $\hat{\mathcal{O}}_\mu$ can be related to the character of the representation obtained by inducing the trivial representation on $P(\mu)$ up to G.

6.2. *A more general statement*

More generally, following a suggestion of Bob Kottwitz, Reid Huntsinger proved the following statement.

Theorem 6.2.1: (Reid Huntsinger) *Fix* $r \in \mathbb{R}$. *If* $T \in J(\mathfrak{g}_r)$, *then* $\hat{T}$ *is represented on* $\mathfrak{g}^{r.s.s.}$ *by*

$$X \mapsto T(\eta_X).$$

Here, for $Y \in \mathfrak{g}$, $\eta_X(Y) := \int_K (\Lambda(\mathrm{tr}(Y \cdot {}^k X)))\, dk$. *(As before,* K *is a compact open subgroup, and* dk *is the normalized Haar measure on* K*.)*

At the heart of the proof of this theorem lies the statement that the map from $\mathfrak{g}^{\mathrm{r.s.s.}}$ to $C^\infty(\mathfrak{g})$ sending X to $\eta_{X,r} := \eta_X \cdot [\mathfrak{g}_r]$ is locally constant. Since the verification of this statement requires some fairly detailed analysis, we shall skip the proof. This statement immediately implies that the function $X \mapsto T(\eta_{X,r}) = T(\eta_X)$ from $\mathfrak{g}^{\mathrm{r.s.s.}}$ to $\mathbb{C}$ is locally constant.

Suppose $f \in C_c^\infty(\mathfrak{g}^{\mathrm{r.s.s.}})$. We need to show

$$\hat{T}(f) = \int_{\mathfrak{g}} f(X) \cdot T(\eta_X)\, dX.$$

From the previous paragraph, there exists a finite collection $\{\omega_i\}_{i=1}^m$ of compact open disjoint subsets of $\mathfrak{g}^{\mathrm{r.s.s.}}$ such that both $X \mapsto T(\eta_X)$ and $X \mapsto f(X)$ are constant on ω_i and the support of f is contained in $\bigcup \omega_i$.

Using Fubini's theorem and the invariance of T, we then have[f]

$$
\begin{aligned}
\int_{\mathfrak{g}} f(X)\cdot T(\eta_X)\,dX &= \sum_{i=1}^{m} \operatorname{meas}_{dX}(\omega_i)\cdot f(X_i)\cdot T(\eta_{X_i}) \\
&= T\Big(\sum_{i=1}^{m} \operatorname{meas}_{dX}(\omega_i)\cdot f(X_i)\cdot \eta_{X_i}\Big) \\
&= T\Big(Y\mapsto \int_{\mathfrak{g}} f(X)\cdot \eta_X(Y)\,dX\Big) \\
&= T\Big(Y\mapsto \int_{\mathfrak{g}} f(X)\int_K \Lambda(\operatorname{tr}({}^kX\cdot Y))\,dk\,dX\Big) \\
&= T\Big(Y\mapsto \int_K \hat{f}({}^kY)\,dk\Big) \\
&= T(\hat{f}).
\end{aligned}
$$

7. Characters

In this section we define the character of an admissible representation and discuss some properties of characters.

7.1. *Admissible representations*

Fix an admissible representation (π, V) of G. Recall, from Gordan Savin's lectures [33], that (π, V) is a representation for which

(1) for all $v \in V$ there exists a compact open subgroup K such that

$$v \in V^K := \{v \in V \mid \pi(k)v = v \text{ for all } k \in K\}$$

and

(2) for all compact open subgroups K of G we have $\dim_{\mathbb{C}} V^K < \infty$.

The second condition is equivalent to saying that for all compact open subgroups K of G and all irreducible representations σ of K, the multiplicity of σ in π is finite.

Example 7.1.1: Suppose $\bar{\sigma}$ is a cuspidal representation of $\mathrm{GL}_n(\mathfrak{f}) \cong K_0/K_1$. Inflate $\bar{\sigma}$ to a representation of K_0 and extend this inflation to a representation σ of $Z(G)\cdot K_0$ where $Z(G)$ denotes the center of G. The

[f]This part of the proof supports David Vogan's adage: "The p-adics: if you can add, you can integrate."

representation (π_σ, V_σ) obtained by (compact) induction of σ from $Z(G)\cdot K_0$ up to G is an (irreducible) admissible representation.

Example 7.1.2: Suppose $m > 0$ and $\bar{\chi}$ is a character of T/T_m. Inflate $\bar{\chi}$ to a character χ of T. Since the unipotent radical N of B is normal in B, we may extend χ to a character of B. The representation (π_χ, V_χ) obtained by (compact) parabolic induction of χ from B to G is an admissible representation of G.

7.2. *The character distribution*

We define the *character distribution*, Θ_π, associated to the representation (π, V) as follows.

For $f \in C_c^\infty(G)$, we define $\pi(f) \in \mathrm{End}(V)$ by

$$\pi(f)v = \int_G f(g) \cdot \pi(g)v \, dg$$

for $v \in V$. Since f is compactly supported and locally constant and (π, V) is admissible, this integral is really just a finite sum. In fact, if K is a compact open subgroup of G such that $f(k_1 g k_2) = f(g)$ for all $k_1, k_2 \in K$ and $g \in G$, then we have

$$\pi(f)v = \pi([K] * f * [K])v = \pi([K])\pi(f)\pi([K])v.$$

(Here $*$ denotes the usual convolution operation.) Thus, $\pi(f)$ is a map from V to V^K. If $e_K := \mathrm{meas}_{dg}(K)^{-1} \cdot \pi([K])$, then e_K is the projection operator from V to V^K, and we have

$$V = V^K \oplus (1 - e_K)V.$$

Consequently, since $\dim_{\mathbb{C}}(V^K) < \infty$, it follows that $\pi(f)$ is a finite-rank operator. We can therefore define the *character distribution*

$$\Theta_\pi : C_c^\infty(G) \to \mathbb{C}$$

by sending f to $\mathrm{tr}(\pi(f))$.

Just as the Fourier transform of an invariant distribution on $\mathfrak{g}$ is represented on $\mathfrak{g}^{\text{r.s.s.}}$ by a function in $C^\infty(\mathfrak{g}^{\text{r.s.s.}})$, so too is the character distribution [15]. We abuse notation and denote this function by $\Theta_\pi \in C^\infty(G^{\text{r.s.s.}})$. ($G^{\text{r.s.s.}}$ denotes the open, dense subset of G consisting of regular semisimple elements, that is, those elements of G whose eigenvalues are distinct.) We call this function the *character* of π.

Remark 7.2.1: Fix $\gamma \in G^{\text{r.s.s.}}$ and $m > 0$ large enough so that $\gamma K_m \subset G^{\text{r.s.s.}}$. Define $f_m = \text{meas}_{dg}(K_m)^{-1} \cdot [\gamma K_m]$. We then have

$$\Theta_\pi(f_m) = \int_G f_m(g) \cdot \Theta_\pi(g)\, dg = \text{meas}_{dg}(K_m)^{-1} \int_{K_m} \Theta_\pi(\gamma k)\, dk.$$

Since the function Θ_π is locally constant on $G^{\text{r.s.s.}}$, the right-hand side becomes $\Theta_\pi(\gamma)$ for all m sufficiently large. Thus, "in the limit" the function Θ_π looks like a character.

7.3. *A calculation*

Fix $j \geq 1$.

Suppose that $\bar{\tau}$ is an irreducible representation of K_j/K_{2j}. Since K_j/K_{2j} is abelian, $\bar{\tau}$ is a character. Let τ denote the corresponding character of K_j; we regard τ as an element of $C_c^\infty(G)$ in the obvious way.

Let $\text{res}_{K_j} \pi$ denote the restriction of π to K_j. Suppose that

$$\text{res}_{K_j} \pi = \oplus_{\sigma \in \widehat{K_j}} m(\sigma, \pi)\sigma$$

where $m(\sigma, \pi)$ denotes the multiplicity of σ in $\text{res}_{K_j} \pi$. We have

$$\begin{aligned}
\Theta_\pi(\tau) &= \text{tr} \int_G \tau(g) \cdot \pi(g)\, dg \\
&= \text{tr} \int_{K_j} \tau(g) \cdot \pi(g)\, dg \\
&= \text{tr} \int_{K_j} \sum_{\sigma \in \widehat{K_j}} \tau(g) \cdot m(\sigma, \pi)\sigma(g)\, dg \\
&= \text{tr} \int_{K_j} \sum_{\bar{\sigma} \in \widehat{K_j/K_{2j}}} \tau(g) \cdot m(\sigma, \pi)\sigma(g)\, dg \\
&= m(\tau^{-1}, \pi) \cdot \text{meas}_{dg}(K_j).
\end{aligned}$$

In other words, the character picks out the multiplicity of τ^{-1} in $\text{res}_{K_j} \pi$ (up to a constant).

7.4. *Depth*

The depth of a representation was introduced in the fundamental papers of Allen Moy and Gopal Prasad [28, 27]. Essentially, the depth of a representation is a rational number which measures the first occurrence of fixed vectors with respect to all filtration subgroups of G that arise naturally from Bruhat-Tits theory.

Recall from §3.4.2 that $\mathcal{O}$ denotes the set of optimal points. Up to conjugation, the set of optimal points is finite, and so the subset $\{r \in \mathbb{R} \mid \mathfrak{g}_{x,r} \neq \mathfrak{g}_{x,r+} \text{ for some } x \in \mathcal{O}\}$ is discrete (and in fact, is a subset of $\mathbb{Q}$). For an admissible representation (π, V), we define $\rho(\pi)$, the depth of (π, V), by

$$\rho(\pi) := \min\{r \in \mathbb{Q}_{\geq 0} \mid \text{there is an } x \in \mathcal{O} \text{ such that } V^{G_{x,r+}} \neq 0\}.$$

Proposition 7.4.1: *$\rho(\pi)$ is the unique rational number satisfying the following statement. If $(x,r) \in \mathcal{A}(T) \times \mathbb{R}_{\geq 0}$ with $V^{G_{x,r+}} \neq \{0\}$, then $r \geq \rho(\pi)$.*

Proof: Suppose $(x,r) \in \mathcal{A}(T) \times \mathbb{R}_{\geq 0}$ with $V^{G_{x,r+}} \neq \{0\}$. From the group version of Lemma 3.4.3 there exists $y \in \mathcal{O}$ such that $G_{x,r+} \supset G_{y,r+}$. Consequently, $\{0\} \neq V^{G_{x,r+}} \subset V^{G_{y,r+}}$. □

All the various things you would want to be true about the depth of a representation are true. For example, if σ is an irreducible representation of a parabolic subgroup of G, and π is an irreducible subquotient of the induced representation, then $\rho(\pi) = \rho(\sigma)$.

Example 7.4.2: Any representation with Iwahori (that is, B_0) fixed vectors has depth zero.

Example 7.4.3: The representation (π_σ, V_σ) defined in Example 7.1.1 does not have Iwahori fixed vectors, but it does have depth zero.

Example 7.4.4: The representation (π_χ, V_χ) defined in Example 7.1.2 has depth $(m-1)$.

7.5. *Elementary Kirillov theory*

Fix an irreducible representation (π, V).

For nilpotent real groups, Kirillov [24] established that the representations were parameterized by coadjoint orbits in the linear dual of the Lie algebra. In our context, the term *Kirillov theory* is used to describe the connection between representations of compact open subgroups occurring in π and coadjoint orbits in the linear dual of $\mathfrak{g}$. Via our nondegenerate trace form, we have identified $\mathfrak{g}$ with its dual. The type of result discussed below was first studied, I believe, by Howe [17].

Fix $r, s \in \mathbb{R}_{\geq 0}$ and $x, y \in \mathcal{B}$.

The map $g \mapsto (g-1)$ induces an isomorphism of the abelian groups $G_{x,r^+}/G_{x,(2r)^+}$ and $\mathfrak{g}_{x,r^+}/\mathfrak{g}_{x,(2r)^+}$. Moreover, every character of $\mathfrak{g}_{x,r^+}/\mathfrak{g}_{x,(2r)^+}$ is of the form $\bar{Y} \mapsto \Lambda(\mathrm{tr}(X \cdot Y)$ for some $\bar{X} \in \mathfrak{g}_{x,-2r}/\mathfrak{g}_{x,-r}$. (That is, the Pontrjagin dual of $G_{x,r^+}/G_{x,(2r)^+} \cong \mathfrak{g}_{x,r^+}/\mathfrak{g}_{x,(2r)^+}$ is isomorphic to $\mathfrak{g}_{x,-2r}/\mathfrak{g}_{x,-r}$.)

Fix a character $\bar{\sigma}$ of $G_{x,r^+}/G_{x,(2r)^+}$ and a character $\bar{\tau}$ of $G_{y,s^+}/G_{y,(2s)^+}$ Let $X_\sigma + \mathfrak{g}_{x,-r} \in \mathfrak{g}_{x,-2r}/\mathfrak{g}_{x,-r}$ represent $\bar{\sigma}$ and $X_\tau + \mathfrak{g}_{y,-s} \in \mathfrak{g}_{y,-2s}/\mathfrak{g}_{y,-s}$ represent $\bar{\tau}$.

Proposition 7.5.1: *If σ and τ both occur in π, then there exists a $g \in G$ such that*

$$^g(X_\tau + \mathfrak{g}_{y,-s}) \cap (X_\sigma + \mathfrak{g}_{x,-r}) \neq \emptyset.$$

Proof: Let $V_\sigma \subset V$ (resp. $V_\tau \subset V$) denote a one-dimensional subspace of V on which $\mathrm{res}_{G_{x,r^+}} \pi$ (resp. $\mathrm{res}_{G_{y,s^+}} \pi$) acts by σ (resp. τ). Since (π, V) is irreducible, there exists a $g \in G$ such that the image of $\pi(g)V_\sigma$ under the projection of V onto V_τ is nonzero. This implies that for all $h \in G_{x,r^+} \cap {}^gG_{y,s^+}$, we have

$$\sigma(h) = \tau(g^{-1}hg).$$

Thus, for all $H \in \mathfrak{g}_{x,r^+} \cap {}^g\mathfrak{g}_{y,s^+}$, we have

$$\Lambda(\mathrm{tr}(X_\sigma \cdot H)) = \Lambda(\mathrm{tr}({}^gX_\tau \cdot H)).$$

This implies that for all $H \in \mathfrak{g}_{x,r^+} \cap {}^g\mathfrak{g}_{y,s^+}$, we have $\mathrm{tr}((X_\sigma - {}^gX_\tau) \cdot H) \in \wp$. Consequently, $(X_\sigma - {}^gX_\tau) \in (\mathfrak{g}_{x,r^+} \cap {}^g\mathfrak{g}_{y,s^+})^* = \mathfrak{g}_{x,-r} + {}^g\mathfrak{g}_{y,-s}$. The proposition follows. □

7.6. *Understanding the distribution* $\mathrm{res}_{C_c^\infty(G_{\rho(\pi)^+})} \Theta_\pi$

Let (π, V) be an irreducible admissible representation.

From our discussion of depth, we know that we can find an $x \in \mathcal{A}(T)$ such that $V^{G_{x,\rho(\pi)^+}} \neq \{0\}$. Thus, the trivial representation of $G_{x,\rho(\pi)^+}$ occurs in π; the associated coset in $\mathfrak{g}$ is $\mathfrak{g}_{x,-\rho(\pi)}$.

Now fix $s > \rho(\pi)$, $y \in \mathcal{B}$, and $\bar{\tau} \in \widehat{G_{y,s}/G_{y,s^+}}$ such that the character τ of $G_{y,s}$ occurs in π. Let $X_\tau + \mathfrak{g}_{y,(-s)^+} \in \mathfrak{g}_{y,(-s)}/\mathfrak{g}_{y,(-s)^+}$ be the coset corresponding to $\bar{\tau}$.

From our discussion of Kirillov theory, there exists a $g \in G$ such that ${}^g\mathfrak{g}_{x,-\rho(\pi)} \cap (X_\tau + \mathfrak{g}_{y,(-s)^+}) \neq \emptyset$. However, from Lemma 4.2.1, we have

$$^g\mathfrak{g}_{x,-\rho(\pi)} \subset \mathfrak{g}_{y,-\rho(\pi)} + \mathcal{N} \subset \mathfrak{g}_{y,(-s)^+} + \mathcal{N}.$$

Consequently, we can assume that X_τ is nilpotent!

We now use this fact to say something about the distribution $\mathrm{res}_{C_c^\infty(G_{\rho(\pi)+})}\Theta_\pi$. For a function $h \in C_c^\infty(\mathfrak{g}_{\rho(\pi)+})$, we define $\tilde{h} \in C_c^\infty(G_{\rho(\pi)+})$ by $\tilde{h}(g) = h(g-1)$. We define the distribution $\Theta_{\pi,\mathfrak{g}}$ on $\mathfrak{g}$ by $\Theta_{\pi,\mathfrak{g}}(f) = \Theta_\pi(\tilde{f}_{\rho(\pi)})$ where $f_{\rho(\pi)} = f \cdot [\mathfrak{g}_{\rho(\pi)+}]$. We let $\hat{\Theta}_\pi$ denote the Fourier transform of $\Theta_{\pi,\mathfrak{g}}$.

From Exercise 4.2.3, we know that the Fourier transform of an element of $D_{-\rho(\pi)}$ belongs to $C_c^\infty(\mathfrak{g}_{\rho(\pi)+})$. Suppose $x \in \mathcal{B}$ and $s > \rho(\pi)$. If $f \in C(\mathfrak{g}_{x,-s}/\mathfrak{g}_{x,-\rho(\pi)})$, then we have $\hat{\Theta}_\pi(f) = \Theta_\pi(\tilde{\hat{f}})$. It follows from our discussion above and a few lines of calculation that

$$\hat{\Theta}_\pi(f) \neq 0 \text{ implies } \mathrm{supp}(f) \cap (\mathfrak{g}_{x,(-s)^+} + \mathcal{N}) \neq \emptyset.$$

This last statement is equivalent to the statement

$$\hat{\Theta}_\pi(f) \neq 0 \text{ implies } \mathrm{supp}(f) \cap \mathfrak{g}_{(-s)^+} \neq \emptyset. \tag{3}$$

7.7. *The Harish-Chandra–Howe local character expansion*

Considerations similar to those above led Roger Howe to make his finiteness conjectures [19] (which we will discuss later) and establish the following remarkable connection between the character of an irreducible representation of G and the Fourier transforms of nilpotent orbital integrals.

Theorem 7.7.1: (Harish-Chandra–Howe local character expansion) *If π is an irreducible admissible representation of G, then there exist constants $c_\mu(\pi)$ indexed by $\mu \in \mathcal{P}(n)$ such that*

$$\Theta_\pi(1+X) = \sum_{\mu \in \mathcal{P}(n)} c_\mu(\pi) \cdot \hat{\mathcal{O}}_\mu(X) \tag{4}$$

for all $X \in \mathfrak{g}^{r.s.s.}$ sufficiently near zero.

Remark 7.7.2: Howe [16] proved this result without any restrictions on the characteristic of k. Later, under the assumption that the characteristic of k is zero, Harish-Chandra [13] generalized the proof to all connected reductive groups. The above theorem and its analogues (local expansions about any semisimple point) play a crucial role in Harish-Chandra's proof that characters are locally integrable on G (not just on $G^{\mathrm{r.s.s.}}$). (The question of integrability is still open for arbitrary groups in positive characteristic. From work of Rodier [30] and Lemaire [25], it is known to be true for $\mathrm{GL}_n(k)$ when k has positive characteristic.)

One of the difficulties with this theorem is that it provides no indication of where equation (4) ought to hold. The conjecture of Hales, Moy, and Prasad [28] is a precise statement about this issue:

Statement 7.7.3: Equation (4) ought to be valid on $\mathfrak{g}^{\text{r.s.s.}}_{\rho(\pi)+}$.

For integral-depth representations this was proved by Waldspurger [35]. With some restrictions on k, it was proved for arbitrary depth representations in [9].

8. An introduction to Howe's conjectures and homogeneity

In his paper, *Two conjectures about reductive p-adic groups*, Howe proposed two remarkable finiteness conjectures that now bear his name. Howe's conjecture for the Lie algebra looks like:

$$\dim_{\mathbb{C}} \operatorname{res}_{C_c(\mathfrak{g}/L)} J(\omega) < \infty.$$

Here $\omega \subset \mathfrak{g}$ is any compactly generated[g], invariant, and closed subset of $\mathfrak{g}$, $J(\omega)$ denotes the space of invariant distributions supported on ω, and L is any lattice in $\mathfrak{g}$. For $T \in J(\omega)$,

$$\operatorname{res}_{C_c(\mathfrak{g}/L)} T$$

denotes the restriction of T to $C_c(\mathfrak{g}/L)$. Howe's conjecture for the group looks very similar:

$$\dim_{\mathbb{C}} \operatorname{res}_{C_c(G/K)} J(\omega) < \infty.$$

Here K is a compact open subgroup of G, and ω is a closed, compactly generated, invariant subset of G, and $J(\omega)$ denotes the space of invariant distributions which are supported on ω.

Of course, although we still refer to these statements as Howe's conjectures, both are known to be true. Howe's conjecture for the Lie algebra was proved by Howe [16] in the 1970s for GL_n and later by Harish-Chandra [13] for general groups (but only in the characteristic zero setting). Waldspurger [37] has also given a proof. Howe's conjecture for the group was proved in the 1980s by Clozel [6, 7] in the characteristic zero setting. In the 1990s, a characteristic free proof of Howe's conjecture for the group was given by Barbasch and Moy in their beautiful paper [2].

[g]That is, ω can be realized as the closure of ${}^{G}C$ for some compact subset C of $\mathfrak{g}$.

8.1. *Homogeneity*

Beginning with Waldspurger's paper [36], much energy has been spent trying to produce optimal versions of Howe's conjectures. By optimal, we mean that we'd like to choose ω and L in such a way so that we can not only describe the dimension of $\mathrm{res}_{C_c(\mathfrak{g}/L)}\, J(\omega)$, but we can also find a good basis for this space in terms of distributions with which we are very familiar, namely, nilpotent orbital integrals.

These optimal versions of Howe's conjectures are referred to as *homogeneity results*. This is an appropriate name, because, according to Webster's Ninth New Collegiate Dictionary, the word homogeneity means

"the state of having identical distribution functions or values".

We restrict our attention to the Lie algebra. Fix $r \in \mathbb{R}$. The role of ω in Howe's conjecture will be played by the G-domain $\mathfrak{g}_{r^+}$ (which is compactly generated). Guided by Harish-Chandra's philosophy and Theorem 7.7.1, it appears that we want something like the following: For all $T \in J(\mathfrak{g}_{r^+})$ and for all $f \in C_c^\infty(\mathfrak{g}_{(-r)})$

$$\hat{T}(f) = \sum_{\mu \in \mathcal{P}(n)} c_\mu(T) \cdot \hat{\mathcal{O}}_\mu(f).$$

This last displayed equation is equivalent to requiring

$$T(\hat{f}) = \sum_{\mu \in \mathcal{P}(n)} c_\mu(T) \cdot \mathcal{O}_\mu(\hat{f}).$$

From Exercise 4.2.3 we have $\hat{f} \in D_{r^+}$. So, the role of $C_c^\infty(\mathfrak{g}/L)$ will be played by D_{r^+}. Putting it all together, we arrive at the homogeneity statement

$$\mathrm{res}_{D_{r^+}}\, J(\mathfrak{g}_{r^+}) = \mathrm{res}_{D_{r^+}}\, J(\mathcal{N}).$$

The fact that $\mathfrak{g}_{r^+} \subset \mathfrak{g}_{x,r^+} + \mathcal{N}$ for all $x \in \mathcal{B}$ gives some feeling as to why this homogeneity statement ought to be true.

8.2. *From Howe's conjectures to the Harish-Chandra–Howe local character expansion*

The homogeneity result discussed in the previous section is actually not strong enough to prove Theorem 7.7.1. As in [16, 13, 9, 35] we need something a bit stronger, but, unfortunately, more complicated.

For $x \in \mathcal{B}$ and $s \le r$, define

$$\tilde{J}_{x,s,r^+} := \{T \in J(\mathfrak{g}) \,|\, \text{for } f \in C(\mathfrak{g}_{x,s}/\mathfrak{g}_{x,r^+}),$$
$$\text{if } \mathrm{supp}(f) \cap (\mathfrak{g}_{s^+}) = \emptyset, \text{ then } T(f) = 0\}.$$

Note that $J(\mathfrak{g}_{r^+}) \subset \tilde{J}_{x,s,r^+}$.

Remark 8.2.1: Although this definition seems a bit unnatural, it arises naturally from our understanding of $\operatorname{res}_{C_c^\infty(G_{\rho(\pi)^+})} \Theta_\pi$. In fact, from Equation (3), we have that $\hat{\Theta}_\pi \in \tilde{J}_{x,s,r^+}$ if $-r > \rho(\pi)$.

The following is the stronger homogeneity statement that needs to be proved in order to recover Statement 7.7.3:

$$\operatorname{res}_{D_{r^+}} \tilde{J}_{r^+} = \operatorname{res}_{D_{r^+}} J(\mathcal{N})$$

where

$$\tilde{J}_{r^+} := \bigcap_{x \in \mathcal{B}} \bigcap_{s \leq r} \tilde{J}_{x,s,r^+}.$$

9. Proving homogeneity results

We recall that for $r \in \mathbb{R}$ we have

$$\mathfrak{g}_{r^+} = \bigcup_{x \in \mathcal{B}} \mathfrak{g}_{x,r^+} = \bigcap_{x \in \mathcal{B}} (\mathfrak{g}_{x,r^+} + \mathcal{N})$$

and

$$D_{r^+} = \sum_{x \in \mathcal{B}} C_c(\mathfrak{g}/\mathfrak{g}_{x,r^+}) = \widehat{C_c^\infty(\mathfrak{g}_{-r})}.$$

We want to show

$$\operatorname{res}_{D_{r^+}} J(\mathfrak{g}_{r^+}) = \operatorname{res}_{D_{r^+}} J(\mathcal{N}). \tag{5}$$

We decided that knowing this, or, in truth, a stronger (but more complicated) statement would tell us that the Harish-Chandra–Howe local character expansion was valid on $\mathfrak{g}^{\text{r.s.s.}}_{\rho(\pi)^+}$.

As we discussed before, $\mathfrak{g}_r = \mathfrak{g}_{r^+}$ unless $r = \frac{k}{n}$ for some $k \in \mathbb{Z}$. Consequently, we only need to verify Equation (5) for $r \in \{\frac{k}{n}\}$.

Moreover, by taking advantage of the fact that nilpotent orbital integrals are homogeneous, we can further restrict our attention to r of the form $\frac{k}{n}$ with $0 \leq k < n$. Indeed, pick $m \in \mathbb{Z}$ such that $m + r \in [0,1)$. For $T \in J(\mathfrak{g})$ and $t \in k^\times$, define T_t by $T_t(f) = T(f_t)$. Note that

$$f \in D_{r^+} \text{ if and only if } f_{\varpi^{-m}} \in D_{(m+r)^+}$$

and

$$T \in J(\mathfrak{g}_{r^+}) \text{ if and only if } T_{\varpi^m} \in J(\mathfrak{g}_{(m+r)^+}).$$

Consequently, if we know

$$\mathrm{res}_{D_{(r+m)+}} J(\mathfrak{g}_{(r+m)+}) = \mathrm{res}_{D_{(r+m)+}} J(\mathcal{N})$$

then for $T \in J(\mathfrak{g}_{r+})$ and $f \in D_{r+}$ we have

$$\begin{aligned} T(f) &= T_{\varpi^m}(f_{\varpi^{-m}}) \\ &= \sum c_\mu(T_{\varpi^m}) \cdot \mathcal{O}_\mu(f_{\varpi^{-m}}) \\ &= \sum c_\mu(T_{\varpi^m}) \cdot q^{\frac{m\cdot\dim(\mathcal{O}_\mu)}{2}} \cdot \mathcal{O}_\mu(f), \end{aligned}$$

and so we may assume $r \in [0,1)$.

For the remainder of this section we discuss how to prove Equation (5). We first prove it for $\mathrm{GL}_1(k)$, we next prove it for $\mathrm{GL}_2(k)$, and finally we discuss a way to go about giving a general proof.

9.1. *A proof for* $\mathrm{GL}_1(k)$

From the reasoning above, we need to show

$$\mathrm{res}_{C_c(k/\wp)} J(\wp) = \mathrm{res}_{C_c(k/\wp)} J(\mathcal{N}).$$

The right-hand side is one-dimensional and spanned by the distribution $f \mapsto f(0)$ for $f \in C_c(k/\wp)$. Moreover, the right-hand side is a vector subspace of the left-hand side.

Since $\mathrm{GL}_1(k)$ is abelian, all distributions are invariant. Note that $J(\wp)$ consists of linear maps from $C_c^\infty(k)$ to $\mathbb{C}$ which are supported on $\wp$. If $f \in C_c(k/\wp)$, then we can write

$$f = \sum_{\bar{X} \in k/\wp} c_{\bar{X}} \cdot [X + \wp]$$

where the $c_{\bar{X}}$ are complex numbers which are almost always zero. For $T \in J(\wp)$, we have

$$T(f) = T(c_{\bar{0}} \cdot [\wp]) = T([\wp]) \cdot f(0).$$

Consequently, the left-hand side is also one-dimensional and so the equality is established.

9.2. *A proof for* $\mathrm{GL}_2(k)$

This is where things begin to become interesting. Thanks to the remarks at the beginning of this section, we only need to verify two statements:

$$\mathrm{res}_{D_{0+}} J(\mathfrak{g}_{0+}) = \mathrm{res}_{D_{0+}} J(\mathcal{N}) \tag{6}$$

and

$$\operatorname{res}_{D_{(1/2)+}} J(\mathfrak{g}_{(1/2)+}) = \operatorname{res}_{D_{(1/2)+}} J(\mathcal{N}). \tag{7}$$

We begin by considering the first statement, Equation (6). We note that D_{0+} may be thought of as an invariant version of $C_c(\mathfrak{g}/\mathfrak{k}_1)$, but, for our purposes, this is not a good way to think about the space D_{0+}.

9.2.1. *Descent and recovery*

Fix $T \in J(\mathfrak{g}_{0+})$. We wish to show that $\operatorname{res}_{D_{0+}} T$ is completely determined by $\operatorname{res}_{C(\mathfrak{k}_0/\mathfrak{k}_1)+C(\mathfrak{b}_0/\mathfrak{b}_{1/2})} T$.

Fix $f \in D_{0+}$. We will demonstrate that $T(f)$ is completely determined by $\operatorname{res}_{C(\mathfrak{k}_0/\mathfrak{k}_1)+C(\mathfrak{b}_0/\mathfrak{b}_{1/2})} T$. We write $f = \sum_i f_i$ with $f_i \in C_c(\mathfrak{g}/\mathfrak{g}_{x_i,0+})$ for some $x_i \in \mathcal{B}$. Since T is linear, without loss of generality we may assume that $f \in C_c(\mathfrak{g}/\mathfrak{g}_{x,0+})$ for some $x \in \mathcal{B}$. We can write

$$f = \sum_{\bar{X} \in \mathfrak{g}/\mathfrak{g}_{x,0+}} c_{\bar{X}} \cdot [X + \mathfrak{g}_{x,0+}]$$

with the $c_{\bar{X}}$ complex constants which are almost always zero. Again, since T is linear, without loss of generality we may assume that $f = [X + \mathfrak{g}_{x,0+}]$.

Now, $T(f) = 0$ if the support of f does not intersect $\mathfrak{g}_{0+}$. Consequently, since $\mathfrak{g}_{0+} \subset \mathfrak{g}_{x,0+} + \mathcal{N}$, we must have

$$(X + \mathfrak{g}_{x,0+}) \cap \mathcal{N} \neq \emptyset.$$

So, without loss of generality, $X \in \mathcal{N}$.

Up to conjugacy, we have two choices for $\mathfrak{g}_{x,0+}$; it is either $\mathfrak{k}_1$ or $\mathfrak{b}_{1/2}$. In what follows, the reader is encouraged to consult Figure 8 to get a more geometric understanding of what is happening.

We first deal with the $\mathfrak{k}_1$ case. Since x_0 is the only point x where $\mathfrak{g}_{x,0+} = \mathfrak{k}_1$, we may suppose that $X \in \mathcal{N} \cap (\mathfrak{g}_{x_0,-m} \setminus \mathfrak{g}_{x_0,(-m)+})$ for some $m > 0$. In other words, $X \in \mathcal{N} \cap (\mathfrak{k}_{-m} \setminus \mathfrak{k}_{1-m})$. Since we are free to conjugate by $G_{x_0,0} = K_0$, we may assume that

$$X = \begin{pmatrix} 0 & \varpi^{-m}u \\ 0 & 0 \end{pmatrix}$$

with $u \in R^\times$. For any point y in the chamber C_0 (that is, a point between x_0 and x' in Figure (8)), we have that X is "closer" to the origin with respect to the y filtration than it is in the x filtration. For example, $X \in \mathfrak{g}_{y_0,(1/2-m)+}$ where y_0 is the barycenter of C_0. The problem is: at the point y we require local constancy not with respect to $\mathfrak{g}_{x_0,0+}$ but with respect to $\mathfrak{g}_{y,0+} = \mathfrak{b}_{1/2}$.

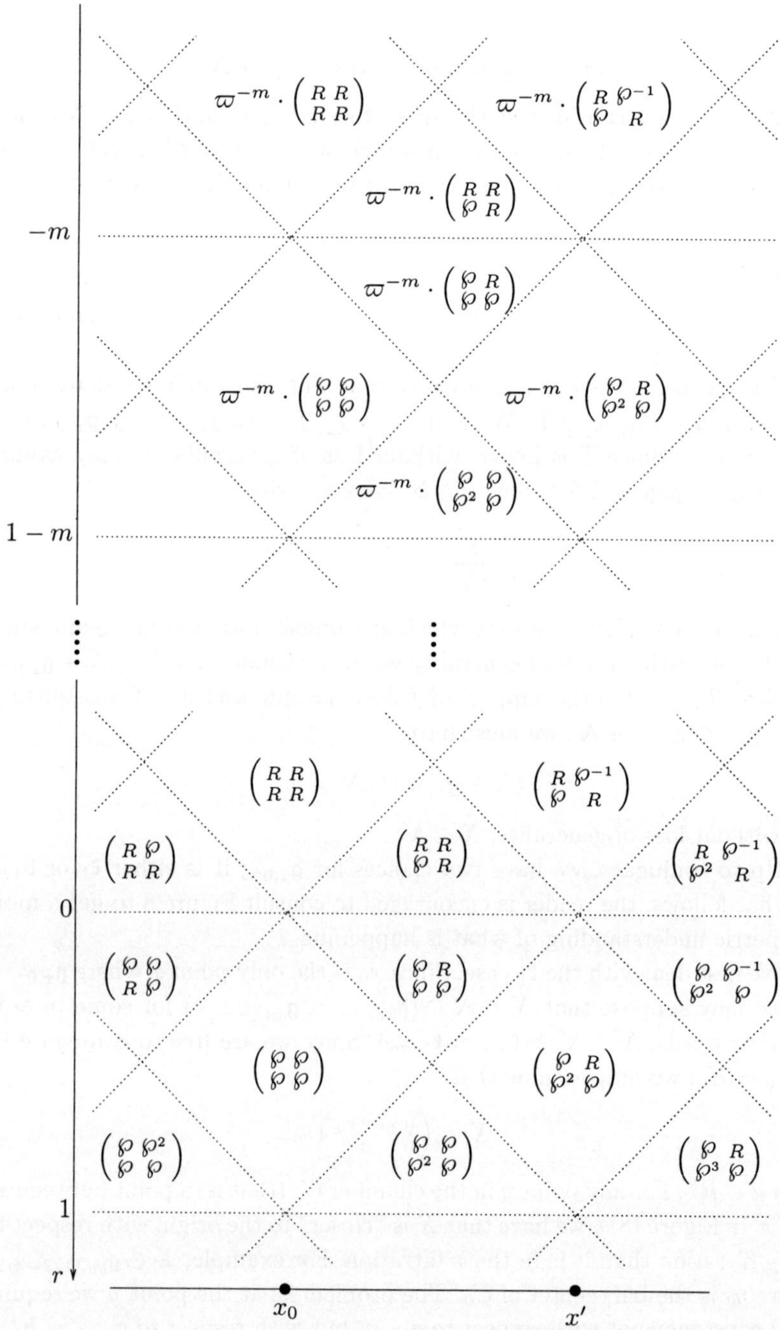

Fig. 8. $(\mathfrak{gl}_2)_{x,r}$

In order to recover the proper type of local constancy, we take advantage of the invariance of T. We write

$$T([X+\mathfrak{k}_1]) = \frac{1}{q^2}\cdot \sum_{\bar{t}\in T_m/T_{m^+}} ([{}^t X + \mathfrak{k}_1])$$

$$= \frac{1}{q^2}\cdot \sum_{\bar{\alpha},\bar{\beta}\in R/\wp} ([X + \begin{pmatrix} 0 & u(\alpha-\beta) \\ 0 & 0 \end{pmatrix} + \mathfrak{k}_1])$$

$$= \frac{1}{q}\cdot T([X+\mathfrak{b}_{1/2}]).$$

We have succeeded in writing $T(f)$ in terms of T evaluated at a function $f' \in D_{0^+}$ which is supported closer to the origin with respect to some other point in the building.

We now examine the $\mathfrak{b}_{1/2}$ case. In this case, we are looking at the coset $X + \mathfrak{g}_{y,0^+}$ where $X \in \mathcal{N}$ and y is any point in C_0. Since we are free to conjugate by $G_{y,0} = B_0$, we may assume that X is either

$$\begin{pmatrix} 0 & \varpi^{-m}u \\ 0 & 0 \end{pmatrix}$$

or

$$\begin{pmatrix} 0 & 0 \\ \varpi^{(1-m)}u & 0 \end{pmatrix}$$

with $m > 0$ and $u \in R^\times$. In the former case, we can write

$$T([X+\mathfrak{g}_{y,0^+}]) = \sum_{\bar{\alpha}\in\wp/\wp^2} T([X + \begin{pmatrix} 0 & 0 \\ \alpha & 0 \end{pmatrix} + \mathfrak{g}_{x_1,0^+}])$$

where $x_1 = {}^{\mathrm{diag}(\varpi^{-1},1)}x_0 = x'$ is the other vertex of $\bar{C}_0$. We have

$$\mathfrak{g}_{x_1,0^+} = \begin{pmatrix} \wp & R \\ \wp^2 & \wp \end{pmatrix}.$$

From Figure 8 it is clear that we have expressed T evaluated at $[X+\mathfrak{g}_{y,0^+}]$ in terms of T evaluated at f' where $f' \in D_{0^+}$ has support closer to the origin with respect to the x_1 filtration than $[X+\mathfrak{g}_{y,0^+}]$ had with respect to the y filtration.

Exercise 9.2.1: Do the analogous analysis for the latter case.

To summarize, the point of *descent and recovery* is as follows. We begin with a simple function $f \in C((\mathfrak{g}_{x,s} \smallsetminus \mathfrak{g}_{x,s^+})/\mathfrak{g}_{x,0^+})$ for some $x \in \mathcal{B}$. From this function, we find a point $y \in \mathcal{B}$ and a function $f' \in C(\mathfrak{g}_{y,s^+}/\mathfrak{g}_{y,0^+})$ so that $T(f) = T(f')$. After a finite number of steps, we will have shown that $T(f)$ is completely determined by $\mathrm{res}_{C(\mathfrak{k}_0/\mathfrak{k}_1)+C(\mathfrak{b}_0/\mathfrak{b}_{1/2})} T$.

9.2.2. *Counting*

We now know the following facts:

(1) Thanks to Harish-Chandra [13], the dimension of the complex vector space $\operatorname{res}_{D_{0+}} J(\mathcal{N})$ is equal to the cardinality of $\mathcal{P}(n)$, which, in this case, is 2.
(2) From §4.4, we have $J(\mathcal{N}) \subset J(\mathfrak{g}_{0+})$ and so $\operatorname{res}_{D_{0+}} J(\mathcal{N}) \subset \operatorname{res}_{D_{0+}} J(\mathfrak{g}_{0+})$.
(3) From the previous section, we know that

$$\dim_{\mathbb{C}} \operatorname{res}_{D_{0+}} J(\mathfrak{g}_{0+}) = \dim_{\mathbb{C}} \operatorname{res}_{C(\mathfrak{k}_0/\mathfrak{k}_1)+C(\mathfrak{b}_0/\mathfrak{b}_{1/2})} J(\mathfrak{g}_{0+}).$$

Consequently, we need only show that

$$\dim_{\mathbb{C}} \operatorname{res}_{C(\mathfrak{k}_0/\mathfrak{k}_1)+C(\mathfrak{b}_0/\mathfrak{b}_{1/2})} J(\mathfrak{g}_{0+}) \leq 2.$$

Since $\mathfrak{g}_{0+} \subset \mathfrak{g}_{x,0+} + \mathcal{N}$ for any $x \in \mathcal{B}$, we have that for $T \in J(\mathfrak{g}_{0+})$, $\operatorname{res}_{C(\mathfrak{k}_0/\mathfrak{k}_1)+C(\mathfrak{b}_0/\mathfrak{b}_{1/2})} T$ is completely determined by

$$T([\left(\begin{smallmatrix} 0 & 1 \\ 0 & 0 \end{smallmatrix}\right) + \mathfrak{k}_1]), \quad T([\mathfrak{k}_1]), \quad \text{and} \quad T([\mathfrak{b}_{1/2}]).$$

Since

$$T([\mathfrak{b}_{1/2}]) = (q-1) \cdot T([\left(\begin{smallmatrix} 0 & 1 \\ 0 & 0 \end{smallmatrix}\right) + \mathfrak{k}_1]) + T([\mathfrak{k}_1]),$$

we are done.

Exercise 9.2.2: Using the above proof as a template, prove Equation (7).

9.3. *The general approach*

In general, the proof is very much like that produced above, only more complicated.

9.3.1. *Descent and recovery*

Suppose $T \in J(\mathfrak{g}_{r+})$. The main point is to show that

$$\operatorname{res}_{D_{r+}} T = 0 \text{ if and only if } \operatorname{res}_{D^r_{r+}} T = 0$$

where

$$D^r_{r+} := \sum_{x \in \mathcal{B}} C(\mathfrak{g}_{x,r}/\mathfrak{g}_{x,r+}).$$

That is, we can find a very small space of functions from which we can choose a dual basis for $\operatorname{res}_{D_{r+}} J(\mathfrak{g}_{r+})$.

Suppose $f \in D_{r^+}$. As before, without loss of generality we may assume that $f = [X + \mathfrak{g}_{x,r^+}]$ for some $X \in \mathcal{N}$ and some $x \in \mathcal{B}$. Under some hypotheses, we can use $\mathfrak{sl}_2(\mathfrak{f})$-theory to find a direction in the building in which to move so that, for some $y \in \mathcal{B}$ in that direction, the support of f is nearer the origin with respect to y than it was with respect to x. Moreover, if necessary, we can use the basics of $\mathfrak{sl}_2(\mathfrak{f})$-representation theory and the invariance of T to "beef-up" the coset $\mathfrak{g}_{x,r^+}$ so that we arrive at a function $f' \in C_c(\mathfrak{g}/\mathfrak{g}_{y,r^+})$ with the properties:

(1) $T(f') = T(f)$ and
(2) the support of f' is "closer" to the origin with respect to y than the support of f with respect to x.

Remark 9.3.1: Invoking $\mathfrak{sl}_2(\mathfrak{f})$-theory requires some restrictions. The theory only works well if the highest weights for the representations one wishes to consider are less than $(p-3)$ where p is the characteristic of $\mathfrak{f}$.

9.3.2. *Counting*

This is the real key. In the end, you need some type of correspondence between "suitable" elements of $D^r_{r^+}$ and $\mathcal{O}(0)$, the set of nilpotent orbits in $\mathfrak{g}$.

When r is zero, the situation is easy to understand. As discussed before, the conjugacy classes of facets in $\mathcal{B}$ are in one-to-one correspondence with the elements of $\mathcal{P}(n)$. Similarly, the nilpotent orbits are in one-to-one correspondence with the elements of $\mathcal{P}(n)$. For $\mu \in \mathcal{P}(n)$, let F_μ be the facet described in §3.4.1. The image of X_μ in $\mathfrak{g}_{F_\mu,0}/\mathfrak{g}_{F_\mu,0^+}$ is distinguished nilpotent (that is, it does not lie in a proper Levi subalgebra of $\mathfrak{g}_{F_\mu,0}/\mathfrak{g}_{F_\mu,0^+}$). We let $[X_\mu + \mathfrak{g}_{F_\mu,0^+}] \in D^0_{0^+}$ denote the characteristic function of the coset $X_\mu + \mathfrak{g}_{F_\mu,0^+}$. This is the correspondence between "suitable" elements of $D^0_{0^+}$ and $\mathcal{O}(0)$ alluded to above. Indeed, for $T \in J(\mathfrak{g}_{0^+})$, the restriction of T to $D^0_{0^+}$ is zero if and only if

$$T([X_\mu + \mathfrak{g}_{F_\mu,0^+}])$$

is zero for all $\mu \in \mathcal{P}(n)$.

When $r \neq 0$, life is much more complicated; but something beautiful is true. We refer the reader to [10].

10. A few comments on the $c_\mu(\pi)$s

Let (π, V) denote an irreducible representation of G. We recall that for all regular semisimple X in some neighborhood of zero we have the Harish-Chandra–Howe local character expansion:

$$\Theta_\pi(1+X) = \sum_{\mu \in \mathcal{P}(n)} c_\mu(\pi) \cdot \hat{\mathcal{O}}_\mu(X).$$

In his paper [16], Howe proves that for irreducible supercuspidal representations (that is, those representations that do not occur as subrepresentations of parabolically induced representations) the coefficients occurring in the Harish-Chandra–Howe local character expansion are all integers. In fact, given a bit of additional information, the same proof shows that this is true for all smooth irreducible representations. We first recall what we know about the $c_\mu(\pi)$s, and we then prove this result of Howe.

10.1. *The coefficient* $c_{(1,1,\ldots,1)}(\pi)$

Suppose that (π, V) is a discrete series representation (that is, the matrix coefficients of (π, V) are square-integrable mod the center of G). In this case, by using Rogawski [32] one may extend a result of Harish-Chandra [13] to show that

$$c_{(1,1,\ldots,1)}(\pi) = \frac{\deg(\pi)}{(-1)^\ell \cdot \deg(\mathrm{St})}.$$

Here $\deg(\pi)$ denotes the formal degree of π, St is the Steinberg representation (see, for example, [5]), and ℓ denotes the semisimple rank of G.

On the other hand, if (π, V) is a tempered representation (that is, it occurs in the Plancherel formula) which is not in the discrete series, then, using results of Kazhdan [23], Huntsinger [20] showed that $c_{(1,1,\ldots,1)}(\pi) = 0$ (see also the paper of Schneider and Stuhler [34]).

10.2. *The leading coefficient*

According to Mœglin and Waldspurger [26], the set

$$\{\mu \in \mathcal{P}(n) \,|\, c_\mu(\pi) \neq 0\}$$

has a unique maximal element; call it μ_π. Moreover, according to Mœglin and Waldspurger [26] and Rodier [31], the coefficient $c_{\mu_\pi}(\pi)$ is an integer which is equal to the dimension of the degenerate Whittaker model corresponding to μ_π.

Example 10.2.1: Suppose $\mu = (n) \in \mathcal{P}(n)$. Recall that $X_{(n)}$ is regular nilpotent. Let $\bar{B}$ be the Borel subgroup consisting of lower triangular matrices in G. Let $\bar{N}$ denote the unipotent radical of $\bar{B}$; that is, the subgroup of $\bar{B}$ formed by the set of lower triangular matrices with ones on the diagonal. Define the character δ of $\bar{N}$ by $\bar{n} \mapsto \Lambda(\mathrm{tr}(X_{(n)} \cdot (\bar{n} - 1)))$. For an irreducible smooth representation (π, V) of G, let $V(\delta)$ denote the subspace of V generated by the set

$$\{(\pi(\bar{n}) - \delta(\bar{n}))v \mid v \in V \text{ and } \bar{n} \in \bar{N}\}.$$

Define $V_{(n)} = V/V(\delta)$. Then according to the results quoted above, we have $c_{(n)}(\pi) = \dim_{\mathbb{C}}(V_{(n)}) = \dim_{\mathbb{C}} \mathrm{Hom}_G(V, \mathrm{Ind}_{\bar{N}}^G \delta)$.

10.3. *The remaining coefficients*

Not much is known about the remaining coefficients. Following an argument of Roger Howe [16], we show that they are all integers.

We begin by making Remark 6.1.4 precise. If $\mu' \in \mathcal{P}(n)$, then

$$\hat{\mathcal{O}}_{\mu'}(X) = \Theta_{\mathrm{Ind}_{P(\mu')}^G 1}(1 + X)$$

for all $X \in \mathfrak{g}_{0+}$. Here $\mathrm{Ind}_{P(\mu')}^G 1$ is the representation of G obtained by inducing the trivial representation of $P(\mu')$ up to G.

Suppose $j > 0$, $\mu \in \mathcal{P}(n)$, and τ_μ is the character of K_j represented by the coset $\varpi^{(1-2j)} X_\mu + \mathfrak{k}_{(1-j)} \in \mathfrak{k}_{1-2j}/\mathfrak{k}_{1-j}$. From §7.3 we have

$$\Theta_\pi(\tau_\mu) = \mathrm{meas}_{dg}(K_j) \cdot m(\tau_\mu^{-1}, \pi) \tag{8}$$

and

$$\Theta_{\mathrm{Ind}_{P(\mu')}^G 1}(\tau_\mu) = \mathrm{meas}_{dg}(K_j) \cdot m(\tau_\mu^{-1}, \mathrm{Ind}_{P(\mu')}^G 1). \tag{9}$$

From the discussion above, Exercise 5.3.1, and Equation (2) we have

$$\hat{\mathcal{O}}_{\mu'}(\tau_\mu) = \mathcal{O}_{\mu'}(\hat{\tau}_\mu) = \mathrm{meas}_{dX}(\mathfrak{k}_j) \cdot \mathcal{O}_{\mu'}([-\varpi^{(1-2j)} \cdot X_\mu + \mathfrak{k}_{(1-j)}])$$
$$= \begin{cases} \mathrm{meas}_{dX}(\mathfrak{k}_j) \cdot m(\tau_\mu^{-1}, \mathrm{Ind}_{P(\mu')}^G 1) & \text{if } \mu < \mu', \\ \mathrm{meas}_{dX}(\mathfrak{k}_j) & \text{if } \mu = \mu', \text{ and} \\ 0 & \text{otherwise.} \end{cases}$$

Thus, if j is sufficiently large, then

$$m(\tau_\mu^{-1}, \pi) = \sum_{\mu \le \mu'} c_{\mu'}(\pi) \cdot m(\tau_\mu^{-1}, \mathrm{Ind}_{P(\mu')}^G 1).$$

We now proceed by induction. From Rodier [31] we have that $c_{(n)}$ is always an integer.

Suppose $\mu \in \mathcal{P}(n)$. By induction, if $\mu' > \mu$, then $c_{\mu'}(\pi) \in \mathbb{Z}$. We have

$$c_\mu(\pi) \cdot m(\tau_\mu^{-1}, \mathrm{Ind}_{P(\mu)}^G 1) = m(\tau_\mu^{-1}, \pi) - \sum_{\mu' > \mu} c_{\mu'}(\pi) \cdot m(\tau_\mu^{-1}, \mathrm{Ind}_{P(\mu')}^G 1) \in \mathbb{Z}.$$

Since $m(\tau_\mu^{-1}, \mathrm{Ind}_{P(\mu)}^G 1) = \mathcal{O}_\mu([-\varpi^{(1-2j)} \cdot X_\mu + \mathfrak{k}_{(1-j)}]) = 1$, we are done.

Acknowledgements

The author was partially supported by National Science Foundation Grant No. 0200542. This paper is based on lectures delivered at the Institute for Mathematical Sciences at the National University of Singapore in 2002. The author thanks the Institute for its support.

References

1. J. Adler and S. DeBacker, *Some applications of Bruhat-Tits theory to harmonic analysis on the Lie algebra of a reductive p-adic group*, Mich. Math. J., **50** (2002), no. 2, pp. 263–286.
2. D. Barbasch and A. Moy, *A new proof of the Howe conjecture*, J. Amer. Math. Soc. **13** (2000), no. 3, pp. 639–650.
3. A. Borel and J. Tits, *Groupes réductifs*, Inst. Hautes Études Sci. Publ. Math., No. 27, 1965, pp. 55–150.
4. R. Carter, *Finite groups of Lie type. Conjugacy classes and complex characters*, Reprint of the 1985 original, Wiley Classics Library, John Wiley & Sons, Ltd., Chichester, 1993.
5. W. Casselman, *Introduction to the theory of admissible representations of p-adic groups*, to appear.
6. L. Clozel, *Orbital integrals on p-adic groups: a proof of the Howe conjecture*, Ann. of Math. (2) **129** (1989), no. 2, pp. 237–251.
7. ———, *Sur une conjecture de Howe. I*, Compositio Math. **56** (1985), no. 1, pp. 87–110.
8. S. DeBacker and P. J. Sally, Jr., *Germs, characters, and the Fourier transforms of nilpotent orbits*, The mathematical legacy of Harish-Chandra (Baltimore, MD, 1998), Proc. Sympos. Pure Math., **68**, Amer. Math. Soc., Providence, RI, 2000, pp. 191–221.
9. S. DeBacker, *Homogeneity results for invariant distributions of a reductive p-adic group*, Ann. Sci. École Norm. Sup., **35** (2002), no. 3, pp. 391–422.
10. ———, *Parametrizing nilpotent orbits via Bruhat-Tits theory*, Ann. of Math., **156** (2002), no. 1, pp. 295–332.
11. ———, *Some applications of Bruhat-Tits theory to harmonic analysis on a reductive p-adic group*, Mich. Math. J., **50** (2002), no. 2, pp. 241–261.

12. R. Kottwitz, *Harmonic analysis on semisimple p-adic Lie algebras*, Proceedings of the International Congress of Mathematicians, Vol. II (Berlin, 1998). Doc. Math. 1998, Extra Vol. II, pp. 553–562.
13. Harish-Chandra, *Admissible invariant distributions on reductive p-adic groups*, Preface and notes by Stephen DeBacker and Paul J. Sally, Jr., University Lecture Series, **16**, American Mathematical Society, Providence, RI, 1999.
14. ———, *The characters of reductive p-adic groups*, Contributions to algebra (collection of papers dedicated to Ellis Kolchin), Academic Press, New York, 1977, pp. 175–182.
15. ———, *A submersion principle and its applications*, Proc. Indian Acad. Sci. Math. Sci., **90** (1981), no. 2, pp. 95–102.
16. R. Howe, *The Fourier transform and germs of characters (case of Gl_n over a p-adic field)*, Math. Ann. **208** (1974), pp. 305–322.
17. ———, *Kirillov theory for compact p-adic groups*, Pacific J. Math. **73** (1977), no. 2, pp. 365–381.
18. ———, *Some qualitative results on the representation theory of* Gl_n *over a p-adic field*, Pacific J. Math. **73** (1977), no. 2, pp. 479–538.
19. ———, *Two conjectures about reductive p-adic groups*, Harmonic analysis on homogeneous spaces, Proceedings of Symposia in Pure Mathematics, vol. 26, American Mathematical Society, 1973, pp. 377–380.
20. R. Huntsinger, *Vanishing of the leading term in Harish-Chandra's local character expansion*, Proc. Amer. Math. Soc. **124** (1996), no. 7, pp. 2229–2234.
21. ———, *Some aspects of invariant harmonic analysis on the Lie algebra of a reductive p-adic group*, Ph.D. Thesis, The University of Chicago, 1997.
22. N. Jacobson, *Basic Algebra, I*, second edition, W.H. Freeman and Company, New York, 1985.
23. D. Kazhdan, *Cuspidal geometry of p-adic groups*, J. Analyse Math. **47** (1986), pp. 1–36.
24. A. A. Kirillov, *Unitary representations of nilpotent Lie groups*, Russ. Math. Surveys, **17** (1962), pp. 53-104.
25. B. Lemaire, *Intégrabilité locale des caractères-distributions de* $\mathrm{GL}_N(F)$ *où* F *est un corps local non-archimédien de caractéristique quelconque*, Compositio Math. **100** (1996), no. 1, pp. 41–75.
26. C. Mœglin and J.-L. Waldspurger, *Modèles de Whittaker dégénérés pour des groupes p-adiques*, Math. Z. **196** (1987), no. 3, pp. 427–452.
27. A. Moy and G. Prasad, *Jacquet functors and unrefined minimal K-types*, Comment. Math. Helvetici **71** (1996), pp. 98–121.
28. ———, *Unrefined minimal K-types for p-adic groups*, Inv. Math. **116** (1994), pp. 393–408.
29. R. Ranga-Rao, *Orbital integrals in reductive groups*, Ann. of Math. (2) **96** (1972), pp. 505–510.
30. F. Rodier, *Intégrabilité locale des caractères du groupe* $\mathrm{GL}(n,k)$ *où* k *est un corps local de caractéristique positive*, Duke Math. J. **52** (1985), no. 3, pp. 771–792.
31. ———, *Modèle de Whittaker et caractères de représentations.* Non-

commutative harmonic analysis (Actes Colloq., Marseille-Luminy, 1974), Lecture Notes in Math., Vol. 466, Springer, Berlin, 1975. pp. 151–171.

32. J. D. Rogawski, *An application of the building to orbital integrals*, Compositio Math. **42** (1980/81), no. 3, pp. 417–423.

33. G. Savin, *Lectures on representations of p-adic groups*, this volume.

34. P. Schneider and U. Stuhler, *Representation theory and sheaves on the Bruhat-Tits building*, Inst. Hautes tudes Sci. Publ. Math. No. 85 (1997), pp. 97–191.

35. J.-L. Waldspurger, *Homogénéité de certaines distributions sur les groupes p-adiques*, Inst. Hautes Études Sci. Publ. Math. No. 81 (1995), pp. 25–72.

36. ———, *Quelques resultats de finitude concernant les distributions invariantes sur les algèbres de Lie p-adiques*, preprint, 1993.

37. ———, *Une formule des traces locale pour les algèbres de Lie p-adiques*, J. Reine Angew. Math. **465** (1995), pp. 41–99.

On Classification of Some Classes of Irreducible Representations of Classical Groups

Marko Tadić

Department of Mathematics
University of Zagreb
Bijenička 30, 10000 Zagreb, Croatia
E-mail: tardic@math.hr

Representation theory of reductive p-adic groups, besides its importance for harmonic analyses, is very important for Langlands program. It also gives us often better understanding of representation theory of reductive Lie groups. In these notes we review some parts of representation theory of reductive groups over local fields, in particular over p-adic fields. We discuss classifications of some families of irreducible representations, which are important for harmonic analysis on these groups. We start with general principles of harmonic analyses on groups (which are given in terms of unitary representations). Then we explain algebraization of the problem of classification of irreducible unitary representations. Langlands classification reduces classification of irreducible representations to tempered representations, which come from square integrable representations by parabolic induction. We present existing classifications of such classes of representations for general linear and classical groups, and discuss connection of this with Langlands correspondences. Special attention is devoted to classification (modulo cuspidal data) of irreducible square integrable representations of classical p-adic groups, which implies parameterization of non-unitary duals. This opens possibility to work on the (very hard) problem of classification of irreducible unitary representation of these group.

Contents

1 Harmonic analysis and unitary duals 97

2 Non-discrete locally compact fields, classical groups, reductive groups 99

3 K_0-finite vectors 103

4 Smooth representations 104

5 Parabolically induced representations 105
6 Jacquet modules 109
7 Filtrations of Jacquet modules 111
8 Square integrable and tempered representations 112
9 Langlands classification 114
10 Geometric lemma and algebraic structures 120
11 Square integrable representations of p-adic general linear groups 123
12 Two simple examples of square integrable representations of classical p-adic groups 125
13 Invariants of square integrable representations of classical p-adic groups 127
14 Reduction to cuspidal lines 135
15 Parameters of $\mathcal{D}(\rho;\sigma)$ 137
16 Integral case 138
17 Non-integral case 141
18 Local Langlands correspondences 143
19 Non-unitary duals of classical p-adic groups 145
20 Unitary duals of general linear groups over local fields 147
21 On the unitarizability problem for classical p-adic groups 154
References 158

Introduction

In these notes of the lectures given during the special period on representation theory of Lie groups in IMS, NUS, Singapore, we shall discuss the problem of classification of some important series of irreducible representations of general linear and classical groups, having in mind unitary representations. We shall discuss more p-adic groups, but a part of notes deals also with real groups.

One of the main goals of the notes is to give an introduction to the classification modulo cuspidal data, of irreducible square integrable representations of classical p-adic groups. After that we shall describe unitary duals of general linear groups over local fields, and describe the proof of the classification theorem in the case of complex general linear groups. We shall finish the notes with a series of questions regarding unitary representations of classical p-adic groups.

In these notes, the fields of real and complex numbers are denoted by $\mathbb{R}$ and $\mathbb{C}$, and the ring of rational integers is denoted by $\mathbb{Z}$ (as usual). Further

$$\mathbb{Z}_+ = \{k \in \mathbb{Z}; k \geq 0\}$$
$$\mathbb{N} = \{k \in \mathbb{Z}; k \geq 1\}.$$

We are thankful to the organizers of the special period for providing a very stimulating atmosphere in which we had opportunity to present the lectures.

1. Harmonic analysis and unitary duals

1.1. One can interpret classical harmonic analysis in terms of unitary representations of $\mathbb{R}^n$ and $(\mathbb{R}/\mathbb{Z})^n$. This point of view opens a possibility of generalizing classical harmonic analysis, and building such a type of theory for a general locally compact group G (in general, neither compact, nor commutative). We shall briefly describe the main problems of harmonic analysis on such a group G. First, we shall introduce a few notions which we shall need for this description.

1.2. A **representation** (π, V) (or simply π or V) of a group G is a group homomorphism π from the group G to the group of all invertible linear operators on a complex vector space V (there is no requirement on continuity in this definition). A representation π on a non-zero vector space V is called **irreducible** (or **algebraically irreducible**) if $\{0\}$ and V are the only vector subspaces of V which are invariant for all $\pi(g), g \in G$.

A representation (π, H) is called unitary if H is a Hilbert space and:

(1) the mapping

$$(g, v) \mapsto \pi(g)v, \quad G \times H \to H$$

is continuous;

(2) each operator $\pi(g), g \in G$ is unitary.

If we omit the second requirement in the above definition, then the representation defined in this way will be called **continuous** (one can consider much more general continuous representations, but we shall not need them in these notes).

A unitary (or only continuous) representation (π, G) is called **irreducible** (or **topologically irreducible**) if $\{0\}$ and H are the only closed subspaces of H which are invariant for all $\pi(g), g \in G$.

1.3. Now we can describe the main goals of harmonic analysis on a locally compact group G (which satisfies some technical requirements, which we shall not discuss here, but which are satisfied for the groups that we shall consider in these notes, i.e. for general linear and classical groups over local fields).

The first problem is to

(1) understand in a convenient way (possibly classify) the set of all the equivalence classes of irreducible unitary representations of G. This set is called the **unitary dual** of G, and it is denoted by

$$\hat{G}.$$

The second problem is to

(2) interpret other important unitary representations of G in terms of $\hat{G}$.

Such important unitary representations are usually given on functional spaces. The most important examples of such representations include representations of G on spaces of square integrable functions (with respect to an invariant measure, assuming that it exists) on a space X where G acts transitively. Then $X \cong H\backslash G$ for some closed group H of G and G acts by right translations on the space $L^2(H\backslash G)$ of the square integrable functions on $H\backslash G$ (in this case $H\backslash G$ carries an invariant measure for right translations of G).

The first example of such representation would be when H is the trivial subgroup of G, i.e. the representation of G on the space $L^2(G)$ of the square integrable functions on G with respect to right invariant measure on G. This representation is very important. A significant portion of Harish-Chandra's work is closely related to this representation in the case of semi simple real Lie groups (among others, he described the representation from $\hat{G}$ necessary for decomposing $L^2(G)$, and found Plancherel measure by which one decomposes $L^2(G)$ in terms of these irreducible representations).

In this lectures we shall be more related to the problem (1) of harmonic analysis, although we shall be also related to the problem (2). Irreducible square integrable representations, which are one of the main topics of our notes, are part of both problems, (1) and (2). They are subrepresentations of $L^2(G)$ if the center of G is compact.

Remark: Some of the most important parts of the Langlands program can be considered as a kind of problems from harmonic analysis on groups in the above sense. For example. the origin of the Langlands program one can view as a kind of problem of harmonic analysis. The program started as a strategy for proving the Artin's conjecture that Artin's L-functions are entire. Roughly, Langlands proposed a strategy that irreducible representations of the absolute Galois group of a number field would parameterize irreducible subrepresentations of adelic general linear groups on the spaces of cuspidal automorphic forms (which are unitary representations on functional spaces), in a way that corresponding L-functions match (this can clearly be regarded as a kind of problem of type (2) of harmonic analysis on groups). Realization of this strategy would imply the Artin's conjecture.

The above philosophy has its local counterpart (with corresponding parameterizations). In the local case of the Langlands program, we are more related to the problem of type (1) of harmonic analysis on groups.

One can extend the above considerations to other reductive groups, and one can consider also different fields. A question may be, can one do the above mentioned parameterizations in a naturally compatible way. Such a question is related to the functoriality problem.

Above we gave only a very rough comments regarding the Langlands program. Details regarding this program can be found in [19] or [36].

2. Non-discrete locally compact fields, classical groups, reductive groups

2.1. Let F be a non-discrete locally compact field. Such field will be called **local field**. If F is connected, then the field is called archimedean. Otherwise, it is called non-archimedean. Non-archimedean fields are totally disconnected. They contain a basis of neighborhoods of 0 consisting of open compact subrings.

If a local field is archimedean, then it is isomorphic to $\mathbb{R}$ or $\mathbb{C}$.

Let p be a prime integer. Ideals $p^k\mathbb{Z}$, $k \in \mathbb{Z}_+$, define a basis of neighborhoods of 0 in $\mathbb{Z}$. The completion of Z with respect to this topology (more precisely, uniform structure defined by this topology) is denoted by $\mathbb{Z}_p$. The field of fractions of $\mathbb{Z}_p$ is denoted by $\mathbb{Q}_p$. This is the field of p-adic numbers.

We can introduce $\mathbb{Q}_p$ also as a completion of $\mathbb{Q}$ with respect to the absolute value

$$\left|\frac{a}{b}p^k\right|_p = p^{-k}, \quad a, b, k \in \mathbb{Z}, b \neq 0, p \not| ab.$$

Any finite extension F of $\mathbb{Q}_p$ is in a natural way a topological space, and with this topology, F is a local non-archimedean field of characteristic 0. One gets each non-archimedean field of characteristic 0 in this way.

Let $\mathbb{F}_q[[X]]$ be the ring of all formal power series $\sum_{k=0}^{\infty} a_n X^n$ over a finite field $\mathbb{F}_q$ (with q elements), and let $\mathbb{F}_q((X))$ be the field of all Laurent power series $\sum_{k=-\infty}^{\infty} a_n X^n$ over $\mathbb{F}_q$ for which there exists $n_0 \in \mathbb{Z}$ such that $a_n = 0$ for all $n \leq n_0$. Then the powers of the ideal $X\mathbb{F}_q[[X]]$ in $\mathbb{F}_q[[X]]$ define a basis of neighborhoods of 0 in $\mathbb{F}_q((X))$, and therefore a topology on $\mathbb{F}_q((X))$. In this way $\mathbb{F}_q((X))$ becomes a local non-archimedean field of positive characteristic. One gets each local non-archimedean field of positive characteristic in this way.

Topology on a local field can be always defined using an absolute value. Moreover, there exists a unique absolute value $|\ |_F$ on a local non-discrete field F such that

$$\int_F f(x)dx = |a|_F^{-1} \int_F f(ax)dx,$$

for any $a \in F^\times$ and for any continuous, compactly supported function f on G, where dx denotes an invariant (for translations) measure on F. We shall always fix such an absolute value on F. Let us note that for $\mathbb{C}$, this absolute value is a square of the standard one.

2.2. We shall recall now of a definition of the classical groups. A **classical group** over a local field F is the group of isomorphisms of either symplectic, or orthogonal or unitary space over F (of finite dimension). For the study of representations of classical groups, it is important to understand the representation theory of general linear groups $GL(n, F)$'s, i.e. of the groups of all isomorphisms of finite dimensional vector spaces over F (soon it will become clear why this is important).

In the study of classical groups, we shall use very convenient language of structure theory of reductive groups without going into this theory. In general, we shall try to keep the technicalities as low as possible.

For the simplicity, we shall consider in these lectures two series of classical groups (these series consist of split and connected groups). The reason for this is only to simplify the notation.

2.3. The first is the series of symplectic groups:

Denote

$$J_n = \begin{bmatrix} 00 \ldots 01 \\ 00 \ldots 10 \\ \vdots \quad\quad \vdots \\ 01 \ldots 00 \\ 10 \ldots 00 \end{bmatrix} \in GL(n, F).$$

Then the symplectic group is

$$Sp(2n, F) = \left\{ S \in GL(2n, F); \ {}^tS \begin{bmatrix} 0 & J_n \\ -J_n & 0 \end{bmatrix} S = \begin{bmatrix} 0 & J_n \\ -J_n & 0 \end{bmatrix} \right\}.$$

Here ${}^tS$ denotes the transposed matrix of S.

2.4. The second series consists of split odd-orthogonal groups:

Denote by I_n the identity matrix in $GL(n, F)$. Let

$$SO(2n+1, F) = \{S \in SL(2n+1, F); \ {}^{\tau}S\, S = I_{2n+1}\}.$$

Here $SL(n, F) = \{g \in GL(n, F); \det(g) = 1\}$ and ${}^{\tau}S$ denotes the transposed matrix of S with respect to the other diagonal.

We could work also with $O(2n+1, F)$ instead of $SO(2n+1, F)$.

In these notes, we shall always deal with matrix forms of classical groups.

2.5. The above groups are connected, split, semi simple algebraic groups over F. They are topological groups in a natural way. In the case when F is a non-archimedean field, these groups are totally disconnected. Then one can write a basis of neighborhoods of identity which consists of open (and closed) compact subgroups. If F is archimedean, then symplectic and odd-orthogonal groups are connected semi simple Lie groups.

If G is $GL(n, F)$, or $Sp(2n, F)$ or $SO(2n+1, F)$, then we shall denote by $P_\emptyset$ the subgroup of all the upper triangular matrices in G. Then $P_\emptyset$ is called standard minimal parabolic subgroup of G. Any subgroup of G containing $P_\emptyset$ is called standard parabolic subgroup of G. There are finitely many of them, and we shall describe them precisely. Any subgroup conjugate to a standard parabolic subgroup is called parabolic subgroup.

Let

$$\alpha = (n_1, ..., n_k)$$

be an ordered partitions of n into positive integers. Consider matrices of $GL(n, F)$ as block matrices with blocks of sizes $n_i \times n_j$. Let P_α^{GL} (resp. M_α^{GL}), be the upper block-triangular matrices (resp. block-diagonal matrices) in $GL(n, F)$. Denote by N_α^{GL} the (block) matrices in P_α^{GL} which have identity matrices on the block-diagonal. Now

$$\alpha \mapsto P_\alpha^{GL}$$

is one-to-one mapping of the set of all ordered partitions of n onto the set of all standard parabolic subgroups of $GL(n, F)$. We have Levi decomposition

$P_\alpha^{GL} = M_\alpha^{GL} N_\alpha^{GL}$. This means that P_α^{GL} is a semi direct product of M_α^{GL} and N_α^{GL}, where N_α^{GL} is a normal subgroup in P_α^{GL}, i.e.

$$P_\alpha^{GL} = M_\alpha^{GL} \ltimes N_\alpha^{GL}.$$

This decomposition of P_α^{GL} is called the **standard Levi decomposition** of P_α^{GL}, where M_α^{GL} is called the **standard Levi factor** of P_α^{GL} and N_α^{GL} is called the **unipotent radical** of P_α^{GL}.

2.6. Standard parabolic subgroups of $Sp(2n, F)$ **and** $SO(2n+1, F)$ are parameterized by ordered partitions $\alpha = (n_1, \dots, n_k)$ of integers m, where $0 \leq m \leq n$. If we consider the group $G = Sp(2n, F)$ set

$$\alpha' = (n_1, ..., n_k, 2n - 2m, n_k, ..., n_1),$$

while in the case of the group $G = SO(2n+1, F)$ set

$$\alpha' = (n_1, ..., n_k, 2n + 1 - 2m, n_k, ..., n_1).$$

Then

$$\alpha \mapsto P_\alpha = P_{\alpha'}^{GL} \cap G$$

gives a parameterization of standard parabolic subgroups in G. In similar way as in the case of general linear groups, one defines standard Levi decompositions in this case, using standard Levi decompositions of $P_{\alpha'}^{GL}$.

In the sequel, we shall denote by G one of the groups $GL(n, F), Sp(2n, F)$ or $SO(2n+1, F)$.

We shall denote by $A_\emptyset$ the subgroup of all diagonal matrices in G. This is a maximal split (over F) torus in G (it is also a maximal torus in G).

2.7. We shall denote by K_0 a maximal compact subgroup of G.

If F is non-archimedean, let

$$\mathcal{O}_F = \{x \in F; |x|_F \leq 1\},$$

$$\mathfrak{p}_F = \{x \in F; |x|_F < 1\}.$$

In the non-archimedean case one can take

$$K_0 = GL(n, \mathcal{O}_F) \cap G.$$

For $F = \mathbb{R}$ (resp. $F = \mathbb{C}$), one can define K_0 in a similar way as above, taking the group $O(n)$ (resp. $U(n)$) of orthogonal matrices in $GL(n, \mathbb{R})$ (resp. unitary matrices in $GL(n, \mathbb{C})$) instead of $GL(n, \mathcal{O}_F)$.

It is important to note that K_0 is an open subgroup if F is non-archimedean field, which is not the case (in general) in the archimedean case.

3. K_0-finite vectors

3.1. Let $(\pi, H) \in \hat{G}$. For $\tau \in \hat{K}_0$ denote by $m(\tau : \pi)$ the multiplicity of τ in π.

The basic property of K_0 is that it is a large subgroup of G (this was proved by Harish-Chandra in the archimedean case, and by J. Bernstein in the non-archimedean case). It means the that the function

$$\pi \mapsto m(\tau : \pi)$$

is a bounded function on $\hat{G}$, for any fixed $\tau \in \hat{K}_0$. This fact has a number of important consequences. Among others, it enables algebraization of the problem of determining of the unitary dual $\hat{G}$ of G.

3.2. Let $(\pi, H) \in \hat{G}$. Denote by H^∞ the set of all vectors $v \in H$ such that

$$\dim_{\mathbb{C}} \operatorname{span}_{\mathbb{C}} \pi(K_0)v < \infty.$$

Then H^∞ is a dense K_0-invariant vector subspace of H^∞.

Suppose that F is non-archimedean. Since for each $g \in G$ the group

$$gK_0g^{-1} \cap K_0$$

has finite index in K_0, H^∞ is G-invariant. It follows easily that the following property holds for H^∞:

> For any $v \in H^\infty$ there exists an open subgroup K of G such that $\pi(k)v = v$ for any $k \in K$.

This follows from the fact that each continuous representation of K_0 is trivial on an open subgroup (since open subgroups in K_0 form a basis of neighborhoods of identity, and $GL(n, \mathbb{C})$ does not contain small (non-trivial) subgroups).

3.3. In the archimedean case, $gK_0g^{-1} \cap K_0$ is not (in general) of finite index in K_0. Because of this, H^∞ is not (in general) G-invariant. But then one can prove that it is invariant for the natural action of Lie algebra $\mathfrak{g}$ of G. Moreover, the action of $\mathfrak{g}$ and K_0 satisfies a natural condition. Such a structure is called $(\mathfrak{g}, K_0)$-module.

3.4. At this point usually archimedean and non-archimedean theory continue to develop separately. In the sequel, we shall more discuss the non-archimedean theory, but a number of topics hold for both theories (these will be examples of Lefschetz principle). We shall usually comment the results which hold in both theories.

4. Smooth representations

We shall assume in the sequel that F is a local non-archimedean field (if it is not otherwise specified).

4.1. A representation (π, V) of G is called **smooth** if it satisfies the following condition:

> For any $v \in V$ there exists an open subgroup K of G such that $\pi(k)v = v$ for any $k \in K$.

Denote by

$$\tilde{G}$$

the set of all equivalence classes of irreducible smooth representations of G. This set is called **non-unitary dual of** G, or **admissible dual of** G.

4.2. The mapping

$$(\pi, H) \mapsto (\pi^\infty, H^\infty); \qquad \hat{G} \to \tilde{G}$$

is injective (here π^∞ denotes the restriction of π to H^∞). Therefore, the unitary dual can be identified with a subset of $\tilde{G}$. We shall assume this identification in further. It can be shown that in this way the unitary dual is identified with the subset of all $(\pi, V) \in \tilde{G}$ such that on V there exists an inner product which is invariant for the action of G.

The problem of classification of $\tilde{G}$ has appeared much more manageable then the problem of classification of $\hat{G}$.

4.3. The problem of classification of unitary dual of G now breaks into two parts:

> problem of classification of $\tilde{G}$, which is called the **problem of non-unitary dual**;
>
> problem of determining the subset $\hat{G}$ of $\tilde{G}$ (in other words, the problem of identifying unitarizable classes in $\tilde{G}$), which is called the **unitarizability problem.**

We shall discuss both problems in these lectures.

4.4. Regarding the problem of non-unitary duals, let us note that there is Langlands classification of non-unitary duals, which reduces the problem of classification of non-unitary duals to the problem of classification of a special kind of irreducible representations of Levi subgroups, namely to the problem of classification of tempered representations, which will be introduced later. In the moment, let us just note that these tempered representations are unitarizable. The problem of classifying of irreducible tempered representations is very far from being easy.

Before we describe Langlands classification, we shall recall of a more simple (and less precise) reduction of the non-unitary duals.

We shall need to have a tool by which we shall be able to produce new representations. This tool is provided by parabolic induction, a construction which generalizes in a natural way induction studied already by Schur and Frobenius in the case of finite groups. Further, we shall need a tool for analyzing induced representations. Jacquet modules will be of great help for this.

5. Parabolically induced representations

5.1. Smooth representations of G and intertwinings form an Abelian category, which will be denoted by $Alg(G)$.

Let (π, V) be a representation of G. Denote

$$V^\infty = \{v \in V; \text{ there exists an open subgroup } K \text{ such that } \pi(k)v = v,\ k \in K\}.$$

The space V^∞ is a G-subrepresentation of G, and it is called the **smooth part** of V. For a compact subgroup K of G let

$$V^K = \{v \in V; \pi(k)v = v \text{ for any } k \in K\}.$$

This vector space is called the space of K-**invariants** of V. Further,

$$(\pi, V) \mapsto V^K$$

is an exact functor on the category $Alg(G)$.

5.2. If (π, V) is a smooth representation of G, then there is a natural representation π' on the space of all linear forms V' on V defined by $(\pi'(g)v')(v) = v'(\pi(g^{-1})v)$. The smooth part of this representation is called the **contragredient** of (π, V). This representation is denoted by

$$(\tilde{\pi}, \tilde{V})$$

(recall $(\tilde{\pi}(g)\tilde{v})(v) = \tilde{v}(\pi(g^{-1})v))$. Then the mapping

$$(v, \tilde{v}) \mapsto < v, \tilde{v} >=: \tilde{v}(v), \qquad V \times \tilde{V} \to \mathbb{C}$$

is called canonical bilinear form. This form is G-invariant.

A function

$$g \mapsto < \pi(g)v, \tilde{v} >$$

is called a **matrix coefficient** of π.

Further,

$$(\pi, V) \mapsto (\tilde{\pi}, \tilde{V})$$

extends to a contravariant functor in a natural way. This functor is exact.

For a representation (π, V) of G, the representation on the complex conjugate vector space $\bar{V}$ of V will be denoted by

$$(\bar{\pi}, \bar{V}).$$

A smooth representation (π, V) will be called **Hermitian** if

$$(\pi, V) \cong (\bar{\tilde{\pi}}, \bar{\tilde{V}}).$$

5.3. A smooth representation (π, V) of G is called **admissible** if

$$\dim_{\mathbb{C}} V^K < \infty$$

for any open compact subgroup K of G.

For a smooth representation (π, V) of G we have always a natural intertwining of V into $\tilde{\tilde{V}}$. If the representation is admissible, then this is an isomorphism. The converse also holds, i.e. if V and $\tilde{\tilde{V}}$ are isomorphic, then (π, V) is admissible.

It is easy to show that each unitarizable admissible representation of G is Hermitian.

5.4. We shall fix the group G of rational points of a connected reductive group defined over a local non-archimedean field F. One of the main examples for us are general linear groups and classical groups. We shall fix a maximal split torus $A_\emptyset$ in G and a minimal parabolic subgroup $P_\emptyset$ of G which contains $A_\emptyset$. Standard parabolic subgroups of G are subgroups of G which contain $P_\emptyset$. For a standard parabolic subgroup P of G, a Levi decomposition of P into semi direct product of a reductive subgroup M and a normal unipotent subgroup N will be called standard if $A_\emptyset \subseteq M$. For standard parabolic subgroups we shall always assume that Levi decompositions are standard.

Parabolic subgroups and their Levi decompositions one gets from standard parabolic subgroups and their standard decompositions by conjugation with elements of G.

We shall fix a maximal compact subgroup K_0 of G for which Iwasawa decomposition

$$G = P_\emptyset K_0$$

holds (such a maximal compact subgroup always exists).

5.5. Let for a moment $\mathcal{G}$ be a locally compact group. Then there always exists a positive measure which is invariant for right translations. Such a measure will be denoted by dg. Right invariance means that

$$\int_{\mathcal{G}} f(gx)\, dg \;=\; \int_{\mathcal{G}} f(g)\, dg$$

for any continuous compactly supported function f on $\mathcal{G}$ and any $x \in \mathcal{G}$. This measure is unique up to a multiplication by a constant, and it is called a **right Haar measure** on $\mathcal{G}$.

A right Haar measure does not need to be left invariant (if it is, then the group is called **unimodular**; reductive groups are unimodular), but there exists a character $\Delta_{\mathcal{G}}$ of $\mathcal{G}$ (which is called the **modular function** or **modular character** of $\mathcal{G}$), such that holds

$$\int_{\mathcal{G}} f(xg)\, dg \;=\; \Delta_{\mathcal{G}}(x)^{-1} \int_{\mathcal{G}} f(g)\, dg$$

holds for any f and x as above.

5.6. Let us return back to the case of a connected reductive group G over a non-archimedean field F. Fix a parabolic subgroup P of G with a Levi decomposition $P = MN$ (more precisely, the group of rational points). Let (σ, U) be a smooth representation of M. Denote by $\underline{\mathrm{Ind}}_P^G(\sigma)$ the space of all functions $f\ G \to U$ which satisfy

$$f(nmg) = \Delta_P(m)^{1/2}\sigma(m)f(g)$$

for each $m \in M$, $n \in N$, $g \in G$. Then G acts on $\underline{\mathrm{Ind}}_P^G(\sigma)$ by right translations $(R_g f)(x) = f(xg)$, $x, g \in G$. The smooth part of the representation $\underline{\mathrm{Ind}}_P^G(\sigma)$ is denoted by

$$\mathrm{Ind}_P^G(\sigma)$$

and called a **parabolically induced representation** of G from P by σ.

Parabolic induction becomes in an obvious way a functor from $Alg(M)$ into $Alg(G)$. The functor of parabolic induction is exact.

5.7. If σ is unitarizable, then $\text{Ind}_P^G(\sigma)$ is also unitarizable. The inner product $(\ ,\ )$ on $\text{Ind}_P^G(\sigma)$ is given by

$$(f_1, f_2) = \int_{K_0} (f_1(k), f_2(k))dk.$$

Further,

$$\text{Ind}_P^G(\sigma)\tilde{} \cong \text{Ind}_P^G(\tilde{\sigma})$$

The canonical bilinear form is given by the same formula as the above inner product:

$$< f_1, f_2 >= \int_{K_0} < f_1(k), f_2(k) > dk.$$

5.8. Suppose that $P = MN$ is a standard parabolic subgroup of G and $P' = M'N'$ another standard parabolic subgroup of G (the above decompositions are considered to be standard Levi decompositions). Let

$$P \subseteq P'.$$

Then

$$\text{Ind}_P^G(\sigma) \cong \text{Ind}_{P'}^G(\text{Ind}_{P\cap M'}^{M'}(\sigma)).$$

This fact is called **induction by stages** (which gives the same result as the original, direct parabolic induction). It is easy to prove it (one writes an explicit isomorphism).

5.9. Iwasawa decomposition implies that $\text{Ind}_P^G(\sigma)$ is an admissible representation if σ is admissible. It is less obvious to prove that if σ is a representation of finite length, then $\text{Ind}_P^G(\sigma)$ is also a representation of finite length (of G).

5.10. Suppose that we have a parabolic subgroup P with Levi decompositions $P = MN$ and $P = M'N'$, which do not need to be the standard one (in the case that really interests us, at least one Levi decomposition should not be the standard one). Suppose

$$M = M'.$$

Let σ be a smooth finite length representation of M. Then

$$\text{Ind}_P^G(\sigma) \text{ and } \text{Ind}_{P'}^G(\sigma) \text{ have the same Jordan-Hölder series.}$$

This is an important fact, called **induction from associate parabolic subgroups**. It is not quite simple to prove it. It relies on the theory of characters. Since we shall not introduce characters in these notes, we shall not comment the proof here.

6. Jacquet modules

In this section we shall introduce a functor which is left adjoint to the functor of parabolic induction.

6.1. Suppose that (π, V) is a smooth representation of G and let $P = MN$ be a parabolic subgroup of G (actually, it is enough to assume that (π, V) is a smooth representation of P only). Let

$$V(N) = \text{span}_{\mathbb{C}}\{\pi(n)v - v; n \in N, v \in V\}.$$

Since N is normal in P, $V(N)$ is P-invariant. In particular, it is M-invariant. We have a natural quotient action of M on

$$r_M^G(V) = V/V(N).$$

We shall consider the action of M on $r_M^G(V)$ which is the quotient action of the action of M (through π) on V, twisted with $\Delta_P^{-1/2}$. This action will be denoted by

$$r_M^G(\pi).$$

The representation

$$(r_M^G(\pi), r_M^G(V))$$

is called the **Jacquet module** of (π, V) with respect to $P = MN$. One defines in a natural way **Jacquet functor** from $Alg(G)$ into $Alg(M)$. Jacquet functor is exact.

6.2. If $P = MN$ and $P' = M'N'$ are standard parabolic subgroup, with standard Levi decompositions, such that

$$P \subseteq P',$$

then

$$r_{M'}^G(r_M^{M'}(\pi)) \cong r_M^G(\pi).$$

This fact will be called **transitivity of Jacquet modules**.

6.3. The fact that Jacquet functor is left adjoint to the functor of parabolic induction means that we have a natural isomorphism

$$\mathrm{Hom}_G(\pi, \mathrm{Ind}_P^G(\sigma)) \cong \mathrm{Hom}_M(r_M^G(\pi), \sigma).$$

The above isomorphism is called **Frobenius reciprocity**. One constructs this isomorphism using evaluation of $f \in \mathrm{Hom}_G(\pi, \mathrm{Ind}_P^G(\sigma))$ at 1.

6.4. A smooth irreducible representation (π, V) of G is called **cuspidal (or supercuspidal)** if all the Jacquet modules for proper parabolic subgroups are trivial modules. It is natural to distinguish these representations, as will become clear very soon. Actually, in the definition of cuspidal representations, it is enough to require triviality Jacquet modules only of proper standard parabolic subgroups.

Cuspidal representations are very special representations, as we shall see later.

6.5. One easily sees that if (π, V) is a finitely generated smooth representation of G, then $r_M^G(\pi)$ is finitely generated representation of M. From this follows that it has an irreducible quotient.

Let (π, V) be an irreducible smooth representation of G. Among all the parabolic subgroups $P = MN$ which satisfy $r_M^G(\pi) \neq \{0\}$, chose a minimal one. Then by the above observation, $r_M^G(\pi)$ has an irreducible quotient, say σ. Minimality of P and transitivity of Jacquet modules (together with exactness) imply that σ is cuspidal. Now Frobenius reciprocity implies that π embeds into $\mathrm{Ind}_P^G(\sigma)$. In this way we have obtained a simple but important

Theorem: *An irreducible smooth representation π of G is cuspidal, or there exists a proper parabolic subgroup $P = MN$ of G and an irreducible cuspidal representation σ of M such that π is isomorphic to a subrepresentation of $\mathrm{Ind}_P^G(\sigma)$.*

In this way the problem of classification of non-unitary dual $\tilde{G}$ breaks into two problems. One problem is to classify cuspidal representations of Levi factors, and the other one is to classify irreducible subrepresentations of representations parabolically induced by cuspidal representations.

From the above theorem it is not clear at all how to classify irreducible subrepresentations of representations parabolically induced by cuspidal representations. Langlands classification will provide a strategy for it. There is another way to describe irreducible cuspidal representations. They can be characterized as those representations which never show up as subquotients of representations parabolically induced from proper parabolic subgroups.

6.6. Remarks: (i) Let G be $GL(2, F)$ and π be an irreducible representation of G which is not cuspidal. Then Schur lemma and the above theorem imply that π is isomorphic to a subrepresentation of $\text{Ind}_{P_\emptyset}^G(\chi)$, for a character χ of the Levi factor $M_\emptyset$ of $P_\emptyset$ (note that $M_\emptyset$ is commutative).

(ii) There is an archimedean version of the above theorem (see [17]).

6.7. One important property of cuspidal representations is that their matrix coefficients are functions which are compactly supported modulo center. W. Casselman has shown that this property characterizes cuspidal representations.

H. Jacquet has proved that each cuspidal representation is admissible (a nice argument for this can be found in [50]). Now 5.9 and Theorem 6.5 imply that each irreducible smooth representation is admissible. This is the reason that $\tilde{G}$ is also called admissible dual.

6.8. Let us note that Jacquet functor carries admissible representations to admissible ones (this is not quite easy to prove). Further, it carries smooth representations of finite length of G to smooth representations of finite length of M. For proofs one can consult [16].

7. Filtrations of Jacquet modules

Jacquet modules are very important in analysis of admissible representations, in particular of the induced ones. In general, it is not easy to determine their structure. There is **geometric lemma**, obtained independently by J. Bernstein and A. V. Zelevinsky, and by W. Casselman, which describes filtrations of Jacquet modules of $\text{Ind}_P^G(\sigma)$ in terms of representations parabolically induced by Jacquet modules of σ. We shall illustrate this lemma on the example of $G = GL(2, F)$. Later we shall describe how one can realize the geometric lemma as a part of an algebraic structure in the case of general linear and classical groups.

In this section we assume that

$$G = GL(2, F).$$

7.1. For $G = GL(2, F)$ we have $P_\emptyset = P_{(1,1)}^{GL}$. Irreducible smooth representations of $M_{(1,1)}^{GL}$ are one dimensional, i.e. characters. Since $M_{(1,1)}^{GL}$ is naturally isomorphic to $F^\times \times F^\times$, each irreducible smooth representations of $M^{GL}_{(1,1)}$ can be written as

$$\chi_1 \otimes \chi_2 \,,$$

where χ_1 and χ_2 are characters of $F^\times$.

We shall consider

$$\mathrm{Ind}^G_{P_\emptyset}(\chi_1 \otimes \chi_2).$$

Denote

$$w_0 = \begin{bmatrix} 0 & -1 \\ 1 & 0 \end{bmatrix}$$

It is not hard to show that the following obvious sequence

$$0 \to \{f \in \mathrm{Ind}^G_{P_\emptyset}(\chi_1 \otimes \chi_2); \mathrm{supp}(f) \subseteq P_\emptyset w_0 P_\emptyset\} \hookrightarrow \mathrm{Ind}^G_{P_\emptyset}(\chi_1 \otimes \chi_2) \xrightarrow{\text{restriction}} \{f|P_\emptyset; f \in \mathrm{Ind}^G_{P_\emptyset}(\chi_1 \otimes \chi_2)\} \to 0. \quad (1)$$

is exact. Further, considering the mapping $f \mapsto f(1)$, one gets easily that

$$r^G_{M_\emptyset}(\{f|P_\emptyset; f \in \mathrm{Ind}^G_{P_\emptyset}(\chi_1 \otimes \chi_2)\}) \cong \chi_1 \otimes \chi_2.$$

It requires a little bit more efforts to show that

$$r^G_{M_\emptyset}(\{f \in \mathrm{Ind}^G_{P_\emptyset}(\chi_1 \otimes \chi_2); \mathrm{supp}(f) \subseteq P_\emptyset w_0 P_\emptyset\}) \cong \chi_2 \otimes \chi_1.$$

7.2. Applying Jacquet functor to the exact sequence (1) (recall that the Jacquet functor is exact), we get the following exact sequence

$$0 \to \chi_2 \otimes \chi_1 \to r^G_{M_\emptyset}(\mathrm{Ind}^G_{P_\emptyset}(\chi_1 \otimes \chi_2)) \to \chi_1 \otimes \chi_2 \to 0.$$

As a consequence of this exact sequence, we can conclude that $\mathrm{Ind}^G_{P_\emptyset}(\chi_1 \otimes \chi_2)$ has at most length 2.

8. Square integrable and tempered representations

8.1. Let (π, V) be an irreducible smooth representation of G. Then Schur lemma implies that the center $Z(G)$ of G acts by scalars. Corresponding character will be denoted by

$$\omega_\pi$$

and called the **central character of** π.

A smooth representation does not need to be irreducible, but the center can act by scalars. Then we shall say that the representation has a central character.

8.2. An admissible representation (π, V) of G will be called **square integrable** (more precisely, **square integrable modulo center**) if it has central character, if the central character is unitary, and if absolute values of all the matrix coefficients

$$g \mapsto | < \pi(g)v, \tilde{v} > |, \quad v \in V, \tilde{v} \in \tilde{V},$$

are square integrable functions modulo center (i.e. square integrable functions on $G/Z(G)$).

In this notes we shall consider only irreducible square integrable representations.

An admissible representation (π, V) of G will be called **essentially square integrable** (or **essentially square integrable modulo center**) if there exists a character χ of G such that $\chi\pi$ is square integrable.

8.3. Each (irreducible) square integrable representation (π, V) of G is unitarizable. To see this, take $\tilde{v}_o \in \tilde{V}$ different from 0. Now for $u, v \in V$ set

$$(u, v) = \int_{G/Z(G)} \tilde{v}_o(\pi(g)u)\overline{\tilde{v}_o(\pi(g)v)}dg.$$

One sees directly that this is a G-invariant inner product on V (if π is not irreducible, one proceeds similarly; see [73]).

8.4. Irreducible square integrable representations are very important. First, they are (very distinguished) elements of the unitary dual. Then, via Langlands classification, they are crucial in the parameterization of non-unitary duals. Further, using matrix coefficients one gets that they are (irreducible) subrepresentations of $L^2(G)$ if G has compact center (in the non-compact case, we have similar situation when one fixes central character). Therefore, they are very important for understanding decomposition of $L^2(G)$.

8.5. An irreducible smooth (which implies admissible) representation (π, V) of G is called **tempered** if there exists a parabolic subgroup $P = MN$ of G (not necessarily proper) and an irreducible square integrable representation δ of M such that π is isomorphic to a subrepresentation of

$$\mathrm{Ind}_P^G(\delta).$$

An irreducible smooth representation (π, V) of G is called **essentially tempered** if there exists a character χ of G such that $\chi\pi$ is tempered.

One usually defines tempered representations without irreducibility requirement, but since we shall work only with irreducible tempered representations in these notes, the above definition is not restrictive for us.

8.6. We shall now introduce notation for general linear groups. The character

$$g \mapsto |\det(g)|_F, \quad GL(n,F) \to \mathbb{R}^\times$$

will be denoted by

$$\nu.$$

If τ is an (irreducible) essentially tempered representation of $GL(n,F)$, then one easily sees there exists a unique

$$e(\tau) \in \mathbb{R}$$

and a unique tempered representation τ^u such that

$$\tau = \nu^{e(\tau)}\tau^u.$$

This requirement uniquely defines $e(\tau)$.

8.7. There are a very useful criteria of Casselman for checking if an irreducible admissible representation is square irreducible or tempered. We shall explain this criterion on the simplest case, on $G = GL(2,F)$.

Let (π, V) be an irreducible essentially square integrable (resp. essentially tempered) representation of $GL(2,F)$. Then for any irreducible subquotient $\chi = \chi_1 \otimes \chi_2$ of $r_{P_\emptyset}^{GL(2,F)}(\pi)$ (χ_1 and χ_2 are characters of $F^\times$) we have

$$e(\chi_1) > e(\chi_2) \qquad (\text{resp.} \quad e(\chi_1) \geq e(\chi_2)).$$

Moreover, the converse also holds for an irreducible admissible representation (π, V).

9. Langlands classification

9.1. Langlands classification parameterizes representations of $\tilde{G}$ by (irreducible) essentially tempered representations of Levi factors of standard parabolic subgroups. These essentially tempered representations need to satisfy certain positiveness condition (which will be discussed later).

Langlands classification claims the following: For an irreducible essentially tempered representation τ of Levi factor M of standard parabolic subgroup P of G, which satisfy the above mentioned positiveness condition, the representation $\text{Ind}_P^G(\tau)$ has a unique irreducible quotient. This irreducible quotient is called the **Langlands quotient** ($\text{Ind}_P^G(\tau)$ is called a **standard module** of G). Each $\pi \in \tilde{G}$ is isomorphic to some Langlands quotient, and moreover π determines uniquely the standard parabolic subgroup P and essentially tempered representation τ.

We shall now explain the positiveness condition for the groups that we consider in these notes. We shall start with general linear groups.

9.2. Each Levi factor of a parabolic subgroup of a general linear group, is a direct product of general linear groups. Because of this, each irreducible essentially tempered representation of a Levi factor of a general linear group is a tensor product of such representations of general linear groups. Therefore, the essentially tempered representations of a Levi factors of a general linear groups are of the form

$$\tau_1 \otimes \tau_2 \otimes \cdots \otimes \tau_k,$$

where $\tau_1, \tau_2, \ldots, \tau_k$ are essentially tempered representations of general linear groups. The positiveness condition here is simply

$$e(\tau_1) > e(\tau_2) > \cdots > e(\tau_k).$$

9.3. We shall see how one gets the Langlands parameterization in the case of the simplest possible example, in the case of $G = GL(2, F)$. Let $\pi \in \tilde{G}$. If π is essentially tempered, then it is its own Langlands parameter. Suppose therefore that π is not essentially tempered. Then, in particular, it is not cuspidal. Therefore

$$\tilde{\pi} \quad \text{is a subquotient of} \quad \text{Ind}_{P_\emptyset}^G(\chi_1 \otimes \chi_2)$$

by Theorem 6.5, for some characters χ_1 and χ_2 of $F^\times$ (see also Remarks 6.6, (i)). Since π is not essentially tempered, $\tilde{\pi}$ is also not essentially tempered. Now by 8.7

$$e(\chi_1) \neq e(\chi_2).$$

Therefore by 7.2

$$r_{P_\emptyset}^G(\tilde{\pi}) \hookrightarrow \chi_1 \otimes \chi_2 \;\oplus\; \chi_2 \otimes \chi_1. \tag{2}$$

Without lost of generality we can assume

$$e(\chi_1) > e(\chi_2),$$

since $\mathrm{Ind}_{P_\emptyset}^G(\chi_1 \otimes \chi_2)$ and $\mathrm{Ind}_{P_\emptyset}^G(\chi_2 \otimes \chi_1)$ have the same Jordan-Hölder series (one sees this using induction from associate parabolic subgroups; see 5.10). Since $\tilde{\pi}$ is not essentially square integrable (recall that π it is not essentially tempered), from criterion for essentially square integrability follows that $\chi_2 \otimes \chi_1$ must be a subquotient of $r_{P_\emptyset}^G(\tilde{\pi})$. Now from (2) we see that there exits a non-trivial homomorphism

$$r_{P_\emptyset}^G(\tilde{\pi}) \to \chi_2 \otimes \chi_1.$$

Frobenius reciprocity implies

$$\tilde{\pi} \hookrightarrow \mathrm{Ind}_{P_\emptyset}^G(\chi_2 \otimes \chi_1).$$

Passing to contragredients we get an epimorphism

$$\mathrm{Ind}_{P_\emptyset}^G(\chi_2^{-1} \otimes \chi_1^{-1}) \to \pi.$$

Note that

$$e(\chi_2^{-1}) = -e(\chi_2) > -e(\chi_1) = e(\chi_1^{-1}).$$

This implies that we have shown the existence of Langlands parameters for irreducible representations of $GL(2, F)$. Their uniqueness follow from the filtration of Jacquet modules (see 7.2).

Thus, we have "proved" Langlands classification for $GL(2, F)$. This case is too simple to illustrate the proof of the Langlands classification in general, but one can get at least some idea from this simplest case how proof goes in general. In any case, we see the importance of the Geometric lemma.

9.4. Since we have defined tempered representations by square integrable representations, it is natural to try to express Langlands classification in terms of square integrable representations, if this is possible.

In the study the representation theory of general linear groups, it is convenient to use notation that was used by Bernstein and Zelevinsky in their work on the representation theory of general linear groups. We shall now recall of (a very small part of) it.

For smooth representations π_1 and π_2 of $GL(n_1, F)$ and $GL(n_2, F)$ denote

$$\pi_1 \times \pi_2 = \mathrm{Ind}_{P^{GL}_{(n_1,n_2)}}^{GL(n_1+n_2,F)}(\pi_1 \otimes \pi_2).$$

Then

$$\pi_1 \times (\pi_2 \times \pi_3) \cong (\pi_1 \times \pi_2) \times \pi_3. \tag{3}$$

This follows from induction by stages. Further, for smooth representation π_1 and π_2 of finite length,

$$\pi_1 \times \pi_2 \text{ and } \pi_2 \times \pi_1 \text{ have the same Jordan-Hölder series.} \tag{4}$$

This follows from induction from associate parabolic subgroups.

Remark: In the case of an archimedean field F, using parabolic induction we define multiplication $\times$ between $(\mathfrak{g}, K_0)$-modules of general linear groups (over F) in the same way as above. Then (3) and (4) hold also in this case.

9.5. A principal result regarding tempered induction for general linear groups is that this induction is irreducible (this fact holds for all the local fields). This fact has been proved independently at several places, but it seems that the first proof in this setting belongs to H. Jacquet, who proved it is for all the local fields (see [29]).

This principal result claims the following:

> If $\pi_1, \pi_2, \ldots, \pi_k$ are (unitarizable) irreducible square integrable representations of general linear groups, then $\pi_1 \times \pi_2 \times \cdots \times \pi_k$ is irreducible.

Either from general facts regarding tempered representations, or from A. V. Zelevinsky paper [76], follows that the tempered representation $\pi_1 \times \pi_2 \times \cdots \times \pi_k$ determines irreducible square integrable representations $\pi_1, \pi_2, \ldots, \pi_k$ up to a permutation.

9.6. Using 9.4 and the fact that for a character χ of $F^\times$ we have

$$[(\chi \circ \det)\pi_1] \times [(\chi \circ \det)\pi_2] \cong (\chi \circ \det)(\pi_1 \times \pi_2)$$

(which one proves directly), we can reformulate the Langlands classification for general linear groups in the following way.

Denote by D the set of all the irreducible essentially square integrable representations of $GL(n, F)$'s for all $n \geq 1$. Let $M(D)$ be the set of all finite multisets in D. These are functions from D into $\mathbb{Z}_+$ with finite supports. We shall write them similar as sets, but repetitions of elements will be allowed. We shall write them as

$$(\delta_1, \delta_2, \ldots, \delta_k), \quad \text{where } \delta_i \in D.$$

Take any $d = (\delta_1, \delta_2, \dots, \delta_k) \in M(D)$. Take a permutation p of $\{1, \dots, k\}$ such that

$$e(\delta_{p(1)}) \geq e(\delta_{p(2)}) \geq \cdots \geq e(\delta_{p(k)}).$$

Now the representation

$$\delta_{p(1)} \times \delta_{p(2)} \times \cdots \times \delta_{p(k)},$$

which will be denoted by

$$\lambda(d),$$

has a unique irreducible quotient (the representation $\lambda(d)$ is determined by $d \in M(d)$ up to an isomorphism.). This quotient will be denoted by

$$L(d).$$

In this way one gets bijection between $M(D)$ and the set of all the irreducible smooth representations of all general linear groups over F.

This is just a reformulation of the Langlands classification for general linear groups.

The Langlands classification has a number of natural properties. Let us mention three:

$$L(\delta_1, \delta_2, \dots, \delta_k)\tilde{} \cong L(\tilde{\delta}_1, \tilde{\delta}_2, \dots, \tilde{\delta}_k), \tag{5}$$

$$L(\delta_1, \delta_2, \dots, \delta_k)^- \cong L(\bar{\delta}_1, \bar{\delta}_2, \dots, \bar{\delta}_k), \tag{6}$$

$$\chi L(\delta_1, \delta_2, \dots, \delta_k) \cong L(\chi\delta_1, \chi\delta_2, \dots, \chi\delta_k) \tag{7}$$

(χ is a multiplicative character of the field, and further, for a representation π of $GL(n, F)$, $\chi\pi$ denotes the representation $(\chi \circ \det)\,\pi$).

Remark: The Langlands classification for general linear groups holds also if the field is archimedean F. In this case the non-unitary dual $GL(n, F)\tilde{}$ of $GL(n, F)$ is the set of all the equivalence classes of irreducible $(\mathfrak{g}, K_0)$-modules of $GL(n, F)$. The irreducible representations (i.e. non-unitary duals) are classified by $M(D)$, where D is the set of all the equivalence classes of irreducible essentially square integrable $(\mathfrak{g}, K_0)$-modules of all $GL(n, F)$'s, $n \geq 1$ (if $F = \mathbb{C}$, then $D = (\mathbb{C}^\times)\tilde{}$, while for $F = \mathbb{R}$ we have $D \subseteq (\mathbb{R}^\times)\tilde{} \cup GL(2, \mathbb{R})\tilde{}$).

9.7. Now we shall describe the Langlands classification for symplectic and odd-orthogonal groups.

It is convenient to introduce for classical groups the following notation, which simplifies notation when one works with parabolically induced representations. Let π be a smooth representation of $GL(n,F)$ and let σ be a smooth representation of $Sp(2m,F)$ (resp. $SO(2m+1,F)$). Denote

$$\pi \rtimes \sigma = \mathrm{Ind}_{P_{(n)}}^{Sp(2(n+m),F)}(\pi \otimes \sigma)$$

$$\left(\text{resp. } \pi \rtimes \sigma = \mathrm{Ind}_{P_{(n)}}^{SO(2(n+m)+1,F)}(\pi \otimes \sigma)\right).$$

From induction by stages follows

$$\pi_1 \rtimes (\pi_2 \rtimes \sigma) \cong (\pi_1 \times \pi_2) \rtimes \sigma.$$

Further, for smooth representations π and σ of finite length,

$$\pi \rtimes \sigma \text{ and } \tilde{\pi} \rtimes \sigma \text{ have the same Jordan-Hölder series.} \tag{8}$$

This follows from induction from associate parabolic subgroups.

9.8. Regarding the Langlands classification for symplectic and odd-orthogonal groups, one can first describe it in terms of essentially tempered representations, and after that pass to a description which include only essentially square integrable representations of general linear groups (and tempered representations of symplectic or odd-orthogonal groups), similarly as we did in the case of Langlands classification for general linear groups. Instead of this, we shall skip over the first description and go directly to the second description.

Set

$$D_+ = \{\delta \in D; e(\delta) > 0\}.$$

Denote by T the set of all equivalence classes of tempered representations of all $Sp(2m,F)$ (resp. $SO(2m+1,F)$), for all $m \geq 0$. Take

$$((\delta_1, \delta_2, \ldots, \delta_k), \tau) \in M(D_+) \times T.$$

After a renumeration, we can assume

$$\delta_1 \geq \delta_2 \geq \cdots \geq \delta_k$$

Then the representation

$$\delta_1 \times \delta_2 \times \cdots \times \delta_k \rtimes \tau$$

has a unique irreducible quotient, which will be denoted by

$$L(\delta_1, \delta_2, \ldots, \delta_k; \tau).$$

Now the mapping

$$((\delta_1, \delta_2, \ldots, \delta_k), \tau) \mapsto L(\delta_1, \delta_2, \ldots, \delta_k; \tau)$$

defines a bijection from the set $M(D_+) \times T$ onto the set of all the equivalence classes of irreducible smooth representations of all $Sp(2m, F)$ (resp. $SO(2m+1, F)$), $m \geq 0$.

This is the Langlands classification for (these) classical groups.

Remark: One can define $\rtimes$ also for the case of archimedean fields. The Langlands classification holds here in the same form.

10. Geometric lemma and algebraic structures

Geometric lemma, which is a technical result describing filtrations of Jacquet modules of induced representations in terms of representations induced by Jacquet modules of inducing representations, is very important for number of purposes. For general linear and classical groups we can "incorporate" it into an algebraic structures on the representations of these groups.

10.1. Let for a moment G be a reductive group over a non-archimedean field F. The Grothendieck group of the category of all smooth G-representations of finite length will be denoted by

$$\mathcal{R}(G).$$

This is just a free $\mathbb{Z}$-module over basis $\tilde{G}$ (it is isomorphic to the group of virtual characters of G). For a finite length representation π, let

$$\text{s.s.}(\pi) = \sum_{\tau \in \tilde{G}} m(\tau : \pi)\tau.$$

This is called semi simplification of π. We consider it as an element of $\mathcal{R}(G)$. There is a natural order on $\mathcal{R}(G)$.

We have

$$\mathcal{R}(G_1 \times G_2) \cong \mathcal{R}(G_1) \otimes \mathcal{R}(G_2). \tag{9}$$

Further, r_M^G factors in a natural way to a homomorphism from $\mathcal{R}(G)$ into $\mathcal{R}(M)$, which is denoted again by r_M^G. This is a homomorphism of ordered groups (i.e. it respects also orders).

We have analogous definition for the parabolic induction: Ind_P^G : $\mathcal{R}(M) \to \mathcal{R}(G)$, which is again a morphism of ordered groups.

10.2. Set

$$R = \oplus_{n \in \mathbb{Z}_+} \mathcal{R}(GL(n,F)).$$

Then one lifts $\times$ to a multiplication on R in a natural way. The mapping $\times : R \times R \to R$ factors in a natural way through a mapping

$$m : R \otimes R \to R.$$

Let $\pi \in GL(n,F)\tilde{}$. We will consider

$$\text{s.s.}\left(r^{GL(n,F)}_{M^{GL}_{(k,n-k)}}(\pi) \right) \in R_k \otimes R_{n-k}$$

using the isomorphism (9). Define

$$m^*(\pi) = \sum_{k=0}^{n} \text{s.s.}\left(r^{GL(n,F)}_{M^{GL}_{(k,n-k)}}(\pi) \right) .$$

We can (and will) consider

$$m^*(\pi) \in R \otimes R ,$$

since each $R_k \otimes R_{n-k} \hookrightarrow R \otimes R$. We can lift m^* to an additive mapping

$$m^* : R \to R \otimes R .$$

This mapping is called comultiplication.

With the multiplication and comultiplication, R is a commutative Hopf algebra (over $\mathbb{Z}$). This algebra was constructed by A. V. Zelevinsky.

The most important part of this Hopf algebra structure is the formula

$$m^*(\pi_1 \times \pi_2) = m^*(\pi_1) \times m^*(\pi_2),$$

which explains how to get composition factors of Jacquet modules (for maximal parabolic subgroups) of induced representations, by induction from Jacquet modules of inducing representations.

10.3. Suppose (only) in this paragraph that F is an archimedean field. One defines $\mathcal{R}(G)$ in the same way as in the non-archimedean case, considering the category of $(\mathfrak{g}, K)$-modules of finite length (G is a connected reductive group over F). This is a free $\mathbb{Z}$-module over basis $\tilde{G}$ (it is isomorphic to the group of virtual characters of G). Now for general linear groups over F one defines R in the same way as in the non-archimedean case. By the same formula as in the non-archimedean case, one defines multiplication $\times$ on R (using parabolic induction). In this way R becomes a commutative ring with identity. In the archimedean case, there is no comodule structure on R as in the non-archimedean case.

Remark: The Kazhdan-Patterson lifting for $GL(n,\mathbb{C})$ has a very nice and natural description in terms of this algebra (see [64]).

10.4. Assume in this section that F is any local field (archimedean or non-archimedean). For $a = (\delta_1, \ldots, \delta_k) \in M(D)$ consider

$$\text{s.s}(\lambda(a)) = \delta_1 \times \cdots \times \delta_k \in R.$$

Then a simple fact regarding Langlands classification implies that

$$\text{s.s}(\lambda(a)), \quad a \in M(D),$$

form a $\mathbb{Z}$-basis of R (this simple fact is usually expressed in the following form: characters of standard modules form a $\mathbb{Z}$-basis of $\mathcal{R}(G)$). In other words, for any local field F holds the following

Proposition: *The ring R is a polynomial $\mathbb{Z}$-algebra over D.*

10.5. We shall denote

$$R(S) = \oplus_{n \in \mathbb{Z}_+} \mathcal{R}(Sp(2n,F))$$

if we consider symplectic groups, and

$$R(S) = \oplus_{n \in \mathbb{Z}_+} \mathcal{R}(SO(2n+1,F))$$

if we consider odd-orthogonal groups.

In this setting, one again lifts $\rtimes$ in a natural way to a multiplication $R \times R(S) \to R(S)$. This multiplication factors through a mapping

$$\mu : R \otimes R(S) \to R(S).$$

For $\pi \in Sp(2n,F)\tilde{}$ (resp. $\pi \in SO(2n+1,F)\tilde{}$) set

$$\mu^*(\pi) = \sum_{k=0}^{n} \text{s.s.}\left(r_{M_{(k)}}^{Sp(2n,F)}(\pi)\right)$$

$$\left(\text{resp. } \mu^*(\pi) = \sum_{k=0}^{n} \text{s.s.}\left(r_{M_{(k)}}^{SO(2n+1,F)}(\pi)\right)\right).$$

Consider $\text{s.s.}\left(r_{M_{(k)}}^{Sp(2n,F)}(\pi)\right)$ $\left(\text{resp. s.s.}\left(r_{M_{(k)}}^{SO(2n+1,F)}(\pi)\right)\right)$ as an element of $R_k \otimes R_{n-k}(S)$ using (9), and consider further

$$\mu^*(\pi) \in R \otimes R(S).$$

Lift μ^* to an additive mapping

$$\mu^* \colon R(S) \to R \otimes R(S) \ ,$$

which will be called comultiplication on $R(S)$.

With the above multiplication and complication, $R(S)$ is a module and a comodule over R. It is not a Hopf module over R, but is also far from this structure as we shall explain now.

Define

$$M^* = (m \otimes 1) \circ (\sim \otimes m^*) \circ s \circ m^* : R \to R \otimes R,$$

where 1 denotes the identity mapping, $\sim$ the contragredient mapping and s the transposition mapping $\sum x_i \otimes y_i \mapsto \sum y_i \otimes x_i$. Then

$$\mu^*(\pi \rtimes \sigma) = M^*(\pi) \rtimes \mu^*(\sigma)$$

($R \otimes R(S)$ is a $R \otimes R$-module in an obvious way). We say that $R(S)$ is an M^*-Hopf module over R.

This is again a (combinatorial) formula from which we can again in a simple way get compositions factors of Jacquet modules of parabolically induced representations for classical group.

11. Square integrable representations of p-adic general linear groups

11.1. Denote by

$$\mathcal{C}$$

the set of all equivalence classes of irreducible cuspidal representations of all $GL(n, F)$, $n \geq 1$.

A **segment in** $\mathcal{C}$ is the set of form

$$\Delta = [\rho, \nu^k \rho] = \{\rho, \nu\rho, \ldots, \nu^k \rho\},$$

where $\rho \in \mathcal{C}, k \in \mathbb{Z}_+$. Denote the set of all such segments by

$$\mathcal{S}.$$

For a segment $\Delta = [\rho, \nu^k \rho] = \{\rho, \nu\rho, \ldots, \nu^k \rho\} \in \mathcal{S}$, the representation

$$\nu^k \rho \times \nu^{k-1} \rho \times \cdots \times \nu\rho \times \rho$$

contains a unique irreducible subrepresentation, which will be denoted by

$$\delta(\Delta).$$

Then $\delta(\Delta)$ is essentially square integrable representation, and in this way one gets a bijection from $\mathcal{S}$ onto D (which is the set of all the irreducible essentially square integrable representations of general linear groups $GL(n, F)$, $n \geq 1$). This is one of the consequences of Bernstein and Zelevinsky theory, which is based on Gelfand-Kazhdan theory of derivatives. One can obtain these results also by different methods.

In applications of square integrable representations of general linear groups, it is important to know what are Jacquet modules of these representations. This tells the following simple formula

$$m^*(\delta([\rho, \nu^k \rho])) = \sum_{i=-1}^{k} \delta([\nu^{i+1}\rho, \nu^k \rho]) \otimes \delta([\rho, \nu^i \rho]) \tag{10}$$

(see [76]).

11.2. As we have seen, each segment of $\mathcal{S}$ determines uniquely essentially square integrable representation. Let us explain how to "read" corresponding segment from an essentially square integrable representation

$$\delta = \delta(\Delta), \Delta \in \mathcal{S}.$$

For this, we shall introduce two natural invariants of δ.

There exists exactly two (inequivalent) $\rho_1, \rho_2 \in \mathcal{C}$ such that $\rho_1 \times \delta$ and $\rho_2 \times \delta$ reduce. We can, after a possible renumeration, assume

$$e(\rho_1) \leq e(\rho_2).$$

Representations ρ_1 and ρ_2 will be called **cuspidal reducibilities of** δ (ρ_1 the lower one and ρ_2 the upper one).

Then

$$\Delta = [\nu\rho_1, \nu^{-1}\rho_2].$$

Thus cuspidal reducibilities of δ determine completely the segment in $\mathcal{C}$ corresponding to δ.

11.3. Let $\delta_1 \in D$ have cuspidal reducibilities ρ_1, and ρ_2. Take $\delta_2 = \delta(\Delta_2) \in D$. Then

$$\delta_1 \times \delta_2$$

reduces if and only if

(1) $\mathrm{card}(\{\rho_1, \rho_2\} \cap \Delta_2) = 1$;
(2) neither ρ_1 nor ρ_2 is a cuspidal reducibility of $\delta(\Delta_2)$.

12. Two simple examples of square integrable representations of classical p-adic groups

In the rest of these notes, we shall fix one of the series of classical groups, symplectic or odd-orthogonal, and denote by S_n either $Sp(2n, F)$ or $SO(2n+1, F)$.

Before we proceed further with description of general square integrable representations, we shall give two examples of square integrable representations of classical groups.

The trivial one-dimensional representation of a group G will be denoted by 1_G.

12.1. Example: In this example we shall describe Steinberg representations for symplectic groups. Steinberg representation can be constructed for any reductive group.

Here we consider the series $S_n = Sp(2n, F)$.

An easy computation of modular character of $P_\emptyset$ in $S_1 = Sp(2, F) = SL(2, F)$ implies that

$$1_{S_1} \hookrightarrow \nu^{-1}\, 1_{F^\times} \rtimes 1_{S_0},$$

since $\nu^{-1} 1_{F^\times} \rtimes 1_{S_0}$ contains constant functions. The length of the Jacquet module of $\nu^{-1}\, 1_{F^\times} \rtimes 1_{S_0}$ for the (standard) minimal parabolic subgroup is 2 (and irreducible subquotients are not isomorphic). This implies that $\nu^{-1}\, 1_{F^\times} \rtimes 1_{S_0}$ is a length two representation. Further, Frobenius reciprocity implies that $\nu^{-1}\, 1_{F^\times} \rtimes 1_{S_0}$ is not completely reducible (i.e. it is not a sum of irreducible subrepresentations). Now passing to contragredients we see that

$$\nu\, 1_{F^\times} \rtimes 1_{S_0}$$

contains a unique irreducible subrepresentation. This representation will be denoted by St_{S_1}, and called Steinberg representation of S_1. We can see (from the algebraic structure of $R(S)$ over R) that the Jacquet module for the minimal parabolic subgroup is $\nu\, 1_{F^\times} \otimes 1_{S_0}$. Now Casselman's square integrability criterion in this situation implies that St_{S_1} is square integrable.

Define St_{S_0} to be 1_{S_0}. Now both

$$\nu^2\, 1_{F^\times} \rtimes \mathrm{St}_{S_1} \quad \text{and} \quad \delta([\nu\, 1_{F^\times}, \nu^2\, 1_{F^\times}]) \rtimes \mathrm{St}_{S_0}$$

embed into

$$\nu^2\, 1_{F^\times} \times \nu\, 1_{F^\times} \rtimes \mathrm{St}_{S_0}.$$

Analyzing Jacquet modules, we would see that these two subrepresentations have exactly one irreducible subquotient in common, and that this subquotient is square integrable. We shall denote it by St_{S_2}. It is a unique irreducible subrepresentation of $\nu^2\, 1_{F^\times} \times \nu\, 1_{F^\times} \rtimes \mathrm{St}_{S_0}$.

Continuing recursively in the above way, we define the **Steinberg representation**

$$\mathrm{St}_{Sp_n}$$

for any S_n. It is a unique irreducible subrepresentation of

$$\nu^n\, 1_{F^\times} \times \nu^{n-1}\, 1_{F^\times} \times \cdots \times \nu^2\, 1_{F^\times} \times \nu\, 1_{F^\times} \rtimes \mathrm{St}_{S_0}$$

(this defines St_{Sp_n}). It is again easy to write what are the Jacquet modules of these representations:

$$\mu^*(\mathrm{St}_{S_n}) = \sum_{k=0}^{n} \delta([\nu^{k+1} 1_{F^\times}, \nu^n 1_{F^\times}]) \otimes \mathrm{St}_{S_k}.$$

12.2. Example: Let we now consider the series $S_n = SO(2n+1, F)$.

An easy computation of modular character of $P_\emptyset$ in $S_1 = SO(3, F)$ implies that

$$1_{S_1} \hookrightarrow \nu^{-1/2}\, 1_{F^\times} \rtimes 1_{S_0}$$

(since $\nu^{-1/2}\, 1_{F^\times} \rtimes 1_{S_0}$ contains constant functions).

Consider the representation

$$\nu^{1/2}\, 1_{F^\times} \times \nu^{-1/2}\, 1_{F^\times} \rtimes 1_{S_0}.$$

Here

$$\delta([\nu^{-1/2}\, 1_{F^\times}, \nu^{1/2}\, 1_{F^\times}]) \rtimes 1_{S_0} \quad \text{and} \quad \nu^{1/2}\, 1_{F^\times} \rtimes 1_{S_1}$$

are subrepresentations. Looking at Jacquet modules, one sees that

$$\delta([\nu^{-1/2}\, 1_{F^\times}, \nu^{1/2}\, 1_{F^\times}]) \rtimes 1_{S_0} \tag{11}$$

reduces. This follows from

$$\text{s.s.}\left(\delta([\nu^{-1/2}\, 1_{F^\times}, \nu^{1/2}\, 1_{F^\times}]) \rtimes 1_{S_0}\right) + \text{s.s.}\left(\nu^{1/2}\, 1_{F^\times} \rtimes 1_{S_1}\right) \not\leq \text{s.s.}\left(\nu^{1/2}\, 1_{F^\times} \times \nu^{-1/2}\, 1_{F^\times} \rtimes 1_{S_0}\right)$$

and

$$\text{s.s.}\left(\delta([\nu^{-1/2}\, 1_{F^\times}, \nu^{1/2}\, 1_{F^\times}]) \rtimes 1_{S_0}\right) \not\leq \text{s.s.}\left(\nu^{1/2}\, 1_{F^\times} \rtimes 1_{S_1}\right),$$

what one checks using the structure of $R(S)$.

The representation (11) is unitarizable, so it is completely reducible. Considering the Jacquet module of this representation for $P_{(2)}$, and applying Frobenius reciprocity, we get that the representation $\delta([\nu^{-1/2}\,1_{F^\times}, \nu^{1/2}\,1_{F^\times}]) \rtimes 1_{S_0}$ reduces into a sum of two inequivalent irreducible subrepresentations, say T_1 and T_2. Thus

$$\delta([\nu^{-1/2}\,1_{F^\times}, \nu^{1/2}\,1_{F^\times}]) \rtimes 1_{S_0} = T_1 \oplus T_2.$$

Now

$$\begin{aligned} &\delta([\nu^{-1/2}\,1_{F^\times}, \nu^{3/2}\,1_{F^\times}]) \rtimes 1_{S_0} \\ &\qquad \hookrightarrow \nu^{3/2}\,1_{F^\times} \times \delta([\nu^{-1/2}\,1_{F^\times}, \nu^{1/2}\,1_{F^\times}]) \rtimes 1_{S_0} \\ &\qquad\qquad = \nu^{3/2}\,1_{F^\times} \times (T_1 \oplus T_2) \\ &\qquad\qquad\qquad \cong \nu^{3/2}\,1_{F^\times} \rtimes T_1 \oplus \nu^{3/2}\,1_{F^\times} \rtimes T_2. \end{aligned}$$

Now the multiplicity of $\nu^{3/2}\,1_{F^\times} \otimes T_i$ in corresponding Jacquet module of $\nu^{3/2}\,1_{F^\times} \rtimes T_i$ is one. This implies that $\nu^{3/2}\,1_{F^\times} \rtimes T_i$ has a unique irreducible subrepresentation. These two irreducible subrepresentations (for $i = 1, 2$) are square integrable.

One can show that $\nu^{3/2}\,1_{F^\times} \rtimes T_i$ are subquotients of corresponding Jacquet module of $\delta([\nu^{-1/2}\,1_{F^\times}, \nu^{3/2}\,1_{F^\times}]) \rtimes 1_{S_0}$. From this we see that

$$\delta([\nu^{-1/2}\,1_{F^\times}, \nu^{3/2}\,1_{F^\times}]) \rtimes 1_{S_0}$$

has exactly two irreducible subrepresentations. They are inequivalent and square integrable.

Next question is how to distinguish these two irreducible square integrable subrepresentations. One can show that they have Jacquet modules of different length. This is one possible way to distinguish them.

13. Invariants of square integrable representations of classical p-adic groups

13.1. Let π be an irreducible square integrable representation of a classical group S_q. C. Mœglin has attached to it a triple

$$(Jord(\pi), \pi_{cusp}, \epsilon_\pi).$$

Each of these three parameters was considered earlier (at least in some form), but C. Mœglin was the first who considered them in this form.

We shall describe each of these parameters. Our goal will be to explain their meaning from the point of harmonic analysis (they have a clear meaning from the point of view of Langlands program, what will be discussed later).

13.2. Jordan block of π: This is probably the most important of the three parameters (this parameter should determine the L-packet in which the representation lies). As we shall see later, the definition of $Jord(\pi)$ is very natural from the point of harmonic analysis (and can be given completely in terms of harmonic analysis).

For $\rho \in \mathcal{C}$ and $a \in \mathbb{N}$ denote

$$\delta(\rho, a) = \delta([\nu^{-(a-1)/2}\rho, \nu^{(a-1)/2}\rho]).$$

We shall start with the first definition of $Jord(\pi)$ (this is a little bit modified definition, to avoid L-functions). $Jord(\pi)$ is called the Jordan block of π and it consists of all $(\rho, a) \in \mathcal{C} \times \mathbb{N}$ such that

(1) ρ is selfdual (i.e. $\tilde{\rho} \cong \rho$; then ρ is unitarizable) and
(2) if $\nu^{1/2}\rho \rtimes 1_{S_0}$ is reducible (resp. irreducible), then a is even (resp. odd) and

$$\delta(\rho, a) \rtimes \pi$$

is irreducible.

Although the above definition is simple, the clear meaning and importance of Jordan blocks is not evident from it. This is the reason that we shall give another description of Jordan blocks, from which will be much more clear importance of Jordan blocks for the harmonic analysis.

13.3. As we have mentioned already above, there is a very natural way to come to Jordan blocks from the point of view of harmonic analysis, and we shall explain it bellow. Besides, because of the importance of Jordan blocks, they deserve to be understood as good as possible.

Before we start to explain it, let us note that the classification of irreducible square integrable representation of classical groups is done under a natural assumption, which will be explained in section 15.1. This assumption shows up in proofs, not in the expression of the parameterization of irreducible square integrable representations. We shall assume that it holds in further.

Now we are going to explain the importance of $Jord(\pi)$ for harmonic analysis.

Once we have an irreducible square integrable representation π of a classical group, having in mind classification of the non-unitary dual via the Langlands classification, the first question that arises is:

Which irreducible tempered representations can be obtained from this π.

In other words, we would like to understand how a representation of the form

$$\delta_1 \times \delta_2 \times \cdots \times \delta_k \rtimes \pi \tag{12}$$

reduces, when δ_i are (unitarizable) irreducible square integrable representations of general linear groups.

If we would know the answer to this question, we would have a reduction of understanding of irreducible tempered representations of the classical groups (and in this way also of all the irreducible representations) to the problem of understanding of irreducible square integrable representations of the classical groups. Therefore, understanding of such a reduction would be of the first class importance.

The theory of R-groups reduces this question to the question when

$$\delta \rtimes \pi \tag{13}$$

reduces, for δ an irreducible (unitarizable) square integrable representation of a general linear group.

Remark: For further discussion of Jordan blocks, one does not need to understand this reduction. But for the classification of the non-unitary duals, one needs it. Therefore, we shall explain the reduction that gives the theory of R-groups, without going deeper in this theory. Consider representation from (12). Denote by ℓ the number of inequivalent δ_i among $\delta_1, \delta_2, \ldots, \delta_k$ such that

$$\delta_i \rtimes \pi$$

reduces. Then $\delta_1 \times \delta_2 \times \cdots \times \delta_k \rtimes \pi$ is a multiplicity one representation and it reduces into a direct sum of

$$2^\ell$$

irreducible (tempered) representations.

If p is a permutation of $\{1, 2, \ldots, k\}$ and $\epsilon_1, \epsilon_2, \ldots, \epsilon_k \in \{\pm 1\}$, then representations $\delta_1 \times \delta_2 \times \cdots \times \delta_k \rtimes \pi$ and $\delta_{p(1)}^{\epsilon_1} \times \delta_{p(2)}^{\epsilon_2} \times \cdots \times \delta_{(k)}^{\epsilon_k} \rtimes \pi$ are equivalent, where $\delta_{p(i)}^{\epsilon_i}$ denotes $\delta_{p(i)}$ if $\epsilon_i = 1$ and it denotes $\tilde{\delta}_{p(i)}$ if $\epsilon_i = -1$.

Let $\delta_1' \times \delta_2' \times \cdots \times \delta_{k'}' \rtimes \pi'$ be another representation, such that δ_j' are (unitarizable) irreducible square integrable representations of general linear groups and π' is an irreducible square integrable representation of a classical group $S_{q'}$. Suppose that $\delta_1 \times \delta_2 \times \cdots \times \delta_k \rtimes \pi$ and $\delta_1' \times \delta_2' \times \cdots \times \delta_{k'}' \rtimes \pi'$ have an irreducible subquotient in common. Then $\pi \cong \pi'$ (and therefore $q = q'$), $k = k'$ and there exists a permutation p of $\{1, 2, \ldots, k\}$ and $\epsilon_1, \epsilon_2, \ldots, \epsilon_k \in \{\pm 1\}$ such that

$$\delta_i' \cong \delta_{p(i)}^{\epsilon_i}$$

for all $i = 1, 2, \ldots, k$.

13.4. As we already have mentioned, to understand irreducible tempered representations, we need to understand when representations $\delta \rtimes \pi$ (from (13) reduce. Having in mind the classification of irreducible square integrable representations of general linear groups, one needs to understand when

$$\delta(\rho, a) \rtimes \pi$$

reduces, for unitarizable $\rho \in \mathcal{C}$ and for $a \in \mathbb{N}$.

When we fix ρ, the reducibility of these representations can be described in a very nice way (a crucial role in this is played by $Jord(\pi)$).

Frobenius reciprocity implies irreducibility if ρ is not selfdual. Therefore, it remains to understand the reducibility for selfdual ρ's.

The following two examples are very simple but nice examples, from which one can get an an idea what happens regarding these reducibilities in general.

13.5. Examples: Let $S_n = Sp(2n, F)$ and $\pi = 1_{S_0}$.

(1) Suppose ψ is a character of order two of $F^\times = GL(1, F)$. Then

$$\delta(\psi, a) \rtimes 1_{S_0} \text{ is irreducible for all even } a;$$

$$\delta(\psi, a) \rtimes 1_{S_0} \text{ is reducible for all odd } a.$$

We see that understanding of reducibility in this case is very simple. One needs only to know the parity of $\mathbb{N}$ for which we have reducibility. Unfortunately, this is not always the case for other square integrable representations.

Now we shall give a simple example of a situation of a slightly different type.

(2) We shall consider now instead of ψ the trivial representation $1_{F^\times}$ of $GL(1, F)$ (on one-dimensional space). Then

$$\delta(1_{F^\times}, a) \rtimes 1_{S_0} \text{ is irreducible for all even } a;$$

$$\delta(1_{F^\times}, a) \rtimes 1_{S_0} \text{ is reducible for all odd } a, \textbf{ except for } a = 1.$$

Now we shall explain what happens in general regarding such reducibility.

Fix selfdual $\rho \in \mathcal{C}$. Then for exactly one parity in $\mathbb{N}$ holds:

(1) $\delta(\rho, a) \rtimes \pi$ is reducible for all a from that parity, with possibly finitely many exceptions;
(2) $\delta(\rho, a) \rtimes \pi$ is irreducible for all a from the other parity.

The parity of $\mathbb{N}$ for which (1) holds, will be called the **parity of reducibility of ρ and π** (note that for this parity we can have finitely many exceptions of reducibility), and the other parity will be called the **parity of irreducibility of ρ and π** (in this parity we have always irreducibility).

Therefore, for understanding tempered representations we need to know which is the parity of reducibility for selfdual $\rho \in \mathcal{C}$, and what are exceptions (if there are exceptions, then clearly they determine the parity of reducibility). Therefore, it is very important to know these exceptions. This is just $Jord(\pi)$:

New definition: $Jord(\pi)$ is the set of all exceptions (ρ, a) in (1), when ρ runs over all selfdual representations in $\mathcal{C}$ ($a \in \mathbb{N}$).

Suppose that π is an irreducible square integrable representation of S_n. C. Mœglin has proved that

$$\sum_{(\rho,a)\in Jord(\pi)} a\, d_\rho \le \begin{cases} 2n & \text{if } S_n = SO(2n+1), \\ 2n+1 & \text{if } S_n = Sp(2n), \end{cases}$$

where d_ρ is determined by the fact that ρ is a representation of $GL(d_\rho, F)$. The above inequality clearly implies that $Jord(\pi)$ is finite.

The above inequality is expected to turn to be an equality.

13.6. Partial cuspidal support of π: In general, (conjugacy class of) an irreducible cuspidal representation τ of a Levi factor M of a parabolic subgroup P in a reductive group G is called **cuspidal support** of π, if π is a subquotient of $\mathrm{Ind}_P^G(\tau)$.

For classical groups, Levi factor M is a direct product of general linear groups and a classical group. Therefore,

$$\tau \cong \rho_1 \otimes \rho_2 \otimes \cdots \otimes \rho_l \otimes \sigma,$$

where $\rho_i \in \mathcal{C}$ and σ is an irreducible cuspidal representation of a classical group.

Now the definition of partial cuspidal support of π is

$$\pi_{cusp} = \sigma.$$

We can define partial cuspidal support of π also in the following way: an irreducible cuspidal representation σ of a classical groups is called partial cuspidal support of π (and denoted by π_{cusp}) if there exists a smooth representation π' of a general linear group, such that

$$\pi \hookrightarrow \pi' \rtimes \sigma.$$

13.7. Partially defined function ϵ_π on $Jord(\pi)$: As we could see already from the simple Example 12.2, the construction of irreducible square integrable representations of classical groups involve reducibility of tempered induction, and thus R-groups (which for classical groups are sums of $\mathbb{Z}/2\mathbb{Z}$). This is roughly behind the fact that parameters of irreducible square integrable representations of classical groups will involve functions with values in $\{\pm 1\}$.

The definition of the domain of partially defined functions ϵ_π on $Jord(\pi)$ is quite technical. Because of this, we shall not give a complete definition of the partially defined functions (besides, from the general definition of partially defined functions, it is not quite easy to understand what are these functions). Rather, we shall try only to explain the main properties of these functions. One can understand pretty well the classification of irreducible square integrable representation of classical groups without knowing all the details of the definition of partially defined functions ϵ_π. Later, we shall give a constructive definition of these functions.

Let X be a free $\mathbb{Z}/2\mathbb{Z}$-module with basis $Jord(\pi)$. We shall denote operation in the module X multiplicatively. Characters of this group are in a natural bijection with functions

$$Jord(\pi) \to \{\pm 1\}.$$

We can think of ϵ_π as a function on a subset of X, in our case on a subset of

$$Jord(\pi) \cup \{x_1x_2 \ ; \ x_1, x_2 \in Jord(\pi), x_1 \neq x_2\},$$

which can be extended to a character of X.

Further, for $x_1, x_2 \in Jord(\pi)$ we shall write $\epsilon(x_1x_2)$, when it is defined, also as

$$\epsilon_\pi(x_1)\epsilon_\pi(x_2)$$

even if $\epsilon_\pi(x_1)$ and $\epsilon_\pi(x_2)$ are not defined.

The fact $\epsilon(x_1x_2) = 1$ (resp. $\epsilon(x_1x_2) = -1$) will be written also as

$$\epsilon_\pi(x_1) = \epsilon_\pi(x_2) \quad (\text{resp. } \epsilon_\pi(x_1) \neq \epsilon_\pi(x_2))$$

even if $\epsilon_\pi(x_1)$ and $\epsilon_\pi(x_2)$ are not defined.

13.8. To give an idea of definition of ϵ_π, we shall describe one important case. The function ϵ_π is always defined on $(\rho, a)(\rho', a')$ if $\rho = \rho'$, $a \neq a'$ (and both $(\rho, a), (\rho', a')$ are from $Jord(\pi)$).

Suppose $(\rho, a_-), (\rho, a)$ are in $Jord(\pi)$, $a_- < a$, and

$$(\rho, a') \notin Jord(\pi) \text{ for any } a_- < a' < a.$$

Then $\epsilon_\pi((\rho, a_-)(\rho, a)) = \epsilon_\pi(\rho, a_-)\epsilon_\pi(\rho, a)$ is defined and

$$\epsilon_\pi(\rho, a_-)\epsilon_\pi(\rho, a) = 1$$

if and only if there exists a smooth representation π' of a classical group such that

$$\pi \hookrightarrow \delta([\nu^{(a_-+1)/2}\rho, \nu^{(a-1)/2}\rho]) \rtimes \pi'.$$

13.9. In general, $\epsilon_\pi(\rho, a)$ is not always defined for $(\rho, a) \in Jord(\pi)$.

It is always defined if a is even. The definition in this case is the following:

Suppose $(\rho, a') \in Jord(\pi)$ with a' even. Chose a minimal a such that $(\rho, a) \in Jord(\pi)$. Then

$$\epsilon_\pi(\rho, a) = 1$$

if and only if there exists a smooth representation π' of a classical group such that

$$\pi \hookrightarrow \delta([\nu^{1/2}\rho, \nu^{(a-1)/2}\rho]) \rtimes \pi'. \tag{14}$$

If a is odd, then $\epsilon_\pi(\rho, a)$ is not always defined. It is not defined if and only if

$$(\rho, b) \in Jord(\pi_{cusp})$$

for some $b \in \mathbb{N}$.

From the above condition for $\epsilon_\pi(\rho, a)$ to be defined in the case of odd a, one can show that if $\epsilon_\pi(\rho, a)$ is defined (for odd a), then

$$\rho \rtimes \pi_{cusp} \tag{15}$$

reduces (this is related to basic assumption under which we consider the classification of irreducible square integrable representations of classical groups; this assumption will be explained later). C. Mœglin has used normalized intertwining operators to define $\epsilon_\pi(\rho, a)$ in this situation.

In the case when $\epsilon_\pi(\rho, a)$ is defined for odd a, as we have mentioned already $\rho \rtimes \pi_{cusp}$ reduces. It reduces into a sum of two inequivalent irreducible subrepresentations. One can chose one of these subrepresentations and attach to it 1, and to the other attach -1. Then one needs to extend in a natural way this choice to other tempered representations coming from inducing representations including $\delta(\rho, a)$ as a factor. One can do this using intertwining operators, but one can also do it without them.

Now we have almost complete definition of ϵ_π.

13.10. C. Mœglin has shown that for an irreducible square integrable representation π of a classical group, the triple

$$(Jord(\pi), \pi_{cusp}, \epsilon_\pi).$$

satisfies some technical conditions. The triples that satisfy these technical conditions she called **admissible triples**. We shall not give in the moment this technical definition. We shall give later a different and more explicit description of admissible triples. Let us just say that admissible triples are combinatorial objects modulo cuspidal data. It will become soon clear what we mean by cuspidal data.

C. Mœglin has proved that the mapping attaching an admissible triple to an irreducible square integrable representation of a classical group, is an injective map from the set of all the equivalence classes of irreducible square integrable representations of classical groups (we fix a series of classical groups and a non-archimedean field F) into the set of all the admissible triples. Jointly, we have proved that this mapping is surjective. This means that we have a bijection between irreducible square integrable representations of classical groups and admissible triples. Since admissible triples are combinatorial objects modulo cuspidal data (as we mentioned already above), we have a classification of irreducible square integrable representations of classical groups modulo cuspidal data.

14. Reduction to cuspidal lines

The classification of irreducible square integrable representations of classical groups modulo cuspidal data will be easier to understand if we pass to cuspidal lines. We shall explain it in this section. This reduction could be important also for some other purposes.

14.1. Fix an irreducible cuspidal representation σ of a classical group S_q and fix inequivalent selfdual irreducible cuspidal representations $\rho_1, \ldots, \rho_k$ of general linear groups. Denote by

$$\mathcal{D}(\rho_1, \ldots, \rho_k; \sigma)$$

the set of all equivalence classes of irreducible square integrable subquotients of representations

$$\nu^{\alpha_1}\tau_1 \times \nu^{\alpha_2}\tau_2 \times \cdots \times \nu^{\alpha_\ell}\tau_\ell \rtimes \sigma,$$

where $\alpha_i \in \mathbb{R}, \tau_i \in \{\rho_1, \ldots, \rho_k\}$. Then there is a natural bijection from $\mathcal{D}(\rho_1, \ldots, \rho_k; \sigma)$ into the Cartesian product

$$\mathcal{D}(\rho_1, \ldots, \rho_k; \sigma) \to \left(\prod_{i=1}^{k} \mathcal{D}(\rho_i; \sigma)\right).$$

This bijection is given in the following way.

Fix $\pi \in \mathcal{D}(\rho_1, \ldots, \rho_k; \sigma)$ and $1 \leq j \leq k$. Then there exists an irreducible representation π_j of some S_{n_j} which is a subquotient of some $\nu^{\beta_1}\rho_j \times \nu^{\beta_2}\rho_j \times \ldots \nu^{\beta_{\ell_j}}\rho_j \rtimes \sigma$, $\beta_i \in \mathbb{R}$, and there exists an irreducible representation θ_j of a general linear group which is a subquotient of $\tau_1 \times \tau_2 \times \cdots \times \tau_{m_j}$ with $\tau_i \in \cup_{i=1, i\neq j}^{k} \{\nu^{\alpha}\rho_i; \alpha \in \mathbb{R}\}$, such that

$$\pi \hookrightarrow \theta_j \rtimes \pi_j.$$

C. Jantzen has proved in [32] that representations $\pi_1, \ldots, \pi_k$ are uniquely determined by π, and that they are square integrable. Further $\pi \mapsto (\pi_1, \ldots, \pi_k)$ defines a bijection from $\mathcal{D}(\rho_1, \ldots, \rho_k; \sigma)$ onto $\prod_{i=1}^{k} \mathcal{D}(\rho_i; \sigma)$.

Each irreducible square integrable representation of a classical group belongs to some $\mathcal{D}(\rho_1, \ldots, \rho_k; \sigma)$. Further, if $\sigma \not\cong \sigma'$, then

$$\mathcal{D}(\rho_1, \ldots, \rho_k; \sigma) \cap \mathcal{D}(\rho_1', \ldots, \rho_{k'}'; \sigma') = \emptyset,$$

and if

$$\{\rho_1, \dots, \rho_k\} \cap \{\rho'_1, \dots, \rho'_{k'}\} = \{\rho''_1, \dots, \rho''_{k''}\}$$

then

$$\mathcal{D}(\rho_1, \dots, \rho_k; \sigma) \cap \mathcal{D}(\rho'_1, \dots, \rho'_{k'}; \sigma) = \mathcal{D}(\rho''_1, \dots, \rho''_{k''}; \sigma)$$

In this way we obtain a reduction of the classification of irreducible square integrable representations of classical groups, to the problem of classification of sets

$$\mathcal{D}(\rho; \sigma)$$

($\rho \in \mathcal{C}$ is selfdual).

14.2. Consider the projection

$$\mathcal{O}_F \to \mathcal{O}_F/\mathfrak{p}_F.$$

Lift it to the level of groups. In this way one gets a natural homomorphism from the maximal compact subgroup K_0 in S_n to the group S_n over the field $\mathcal{O}_F/\mathfrak{p}_F$. The preimage in K_0 of the standard minimal parabolic subgroup in S_n over the field $\mathcal{O}_F/\mathfrak{p}_F$ is denoted by

$$I.$$

This open compact subgroup is called **Iwahori subgroup** of S_n.

The study of irreducible smooth representations with Iwahori fixed vectors has attracted a lot of attention. In this case, for building the representation theory, one does not need non-trivial cuspidal representations (i.e. other than characters). Also, corresponding group algebras for this setting have a nice geometric realization. Using this, one can obtain construction of irreducible representations by geometric methods, what was done by D. Kazhdan and G. Lusztig.

For classical groups, to determine irreducible square integrable representations with Iwahori fixed vectors, it is equivalent to determining of

$$\mathcal{D}(1_{F^\times}; 1_{S_0}) \cup \mathcal{D}(\psi; 1_{S_0}),$$

where ψ is a (unique) character of order 2, which is unramified (i.e. which is trivial on $\mathcal{O}_F^\times$). Irreducible square integrable representations with Iwahori fixed vectors are parameterized by the Cartesian product $\mathcal{D}(1_{F^\times}; 1_{S_0}) \times \mathcal{D}(\psi; 1_{S_0})$.

15. Parameters of $\mathcal{D}(\rho;\sigma)$

15.1. For selfdual $\rho \in \mathcal{C}$ and an irreducible cuspidal representation σ of S_q, A. Silberger has proved there exists a unique

$$\alpha_{\rho,\sigma} \geq 0$$

such that

$$\nu^{\alpha_{\rho,\sigma}}\rho \rtimes \sigma$$

reduces.

Now we shall say what is the basic assumption (for ρ and σ), under which $\mathcal{D}(\rho,\sigma)$ is classified:

(BA) for ρ and σ $$\alpha_{\rho,\sigma} - \alpha_{\rho,1_{S_0}} \in \mathbb{Z}.$$

This assumption is needed (essentially only) in proofs.

F. Shahidi has proved that (BA) holds if σ is generic. It is also known that (BA) holds in some other cases. In general, (BA) would follow from the truth of some general Arthur's conjectures.

F. Shahidi has proved that

$$\alpha_{\rho,1_{S_0}} \in (1/2)\mathbb{Z}. \tag{16}$$

15.2. We shall fix ρ and σ as above, and assume that (BA) holds for ρ and σ. Denote in the sequel

$$\alpha = \alpha_{\rho,\sigma}.$$

Now since we assume that (BA) holds for ρ and σ, (16) implies

$$\alpha = \alpha_{\rho,\sigma} \in (1/2)\mathbb{Z}_+.$$

Note that for $\pi \in \mathcal{D}(\rho,\sigma)$,

$$\pi_{cusp} = \sigma.$$

Therefore, since σ is fixed, for classification of $\mathcal{D}(\rho,\sigma)$ it is enough to consider instead of triples

$$(Jord(\pi), \pi_{cusp}, \epsilon_\pi)$$

pairs

$$(Jord(\pi), \epsilon_\pi)$$

(which form with σ admissible triples).

Further, for classification of $\mathcal{D}(\rho, \sigma)$ it is convenient (and enough) to work with

$$Jord_\rho(\pi) = \{a \in \mathbb{N}; (\rho, a) \in Jord(\pi)\}.$$

instead of $Jord(\pi)$. Now pairs

$$(Jord_\rho(\pi), \epsilon_\pi)$$

will be parameters of representation in $\mathcal{D}(\rho, \sigma)$. Note that $Jord_\rho(\pi)$ is a finite subset of either $2\mathbb{N}$ or $2\mathbb{N} - 1$ and ϵ_π is now regarded as a partially defined function on $Jord_\rho(\pi)$. Because of this, the parameters of $\mathcal{D}(\rho, \sigma)$ are now simpler than before.

There are two possibilities for α. The first is

$$\alpha \in \mathbb{Z}_+,$$

which will be called **integral case**, and the second is

$$\alpha \in ((1/2)\mathbb{Z}_+ \setminus \mathbb{Z}_+),$$

which will be called **non-integral case**.

16. Integral case

We shall suppose in this section that

$$\alpha \in \mathbb{Z}_+,$$

and describe parameters $(Jord_\rho, \epsilon)$ of elements of $\mathcal{D}(\rho, \sigma)$ in this case.

16.1. In the integral case we have always

$$Jord_\rho \subseteq 2\mathbb{N} - 1.$$

Here partially defined function is defined on elements of $Jord_\rho$ if and only if $\alpha = 0$. If it is defined on $Jord_\rho$, then the values on $Jord_\rho$ completely determine the partially defined function.

If $\alpha \geq 1$, then ϵ is defined only on pairs from $Jord_\rho$ (and this partially defined function can be extended to a character of a free $\mathbb{Z}/2\mathbb{Z}$-module with basis $Jord_\rho$).

16.2. In the integral case, $Jord_\rho$ will be called of **alternated type** if

$$\text{card}(Jord_\rho) = \alpha.$$

Here always exists a unique partially defined function ϵ such that $Jord_\rho$, ϵ and σ form an admissible triple. If $\alpha = 0$, then there is nothing to define. If $\alpha \geq 1$, then ϵ is not defined on elements on $Jord_\rho$, but it is defined on pairs. It is completely defined by the following property:

For each $a_-, a \in Jord_\rho, a_- < a$, such that

$$[a_-, a] \cap Jord_\rho = \{a_-, a\}$$

we have

$$\epsilon(a_-) \neq \epsilon(a)$$

(this is where the name alternated type comes from).

Now we shall define the representation corresponding to alternated $Jord_\rho$ (and ϵ). Write $Jord_\rho = \{a_1, a_2, \ldots, a_\alpha\}$. After a renumeration we can assume

$$a_1 < a_2 < \cdots < a_\alpha.$$

Now the representation

$$\left(\prod_{i=1}^{\alpha} \delta([\nu^i \rho, \nu^{(a_i-1)/2}\rho])\right) \rtimes \sigma$$

has a unique irreducible subrepresentation, which will be denoted by

$$\pi_{(Jord_\rho, \sigma, \epsilon)}.$$

This representation is square integrable.

An example of such representations are Steinberg representations for symplectic groups.

16.3. We shall describe now general parameters of elements of $\mathcal{D}(\rho; \sigma)$. Take $Jord_\rho$ (and ϵ) of alternated type. Take any $a_-, a \in 2\mathbb{N} - 1$, $a_- < a$, such that

$$[a_-, a] \cap Jord_\rho = \emptyset.$$

Denote

$$Jord_\rho^{(1)} = Jord_\rho \cup \{a_-, a\}.$$

Extend ϵ in a way that

$$\epsilon(a_-) = \epsilon(a).$$

It is easy to see that there are exactly two such extensions. Denote them by ϵ_1, ϵ_2.

Now $(Jord_\rho^{(1)}, \sigma, \epsilon_i)$, $i = 1, 2$, are (new) admissible triples (in this setting). These triples are no more of alternated type. We can continue this construction, but now starting from $Jord_\rho^{(1)}, \sigma, \epsilon_i$.

Continuing this process, we construct $Jord_\rho^{(2)}$, $Jord_\rho^{(3)}$, ... (and corresponding partially defined functions). In this way, we shall get all the parameters of $\mathcal{D}(\rho; \sigma)$.

16.4. We shall describe now representations corresponding to these (new) parameters. Let $(Jord_\rho, \epsilon)$, a_-, a and $(Jord_\rho^{(1)}, \sigma, \epsilon_i)$ be as in 16.3. To alternated $(Jord_\rho, \epsilon)$ we have already attached in 16.2 a square integrable representation

$$\pi_{(Jord_\rho, \sigma, \epsilon)}.$$

Now the representation

$$\delta([\nu^{-(a_- -1)/2}\rho, \nu^{(a-1)/2}\rho]) \rtimes \pi_{(Jord_\rho, \sigma, \epsilon)}$$

contains exactly two irreducible subrepresentations. One shows this using the strategy that we have used in Example 12.2. These irreducible subrepresentations are square integrable, and their parameters are $(Jord_\rho^{(1)}, \sigma, \epsilon_1)$ and $(Jord_\rho^{(1)}, \sigma, \epsilon_2)$.

If $Jord_\rho \neq \emptyset$, one determines from 13.8 which subrepresentation corresponds to which ϵ_i.

16.5. It remains to say which subrepresentation to attach to which ϵ_i if $Jord_\rho = \emptyset$. C. Mœglin has used normalized intertwining operators for this attaching.

One possibility would be to proceed in the following way. Suppose

$$Jord_\rho = \emptyset.$$

Then $\alpha = 0$. Write

$$\rho \rtimes \sigma = \tau_1 \oplus \tau_{-1}$$

as a sum of (two inequivalent) irreducible representations. The representations

$$\delta([\nu^{-(a_- -1)/2}\rho, \nu^{(a_- -1)/2}\rho]) \rtimes \sigma$$

and

$$\delta([\nu\rho, \nu^{(a_- -1)/2}\rho]) \times \delta([\nu\rho, \nu^{(a_- -1)/2}\rho]) \rtimes \tau_i$$

have exactly one irreducible subquotient in common (for each $i = -1, 1$). Denote it by T_i. Now

$$\delta([\nu^{(a_-+1)/2}\rho, \nu^{(a-1)/2}\rho]) \rtimes T_i$$

contains a unique irreducible subrepresentation. Denote it by π_i. Then π_1 and π_{-1} are two inequivalent irreducible subrepresentations of $\delta([\nu^{-(a_--1)/2}\rho, \nu^{(a-1)/2}\rho]) \rtimes \sigma$ (here $\pi_{(Jord_\rho,\sigma,\epsilon)} = \sigma$).

One natural possibility to distinguish irreducible subrepresentations of

$$\delta([\nu^{-(a_--1)/2}\rho, \nu^{(a-1)/2}\rho]) \rtimes \sigma$$

is to attach ϵ_i to $\pi_{\epsilon_i(a)}$ (for $i = -1, 1$). Let us note that we have not checked that this choice is the same as the choice that C. Mœglin made using normalized intertwining operators.

16.6. One proceeds further from $Jord_\rho^{(1)}$ to $Jord_\rho^{(2)}$, $Jord_\rho^{(2)}$ to $Jord_\rho^{(3)}$, ...recursively in the same way as we did in passing from $Jord_\rho$ to $Jord_\rho^{(1)}$ (it is even less complicated here, since always $Jord_\rho^{(1)} \neq \emptyset, Jord_\rho^{(2)} \neq \emptyset, \ldots$ and therefore we do not need to make choices as in 16.5).

17. Non-integral case

Now we shall assume that

$$\alpha \in ((1/2)\mathbb{Z}_+ \setminus \mathbb{Z}_+).$$

17.1. In this case.

$$Jord_\rho \subseteq 2\mathbb{N}.$$

Partially defined functions are defined on elements of $Jord_\rho$'s (and values on $Jord_\rho$'s completely determine partially defined functions).

17.2. In the non-integral case, $Jord_\rho$ will be called of **alternated type** if $\text{card}(Jord_\rho) = \alpha \pm 1/2$, i.e. if

$$\text{card}(Jord_\rho) = \alpha - 1/2 \quad \text{or} \quad \text{card}(Jord_\rho) = \alpha + 1/2.$$

In this case, there exists a unique partially defined function ϵ on $Jord_\rho$ such that $Jord_\rho$, σ and ϵ form an admissible triple. This partially defined function ϵ is defined (and uniquely determined) by the following conditions:

For each $a_-, a \in Jord_\rho, a_- < a$, such that

$$[a_-, a] \cap Jord_\rho = \{a_-, a\}$$

we have

$$\epsilon(a_-) \neq \epsilon(a),$$

and

$$\epsilon(\min(Jord_\rho)) = \begin{cases} 1 & \text{if } \operatorname{card}(Jord_\rho) = \alpha + 1/2; \\ -1 & \text{if } \operatorname{card}(Jord_\rho) = \alpha - 1/2. \end{cases}$$

17.3. We shall now describe the representation corresponding to the above alternated $Jord_\rho$. Let $Jord_\rho = \{a_1, a_2, \ldots, a_{\alpha\pm1/2}\}$. After a renumeration we can assume

$$a_1 < a_2 < \cdots < a_{\alpha\pm1/2}.$$

Consider first the case

$$\operatorname{card}(Jord_\rho) = \alpha - 1/2.$$

Then the representation

$$\left(\prod_{i=1}^{\alpha-1/2} \delta([\nu^{i+1/2}\rho, \nu^{(a_i-1)/2}\rho]) \right) \rtimes \sigma$$

has a unique irreducible subrepresentation. This subrepresentation will be denoted by

$$\pi_{(Jord_\rho,\sigma,\epsilon)}.$$

It is a square integrable representation which corresponds to $(Jord_\rho, \epsilon)$.

Now assume

$$\operatorname{card}(Jord_\rho) = \alpha + 1/2.$$

Then the representation

$$\left(\prod_{i=1}^{\alpha+1/2} \delta([\nu^{i-1/2}\rho, \nu^{(a_i-1)/2}\rho]) \right) \rtimes \sigma$$

has a unique irreducible subrepresentation. We again denote this subrepresentation by

$$\pi_{(Jord_\rho,\sigma,\epsilon)}.$$

This is a square integrable representation which corresponds to $(Jord_\rho, \epsilon)$.

We have described above how one attaches square integrable representations to alternated parameters.

17.4. Now we define general parameters of $\mathcal{D}(\rho;\sigma)$ in the same way as in the integral case. We also attach square integrable representations in the same way.

The only difference which occurs between integral and non-integral case is in passing from $Jord_\rho$ to $Jord_\rho^{(1)}$ when $Jord_\rho = \emptyset$. In the non-integral case one uses (14) to determine which irreducible subrepresentation corresponds to which ϵ_i.

18. Local Langlands correspondences

18.1. Denote by W_F the Weil group of F. This is a dense subgroup of the Galois group of the separable algebraic closure of F over F (the topology is not the induced one from the Galois group, but a slightly modified one).

Let G be a split connected reductive group over F (as $GL(n,F), Sp(2n,F)$ or $SO(2n+1,F)$). By the Langlangs program, there should exist a natural partition of $\tilde{G}$ into finite subsets, called L-packets, which are indexed by (conjugacy classes of) admissible homomorphism of $W_F \times SL(2,\mathbb{C})$ into the complex dual group ${}^LG^0$ of G (admissible here means that the homomorphisms are continuous, that they carry W_F into semi simple elements and that they are algebraic on $SL(2,\mathbb{C})$).

For the cases of the groups that we consider, complex dual groups are as follows:

$$\begin{aligned} {}^LGL(n,F)^0 &= GL(n,\mathbb{C}), \\ {}^LSp(2n,F)^0 &= SO(2n+1,\mathbb{C}), \\ {}^LSO(2n+1,F)^0 &= Sp(2n,\mathbb{C}) \end{aligned}$$

(a property of the complex dual group ${}^LG^0$ is that it has the root system dual to the root system of G).

We shall now concentrate our attention regarding the above aspect of the Langlands program to irreducible square integrable representations. In this case square integrable L-packets should be parameterized by admissible homomorphisms whose image is not contained in any proper Levi factor. Further, elements of a square integrable L-packet, which is indexed by an admissible homomorphism ψ, should be parameterized by irreducible representations of the component group of ψ (which is the quotient of the centralizer of the image of ψ by the connected component).

The correspondence that one would get in this way is called local Langlands correspondence for G.

18.2. In the case of general linear groups, by the Langlands program there should be a bijection of irreducible square integrable representations of $GL(n,F)$ and n-dimensional irreducible representations of $W_F \times SL(2,\mathbb{C})$ (which are admissible homomorphisms). Here component groups are trivial. In this bijection irreducible cuspidal representations of $GL(n,F)$ should correspond to irreducible representations of W_F

The work of Bernstein and Zelevinsky, which gave classification of irreducible square integrable representations of general linear groups modulo cuspidal representations, resulted in a reduction of establishing of local Langlands correspondence to the cuspidal case, i.e. to establishing a correspondence between (classes of) irreducible cuspidal representations of $GL(n,F)$ and (classes of) irreducible representations of W_F. More precisely, suppose that

$$\varphi$$

is such a correspondence for general linear groups between irreducible cuspidal representations of general linear groups and irreducible representations of W_F (we consider all general linear groups over F together). From the representation theory of $SL(2,\mathbb{C})$ one knows that for each $a \in \mathbb{N}$ there exists a unique irreducible algebraic representation

$$E_a$$

of $SL(2,\mathbb{C})$ on a-dimensional complex vector space (up to an equivalence). Then the formula for local Langlands correspondence on the set of (classes of) irreducible square integrable representations of general linear groups, which we shall denote also by φ, would be

$$\varphi(\delta(\rho,a)) = \varphi(\rho) \otimes E_a.$$

Local Langlands conjecture for $GL(n,F)$ has been recently proved in full generality (by M. Harris and R. Taylor in [25], and by G. Henniart in [26]).

18.3. One may ask does the classification of irreducible square integrable representations of classical groups modulo cuspidal data give also a similar reduction.

The natural candidate for the Langlands correspondence Φ for classical groups is

$$\Phi(\pi) = \oplus_{(\rho,a)\in Jord(\pi)} \varphi(\rho) \otimes E_a$$

(φ is the local Langlands correspondence for general linear groups, which we have considered before). But it remains a number of facts to prove even to see that it is a good candidate (for the beginning, it is not clear at all that $\Phi(\pi)$ goes in the right group).

18.4. For classical groups, the centalizers of images of admissible homomorphism of $W_F \times SL(2,\mathbb{C})$ into the complex dual group, whose images are not contained in any proper Levi factor, are finite groups which are $\mathbb{Z}/2\mathbb{Z}$-modules (these are component groups). Therefore, after choosing a basis, irreducible representations of the component group correspond to functions from the basis to $\{\pm 1\}$.

Now ϵ_π should give a part of the irreducible representation (i.e. character) of the component group corresponding to π. The rest should come from $\epsilon_{\pi_{cusp}}$ (once the local Langlands correspondence is established for cuspidal representations of classical groups).

Complete discussion regarding this reduction one can find in [41].

19. Non-unitary duals of classical p-adic groups

19.1. The classification of irreducible square integrable representations of classical groups modulo cuspidal data implies also a classification of all the irreducible smooth representations of classical groups modulo cuspidal data (by cuspidal data we mean irreducible cuspidal representations of general linear and classical groups, and cuspidal reducibilities).

Suppose that a selfdual $\rho \in \mathcal{C}$ and irreducible square integrable representation π of a classical group are given. For understanding tempered representations, one needs to know the parity of reducibility.

If $Jord_\rho(\pi) \neq \emptyset$, then the parity which shows up in $Jord_\rho(\pi)$ is the parity of reducibility of ρ and π.

If $Jord_\rho(\pi) = \emptyset$, then $Jord_\rho(\pi_{cusp}) = \emptyset$ and then the reducibility of ρ and π_{cusp} (and also π) is at 0 or 1/2. If the reducibility is at 0 (resp 1/2), then the parity of reducibility of ρ and π is odd (resp. even).

19.2. We can describe also the non-unitary dual by reduction to cuspidal lines.

Let σ be an irreducible cuspidal representation of a classical group S_q and let $\rho_1, \dots, \rho_k \in \mathcal{C}$ be unitarizable such that for $i \neq j$, sets $\{\rho_i, \tilde{\rho}_i\}$ and $\{\rho_j, \tilde{\rho}_j\}$ have no equivalent representations (i.e. $\rho_i \not\cong \rho_j$ and $\rho_i \not\cong \tilde{\rho}_j$).

Denote by

$$\mathcal{I}(\rho_1, \ldots, \rho_k; \sigma)$$

the set of all equivalence classes of irreducible subquotients of

$$\nu^{\alpha_1}\tau_1 \times \nu^{\alpha_2}\tau_2 \times \cdots \times \nu^{\alpha_\ell}\tau_\ell \rtimes \sigma,$$

where

$$\alpha_i \in \mathbb{R}, \tau_i \in \{\rho_1, \ldots, \rho_k, \tilde{\rho}_1, \ldots, \tilde{\rho}_k\}.$$

Then by [32] there exists a bijection

$$\mathcal{I}(\rho_1, \ldots, \rho_k; \sigma) \to \left(\prod_{i=1}^{k} \mathcal{I}(\rho_i; \sigma)\right),$$

$$\pi \to (\pi_1, \ldots, \pi_k),$$

similarly as in 14.1 (for complete definition of the bijection one should consult [32]). The classification of the non-unitary duals of classical groups reduces to the classification of the sets $\mathcal{I}(\rho_i; \sigma)$ in a similar way as the classification of the irreducible square integrable representations in 14.1 reduces to cuspidal lines (we shall not write details here, but the reduction is analogous).

19.3. Fix a unitarizable $\rho \in \mathcal{C}$ and fix an irreducible cuspidal representation σ of a classical group.

If ρ is not selfdual, then the tempered induction in $\mathcal{I}(\rho; \sigma)$ is always irreducible. Now irreducible tempered representations which show up as Langlands parameters of representations in $\mathcal{I}(\rho; \sigma)$ is easy to write down (using Remark 13.3 to know equivalences among them).

Suppose now that ρ is selfdual. Let the reducibility of ρ and σ be $\alpha = \alpha_{\rho,\sigma} \geq 0$. Now the parity of reducibility of ρ and σ (and also each square integrable representation in $\mathcal{I}(\rho; \sigma)$) is odd (resp. even) if $\alpha \in \mathbb{Z}$ (resp. $\alpha \notin \mathbb{Z}$). Further, one can describe easily irreducible tempered representations which show up as Langlands parameters of representations in $\mathcal{I}(\rho; \sigma)$, since we know the parity of reducibility of ρ and σ (one needs also to use Remark 13.3).

20. Unitary duals of general linear groups over local fields

20.1. Denote

$$D^u = \{\delta \in D; e(\delta) = 0\}.$$

These are just all the square integrable classes in D.

The following theorem describes the unitary duals of general linear groups over any local field (archimedean or non-archimedean).

Theorem: *For a representation $\delta \in D^u$ and $m \geq 1$ denote*

$$u(\delta, m) = L(\nu^{(m-1)/2}\delta, \nu^{(m-3)/2}\delta, \ldots, \nu^{-(m-1)/2}\delta)$$

For $0 < \alpha < 1/2$ and δ and m as above, denote

$$\pi(u(\delta, m), \alpha) = \nu^{\alpha} u(\delta, m) \times \nu^{-\alpha} u(\delta, m).$$

Let B be the set of all possible $u(\delta, m)$ and $\pi(u(\delta, m), \alpha)$ with δ, m, α as above. Then

(i) *If $\tau_1, \tau_2, \ldots, \tau_n \in B$, then the representation*

$$\tau_1 \times \tau_2 \times \cdots \times \tau_n$$

is an irreducible unitarizable representation of a general linear group.

(ii) *Let $\tau_1, \tau_2, \ldots, \tau_n, \tau_1', \tau_2', \ldots, \tau_{n'}' \in B$, Then*

$$\tau_1 \times \tau_2 \times \cdots \times \tau_n \cong \tau_1' \times \tau_2' \times \cdots \times \tau_{n'}'$$

if and only if $n = n'$ and if one can obtain the sequence $(\tau_1, \tau_2, \ldots, \tau_n)$ from $(\tau_1', \tau_2', \ldots, \tau_n')$ by a permutation.

(iii) *Each irreducible unitary representation of a general linear group is isomorphic to a representation*

$$\tau_1 \times \tau_2 \times \cdots \times \tau_n,$$

for some $\tau_1, \tau_2, \ldots, \tau_n \in B$.

The above classification theorem is the same for all the local fields. The difference in the form of unitary duals comes from the difference of the sets D^u for different fields.

20.2. This theorem is proved in [60] in the non-archimedean case. As we already mentioned, the theorem holds for archimedean fields in the same form (using the notion of $(\mathfrak{g}, K)$ modules), with the proof along the same strategy as in the non-archimedean case (see [71]). D. Vogan in [72] has made quite different approach to the classification of unitary duals of general linear groups over archimedean fields.

Not to deal all the time with non-archimedean fields, we shall now describe the proof of the above theorem for $F = \mathbb{C}$. Since the proofs in the archimedean and non-archimedean case are along the same strategy, one will be able to get from this description quite good idea of the proof in the non-archimedean case.

20.3. In the sequel of this section, by a representation we shall mean corresponding $(\mathfrak{g}, K_0)$-module. Denote by

$$Irr^u = \cup_{n=0}^{\infty} GL(n, \mathbb{C})\hat{}.$$

Consider algebra R for complex general linear groups (constructed in 10.3). We shall consider

$$Irr^u \subseteq R.$$

Recall that R is a polynomial ring over D (Proposition 10.7). In particular, R is a factorial ring. Therefore, we can talk about prime elements in R.

In the complex case we have

$$D = GL(1, \mathbb{C})\tilde{}.$$

Note also that $| \ |_{\mathbb{C}}$ is the square of the usual absolute value in $\mathbb{C}$.

We shall now introduce several claims, whose proofs shall be discussed later:

(U0) $\sigma, \tau \in \mathrm{Irr}^u \implies \sigma \times \tau \in \mathrm{Irr}^u$.
(U1) $\delta \in D^u$ and $n \in \mathbb{N} \implies u(\delta, n) \in \mathrm{Irr}^u$.
(U2) $\delta \in D^u$, $n \in \mathbb{N}$ and $0 < \alpha < 1/2 \implies \pi(u(\delta, n), \alpha) \in \mathrm{Irr}^u$.
(U3) $\delta \in D$ and $n \in \mathbb{N} \implies u(\delta, n)$ is prime in R.
(U4) $a, b \in M(D) \implies L(a) \times L(b)$ contains $L(a + b)$ as a subquotient.

The addition of multisets (which shows up in (U4)) is defined in obvious way:

$$(x_1, \ldots, x_u) + (y_1, \ldots, y_v) = (x_1, \ldots, x_u, y_1, \ldots, y_v).$$

Proposition: *Claims (U0) - (U4) imply Theorem 20.1.*

Proof: First observe that (U1), (U2) and (U0) imply (i) of the theorem.

Further, commutativity of R give implication $\Longleftarrow$ in (ii). The implication $\Longrightarrow$ follows from (U3).

It remains to prove (iii) (i.e. exhaustion). Suppose $\pi \in Irr^u$. Then

$$\pi = L(\gamma_1, \gamma_2, \dots, \gamma_\ell)$$

for some $\gamma_1, \gamma_2, \dots, \gamma_\ell \in D$ (see Remark 9.6). Note that π is hermitian (since it is unitary). This, together with (5) and (6), implies that

$$\pi = L(\nu^{\alpha_1}\delta_1, \nu^{-\alpha_1}\delta_1, \dots, \nu^{\alpha_k}\delta_k, \nu^{-\alpha_k}\delta_k, \delta_{k+1}, \dots, \delta_s),$$

for some $\alpha_i > 0$, $1 \le i \le k$ and $\delta_j \in D^u$, $1 \le j \le s$.

We shall use in the sequel the following simple fact

$$\begin{aligned}&\text{If } a_1, a_2, \dots, a_m \in M(D) \text{ and } L(a_1), L(a_2), \dots, L(a_m) \in Irr^u,\\ &\text{then } L(a_1) \times L(a_2) \times \dots \times L(a_m) = L(a_1 + a_2 + \dots + a_m).\end{aligned} \tag{17}$$

This follows directly from (U4) and (U0) (by induction).

To get an idea of proof of (iii), we shall now give a proof of it in the rigid case, i.e. when all $\alpha_i \in (1/2)\mathbb{Z}$.

Denote

$$a(\delta, m) = (\nu^{(m-1)/2}\delta, \nu^{(m-3)/2}\delta, \dots, \nu^{-(m-1)/2}\delta).$$

Then obviously

$$u(\delta, m) = L(a(\delta, m)).$$

Using the fact that

$$(\nu^{\alpha_i}\delta_i, \nu^{-\alpha_i}\delta_i) + a(\delta_i, 2\alpha_i - 1) = a(\delta_i, 2\alpha_i + 1)$$

and (17) (several times), from (U1) we get that

$$\begin{aligned}&\pi \times u(\delta_i, 2\alpha_1 - 1) \times \dots \times u(\delta_k, 2\alpha_k - 1)\\ &\qquad\qquad \cong u(\delta_i, 2\alpha_1 + 1) \times \dots \times u(\delta_k, 2\alpha_k + 1) \times \delta_{k+1} \times \dots \times \delta_s.\end{aligned}$$

By (U3), on the right hand side we have prime elements (from B). Since R is factorial, π must be a subproduct of the right hand side (up to a sign). So, π must be a product of elements from B. This proves (iii) in the rigid case.

The proof of (iii) in the non-rigid case proceeds along a similar idea, but it is slightly technically more complicated in this case. □

By the above proposition, to prove the theorem, it is enough to prove (U0) - (U4). Now we shall explain how the proofs of each of these claims go.

20.4. (U1): Considering the modular function of the standard minimal parabolic subgroup in $GL(m,\mathbb{C})$, we get easily that for $\delta \in D = GL(1,\mathbb{C})\hat{}$ (i.e. a character of $\mathbb{C}^\times$) we have

$$u(\delta, m) = \delta \circ \det : GL(m,\mathbb{C}) \to \mathbb{C}^\times.$$

Thus, $u(\delta, m)$ is unitarizable. Therefore, (U1) holds.

20.5. (U2): The restriction of the representation $\pi(u(\delta, m), \alpha)$ (which we consider in (U2)) to $SL(2m,\mathbb{C})$, is a Stein's complementary series representation from [58] (if $m > 1$; if $m = 1$, then this is a well known complementary series representation of $SL(2,\mathbb{C})$). Therefore, it is unitarizable. From this one gets directly that $\pi(u(\delta, m), \alpha)$ is unitarizable as a representations of $GL(2m,\mathbb{C})$ (since it has unitarizable central character).

One can get the unitarizability of representations $\pi(u(\delta, m), \alpha)$ by standard construction of complementary series representations (which are unitarizable). For this, see 20.11 bellow.

20.6. (U3): We shall illustrate the proof of (U3) on the example of $u(\delta, 2)$.

Note first that R is a graded ring (by definition). The degree of $u(\delta, 2)$ is two.

The representation theory of $SL(2,\mathbb{C})$ implies that

$$u(\delta, 2) = X_1 \times X_2 - X_3 \times X_4 \tag{18}$$

for some $X_i \in D$, $i = 1, 2, 3, 4$, where all X_i are different. Suppose that $u(\delta, 2)$ is not prime. Since it is primitive (the greatest common divisor of coefficients is 1), it must be a product of homogeneous elements f_1 and f_2 of degree one. Write

$$f_i = c_1^{(i)} X_1 + c_2^{(i)} X_2 + c_3^{(i)} X_3 + c_4^{(i)} X_4, \quad i = 1, 2.$$

Since $X_1 \times X_2$ shows up in $u(\delta, 2)$ (see (18)), it follows that $c_1^{(1)} \neq 0$ and $c_2^{(2)} \neq 0$ (after possible changing indexes of f_1 and f_2). Since $X_3 \times X_4$ shows up in $u(\delta, 2)$, it follows that $c_3^{(1)} \neq 0$ and $c_4^{(2)} \neq 0$ (after possible changing indexes of X_3 and X_4). These observations imply that the total degree of $u(\delta, 2)$ in variables X_1 and X_4 is 2. This obviously contradicts to the expression (18). This contradiction completes the proof that $u(\delta, 2)$ is prime (in R).

The proof of (U3) in the general case follows the same strategy and uses only very basic facts about composition series of standard modules, i.e. principal series (which are standard facts of Langlands classification).

20.7. (U4): This claim follows from basic properties of composition series of principal series (which are standard facts of Langlands classification). We shall explain now how it follows.

We shall consider s.s.$(\lambda(a)) \in R$ for $a \in M(D)$. For simplicity, we shall write s.s.$(\lambda(a))$ as an element of R simply as $\lambda(a) \in R$.

Let $a_1, a_2 \in M(D)$. There exists a partial order $\leq$ on $M(D)$ (which is quite explicit and which is simple to describe), such that we have

$$L(a_i) = \lambda(a_i) + \sum_{b^{(i)} < a_i} m^{(i)}_{b^{(i)}} \lambda(b^{(i)}), \quad i = 1, 2,$$

in R (here $m^{(i)}_{b^{(i)}} \in \mathbb{Z}$). Now

$$\begin{aligned} &L(a_1) \times L(a_2) \\ &= \lambda(a_1) \times \lambda(a_2) + \sum_{b^{(1)} < a_1} m^{(1)}_{b^{(1)}} \lambda(b^{(1)}) \times \lambda(a_2) + \sum_{b^{(2)} < a_2} m^{(2)}_{b^{(2)}} \lambda(b^{(2)}) \times \lambda(a_1) \\ &\qquad + \sum_{b^{(1)} < a_1} m^{(1)}_{b^{(1)}} \lambda(b^{(1)}) \times \sum_{b^{(2)} < a_2} m^{(2)}_{b^{(2)}} \lambda(b^{(2)}). \end{aligned}$$

We know that $L(a_1 + a_2)$ is a subquotient of $\lambda(a_1) \times \lambda(a_2)$. Standard properties of the Langlands classification imply that $L(a_1 + a_2)$ is not a subquotient of any of three sums on the right hand side of the above equality. This proves (U4).

20.8. (U0): Let P_n be the subgroup of all the matrices in $GL(n, \mathbb{C})$ which have bottom row equal to $(0, 0, \ldots, 0, 1) \in \mathbb{C}^n$. Suppose that we know that

(K) $$\pi \in GL(n, \mathbb{C})\hat{} \implies \pi | P_n \text{ is irreducible}$$

(in the above claim, we consider π as an irreducible unitary representation on a Hilbert space, not as a $(\mathfrak{g}, K_0)$-module).

It has been known for a long time that (K) implies (U0). The implication follows using small Mackey theory. For the implication, for irreducible unitary representations π_1 and π_2 of $GL(n_1, \mathbb{C})$ and $GL(n_2, \mathbb{C})$, using (K) one shows that $(\pi_1 \times \pi_2)|P_{n_1+n_2}$ is an induced unitary representation, which is irreducible by small Mackey theory. A complete proof of this implication can be found in [49] (but the proof is implicit already in [22]).

M. Baruch proved (K) in [7].

Additional comments

20.9. We shall give here a little bit more explanations regarding Baruch's proof of (K) and the history of proving of (K).

A.A. Kirillov observed in [35] that on a dense open subset of $GL(n,\mathbb{C})$, in each $GL(n,\mathbb{C})$-conjugacy classes there exists an open dense P_n-conjugacy class. This clearly implies that each continuous function on $GL(n,\mathbb{C})$, which is constant on P_n-conjugacy classes, is constant on $GL(n,\mathbb{C})$-conjugacy classes. Further, the last observation implies that if a P_n-invariant distribution on $GL(n,\mathbb{C})$ is represented by a continuous function with respect to the Haar measure, then it is an invariant distribution, i.e. invariant for conjugation by elements of $GL(n,\mathbb{C})$ (actually, this conclusion holds for wider class of functions than the continuous ones).

Kirillov expected that this property holds for any P_n-invariant distribution on $GL(n,\mathbb{C})$ (not only for those one which are give by integration against continuous functions). He observed that this property would imply (K) in the following way. Take any T in the commutator of the representation $\pi|P_n$. For proof of (K), by Schur lemma it is enough to show that T is scalar.

Kirillov considered the distribution

$$\Lambda_T : \varphi \mapsto \mathrm{Trace}(T\pi(\varphi)), \tag{19}$$

which is P_n-invariant (T is in the commutator of $\pi|P_n$). Now the property that Kirillov expected for general P_n-invariant distributions, would imply that the above distribution is invariant for the whole group. Using irreducibility of π, it is easy to show that this would imply that T must be a scalar operator.

Note that for proving (K), it is enough to prove Kirillov expectation only for P_n-invariant eigen-distributions (since Λ_T is an eigen-distribution).

At this point, let we recall of the Harish-Chandra regularity theorem for invariant eigen-distributions. He showed that such a distribution is represented by a locally integrable function, which is analytic on regular semi simple elements. If one could prove such a type of result for P_n-invariant eigen-distributions, then the Kirillov's argument for P_n-invariant distributions represented by continuous functions could be used to see the invariance for the whole group (and we would prove (K) in this way). Since the geometry of P_n and $GL(n,\mathbb{C})$-conjugacy classes is the same on a big open set in $GL(n,\mathbb{C})$, it make sense to try to follow Harish-Chandra's strategy of proof of the regularity theorem, to try to prove such a type of result for P_n-invariant eigen-distributions.

As we already have mentioned, M. Baruch proved (K) in [7]. The strategy of his proof may be considered as a further development of the ideas that we discussed above (there is a plenty of new moments).

20.10. J. Bernstein proposed the following strategy for proving (K) (and also (U0)). Consider $GL(n-1,\mathbb{C}) \subseteq GL(n,\mathbb{C})$ in obvious way. Bernstein asked if each $GL(n-1,\mathbb{C})$-invariant distribution on $GL(n,\mathbb{C})$ is invariant for transposition (for our purpose, it is enough to consider only $GL(n-1,\mathbb{C})$-invariant eigen-distributions). Positive answer to this question would also imply (K).

One can consider also other local fields regarding the above question. As far as we remember, in a discussion with D. Miličić we saw that it is an easy exercise to show that the answer to this question is positive for $GL(2,\mathbb{R})$. Unfortunately, this argument cannot be extended to higher $GL(n,\mathbb{R})$.

20.11. Now we shall explain how one can prove (U2), using the standard construction of complementary series representations. For this one needs a simple

Lemma: *Let* $\gamma_1,\ldots,\gamma_u,\delta_1,\ldots,\delta_v \in D$. *Suppose that* $\delta_i \times \gamma_j$ *is irreducible for all indexes* $1 \le i \le u$, $1 \le j \le v$. *Then*

$$L(\gamma_1,\ldots,\gamma_u) \times L(\delta_1,\ldots,\delta_v)$$

is irreducible.

Proof: Using the fact that all $\delta_i \times \gamma_j$ are irreducible (which implies $\delta_i \times \gamma_j \cong \gamma_j \times \delta_i$), and associativity of operation $\times$ among representations (see Remark 9.4), we get

$$\lambda(\gamma_1,\ldots,\gamma_u,\delta_1,\ldots,\delta_v) \cong \lambda(\gamma_1,\ldots,\gamma_u) \times \lambda(\delta_1,\ldots,\delta_v)$$

(for the definition of $\lambda(a)$, see 9.6). From the above isomorphism, one concludes that $L(\gamma_1,\ldots,\gamma_u) \times L(\delta_1,\ldots,\delta_v)$ has a unique irreducible quotient, and that this quotient is isomorphic to $L(\gamma_1,\ldots,\gamma_u,\delta_1,\ldots,\delta_v)$.

Since $\tilde{\delta}_1 \times \tilde{\gamma}_j$ are also all irreducible, the above conclusion hold also for them: the representation $L(\tilde{\gamma}_1,\ldots,\tilde{\gamma}_u) \times L(\tilde{\delta}_1,\ldots,\tilde{\delta}_v)$ has a unique irreducible quotient, which is $L(\tilde{\gamma}_1,\ldots,\tilde{\gamma}_u,\tilde{\delta}_1,\ldots,\tilde{\delta}_v)$. Now passing to contragredients (and using (5)), we shall get that $L(\gamma_1,\ldots,\gamma_u) \times L(\delta_1,\ldots,\delta_v)$ has a unique irreducible subrepresentation, which is isomorphic to the representation $L(\gamma_1,\ldots,\gamma_u,\delta_1,\ldots,\delta_v)$.

A basic property of Langlands classification is that the multiplicity of a Langlands quotient in corresponding standard module is one. This implies irreducibility of $L(\gamma_1, \ldots, \gamma_u) \times L(\delta_1, \ldots, \delta_v)$. □

The above lemma and the representations theory of $SL(2, \mathbb{C})$ imply that the continuous family of representations

$$\pi(u(\delta, m), \alpha), \quad 0 \leq \alpha < 1/2,$$

is irreducible (irreducibility at 0 follows from (U0)). From this, using standard integral intertwining operators we can get on these representations (non-degenerate) hermitian forms, which depend continuously on α, and which make these representations hermitian. Since we have positive definiteness in 0 and we have continuous family of hermitian representations, this implies positive definiteness for all $0 \leq \alpha < 1/2$ (this easily follows from a finite dimensional argument; see [65]). Therefore, representations $\pi(u(\delta, m), \alpha)$ from (U2) are unitarizable.

21. On the unitarizability problem for classical p-adic groups

21.1. Constructing new irreducible unitarizable representations is a very interesting and puzzling problem. It may be related to a number of other problems.

The most interesting (and hardest) question is constructing of irreducible unitarizable representations which show up from "nowhere", i.e., of isolated irreducible unitarizable representations. Namely, there is a natural topology on unitary duals (defined by approximation of matrix coefficients) and isolated representations are those ones for which $\{\pi\}$ is an open set (if the center is not compact, one defines isolated representations modulo center; these representations play the role of isolated representations in this case).

As we mentioned above, the construction of isolated irreducible unitarizable representations can be related to a number of other questions. Let us mention some of them: representations in residual spectrum of the group over adels, θ-correspondences, (conjecturaly) involution (which we mention bellow in 21.2) of square integrable representations. A natural question is: how big portion of isolated representations is in a range of each of these methods, combined with some standard constructions of irreducible unitarizable representations (see section 3. of [65] for standard constructions). We plan to address these questions in the future.

Now we shall formulate some other (precise) questions regarding the unitarizability problem (the first question is around for a long time and we do not know who posed it for the first time in full generality). These questions may provide a strategy (or may be considered as a part of a general strategy) for attacking unitarizability problem for classical p-adic groups.

It may happen that answers to (at least some of these) questions will be obtained in the same time as we will get the solution of the unitarizability problem for classical p-adic groups. Nevertheless, as we already mentioned above, some of these questions may be useful guidance on the way to the solution of the unitarizability problem for classical p-adic groups. This is the reason that we collect them here.

21.2. A.-M. Aubert and also P. Schneider and U. Stuhler ([3], [52]) defined an involution on irreducible representations of connected reductive p-adic groups. This involution carries irreducible unitarizable representations of general linear groups to the unitarizable ones. This was conjectured by J. Bernstein in [9] (and shown by this author; see for example [60]). Further, this involution in the case of irreducible representations with Iwahori fixed vectors carries unitarizable representations to the (irreducible) unitarizable ones. This was proved by D. Barbasch and A. Moy in [6].

It is natural to ask if this is the case in general. It would be very important to show this (if this is the case).

21.3. Fix an irreducible cuspidal representation σ of a classical group S_q and fix unitarizable $\rho_1, \dots, \rho_k \in \mathcal{C}$ such that for $i \neq j$, sets $\{\rho_i, \tilde{\rho}_i\}$ and $\{\rho_j, \tilde{\rho}_j\}$ have no equivalent representations (i.e. $\rho_i \not\cong \rho_j$ and $\rho_i \not\cong \tilde{\rho}_j$). We have already observed in 19.2 that there exists a bijection

$$\mathcal{I}(\rho_1, \dots, \rho_k; \sigma) \rightarrow \left(\prod_{i=1}^{k} \mathcal{I}(\rho_i; \sigma) \right),$$

$$\pi \rightarrow (\pi_1, \dots, \pi_k)$$

(for details see [32])

The question is:

Is π unitarizable if and only if all $\pi_1, \pi_2, \dots, \pi_k$ are unitarizable?

21.4. Suppose

$$\rho \not\cong \tilde{\rho},$$

where $\rho \in \mathcal{C}$ is unitarizable. Take an irreducible unitarizable representation π of a general linear group which is an irreducible subquotients of

$$\nu^{\alpha_1}\tau_1 \times \nu^{\alpha_2}\tau_2 \times \cdots \times \nu^{\alpha_\ell}\tau_\ell,$$

where $\alpha_i \in \mathbb{R}, \tau_i \in \{\rho, \tilde{\rho}\}$. Then $\pi \rtimes \sigma$ is an irreducible unitarizable representation.

The question is:

Does every unitarizable representation in $\mathcal{I}(\rho, \sigma)$ come in this way?

It is not hard to show that the answer to this question is postive.

This gives a reduction of unitarizability problem in $\mathcal{I}(\rho, \sigma)$ to the case of general linear groups (where the unitarizability problem has been solved).

21.5. Suppose that $\delta_1, \ldots, \delta_k \in D$ are unitarizable, and π is an irreducible square integrable representation of a classical group. Let τ_1 and τ_2 be irreducible subrepresentations of

$$\delta_1 \times \cdots \times \delta_k \rtimes \sigma.$$

Let $\delta'_1, \ldots, \delta'_\ell \in D_+$. One may ask the following question:

Is $L(\delta'_1, \ldots, \delta'_\ell, \tau_1)$ unitarizable if and only if $L(\delta'_1, \ldots, \delta'_\ell, \tau_2)$ is unitarizable?

The answer to this question is negative (the first example that we know, which shows that the answer is negative, is for $SO(7, F)$).

21.6. Consider $\mathcal{I}(\rho; \sigma)$, with ρ selfdual. The question is

Can the description of unitarizable representations in $\mathcal{I}(\rho; \sigma)$ be expressed only in terms of the reducibility point $\alpha_{\rho,\sigma}$ (similarly as for $\mathcal{D}(\rho; \sigma)$)?

21.7. In the case of general linear groups, solution of the unitarizability problem may be expressed independently of the nature of the (local) field F (see 20.1). This may be viewed as a (very strong) example of Lefschetz principle. The question on the same line is:

Can one get also a descriptions of unitary duals for each of series of classical groups independent of the nature of the field?

21.8. W. Casselman has proved that in the induced representation from minimal parabolic subgroup, where the Steinberg representation shows up, only two irreducible subquotients are unitarizable, the Steinberg and the trivial one (supposing that the group is not compact modulo center). This fact is important not just for unitarizability problem (see [13]). Let us recall that this fact can be also derived from Howe-Moore theorem on asymptotic behavior of infinite dimensional irreducible unitary representations of reductive groups over local fields.

This fact concerns only finitely many irreducible representations regarding unitarizability (for a fixed group). Nevertheless, its importance much overcomes this finite set. Namely, it implies that Steinberg and trivial representation are isolated in unitary dual if the group has compact center and if its split rank is greater than 1. It also implies that the induced representations around the induced representation where the Steinberg and the trivial representation show up, does not have unitarizable subquotients (these regions are determined by certain irreducibility conditions). Using this fact and some simple standard results, one can solve the unitarizability problem for rank two groups (let us note that to apply this result, one needs to understand reducibility of parabolically induced representations).

Such a fact about exactly two unitarizable subquotients in the whole induced representation where the arbitrary irreducible square integrable representation shows up, holds in the case of general linear groups (the induced representation is always multiplicity one, and further, there is exactly one irreducible square integrable subquotient there).

One may ask if this holds for other groups.

Already the first example other than general linear group, the example $Sp(4, F)$, tells that neither one of the nice properties discussed above regarding the whole induced representation, where the arbitrary irreducible square integrable representation shows up hold in this case. Namely, already for $Sp(4, F)$ there is an example of such induced representation which is not multiplicity one, which contains two inequivalent irreducible square integrable subquotients, which has total length 6, and where all the irreducible subquotients are unitarizable.

In general, for classical groups we have quite often plenty of unitarizable subquotients in whole induced representations where the irreducible square integrable representations show up (for the beginning, we can have as many square integrable subquotients as we want). In the moment, we shall ask only the following question:

Suppose that $\rho \in \mathcal{C}$ is selfdual. Let σ be an irreducible cuspidal representation of a classical group. Suppose that $\nu^\alpha \rho \rtimes \sigma$ reduces for some $\alpha > 0$. Then

$$\nu^{\alpha+n}\rho \times \nu^{\alpha+n-1}\rho \times \cdots \times \nu^\alpha \rho \rtimes \sigma$$

is a multiplicity one representation. It contains exactly one irreducible square integrable subquotient (actually, it is a unique subrepresentation in the above representation; such representations we call square integrable representations of Steinberg type). The question is:

> Does the above induced representation have exactly two irreducible subquotients which are unitarizable (a weaker question is, does it have at most two)?

References

1. Arthur, J., "On some problems suggested by the trace formula", Lie Group Representations II, Proceedings, University of Maryland 1982-83, 1–49, Lecture Notes in Math. 1041, Springer-Verlag, Berlin, 1984.
2. Arthur, J., "Unipotent automorphic representations: conjectures", *Astérisque*, **171-172** (1989), 13–71.
3. Aubert, A.-M., "Dualité dans le groupe de Grothendieck de la catégorie des représentations lisses de longueur finie d'un groupe réductif p-adique", *Trans. Amer. Math. Soc.*, **347** (1995), 2179–2189 (and *Erratum, Trans. Amer. Math. Soc.*, **348** (1996), 4687–4690).
4. Barbasch, D., "The unitary dual for complex classical groups", *Invent. Math.*, **96** (1989), 103–176.
5. Barbasch, D., "Unitary spherical spectrum for split classical groups", preprint, 2002.
6. Barbasch, D. and Moy, A., "A unitarity criterion for p-adic groups", *Invent. Math.*, **98** (1989), 19–37.
7. Baruch, E.M., "A proof of Kirillov's conjecture", *Ann. of Math.*, **158**(1) (2003), 207–252.
8. Bernstein, J.N., "All reductive p-adic groups are tame", *Jour. Functional Anal. Appl.*, **8** (1974), 91–93.
9. Bernstein, J., "P-invariant distributions on $GL(N)$ and the classification of unitary representations of $GL(N)$ (non-archimedean case)", Lie Group Representations II, Proceedings, University of Maryland 1982-83, 50–102, Lecture Notes in Math. 1041, Springer-Verlag, Berlin, 1984.
10. Bernstein, J. (written by K. Rumelhart), "Draft of: Representations of p-adic groups", preprint.
11. Bernstein, I. N. and Zelevinsky, A.V., "Representations of the group $GL(n, F)$, where F is a local non-Archimedean field", *Uspekhi Mat. Nauk.*, **31** (1976), 5–70.
12. Bernstein, J. and Zelevinsky, A.V., "Induced representations of reductive p-adic groups I", *Ann. Sci. École Norm Sup.*, **10** (1977), 441–472.

13. Borel, A. and Wallach, N., *Continuous cohomology, discrete subgroups, and representations of reductive groups*, Princeton University Press, Princeton, 1980.
14. Bushnell, C.J. and Kutzko, P., *The admissible dual of $GL(N)$ via compact open subgroups*, Princeton University Press, Princeton, 1993.
15. Cartier, P., "Representations of p-adic groups; a survey", Symp. Pure Math. 33, part 1, 111–155, Amer. Math. Soc., Providence, Rhone Island, 1979.
16. Casselman, W., "Introduction to the theory of admissible representations of p-adic reductive groups", preprint.
17. Casselman W. and Miličić D., "Asymptotic behavior of matrix coefficients of admissible representations", *Duke Math. J.*, **49** (No. 4) (1982), 869–930.
18. Clozel L., "Progrès récents vers la classification du dual unitaire des groupes réductifs réels", Séminaire Bourbaki, no 681 (1986-87), *Astérisque.*
19. Gelbart, S.S., "An elementary introduction to the Langlands program", *Bulletin Amer. Math. Soc.*, **10** (1984), 177–219.
20. Gelfand, I.M., Graev, M. and Piatetski-Shapiro, *Representation theory and automorphic functions*, Saunders, Philadelphia, 1969.
21. Gelfand, I.M. and Kazhdan, D.A., "Representations of $GL(n,k)$", Lie groups and their Representations, 95–118, Halstead Press, Budapest, 1974.
22. Gelfand, I.M. and Naimark, M.A., "Unitäre Darstellungen der Klassischen Gruppen", Akademie Verlag, Berlin, 1957, (German translation).
23. Goldberg, D., "Reducibility of induced representations for $Sp(2n)$ and $SO(n)$", *Amer. J. Math.*, **116** (5) (1994), 1101–1151.
24. Harish-Chandra, *Collected papers*, Springer-Verlag, Berlin, 1983.
25. Harris, M. and Taylor, R., *On the geometry and cohomology of some simple Shimura varieties*, Princeton University Press, Annals of Math. Studies 151, 2001.
26. Henniart, G., "Une preuve simple des conjectures de Langlands pour GL(n) sur un corps p-adique", *Invent. Math.*, **139** (2) (2002), 439–455.
27. Howe, R., "θ-series and automorphic forms", *Symp. Pure Math.* **33**, part 1, 275–286, Amer. Math. Soc., Providence, Rhone Island, 1979.
28. Humpreys, J.E., *Linear algebraic groups*, Springer-Verlag, New York, 1975.
29. Jacquet, H., "Generic representations", *Non-Commutative Harmonic Analysis*, 91–101, Lecture Notes in Math. 587, Spring Verlag, Berlin, 1977.
30. Jacquet, H., "Principal L-functions of the linear group", *Symp. Pure Math.* **33**, part 2, 63–86, Amer. Math. Soc., Providence, Rhone Island, 1979.
31. Jacquet, H., "On the residual spectrum of $GL(n)$", *Lie Group Representations II*, 185–208, Proceedings, University of Maryland, 1982-83. Lecture Notes in Math. 1041, Springer-Verlag, Berlin, 1984.
32. Jantzen, C., "On supports of induced representations for symplectic and odd-orthogonal groups", *Amer J. Math.*, **119** (1997), 1213–1262.
33. Jantzen, C., "On square integrable representations of classical p-adic groups II", *Representation Theory*, **4** (2000), 127–180.
34. Kazhdan, D. and Lusztig, G., "Proof of the Deligne-Langlands conjecture for Hecke algebras", *Invent. Math.*, **87** (1987), 153–215.
35. Kirillov, A. A., "Infinite dimensional representations of the general linear

group", *Dokl. Akad. Nauk SSSR*, **114** (1962), 37–39. *Soviet Math. Dokl.*, **3** (1962), 652–655.

36. Knapp, A. W., "Introduction to the Langlands program", *Representation Theory and Automorphic Forms*; An Instructional Conference at the International Centre for Mathematical Sciences, Edinburgh, 1996, 245–302, American Mathematical Society, Proceedings of Symposia in Pure Mathematics 61, Providence, Rhode Island, 1997.
37. Langlands, R.P., "On the classification of irreducible representations of real algebraic groups", *Representation Theory and Harmonic Analysis on Semisimple Lie Groups*, P.J. Sally, Jr. and D. A. Vogan, Jr. editors, Amer. Math. Soc., Providence, 1989.
38. Lapid, E., Goran, M. and Tadić, M., "On the generic unitary dual of quasi-split classical groups", preprint, 2003.
39. Miličić, D., "On C^*-algebras with bounded trace", *Glasnik Mat.*, **8** (28) (1973), 7–21.
40. Mœglin C., "Normalisation des opérateurs d'entrelacement et réductibilité des induites de cuspidales; le cas des groupes classiques p-adiques", *Annals of Math.*, **151** (2) (2000), 817–847.
41. Mœglin, C., "Sur la classification des séries discrètes des groupes classiques p-adiques: paramètres de Langlands et exhaustivité", to appear in *Journal of the European Mathematical Society.*
42. Mœglin, C., "Représentations quadratiques unipotentes des groupes classiques p-adiques.", *Duke Math. Jour.*, **84** (1996), 267–332.
43. Moeglin, C., "Points de réducibilité pour les induites de cuspidales" preprint, Institute de mathématiques de Jussieu, 2001.
44. Mœglin, C. and Tadić, M., "Construction of discrete series for classical p-adic groups", *J. Amer. Math. Soc.*, **15** (2002), 715–786.
45. Mœglin, C., Vignéras, M.-F. and Waldspurger J.-L., *Correspondance de Howe sur un corps p-adique*, Lecture Notes in Math 1291, Springer-Verlag, Berlin, 1987.
46. Moy, A., "Representations of $GSp(4)$ over a p-adic field: parts 1 and 2", *Compositio Math.*, **66** (1988), 237–328.
47. Muić, G., "On generic irreducible representations of $Sp(n, F)$ and $SO(2n+1, F)$", *Glasnik Mat.*, **33**(53) (1998), 19–31.
48. Rodier, F., "Représentations de $GL(n, k)$ où k est un corps p-adique", Séminaire Bourbaki no 587 (1982), *Astérisque*, **92-93** (1982), 201–28.
49. Sahi, S., "On Kirillov's conjecture for archimedean fields", *Compos. Math.*, **72** (1989), 67–86.
50. Savin, G., "Lectures on representations of p-adic groups", in this volume.
51. Sally, P.J. and Tadić, M., "Induced representations and classifications for $\mathrm{GSp}(2, F)$ and $\mathrm{Sp}(2, F)$", *Mémoires Soc. Math. France*, **52** (1993), 75–133.
52. Schneider, P. and Stuhler, U., "Representation theory and sheaves on the Bruhat-Tits building", *Publ. Math. IHES*, (1997), 97–191.
53. Shahidi, F., "A proof of Langlands conjecture on Plancherel measures; complementary series for p-adic groups", *Ann. of Math.*, **132** (1990), 273–330.
54. Shahidi, F., "Twisted endoscopy and reducibility of induced representations

for p-adic groups", *Duke Math. Jour.*, **66** (1992), 1–41.
55. Silberger, A., "The Langlands quotient theorem for p-adic groups", *Math. Ann.*, **236** (1978), 95–104.
56. Silberger, A., "Special representations of reductive p-adic groups are not integrable", *Ann. of Math.*, **111** (1980), 571–587.
57. Speh B., "Unitary representations of $GL(n,\mathbb{R})$ with non-trivial $(\mathfrak{g},K)$-cohomology", *Invent. Math.*, **71** (1983), 443–465.
58. Stein, E.M., "Analysis in matrix spaces and some new representations of $SL(N,\mathbb{C})$", *Ann. of Math.*, **86** (1967), 461–490.
59. Steinberg, R., *Lectures on Chevalley groups*, Yale University, 1968.
60. Tadić, M., "Classification of unitary representations in irreducible representations of general linear group (non-archimedean case)", *Ann. Sci. École Norm. Sup.*, **19** (1986), 335–382.
61. Tadić, M., "Topology of unitary dual of non-archimedean $GL(n)$", *Duke Math. Jour.*, **55** (1987), 385–422.
62. Tadić, M., "Representations of p-adic symplectic groups", *Compositio Math.*, **90** (1994), 123–181.
63. Tadić, M., "Structure arising from induction and Jacquet modules of representations of classical p-adic groups", *Journal of Algebra*, **177** (1) (1995), 1–33.
64. Tadić, M., "Correspondence on characters of irreducible unitary representations of $GL(n,\mathbb{C})$", *Mathematischen Annalen*, **305** (1996), 419–438.
65. Tadić, M., "An external approach to unitary representations", *Bulletin Amer. Math. Soc.*, **28**(2) (1993), 215–252.
66. Tadić, M., "Representations of classical p-adic groups", *Representations of Lie groups and quantum groups*, 129–204, Pitman Research Notes in Mathematics, Series 311, Longman, Essex, 1994.
67. Tadić, M., "On regular square integrable representations of p-adic groups", *Amer. J. Math.*, **120**(1) (1998), 159–210.
68. Tadić, M., "On reducibility of parabolic induction", *Israel J. Math.*, **107** (1998), 29–91.
69. Tadić, M., "Square integrable representations of classical p-adic groups corresponding to segments", *Representation Theory*, **3** (1999), 58–89.
70. Tadić, M., "A family of square integrable representations of classical p-adic groups", *Glasnik Mat.*, **37**(57) (2002), 21–57.
71. Tadić, M., "$GL(n,\mathbb{C})\hat{}$ and $GL(n,\mathbb{R})\hat{}$", preprint, 2002.
72. Vogan, D. A., "The unitary dual of $GL(n)$ over an archimedean field", *Invent. Math.*, **82** (1986), 449–505.
73. Waldspurger, J.-L., "La formule de Plancherel pour les groupes p-adiques, d'après Harish-Chandra", *Journal de l'Institut de Math. de Jussieu*, **2** (2) (2003), 235–333.
74. Warner, G., *Harmonic analysis on Semi-Simple Lie Groups* I, II, Springer-Verlag, Berlin, 1972.
75. Weil, A., *Basic number theory*, Springer-Verlag, New York, 1974.
76. Zelevinsky, A.V., "Induced representations of reductive p-adic groups II, On irreducible representations of $GL(n)$", *Ann. Sci École Norm Sup.*, **13** (1980),

165–210.

77. Zelevinsky, A. V., *Representations of Finite Classical Groups, A Hopf Algebra Approach*, Lecture Notes in Math 869, Springer-Verlag, Berlin, 1981.

Dirac Operators in Representation Theory

Jing-Song Huang and Pavle Pandžić

Department of Mathematics
The Hong Kong University of Science and Technology
Clear Water Bay, Kowloon, Hong Kong SAR, China
Email: mahuang@ust.hk

Department of Mathematics
University of Zagreb
Bijenička 30, 10000 Zagreb, Croatia
E-mail: pandzic@math.hr

Vogan's conjecture on Dirac cohomology reveals an algebraic nature of Dirac operators. In these notes we explain our proof of the Vogan's conjecture. As applications we describe how to simplify the proof of Atiyah-Schmid on geometric construction of the discrete series and sharpen the Langlands-Hotta-Parthasarathy formula on automorphic forms. We also explore the relationship between Dirac cohomology and Lie algebra cohomology. To make these notes more accessible, we include some well known background material on representations and Lie algebra cohomology.

1. $(\mathfrak{g}, K)$-modules

1.1. *Lie groups and algebras*

A Lie group G is a group which is also a smooth manifold, in such a way that the group operations are smooth. In more words, the multiplication map from $G \times G$ into G and the inverse map from G into G are required to be smooth.

The Lie algebra $\mathfrak{g}$ of G consists of left invariant vector fields on G. The left invariance condition means the following: let $l_g : G \to G$ be the left translation by $g \in G$, i.e., $l_g(h) = gh$. A vector field X on G is left invariant if $(dl_g)_h X_h = X_{gh}$ for all $g, h \in G$.

It is clear that $\mathfrak{g}$ is a vector space. It can be identified with the tangent space to G at the unit element e. Namely, to any left invariant vector field one can attach its value at e. Conversely, a tangent vector at e can be translated to all other points of G to obtain a left invariant vector field.

Note that we did not require our vector fields to be smooth; it is however a fact that a left invariant vector field is automatically smooth.

The operation making $\mathfrak{g}$ into a Lie algebra is the bracket of vector fields:

$$[X,Y]f = X(Yf) - Y(Xf),$$

for $X, Y \in G$ and f a smooth function on G. Here we identify vector fields with derivations of the algebra $C^\infty(G)$, i.e., think of them as first order differential operators.

The operation $[\,,]$ satisfies the well known properties of a Lie algebra operation: it is bilinear and anticommutative, and it satisfies the Jacobi identity:

$$[[X,Y],Z] + [[Y,Z],X] + [[Z,X],Y] = 0$$

for any $X, Y, Z \in \mathfrak{g}$.

The main examples are various matrix groups. They fall into several classes. We are primarily interested in the semisimple connected groups, like the group $SL(n,\mathbb{R})$ of $n\times n$ real matrices with determinant 1. Its Lie algebra is $\mathfrak{sl}(n,\mathbb{R})$, consisting of the $n \times n$ matrices with trace 0. There are also familiar series of compact groups: $SU(n)$, the group of unitary (complex) matrices, with Lie algebra $\mathfrak{su}(n)$ consisting of skew Hermitian matrices, and $SO(n)$, the group of orthogonal (real) matrices with Lie algebra $\mathfrak{so}(n)$ consisting of antisymmetric matrices. Further examples of matrix groups are the groups $Sp(n)$, $SU(p,q)$ and $SO(p,q)$; they are all defined as groups of operators preserving certain forms.

Other classes of Lie groups one needs to study are solvable groups, like the groups of upper triangular matrices; nilpotent groups like the groups of unipotent matrices and abelian groups like $\mathbb{R}^n$ or the groups of diagonal matrices. They are not of our primary interest, but they show up as subgroups of our semisimple groups and therefore have to be understood.

One also often considers reductive groups, which include semisimple groups but are allowed to have a larger center, like $GL(n,\mathbb{R})$ or $U(n)$.

The definitions are easier to formulate for Lie algebras $\mathfrak{g}$: define $C^0\mathfrak{g} = D^0\mathfrak{g} = \mathfrak{g}$, and inductively $C^{i+1}\mathfrak{g} = [\mathfrak{g}, C^i\mathfrak{g}]$, $D^{i+1}\mathfrak{g} = [D^i\mathfrak{g}, D^i\mathfrak{g}]$. Then $\mathfrak{g}$ is nilpotent if $C^i\mathfrak{g} = 0$ for large i, $\mathfrak{g}$ is solvable if $D^i\mathfrak{g} = 0$ for large i, $\mathfrak{g}$ is semisimple if it contains no nonzero solvable ideals, and $\mathfrak{g}$ is reductive if it contains no nonabelian solvable ideals. Abelian means that $[\,,] = 0$.

Instead of checking that each of the above mentioned groups is a Lie group, on can refer to a theorem of Cartan, which asserts that every closed subgroup of a Lie group is automatically a Lie subgroup in a unique way. Since each of the groups we mentioned is contained in $GL(n,\mathbb{C})$ we only need to see that $GL(n,\mathbb{C})$ is a Lie group. But $GL(n,\mathbb{C})$ is an open subset of $\mathbb{C}^{n^2}$, hence is a manifold, and the matrix multiplication and inverting are clearly smooth. Without using Cartan's theorem, one can apply implicit/inverse function theorems, as each of the above groups is given by certain equations that the matrix coefficients must satisfy.

Let us note some common features of all the mentioned examples:

• $\mathfrak{g}$ is contained in the matrix algebra $M_n(\mathbb{C})$, and $[X,Y] = XY - YX$, the commutator of matrices;

• G acts on $\mathfrak{g}$ by conjugation: $Ad(g)X = gXg^{-1}$. This is called the adjoint action. The differential of this action with respect to g gives an action of $\mathfrak{g}$ on itself, $ad(X)Y = [X,Y]$, which is also called the adjoint action.

• There is an exponential map $\exp : \mathfrak{g} \to G$ mapping X to e^X. This is a local diffeomorphism around 0, i.e., sends a neighborhood of 0 in $\mathfrak{g}$ diffeomorphically onto a neighborhood of e in G.

1.2. *Finite dimensional representations*

Let V be a complex n-dimensional vector space. A representation of G on V is a continuous homomorphism

$$\pi : G \to GL(V).$$

Any such homomorphism is automatically smooth; this is a version of the already mentioned Cartan's theorem.

Given a representation of G as above, we can differentiate it at e and obtain a homomorphism

$$d\pi = \pi : \mathfrak{g} \to \mathfrak{gl}(V)$$

of Lie algebras.

An important special case is the case of a unitary representation. This means V has an inner product such that all the operators $\pi(g)$, $g \in G$, are unitary. Then all $\pi(X)$, $X \in \mathfrak{g}$, are skew-hermitian.

The main idea of passing from G to $\mathfrak{g}$ is turning a harder, analytic problem of studying representations of G into an easier, purely algebraic (or even combinatorial in some sense) problem of studying representations of $\mathfrak{g}$. Actually, since we are only considering complex representations, we can as well complexify $\mathfrak{g}$ and study representations of $\mathfrak{g}_{\mathbb{C}}$.

Example: The most basic example and the first one to study is the representations of $\mathfrak{g} = \mathfrak{sl}(2, \mathbb{C})$. There is an obvious basis for $\mathfrak{g}$: take

$$h = \begin{pmatrix} 1 & 0 \\ 0 & -1 \end{pmatrix}; \quad e = \begin{pmatrix} 0 & 1 \\ 0 & 0 \end{pmatrix}; \quad f = \begin{pmatrix} 0 & 0 \\ 1 & 0 \end{pmatrix}.$$

Then the commutator relations are

$$[h, e] = 2e; \quad [h, f] = -2f; \quad [e, f] = h.$$

Let V be an irreducible representation of $\mathfrak{g}$ of dimension $n+1$. Then there is a basis

$$\{v_k \big| k = -n, -n+2, \dots, n-2, n\}$$

of V, such that the action of h is diagonal in this basis:

$$\pi(h)v_k = kv_k.$$

The integers k are called the weights of V. The operator $\pi(e)$ raises the weight by 2, i.e., $\pi(e)v_k$ is proportional to v_{k+2} (which is taken to be 0 if $k = n$). The operator $\pi(f)$ lowers the weight by 2. There is an obvious ordering on the set of weights (usual ordering of integers), and n is the highest weight of V; this corresponds to the fact that $\pi(e)v_n = 0$. Here is a picture of our representation V:

$$v_{-n} \underset{\underset{f}{\longleftarrow}}{\overset{e}{\longrightarrow}} v_{-n+2} \underset{\underset{f}{\longleftarrow}}{\overset{e}{\longrightarrow}} \cdots \underset{\underset{f}{\longleftarrow}}{\overset{e}{\longrightarrow}} v_{n-2} \underset{\underset{f}{\longleftarrow}}{\overset{e}{\longrightarrow}} v_n$$

Finally, the actions of e and f are also completely determined: let us normalize the choice of v_k's by taking v_n to be an arbitrary $\pi(h)$-eigenvector for eigenvalue n, and then taking $v_{n-2j} = \pi(f)^j v_n$. Then

$$(*) \qquad \pi(e)v_{n-2j} = j(n-j+1)v_{n-2j+2}$$

All of this is easy to prove; we give an outline here and encourage the reader to complete the details.

First, it is a general fact that the action of h diagonalizes in any finite dimensional representation, as h is a semisimple element of $\mathfrak{g}$. One can however avoid using this, and just note that the action of h will have an eigenvalue, say $\lambda \in \mathbb{C}$, with an eigenvector v_λ. For any $\mu \in \mathbb{C}$, let us denote by V_μ the μ-eigenspace for $\pi(h)$; of course, V_μ will be zero for all but finitely many μ.

From the commutator relations, it is immediate that $\pi(e)$ sends V_μ to $V_{\mu+2}$, while $\pi(f)$ sends V_μ to $V_{\mu-2}$ for any μ. In particular, the sum of all V_μ for $\mu \in \lambda + 2\mathbb{Z}$ is $\mathfrak{g}$-invariant, hence it is all of V by irreducibility. Moreover, the subspace generated by v_λ will already be everything, so all V_μ are one-dimensional.

We can now replace λ by the highest weight in the obvious ordering coming from $\mathbb{Z}$, and fix a highest weight vector v_λ. Then we can define (as above) $v_{\lambda-2j} = \pi(f)^j v_\lambda$, and prove the analog of the formula (*) with λ replacing n, by induction from the commutator relations. From (*), finite dimensionality and irreducibility it now follows that λ must be a nonnegative integer, say n, and that V is the span of the (nonzero) vectors $v_n, v_{n-2}, \dots, v_{-n}$.

Thus we have described all irreducible finite dimensional $\mathfrak{sl}(2,\mathbb{C})$-modules. Other finite dimensional representations are direct sums of irreducible ones; this is a special case of Weyl's theorem, which says this is true for any semisimple Lie algebra $\mathfrak{g}$. One way to prove this theorem is the so called unitarian trick of Weyl: one shows that there is a compact group G with the same representations as $\mathfrak{g}$. E.g., for $\mathfrak{sl}(2,\mathbb{C})$, $G = SU(2)$. Now using invariant integration one shows that every representation of a compact group is unitary. On the other hand, unitary (finite dimensional) representations are easily seen to be direct sums of irreducibles; namely, for every invariant subspace, its orthogonal is an invariant complement.

We now pass on to describe finite dimensional representations of a general semisimple Lie algebra $\mathfrak{g}$ over $\mathbb{C}$. Instead of just one element h to diagonalize, we can now have a bunch of them. They comprise a Cartan subalgebra $\mathfrak{h}$ of $\mathfrak{g}$, which is by definition a maximal abelian subalgebra consisting of semisimple elements. Elements of $\mathfrak{h}$ can be simultaneously diagonalized in any (finite dimensional) representation; for each joint eigenspace, the eigenvalues for various $X \in \mathfrak{h}$ are described by a functional $\lambda \in \mathfrak{h}^*$, a weight of the representation under consideration. All possible weights of finite dimensional representations form a lattice in $\mathfrak{h}^*$, called the weight lattice of $\mathfrak{g}$. The nonzero weights of the adjoint representation of $\mathfrak{g}$ on itself have a prominent role in the theory; they are called the roots of $\mathfrak{g}$, and satisfy a number of nice symmetry properties. For example, the roots of $\mathfrak{sl}(3,\mathbb{C})$ form a regular hexagon in the plane. The roots of $\mathfrak{sl}(2,\mathbb{C})$ are 2 and -2 (upon identifying $\mathfrak{h}^* = (\mathbb{C}h)^*$ with $\mathbb{C}$).

One can divide up roots into positive and negative roots, which gives an ordering on $\mathfrak{h}^*_{\mathbb{R}}$, the real span of roots, and also a notion of positive ("dominant") weights. The irreducible finite dimensional representations of $\mathfrak{g}$ are classified by their highest weights. The possible highest weights are precisely the dominant weights.

If we denote by $\mathfrak{n}^+$ (respectively $\mathfrak{n}^-$) the subalgebra of $\mathfrak{g}$ spanned by all positive (respectively negative) root vectors, then we have a triangular decomposition

$$\mathfrak{g} = \mathfrak{n}^- \oplus \mathfrak{h} \oplus \mathfrak{n}^+.$$

For example, if $\mathfrak{g} = \mathfrak{sl}(n, \mathbb{C})$, one can take $\mathfrak{h}$ to be the diagonal matrices in $\mathfrak{g}$, $\mathfrak{n}^+$ the strictly upper triangular matrices and $\mathfrak{n}^-$ the strictly lower triangular matrices.

Let V be an irreducible finite dimensional representation with highest weight λ. The highest weight vector is unique up to scalar, and is characterized by being an eigenvector for each $X \in \mathfrak{h}$, with eigenvalue $\lambda(X)$, and by being annihilated by all elements of $\mathfrak{n}^+$.

We now address the question of "going back", i.e., getting representations of G from representations of $\mathfrak{g}$ which we have just described. This is called "integrating" or "exponentiating" representations.

It turns out there is a topological obstacle to integrating representations; this can already be seen in the simplest case of 1-dimensional (abelian) Lie groups. There are two connected 1-dimensional groups: the real line $\mathbb{R}$ and the circle group $\mathbb{T}^1$. Both have the same Lie algebra $\mathbb{R}$.

Consider the one-dimensional representations of the Lie algebra $\mathbb{R}$; each of them is given by $t \mapsto t\lambda$ for some $\lambda \in \mathbb{C}$ (we identify 1×1 complex matrices with complex numbers). All of these representations exponentiate to the group $\mathbb{R}$, and give all possible characters of $\mathbb{R}$:

$$t \longmapsto e^{t\lambda}, \quad \lambda \in \mathbb{C}.$$

However, of these characters only the periodic ones will be well defined on $\mathbb{T}^1$, and $e^{t\lambda}$ is periodic if and only if $\lambda \in 2\pi i\mathbb{Z}$.

In general, when G is connected and simply connected, then all finite dimensional representations of $\mathfrak{g}$ integrate to G. Any connected G will be covered by a simply connected $\tilde{G}$, and the representations of G are those representations of $\tilde{G}$ which are trivial on the kernel of the covering.

If G is semisimple connected with finite center, then there is a decomposition

$$G = KP$$

called the Cartan decomposition; here K is the maximal compact subgroup of G, while P is diffeomorphic to a vector space. For example, if $G =$

$SL(n,\mathbb{R})$, then $K = SO(n)$, and P consists of positive matrices in $SL(n,\mathbb{R})$; so P is diffeomorphic to the vector space of all symmetric matrices of trace zero via the exponential map. It follows that the topology of G is the same as the topology of K, and a (finite dimensional) representation of $\mathfrak{g}$ will exponentiate to G if and only if it exponentiates to K.

1.3. *Infinite dimensional representations*

In general, a representation of G is a continuous linear action on a topological vector space $\mathcal{H}$. Some typical choices for $\mathcal{H}$ are: a Hilbert space, a Banach space, or a Fréchet space.

The first question would be: can we differentiate a representation to get a representation of $\mathfrak{g}$? The answer is: not quite. Actually, $\mathfrak{g}$ acts, but only on the dense subspace $\mathcal{H}_\infty$ of smooth vectors (a vector $v \in \mathcal{H}$ is smooth if the map $g \mapsto \pi(g)v$ from G into $\mathcal{H}$ is smooth).

For semisimple G with maximal compact K, there is a better choice than $\mathcal{H}_\infty$: we can consider the dense subspace $\mathcal{H}_K$ of K-finite vectors in $\mathcal{H}$ (a vector $v \in \mathcal{H}$ is K-finite if the set $\pi(K)v$ spans a finite dimensional subspace of $\mathcal{H}$). One shows that K-finite vectors are all smooth, and so $\mathcal{H}_K$ becomes a Harish-Chandra module, or a $(\mathfrak{g}, K)$ module. A vector space V is a $(\mathfrak{g}, K)$-module if it has:

(1) an action of $\mathfrak{g}$;
(2) a finite action of K;
(3) the two $\mathfrak{k}$-actions obtained from (1) and (2) agree.

In case we want to allow disconnected groups, we also need

(4) the $\mathfrak{g}$-action is K-equivariant, i.e., $\pi(k)\pi(X)v = \pi(Ad(k)X)\pi(k)v$, for all $k \in K$, $X \in \mathfrak{g}$ and $v \in V$.

One usually also puts some finiteness conditions, like finite generation, or admissibility defined below.

Now a basic fact about (unitary) representations of compact groups is that they can be decomposed into Hilbert direct sums of irreducibles, which are all finite dimensional. This can be proved using the basic facts about compact operators. Thus our $\mathcal{H}$ decomposes into a Hilbert direct sum of K-irreducibles, and Harish-Chandra modules, being K-finite, decompose into algebraic direct sums of K-irreducibles. For $\delta \in \widehat{K}$, the unitary dual of K, we denote by $V(\delta)$ the δ-isotypic component of a Harish-Chandra module, i.e., the largest K-submodule of V which is isomorphic to a sum of copies of

δ. We say V is admissible if each $V(\delta)$ is finite dimensional. In other words, every $\delta \in \widehat{K}$ occurs in V with finite multiplicity. Harish-Chandra proved that for each irreducible unitary representation $\mathcal{H}$ of G, the $(\mathfrak{g}, K)$-module $\mathcal{H}_K$ is admissible.

Going back from $(\mathfrak{g}, K)$-modules to representations of G is hard. We call a representation $(\pi, \mathcal{H})$ of G a globalization of a Harish-Chandra module V, if V is isomorphic to $\mathcal{H}_K$. Every irreducible V has globalizations. In fact there are many of them; one can choose e.g. a Hilbert space globalization (not necessarily unitary), or a smooth globalization. There are also notions of minimal and maximal globalizations. A few names to mention here are Harish-Chandra, Lepowsky, Rader, Casselman, Wallach and Schmid. We will not need any globalizations, but will from now on work only with $(\mathfrak{g}, K)$-modules.

1.4. *Infinitesimal characters*

Recall that a representation of a Lie algebra $\mathfrak{g}$ on a vector space V is a Lie algebra morphism from $\mathfrak{g}$ into the Lie algebra $\mathrm{End}(V)$ of endomorphisms of V. Now $\mathrm{End}(V)$ is actually an associative algebra, which is turned into a Lie algebra by defining $[a, b] = ab - ba$; this can be done for any associative algebra. What we want is to construct an associative algebra $U(\mathfrak{g})$ containing $\mathfrak{g}$, so that representations of $\mathfrak{g}$ extend to morphisms $U(\mathfrak{g}) \to \mathrm{End}(V)$ of associative algebras. The construction goes like this: consider first the tensor algebra $T(\mathfrak{g})$ of the vector space $\mathfrak{g}$. Then define

$$U(\mathfrak{g}) = T(\mathfrak{g})/I,$$

where I is the two-sided ideal of $T(\mathfrak{g})$ generated by elements $X \otimes Y - Y \otimes X - [X, Y]$, $X, Y \in \mathfrak{g}$. It is easy to see that $U(\mathfrak{g})$ satisfies a universal property with respect to maps of $\mathfrak{g}$ into associative algebras; in particular, representations of $\mathfrak{g}$ (i.e., $\mathfrak{g}$-modules) are the same thing as $U(\mathfrak{g})$-modules. Some further properties are

- There is a filtration by degree on $U(\mathfrak{g})$, coming from $T(\mathfrak{g})$;
- The graded algebra associated to the above filtration is the symmetric algebra $S(\mathfrak{g})$;
- One can get a basis for $U(\mathfrak{g})$ by taking monomials over an ordered basis of $\mathfrak{g}$.

The last two properties are closely related and are the content of the Poincaré-Birkhoff-Witt theorem. Loosely speaking, one can think of $U(\mathfrak{g})$ as "noncommutative polynomials over $\mathfrak{g}$", with the commutation laws given by the bracket of $\mathfrak{g}$. If we think of elements of $\mathfrak{g}$ as left invariant vector fields on G, then $U(\mathfrak{g})$ consists of left invariant differential operators on G.

Here are some benefits of introducing the algebra $U(\mathfrak{g})$.

(1) One can use constructions from the associative algebra setting. For example, there is a well known notion of "extension of scalars": let $B \subset A$ be associative algebras and let V be a B-module. One can consider A as a right B-module for the right multiplication and form the vector space $A \otimes_B V$. This vector space is an A-module for the left multiplication in the first factor. So we get a functor from B-modules to A-modules. Another functor like this is obtained by considering $\mathrm{Hom}_B(A, V)$; now the Hom is taken with respect to the left multiplication action of B on A (and the given action on V), and the (left!) A-action on the space $\mathrm{Hom}_B(A, V)$ is given by right multiplication on A.

(2) Since $\mathfrak{g}$ is semisimple, it has no center. $U(\mathfrak{g})$ however has a nice center $Z(\mathfrak{g})$. It is a finitely generated polynomial algebra (e.g. for $\mathfrak{g} = \mathfrak{sl}(n, \mathbb{C})$ there are $n-1$ generators and their degrees are $2, 3, \dots, n$). The importance of the center follows from a simple observation that is often used in linear algebra: if two operators commute, then an eigenspace for one of them is invariant for the other. This means that we can reduce representations by taking a joint eigenspace for $Z(\mathfrak{g})$.

Let us examine the center $Z(\mathfrak{g})$ and its use in representation theory in more detail. First, there is an element that can easily be written down; it is the simplest and most important element of $Z(\mathfrak{g})$ called the Casimir element. To define it we need a little more structure theory.

There is an invariant bilinear form on $\mathfrak{g}$, the Killing form B. It is defined by

$$B(X, Y) = \mathrm{tr}(\mathrm{ad}\, X\, \mathrm{ad}\, Y), \quad X, Y \in \mathfrak{g}.$$

The invariance condition means that for a Lie group G with Lie algebra $\mathfrak{g}$, one has

$$B(\mathrm{Ad}(g)X, \mathrm{Ad}(g)Y) = B(X, Y), \quad g \in G,\ X, Y \in \mathfrak{g}.$$

The Lie algebra version of this identity is

$$B([Z, X], Y) + B(X, [Z, Y]) = 0, \quad X, Y, Z \in \mathfrak{g}.$$

Furthermore, B is nondegenerate on $\mathfrak{g}$; this is actually equivalent to $\mathfrak{g}$ being semisimple. In many cases like for $\mathfrak{sl}(n)$ over $\mathbb{R}$ or $\mathbb{C}$, one can instead of B use a simpler form $\mathrm{tr}(XY)$, which is equal to B up to a scalar.

There is a Cartan decomposition of $\mathfrak{g}$:

$$\mathfrak{g} = \mathfrak{k} \oplus \mathfrak{p}.$$

$\mathfrak{k}$ and $\mathfrak{p}$ can be defined as the eigenspaces for the so called Cartan involution θ for the eigenvalues 1 respectively -1. There is also a Cartan decomposition $G = KP$ on the group level that we already mentioned; here K is the maximal compact subgroup of G (if G is connected with finite center), and $\mathfrak{k}$ is the Lie algebra of K.

Rather than defining the Cartan involution in general, let us note that for all the matrix examples in 1.1, $\theta(X)$ is minus the (conjugate) transpose of X. So e.g. for $\mathfrak{g} = \mathfrak{sl}(n, \mathbb{R})$, $\mathfrak{k}$ is $\mathfrak{so}(n)$ and $\mathfrak{p}$ is the space of symmetric matrices in $\mathfrak{g}$.

Some further properties of the Cartan decomposition: $\mathfrak{k}$ is a subalgebra, $[\mathfrak{k}, \mathfrak{p}] \subset \mathfrak{p}$ and $[\mathfrak{p}, \mathfrak{p}] \subset \mathfrak{k}$. The Killing form B is negative definite on $\mathfrak{k}$ and positive definite on $\mathfrak{p}$.

To define the Casimir element, we choose orthonormal bases W_k for $\mathfrak{k}$ and Z_i for $\mathfrak{p}$, so that

$$B(W_k, W_l) = -\delta_{kl}; \quad B(Z_i, Z_j) = \delta_{ij}.$$

The Casimir element is then

$$\Omega = -\sum_k W_k^2 + \sum_i Z_i^2.$$

It is an element of $U(\mathfrak{g})$, and one shows by an easy calculation (which we leave as an exercise) that it commutes with all elements of $\mathfrak{g}$ and thus is an element of $Z(\mathfrak{g})$. Furthermore, it does not depend on the choice of bases W_k and Z_i.

Assume now X is an irreducible $(\mathfrak{g}, K)$-module. Then every element of $Z(\mathfrak{g})$ acts on X by a scalar. For finite dimensional X it is the well known and obvious Schur's lemma: let $z \in Z(\mathfrak{g})$ and take an eigenspace for z in X. This eigenspace is a submodule, hence has to be all of X. For infinite dimensional X the same argument is applied to a fixed K-type in X; see [28], 0.3.19, or [15], Ch. VIII, §3.

Now all the scalars coming from the action of $Z(\mathfrak{g})$ on X put together give a homomorphism

$$\chi_X : Z(\mathfrak{g}) \to \mathbb{C}$$

of algebras, which is called the infinitesimal character of X.

By a theorem of Harish-Chandra, $Z(\mathfrak{g})$ is isomorphic (as an algebra) to $S(\mathfrak{h})^W$, the Weyl group invariants in the symmetric algebra of a Cartan subalgebra $\mathfrak{h}$ in $\mathfrak{g}$. (The Weyl group W is a finite reflection group generated by reflections with respect to roots.) This is obtained by taking the triangular decomposition $\mathfrak{g} = \mathfrak{n}^- \oplus \mathfrak{h} \oplus \mathfrak{n}^+$ mentioned before and building a Poincaré-Birkhoff-Witt basis from bases in $\mathfrak{n}^-$, $\mathfrak{h}$ and $\mathfrak{n}^+$. We now get a linear map from $U(\mathfrak{g})$ into $U(\mathfrak{h}) = S(\mathfrak{h})$ by projecting along the span of all monomials that contain a factor which is not in $\mathfrak{h}$. This turns out to be an algebra homomorphism when restricted to $Z(\mathfrak{g})$, and this gives the required isomorphism.

Now we can identify $S(\mathfrak{h})$ with the algebra $P(\mathfrak{h}^*)$ of polynomials on $\mathfrak{h}^*$, and recall that any algebra homomorphism from $P(\mathfrak{h}^*)$ into $\mathbb{C}$ is given by evaluation at some $\lambda \in \mathfrak{h}^*$. It follows that the homomorphisms from $Z(\mathfrak{g})$ into $\mathbb{C}$ correspond to W-orbits $W\lambda$ in $\mathfrak{h}^*$. So, infinitesimal characters are parametrized by the space $\mathfrak{h}^*/W$. They are important parameters for classifying irreducible $(\mathfrak{g}, K)$-modules. It turns out that for every fixed infinitesimal character, there are only finitely many irreducible $(\mathfrak{g}, K)$-modules with this infinitesimal character. More details about Harish-Chandra isomorphism can be found e.g. in [16].

To finish, let us describe an example where it is easy to explicitly write down all irreducible $(\mathfrak{g}, K)$-modules. This is the case $G = SL(2, \mathbb{R})$, whose representations correspond to $(\mathfrak{sl}(2, \mathbb{C}), SO(2))$-modules.

To see the action of $K = SO(2)$ better, we change basis of $\mathfrak{sl}(2, \mathbb{C})$ and instead of h, e, f used earlier we now use

$$W = \begin{pmatrix} 0 & -i \\ i & 0 \end{pmatrix}, \quad X = \frac{1}{2}\begin{pmatrix} 1 & i \\ i & -1 \end{pmatrix}, \quad Y = \frac{1}{2}\begin{pmatrix} 1 & -i \\ -i & -1 \end{pmatrix}$$

Note that $W \in \mathfrak{k}_{\mathbb{C}}$. The elements W, X and Y satisfy the same relations as before:

$$[W, X] = 2X; \quad [W, Y] = -2Y; \quad [X, Y] = W.$$

Fix $\lambda \in \mathbb{C}$ and $\epsilon \in \{0, 1\}$. Define an $(\mathfrak{sl}(2, \mathbb{C}), SO(2))$-module $V_{\lambda,\epsilon}$ as follows:

- a basis of $V_{\lambda,\epsilon}$ is given by v_n, $n \in \mathbb{Z}$, n congruent to ϵ modulo 2;
- $\pi\begin{pmatrix} \cos\theta & \sin\theta \\ -\sin\theta & \cos\theta \end{pmatrix} v_n = e^{in\theta} v_n$;
- $\pi(W)v_n = nv_n$;
- $\pi(X)v_n = \frac{1}{2}(\lambda + (n+1))v_{n+2}$;
- $\pi(Y)v_n = \frac{1}{2}(\lambda - (n-1))v_{n-2}$.

The picture is similar to the one that we had for finite dimensional representations of $\mathfrak{sl}(2, \mathbb{C})$, but now it is infinite:

$$\cdots \underset{\overleftarrow{\ f\ }}{\overset{e}{\rightarrow}} v_{n-2} \underset{\overleftarrow{\ f\ }}{\overset{e}{\rightarrow}} v_n \underset{\overleftarrow{\ f\ }}{\overset{e}{\rightarrow}} v_{n+2} \underset{\overleftarrow{\ f\ }}{\overset{e}{\rightarrow}} \cdots$$

Also, note that we changed normalization for the $v_n' s$; now we do not have a natural place to start, like a highest weight vector, so it is best to make $\pi(X)$ and $\pi(Y)$ as symmetric as possible.

The following facts are not very difficult to check:

- $V_{\lambda,\epsilon}$ is irreducible unless $\lambda+1$ is an integer congruent to ϵ modulo 2;
- The Casimir element Ω acts by the scalar λ^2-1 on $V_{\lambda,\epsilon}$;
- If $\lambda = k-1$ where $k \geq 1$ is an integer congruent to ϵ modulo 2, then $V_{\lambda,\epsilon}$ contains two irreducible submodules, one with weights $k, k+2, \dots$ and the other with weights $\dots, -k-2, -k$. If $k > 1$, these are called discrete series representations, as they occur discretely in the decomposition of the representation $L^2(G)$. The quotient of $V_{\lambda,\epsilon}$ by the sum of these two submodules is an irreducible module of dimension $k-1$. For $k=1$ the two submodules are called the limits of discrete series, and their sum is all of $V_{0,1}$.

All this can be found with many more details and proofs in Vogan's book [28], Chapter 1. Actually, Chapters 0 and 1 of that book contain a lot of material from this section (plus more) and comprise a good introductory reading. Other books where a lot about $(\mathfrak{g}, K)$-modules can be found are [15] and [33].

2. Clifford algebras, spinors and Dirac operators

2.1. *Clifford algebras*

From now on we will adopt the convention to denote real Lie algebras with the subscript 0 and to refer to complexified Lie algebras with the same letter but no subscript. For example, $\mathfrak{g}_0 = \mathfrak{k}_0 \oplus \mathfrak{p}_0$ will denote the Cartan decomposition of the real Lie algebra $\mathfrak{g}_0$, while $\mathfrak{g} = \mathfrak{k} \oplus \mathfrak{p}$ will be the complexified Cartan decomposition.

We saw in 1.4 that $Z(\mathfrak{g})$ has an important role: it reduces representations, and defines a parameter (infinitesimal character) for the irreducible representations. We also defined the "smallest" non-constant element of $Z(\mathfrak{g})$, the Casimir element Ω.

Let us imagine for a moment that $Z(\mathfrak{g})$ contains a smaller (degree one) element T, such that $T^2 = \Omega$. Then T would (in principle) have more eigenvalues than Ω, as two opposite eigenvalues for T would square to the same eigenvalue for Ω. As a consequence, we would get a better reduction of representations, and infinitesimal character would be a finer invariant. Of course, such a T does not exist; degree one elements of $U(\mathfrak{g})$ are the elements of $\mathfrak{g}$ and $\mathfrak{g}$ has no center being semisimple.

The idea is then to twist $U(\mathfrak{g})$ by a finite dimensional algebra, the Clifford algebra $C(\mathfrak{p})$. The algebra $U(\mathfrak{g}) \otimes C(\mathfrak{p})$ will contain an element D (the Dirac element) whose square is close to $\Omega \otimes 1$.

One can define the Clifford algebra $C(\mathfrak{p})$ as an associative algebra with unit, generated by an orthonormal basis Z_i of $\mathfrak{p}$ (with respect to the Killing form B), subject to the relations

$$Z_i Z_j = -Z_j Z_i \quad (i \neq j); \qquad Z_i^2 = -1.$$

(There are variants obtained by replacing the -1 in the second relation by 1 or by 1/2.) This definition involves choosing a basis, so let us give a nicer one:

$$C(\mathfrak{p}) = T(\mathfrak{p})/I,$$

where $T(\mathfrak{p})$ is the tensor algebra of the vector space $\mathfrak{p}$ and I is the two-sided ideal of $T(\mathfrak{p})$ generated by elements of the form

$$X \otimes Y + Y \otimes X + 2B(X,Y).$$

This definition resembles the definition of $U(\mathfrak{g})$; the similarity extends to an analog of the Poincaré-Birkhoff-Witt theorem. Namely, $C(\mathfrak{p})$ inherits a filtration by degree from $T(\mathfrak{p})$. The associated graded algebra is the exterior algebra $\bigwedge(\mathfrak{p})$. One can obtain a basis for $C(\mathfrak{p})$ by taking an (orthonormal) ordered basis Z_i for $\mathfrak{p}$ and forming monomials over it to obtain

$$Z_{i_1} \dots Z_{i_k}, \quad i_1 < \dots < i_k.$$

Note that no repetitions are allowed here, as Z_i^2 is of lower order. Together with the empty monomial 1, the above monomials form a basis of $C(\mathfrak{p})$ which thus has dimension equal to $2^{\dim \mathfrak{p}}$.

Finally, let us note that one can analogously construct a Clifford algebra for any vector space with a symmetric bilinear form.

2.2. *Dirac operator*

Using again our orthonormal basis Z_i of $\mathfrak{p}$, we define the Dirac operator

$$D = \sum_i Z_i \otimes Z_i \quad \in \quad U(\mathfrak{g}) \otimes C(\mathfrak{p}).$$

It is easy to show that D is independent of the choice of basis Z_i and K-invariant (for the adjoint action of K on both factors).

The adjoint action of $\mathfrak{k}_0$ on $\mathfrak{p}_0$ gives a map from $\mathfrak{k}_0$ into $\mathfrak{so}(\mathfrak{p}_0)$. On the other hand, there is a well known embedding of $\mathfrak{so}(\mathfrak{p}_0)$ into $C(\mathfrak{p}_0)$, given by

$$E_{ij} - E_{ji} \mapsto -\frac{1}{2} Z_i Z_j,$$

where E_{ij} denotes the matrix with all matrix entries equal to zero except the ij entry which is 1. One can check directly that this map is a Lie algebra morphism (where $C(\mathfrak{p}_0)$ is considered a Lie algebra in the usual way, by $[a,b] = ab - ba$). Combining these two maps, we get a map $\alpha : \mathfrak{k}_0 \to C(\mathfrak{p}_0) \subset C(\mathfrak{p})$, and we use it to produce a diagonal embedding of $\mathfrak{k}_0$ into $U(\mathfrak{g}) \otimes C(\mathfrak{p})$, by

$$X \mapsto X \otimes 1 + 1 \otimes \alpha(X), \quad X \in \mathfrak{k}_0.$$

Complexifying this we get a map of $\mathfrak{k}$, hence also of $U(\mathfrak{k})$ and $Z(\mathfrak{k})$ into $U(\mathfrak{g}) \otimes C(\mathfrak{p})$. We denote the images by $\mathfrak{k}_\Delta$, $U(\mathfrak{k}_\Delta)$ and $Z(\mathfrak{k}_\Delta)$ (Δ for diagonal). In particular, we get $\Omega_{\mathfrak{k}_\Delta}$, the image of the Casimir element $\Omega_\mathfrak{k}$ of $Z(\mathfrak{k})$. Since $\Omega_\mathfrak{k} = -\sum_k W_k^2$, we see

$$\Omega_{\mathfrak{k}_\Delta} = -\sum_k (W_k \otimes 1 + 1 \otimes \alpha(W_k))^2.$$

Here is now the announced relationship between D^2 and the Casimir element $\Omega_\mathfrak{g}$:

Lemma: (Parthasarathy)

$$D^2 = -\Omega_\mathfrak{g} \otimes 1 + \Omega_{\mathfrak{k}_\Delta} + C,$$

where C is a constant that can be computed explicitly ($C = ||\rho_c||^2 - ||\rho||^2$, where ρ is the half sum of positive roots and ρ_c is the half sum of compact positive roots).

We just start the calculation and invite the reader to continue. Using the relations in $C(\mathfrak{p})$, we see the left hand side is

$$D^2 = \sum_{i,j} Z_i Z_j \otimes Z_i Z_j = -\sum_i Z_i^2 \otimes 1 + \sum_{i<j} [Z_i, Z_j] \otimes Z_i Z_j.$$

On the other hand, the right hand side is

$$(\sum_k W_k^2 \otimes 1 - \sum_i Z_i^2 \otimes 1) - \sum_k (W_k^2 \otimes 1 + 2W_k \otimes \alpha(W_k) + 1 \otimes \alpha(W_k)^2) + C.$$

One now shows easily that

$$\sum_{i<j} [Z_i, Z_j] \otimes Z_i Z_j = -2 \sum_k W_k \otimes \alpha(W_k),$$

and somewhat less easily that $\sum_k \alpha(W_k)^2$ is a constant (actually, exactly the one mentioned above).

2.3. *Spinors*

Let X be a $(\mathfrak{g}, K)$-module, so $U(\mathfrak{g})$ acts on X. We want to get the Dirac operator to act, so we have to tensor X with a $C(\mathfrak{p})$-module S; then $U(\mathfrak{g}) \otimes C(\mathfrak{p})$ will act on $X \otimes S$, in particular D will act. Since we also want to stay as close as possible to X, we want to take minimal, i.e., simple S. It turns out there are only one or two choices for S, depending on whether $\dim \mathfrak{p}$ is even or odd. A simple $C(\mathfrak{p})$-module is called the spin module, or a space of spinors.

Let us first consider the case when $\dim \mathfrak{p}$ is even, say $2r$. The construction of S involves taking a maximal isotropic subspace $\mathfrak{u}$ of $\mathfrak{p}$ with respect to B. We can for example construct one such subspace starting from our orthonormal basis Z_i of $\mathfrak{p}_0$ and dividing it into two groups, $Z_1, \ldots, Z_r$ and $Z_{r+1}, \ldots, Z_{2r}$. Now we can take for $\mathfrak{u}$ the span of all $Z_s + iZ_{r+s}$, $s = 1, \ldots, r$. Clearly, $\bar{\mathfrak{u}}$ will be a complementary isotropic subspace (the conjugation is with respect to $\mathfrak{p}_0$), and we have

$$\mathfrak{p} = \mathfrak{u} \oplus \bar{\mathfrak{u}}.$$

Since the restriction of B to both $\mathfrak{u}$ and $\bar{\mathfrak{u}}$ is zero, it follows that B identifies $\bar{\mathfrak{u}}$ with the dual space $\mathfrak{u}^*$ of $\mathfrak{u}$. Furthermore, the Clifford algebras over $\mathfrak{u}$ and $\bar{\mathfrak{u}}$ are equal to the exterior algebras, since $B = 0$. We take a basis u_i of $\mathfrak{u}$ and the dual basis $\bar{u}_i$ of $\bar{\mathfrak{u}}$. Let $\bar{u}_* = \bar{u}_1 \ldots \bar{u}_r$ be the corresponding basis element of $\bigwedge^{top} \bar{\mathfrak{u}}$. We define S to be the left ideal in $C(\mathfrak{p})$ generated by $\bar{u}_*$.

More explicitly, one can identify

$$S \cong \bigwedge(\mathfrak{u})\bar{u}_* \cong \bigwedge(\mathfrak{u}),$$

(since $\bar{\mathfrak{u}}$ annihilates $\bar{u}_*$). The action of $\mathfrak{p}$ is now given by

$$u \cdot (u_{i_1} \wedge \cdots \wedge u_{i_s}) = u \wedge u_{i_1} \wedge \cdots \wedge u_{i_s}$$

$$\bar{u} \cdot (u_{i_1} \wedge \cdots \wedge u_{i_s}) = -2 \sum_k B(\bar{u}, u_{i_k}) u_{i_1} \wedge \ldots \widehat{u}_{i_k} \cdots \wedge u_{i_s}$$

for $u \in \mathfrak{u}$ and $\bar{u} \in \bar{\mathfrak{u}}$. Namely, u just Clifford multiplies, or equivalently wedge multiplies, from the left, while $\bar{u}$ has to commute through, and then eventually gets killed upon meeting $\bar{u}_*$.

It is quite easy to show that S is a simple $C(\mathfrak{p})$-module, and not too difficult that it is the only one up to isomorphism. See e.g. [33], 9.2.1.

Let us now briefly examine the case when $\dim \mathfrak{p}$ is odd, say $2r+1$. We can still consider maximal isotropic $\mathfrak{u}$ and $\bar{\mathfrak{u}}$ as before, but now there is an additional element $Z \in \mathfrak{p}$, orthogonal to both $\mathfrak{u}$ and $\bar{\mathfrak{u}}$, and we take it to be of norm 1. We again take S to be $\bigwedge \mathfrak{u}$, and we want to define an action of Z on S. From the relations it follows that Z must be scalar on every homogeneous component of $\bigwedge \mathfrak{u}$, and that these scalars must alternate as the degree changes. To see this, start by observing that all elements of $\mathfrak{u}$ kill the top wedge $\bigwedge^r \mathfrak{u}$, and that there are no other elements of $\bigwedge \mathfrak{u}$ killed by all elements of $\mathfrak{u}$. Since $\bigwedge^r \mathfrak{u}$ is one dimensional, Z must act on it by a scalar. Now use anticommuting of Z and u_i^*'s to see that Z acts on $\bigwedge^{r-1} \mathfrak{u}$ by the opposite scalar, etc. From $Z^2 = -1$, the scalars by which Z can act are i and $-i$. We thus have two choices: Z can be i on $\bigwedge^{even} \mathfrak{u}$ and $-i$ on $\bigwedge^{odd} \mathfrak{u}$, or vice versa. This gives us two nonisomorphic simple $C(\mathfrak{p})$-modules S_1 and S_2. One shows by a similar argument as above that these are the only simple $C(\mathfrak{p})$-modules up to isomorphism.

2.4. *Spin group*

Let us again first assume that $\dim \mathfrak{p}$ is even. Let $v \in \mathfrak{p}_0$ be of length 1, i.e., $B(v,v) = 1$. Then v is clearly invertible in $C(\mathfrak{p}_0)$, as $v^{-1} = -v$. Consider now the action of v on $\mathfrak{p}_0$ by conjugation in $C(\mathfrak{p}_0)$:

$$r_v(X) = vXv^{-1} = -vXv, \quad X \in \mathfrak{p}_0.$$

If X is orthogonal to v, then v and X anticommute, and hence $r_v(X) = -X$. If X is proportional to v, then $r_v(X) = X$. So we see that r_v is minus the reflection with respect to the hyperplane orthogonal to v.

All v as above generate a subgroup of the invertible elements of $C(\mathfrak{p}_0)$ which we denote by $Pin(\mathfrak{p}_0)$. The above discussion shows that we have a map

$$Pin(\mathfrak{p}_0) \longrightarrow O(\mathfrak{p}_0),$$

which is surjective and has kernel $\{1,-1\}$. The connected component of $Pin(\mathfrak{p}_0)$ is the spin group $Spin(\mathfrak{p}_0)$; it coincides with the products of an even number of vectors v as above. It is a compact, semisimple group which is a double cover of the group $SO(\mathfrak{p}_0)$.

In case $\dim \mathfrak{p}$ is odd, one can do a similar construction. Instead of doing this, let us give a uniform description of $Spin(\mathfrak{p}_0)$ valid regardless of the parity of $\dim \mathfrak{p}$ (and also for forms which are not necessarily positive definite); see [17], p. 282. Let α be the antiautomorphism of $C(\mathfrak{p})$ given by the identity on $\mathfrak{p}$. Then $Spin(\mathfrak{p}_0)$ is the group of all even elements g of $C(\mathfrak{p}_0)$

such that $g\alpha(g) = 1$ and $gx\alpha(g) \in \mathfrak{p}_0$ for all $x \in \mathfrak{p}_0$. For $g \in Spin(\mathfrak{p}_0)$, define $T(g) \in GL(\mathfrak{p}_0)$ by $T(g)x = gx\alpha(g)$. Then $T : Spin(\mathfrak{p}_0) \to SO(\mathfrak{p}_0)$ is a double covering. In particular, $Spin(\mathfrak{p}_0)$ is compact (since B is positive definite on $\mathfrak{p}_0$).

Since $Spin(\mathfrak{p}_0)$ is contained in $C(\mathfrak{p})$, it acts on any $C(\mathfrak{p})$-module. For $\dim \mathfrak{p}$ even, there is only one such simple module, S. Since $Spin(\mathfrak{p}_0)$ consists of even elements of $C(\mathfrak{p})$, it preserves the subspaces

$$S^+ = \bigwedge\nolimits^{even} \mathfrak{u}; \qquad S^- = \bigwedge\nolimits^{odd} \mathfrak{u}$$

of S. These are actually irreducible, and they are called spin representations.

For $\dim \mathfrak{p}$ odd, the two spaces of spinors S_1 and S_2 are irreducible for $Spin(\mathfrak{p}_0)$ and equivalent to each other. This representation is also called the spin representation.

2.5. *Dirac cohomology*

We consider the spin double cover $\tilde{K}$ of K, constructed from the following pullback diagram:

$$\begin{array}{ccc} \tilde{K} & \longrightarrow & Spin(\mathfrak{p}_0) \\ \downarrow & & \downarrow \\ K & \longrightarrow & SO(\mathfrak{p}_0) \end{array}$$

Now if X is a $(\mathfrak{g}, K)$-module, then $\tilde{K}$ acts on $X \otimes S$ by acting on both factors: on X through K and on S through $Spin(\mathfrak{p}_0)$. Moreover, it is easy to show that $X \otimes S$ is a $(U(\mathfrak{g}) \otimes C(\mathfrak{p}), \tilde{K})$-module. Such modules are defined analogously as $(\mathfrak{g}, K)$-modules; here $\tilde{K}$ acts on $U(\mathfrak{g}) \otimes C(\mathfrak{p})$ through the adjoint action of K on both factors, and the Lie algebra of $\tilde{K}$ embeds into $U(\mathfrak{g}) \otimes C(\mathfrak{p})$ as the diagonal $\mathfrak{k}_\Delta$ described earlier.

Dirac operator D acts on $X \otimes S$ and we define the Dirac cohomology of X to be the $\tilde{K}$ module

$$H_D(X) = \operatorname{Ker} D / \operatorname{Ker} D \cap \operatorname{Im} D.$$

If we suppose X is a unitary $(\mathfrak{g}, K)$-module, then we can define a positive definite hermitian form such that D is symmetric with respect to this form. For this we use the usual form on S for which all elements of $\mathfrak{p}_0$ are skew-hermitian (see [33], 9.2.3, or [7]). It now follows that if X is unitary, then $\operatorname{Ker} D \cap \operatorname{Im} D = 0$, and the Dirac cohomology of X is just $\operatorname{Ker} D$. More or less all of the above can be found in any of the following references: [7], [17], [33] or [15] (even case).

3. Dirac operators and group representations

3.1. *Paul Dirac and his operator*

Paul Dirac (1902-1984) was one of the greatest physicists of the 20th century. His research interests were mainly quantum mechanics and elementary particles. A free particle T in $\mathbb{R}^3$ is described by a state function $\psi(t,x)$ with $t \in \mathbb{R}$ and $x \in \mathbb{R}^3$. To understand this function, one needs to understand the square root of the wave operator

$$\square = \frac{\partial^2}{\partial x_0^2} - \frac{\partial^2}{\partial x_1^2} - \frac{\partial^2}{\partial x_2^2} - \frac{\partial^2}{\partial x_3^2}.$$

In 1928, Dirac found the square root D of $\square$. It is a matrix valued first order differential operator, and it is now called the Dirac operator.

3.2. *Square root of Laplace operator*

We first describe the square root of the Laplace operator in $\mathbb{R}^n$:

$$\Delta = -\frac{\partial^2}{\partial x_1^2} - \frac{\partial^2}{\partial x_2^2} - \cdots - \frac{\partial^2}{\partial x_n^2}.$$

For $n = 1$, the Laplacian operator has the form $\Delta = -\frac{\partial^2}{\partial x^2}$ and the Dirac operator $D = i\frac{\partial}{\partial x}$ acts on the space of smooth functions $f\colon \mathbb{R} \to \mathbb{C}$.

For $n = 2$, the Laplacian operator has the form $\Delta = -\frac{\partial^2}{\partial x^2} - \frac{\partial^2}{\partial y^2}$. We define

$$D = \begin{pmatrix} 0 & i \\ i & 0 \end{pmatrix} \frac{\partial}{\partial x} + \begin{pmatrix} 0 & 1 \\ -1 & 0 \end{pmatrix} \frac{\partial}{\partial y} = \gamma_x \frac{\partial}{\partial x} + \gamma_y \frac{\partial}{\partial y},$$

which acts on the space of smooth functions $f\colon \mathbb{R}^2 \to \mathbb{C}^2$. It follows from

$$\gamma_x^2 = \gamma_y^2 = -I, \ \gamma_x\gamma_y + \gamma_y\gamma_x = 0$$

that

$$D^2 = \Delta \begin{pmatrix} 1 & 0 \\ 0 & 1 \end{pmatrix} = \Delta I.$$

For $n = 3$, the Laplacian operator has the form $\Delta = -\frac{\partial^2}{\partial x_1^2} - \frac{\partial^2}{\partial x_2^2} - \frac{\partial^2}{\partial x_3^2}$. We define

$$\gamma_1 = \begin{pmatrix} i & 0 \\ 0 & -i \end{pmatrix}, \gamma_2 = \begin{pmatrix} 0 & -1 \\ 1 & 0 \end{pmatrix}, \gamma_3 = \begin{pmatrix} 0 & i \\ i & 0 \end{pmatrix}.$$

Then one has

$$\gamma_i^2 = -I, \gamma_i\gamma_j + \gamma_j\gamma_i = 0 (i \neq j), i, j \in \{1, 2, 3\}.$$

Set

$$D = \gamma_1 \frac{\partial}{\partial x_1} + \gamma_2 \frac{\partial}{\partial x_2} + \gamma_3 \frac{\partial}{\partial x_3},$$

which acts on the space of smooth functions $f\colon \mathbb{R}^3 \to \mathbb{C}^2$. It follows that

$$D^2 = \Delta \begin{pmatrix} 1 & 0 \\ 0 & 1 \end{pmatrix} = \Delta I.$$

3.3. *The original Dirac operator and its generalizations*

Dirac defined the square root of the wave operator $\Box$ in terms of Pauli matrices:

$$\sigma_0 = \begin{pmatrix} 1 & 0 \\ 0 & 1 \end{pmatrix}, \sigma_1 = \begin{pmatrix} 1 & 0 \\ 0 & -1 \end{pmatrix}, \sigma_2 = \begin{pmatrix} 0 & -i \\ i & 0 \end{pmatrix}, \sigma_3 = \begin{pmatrix} 0 & 1 \\ 1 & 0 \end{pmatrix}.$$

Then one has

$$\sigma_i^2 = I, \sigma_j \sigma_k = -\sigma_k \sigma_j = -i\sigma_l, \text{ if } j < k \text{ and } \{j, k, l\} = \{1, 2, 3\}.$$

Define 4×4 matrices

$$\gamma_0 = \begin{pmatrix} 0 & \sigma_0 \\ \sigma_0 & 0 \end{pmatrix}, \gamma_j = \begin{pmatrix} 0 & \sigma_j \\ -\sigma_j & 0 \end{pmatrix}, \quad j = 1, 2, 3$$

and

$$D = \gamma_0 \frac{\partial}{\partial x_0} + \gamma_1 \frac{\partial}{\partial x_1} + \gamma_2 \frac{\partial}{\partial x_2} + \gamma_3 \frac{\partial}{\partial x_3}.$$

It follows from

$$\gamma_0^2 = I, \ \gamma_1^2 = \gamma_2^2 = \gamma_3^2 = -I, \ \ \gamma_j \gamma_k = -\gamma_k \gamma_j \ (j \neq k)$$

that

$$D^2 = \Box I.$$

Brauer and Weyl afterwards generalized the definition of Dirac operator to arbitrary finite-dimensional (quadratic) space of arbitrary signature. The Dirac operator defined in 2.2 is a much more recent analogue for semisimple Lie algebras; it was introduced by Parthasarathy in 1972.

3.4. *Group representations*

Representations of finite groups were studied by Dedekind, Frobenius, Hurwitz and Schur at the beginning of the 20th century. In 1920's, the focus of investigations was representation theory of compact Lie groups and its relations to invariant theory. Cartan and Weyl obtained the well-known classification of equivalence classes of irreducible unitary representations of compact Lie groups in terms of highest weights. In 1930's, Dirac and Wigner started the investigation of infinite-dimensional representations of noncompact Lie groups.

Harish-Chandra (1923-1983) was a Ph.D. student at the University of Cambridge under supervision of Dirac during the years 1945-1947. It was Harish-Chandra who began a systematical investigation of infinite-dimensional representations of semisimple Lie groups after his Ph.D. thesis. He laid down the foundation for further development of the theory for the last half century. In 1964 and 1965, Harish-Chandra published two papers which gave a complete parametrization of discrete series representations. Later he used this classification to prove the Plancherel formula. This classification of discrete series is also crucial to Langlands classification of admissible representations. However, Harish-Chandra did not give explicit construction of discrete series. His work was parallel to that of Cartan-Weyl for irreducible unitary representations of compact Lie groups.

In 1955, Borel and Weil gave explicit realization of irreducible unitary representations of compact Lie groups. In 1957, Bott generalized Borel-Weil theorem by considering Dolbeault cohomology of line bundles over G/T, where T is a maximal torus. Soon thereafter, Kostant and Langlands conjectured that discrete series representations are equivalent to certain Dolbeault cohomology of line bundles over G/T for noncompact G. In his 1967 thesis [25], Schmid started a proof of the conjecture of Kostant and Langlands. This work was completed in his Annals of Mathematics paper [26].

After Hotta's work on construction of holomorphic discrete series and Parthasarathy's construction of most discrete series representations by Dirac operators, Atiyah and Schmid in 1977 proved that all discrete series can be constructed as kernels of the Dirac operator over twisted spinor bundles.

The definition of the Dirac operator for a spin bundle was a major accomplishment of Atiyah-Singer, who obtained the celebrated index theorem. The Dirac cohomology is a far reaching generalization of the idea of index theory to representation theory. In the rest of the sections we see how this simplifies proofs of many classical results such as Bott-Borel-Weil theorem and Atiyah-Schmid theorem and how it sharpens a result of Langlands and Hotta-Parthasarthy on multiplicities of automorphic forms. An important tool in our approach is the theory of $A_{\mathfrak{q}}(\lambda)$-modules which we review in the next section. One should bear in mind that this theory was not available at the time when the above mentioned classical results were proved.

4. Introduction to $A_{\mathfrak{q}}(\lambda)$-modules

Mackey's construction of induction is based on real analysis, and Zuckerman's construction of induction which is needed for $A_{\mathfrak{q}}(\lambda)$-modules is based on complex analysis. It amounts to the geometric construction of representations by using Dolbeault cohomology sections of vector bundles over a noncompact complex homogeneous spaces then passing to the Taylor coefficients.

4.1. *θ-stable parabolic subalgerbas*

Let $H = TA$ be a fundamental Cartan in G. Let $\mathfrak{h}_0 = \mathfrak{t}_0 + \mathfrak{a}_0$ be the corresponding θ-stable Cartan subalgebra. Then $\mathfrak{t}_0 = \mathfrak{h}_0 \cap \mathfrak{k}_0$ is a Cartan subalgebra of $\mathfrak{k}_0$. As usual we drop the subscript 0 for the complexified Lie algebras. Let $X \in i\mathfrak{t}_0$ be such that $\mathrm{ad}(X)$ is semisimple with real eigenvalues. We define

(i) $\mathfrak{l}$ to be the zero eigenspace of $\mathrm{ad}(X)$,
(ii) $\mathfrak{u}$ to be the sum of positive eigenspaces of $\mathrm{ad}(X)$,
(iii) $\mathfrak{q}$ to be the sum of non-negative eigenspaces of $\mathrm{ad}(X)$.

Then $\mathfrak{q}$ is a parabolic subalgebra of $\mathfrak{g}$ and $\mathfrak{q} = \mathfrak{l} + \mathfrak{u}$ is a Levi decomposition. Furthermore, $\mathfrak{l}$ is the complexification of $\mathfrak{l}_0 = \mathfrak{q} \cap \mathfrak{g}_0$. We write L for the connected subgroup of G with Lie algebra $\mathfrak{l}_0$. Since $\theta(X) = X$, $\mathfrak{l}$,$\mathfrak{u}$ and $\mathfrak{q}$ are all invariant under θ, so

$$\mathfrak{q} = \mathfrak{q} \cap \mathfrak{k} + \mathfrak{q} \cap \mathfrak{p}.$$

In particular, $\mathfrak{q} \cap \mathfrak{k}$ is a parabolic subalgebra of $\mathfrak{k}$ with Levi decomposition

$$\mathfrak{q} \cap \mathfrak{k} = \mathfrak{l} \cap \mathfrak{k} + \mathfrak{u} \cap \mathfrak{k}.$$

We call such a $\mathfrak{q}$ a θ-stable parabolic subalgebra.

Let $\mathfrak{f} \subset \mathfrak{q}$ be any subspace stable under $\mathrm{ad}(\mathfrak{t})$. Then there is a subset $\{\alpha_1, \ldots, \alpha_r\}$ of $\mathfrak{t}^*$ and subspaces $\mathfrak{f}_{\alpha_i}$ of $\mathfrak{f}$ such that if $y \in \mathfrak{t}$ and $v \in \mathfrak{f}_{\alpha_i}$, then

$$\mathrm{ad}(y)v = \alpha_i(y)v.$$

We write

$$\Delta(\mathfrak{f},\mathfrak{t}) = \Delta(\mathfrak{f}) = \{\alpha_1,\ldots,\alpha_r\},$$

the weights or roots of $\mathfrak{t}$ in $\mathfrak{f}$. Here $\Delta(\mathfrak{f})$ is a set with multiplicities, with α_i having multiplicity $\dim \mathfrak{f}_{\alpha_i}$. Then if

$$\rho(\mathfrak{f}) = \rho(\Delta(\mathfrak{f})) = \frac{1}{2}\sum_{\alpha_i\in\Delta(\mathfrak{f})} \alpha_i \in \mathfrak{t}^*,$$

we have

$$\rho(\mathfrak{f})(y) = \frac{1}{2}\operatorname{tr}(\operatorname{ad}(y)|_{\mathfrak{f}}) \qquad (y\in\mathfrak{t}).$$

Fix a system $\Delta^+(\mathfrak{l}\cap\mathfrak{k})$ of positive roots in the root system $\Delta(\mathfrak{l}\cap\mathfrak{k},\mathfrak{t})$. (Note that we extend the meaning of root system to include the zero weights.) Then

$$\Delta^+(\mathfrak{k}) = \Delta^+(\mathfrak{l}\cap\mathfrak{k})\cup\Delta(\mathfrak{u}\cap\mathfrak{k})$$

is a positive root system for $\mathfrak{t}$ in $\mathfrak{k}$.

If Z is an $(\mathfrak{l}, L\cap K)$-module, we write $Z^{\#}$ for $Z\otimes\wedge^{\mathrm{top}}\mathfrak{u}$. We set

$$\mathrm{pro}(Z^{\#}) = \mathrm{Hom}_{U(\mathfrak{q})}(U(\mathfrak{g}), Z^{\#})_{L\cap K-finite} \text{ and } \mathrm{ind}(Z^{\#}) = U(\mathfrak{g})\otimes_{U(\bar{\mathfrak{q}})} Z^{\#}.$$

Then both *pro*$(Z^{\#})$ and *ind*$(Z^{\#})$ are $(\mathfrak{g}, L\cap K)$-modules.

4.2. *Zuckerman and Bernstein functors* Γ

Let V be a $(\mathfrak{g}, L\cap K)$-module. The Zuckerman functor $\Gamma = \Gamma^{\mathfrak{g},K}_{\mathfrak{g},L\cap K}$ can be defined as follows (if K is connected): Let $\Gamma(V)$ be the sum of all finite-dimensional $\mathfrak{k}_0$-invariant subspaces of V such that the $\mathfrak{k}_0$-action can be lifted to K. For a morphism $\varphi: V\to W$ of $(\mathfrak{g}, L\cap K)$-modules, let $\Gamma(\varphi)$ be the restriction of φ to $\Gamma(V)$.

The functor Γ is not exact, but only left exact. Therefore one also needs to study the right derived functors Γ^j. For the study of unitarity, it is useful to consider also a "left analog" of Γ, the Bernstein functor $\Pi = \Pi^{\mathfrak{g},K}_{\mathfrak{g},L\cap K}$ and its left derived functors Π_j. Π is not so easily defined directly; however, there is a uniform way to describe Zuckerman and Bernstein functors in terms of Hecke algebras, or in terms of relative Lie algebra (co)homology.

Let $R(K)$ be the space of distributions on K. Let $R(\mathfrak{g},K)$ be the space of left and right K-finite distributions on G with support in K. Then

$$R(\mathfrak{g},K)\cong R(K)\otimes_{U(\mathfrak{k})} U(\mathfrak{g}) \cong U(\mathfrak{g})\otimes_{U(\mathfrak{k})} R(K).$$

Then one has that the Zuckerman functor on V is

$$\Gamma(V) = \mathrm{Hom}_{R(\mathfrak{g},L\cap K)}(R(\mathfrak{g},K),V)_{K-finite},$$

and the Bernstein functor is

$$\Pi(V) = R(\mathfrak{g},K)\otimes_{R(\mathfrak{g},L\cap K)} V.$$

We set

$$\mathcal{R}^j(Z) = \Gamma^j(\mathrm{pro}\, Z^\#) \text{ and } \mathcal{L}_j(Z) = \Pi_j(\mathrm{ind}\, Z^\#).$$

4.3. *Irreducibility and unitarity of cohomologically induced modules*

The hermitian inner product is not obvious for cohomological parabolic induction. This makes studying unitarity a very difficult problem for cohomologically induced modules. Nevertheless, Vogan proved the following powerful theorem.

Theorem: ([29]) *Suppose $\mathfrak{q}$ is a θ-stable parabolic subalgebra of $\mathfrak{g}$ and Z is an $(\mathfrak{l}, L\cap K)$-module with infinitesimal character λ. If Z is weakly good (i.e. $\mathrm{Re}\langle\lambda+\rho(\mathfrak{u}),\alpha\rangle \geq 0$ for any $\alpha\in\Delta(\mathfrak{u})$), then*

(i) *$\mathcal{L}_j(Z) = \mathcal{R}^j(Z) = 0$ for $j\neq s$ ($s = \dim\mathfrak{u}\cap\mathfrak{k}$).*

(ii) *$\mathcal{L}_s(Z)\cong\mathcal{R}^s(Z)$.*

(iii) *If Z is irreducible, then $\mathcal{L}_s(Z)$ is irreducible or zero.*

(iv) *If Z is irreducible and in addition is good (i.e. $\mathrm{Re}\langle\lambda+\rho(\mathfrak{u}),\alpha\rangle > 0$ for any $\alpha\in\Delta(\mathfrak{u})$), then $\mathcal{L}_s(Z)$ is irreducible and nonzero.*

(v) *If Z is unitary, then $\mathcal{L}_s(Z)$ is unitary.*

Remark: There are two dualities:

$$(\Pi^{\mathfrak{g},K}_{\mathfrak{g},L\cap K})_j \cong (\Gamma^{\mathfrak{g},K}_{\mathfrak{g},L\cap K})^{2s-j} \quad \text{and} \quad \Pi_j(W)^h \cong \Gamma^j(W^h)$$

where W^h is the hermitian dual of W.

4.4. $A_q(\lambda)$*-modules*

Now we consider Z to be a one-dimensional representation. $\lambda\colon\mathfrak{l}\to\mathbb{C}$ is called admissible if it satisfies the following conditions:

(i) λ is the differential of a unitary character of L

(ii) if $\alpha\in\Delta(\mathfrak{u})$, then $\langle\alpha,\lambda|_{\mathfrak{t}}\rangle\geq 0$.

Given $\mathfrak{q}$ and an admissible λ, define

$$\mu(\mathfrak{q},\lambda) = \text{representation of } K \text{ of highest weight } \lambda|_{\mathfrak{t}} + 2\rho(\mathfrak{u}\cap\mathfrak{p}). \qquad (1)$$

The following theorem is due to Vogan and Zuckerman.

Theorem: ([32], [29]) *Suppose* $\mathfrak{q}$ *is a* θ*-stable parabolic subalgebra of* $\mathfrak{g}$ *and* $\lambda: \mathfrak{l} \to \mathbb{C}$ *is admissible as defined above. Then there is a unique unitary* $(\mathfrak{g}, K)$*-module* $A_\mathfrak{q}(\lambda)$ *with the following properties:*

(i) *The restriction of* $A_\mathfrak{q}(\lambda)$ *to* $\mathfrak{k}$ *contains* $\mu(\mathfrak{q}, \lambda)$ *as defined in (4.1);*

(ii) $A_\mathfrak{q}(\lambda)$ *has infinitesimal character* $\lambda + \rho$*;*

(iii) *If the representation of* $\mathfrak{k}$ *of the highest weight* δ *occurs in* $A_\mathfrak{q}(\lambda)$*, then*

$$\delta = \mu(\mathfrak{q}, \lambda) + \sum_{\beta \in \Delta(\mathfrak{u} \cap \mathfrak{p})} n_\beta \beta$$

with $\mathfrak{n}_\beta$ *non-negative integers. In particular,* $\mu(\mathfrak{q}, \lambda)$ *is the lowest* K*-type of* $A_\mathfrak{q}(\lambda)$.

We note that the unitarity of $A_\mathfrak{q}(\lambda)$ in the above theorem was proved in [29]. In the context of definition of θ-stable parabolic subalgebras, if we take X to be a regular element, then we obtain a minimal θ-stable subalgebra $\mathfrak{b} = \mathfrak{h} + \mathfrak{n}$. We call such a subalgebra $\mathfrak{b}$ a θ-stable Borel subalgebra. The corresponding representation $A_\mathfrak{b}(\lambda)$ is called a fundamental series representation. It is the $(\mathfrak{g}, K)$-module of a tempered representation of G. If G has a compact Cartan subgroup, then $A_\mathfrak{b}(\lambda)$ is the $(\mathfrak{g}, K)$-module of a discrete series representation of G. Moreover, all $(\mathfrak{g}, K)$-modules of discrete series representations of G are of this form; this will be important in Section 6. For the proof of the main result of [12] (Section 5) it is only needed that $A_\mathfrak{b}(\lambda)$ has infinitesimal character $\lambda + \rho$ and the lowest K-type $\mu(\mathfrak{b}, \lambda) = \lambda + 2\rho_n$, where $\rho_n = \rho(\mathfrak{n} \cap \mathfrak{p})$. These facts are contained in Theorem 4.4.

4.5. *Salamanca-Riba's classification of the unitary dual with strongly regular infinitesimal characters*

As before, let $\mathfrak{h}$ be the complexification of a fundamental Cartan subalgebra $\mathfrak{h}_0$ of $\mathfrak{g}_0$. Given any weight $\Lambda \in \mathfrak{h}^*$, fix a choice of positive roots $\Delta^+(\Lambda, \mathfrak{h})$ for Λ so that

$$\Delta^+(\Lambda, \mathfrak{h}) \subset \{\alpha \in \Delta(\mathfrak{g}, \mathfrak{h}) \mid \mathrm{Re}\langle \Lambda, \alpha\rangle \geq 0\}.$$

Set

$$\rho(\Lambda) = \frac{1}{2} \sum_{\alpha \in \Delta^+(\Lambda, \mathfrak{h})} \alpha.$$

Definition: A weight $\Lambda \in \mathfrak{h}^*$ is said to be real if

$$\Lambda \in i\mathfrak{t}_0^* + \mathfrak{a}_0^*,$$

and to be strongly regular if it is real and

$$\langle \Lambda - \rho(\Lambda), \alpha \rangle \geq 0, \quad \forall \alpha \in \Delta^+(\Lambda, \mathfrak{h}).$$

Salamanca-Riba [24] proved:

Theorem: (Salamanca-Riba) *Suppose that X is an irreducible unitary $(\mathfrak{g}, K)$-module with strongly regular infinitesimal character $\Lambda \in \mathfrak{h}^*$. Then there exist a θ-stable parabolic subalgebra $\mathfrak{q} = \mathfrak{l} + \mathfrak{u}$ and an admissible character λ of L such that X is isomorphic to $A_\mathfrak{q}(\lambda)$.*

5. Vogan's conjecture and its proof

In this section we explain Vogan's conjecture on Dirac cohomology and a proof of this conjecture. The presentation mostly follows [12].

5.1. *Vogan's conjecture*

Let T be a maximal torus in K, with Lie algebra $\mathfrak{t}_0$. Let $\mathfrak{h}$ be the centralizer of $\mathfrak{t}$ in $\mathfrak{g}$; it is a θ-stable Cartan subalgebra of $\mathfrak{g}$ containing $\mathfrak{t}$. Since $\mathfrak{h} = \mathfrak{t} \oplus \mathfrak{p}^\mathfrak{t}$, we get an embedding of $\mathfrak{t}^*$ into $\mathfrak{h}^*$. Therefore any element of $\mathfrak{t}^*$ determines a character of the center $Z(\mathfrak{g})$ of $U(\mathfrak{g})$. Here we are using the standard identification $Z(\mathfrak{g}) \cong S(\mathfrak{h})^W$ via the Harish-Chandra homomorphism (W is the Weyl group), by which the characters of $Z(\mathfrak{g})$ correspond to the W-orbits in $\mathfrak{h}^*$.

We fix a positive root system $\Delta^+(\mathfrak{k}, \mathfrak{t})$ for $\mathfrak{t}$ in $\mathfrak{k}$; let $\rho_c = \rho(\Delta^+(\mathfrak{k}, \mathfrak{t}))$ be the corresponding half sum of the positive roots. For any finite dimensional irreducible representation (γ, E_γ) of $\mathfrak{k}$, we denote its highest weight in $\mathfrak{t}^*$ by γ again.

Vogan [30] made a conjecture which was proved as the following theorem:

Theorem: ([12]) *Let X be an irreducible $(\mathfrak{g}, K)$-module, such that the Dirac cohomology of X is non-zero. Let γ be a $\tilde{K}$-type contained in the Dirac cohomology. Then the infinitesimal character of X is given by $\gamma + \rho_c$.*

In view of the remarks in 2.5, in case X is unitarizable, we get the following consequence:

Corollary: *Let X be an irreducible unitarizable $(\mathfrak{g}, K)$-module, such that* $\operatorname{Ker} D \neq 0$. *Let γ be a $\tilde{K}$-type contained in* $\operatorname{Ker} D$. *Then the infinitesimal character of X is given by $\gamma + \rho_c$.*

In fact, Vogan first conjectured the above corollary and then he saw that the above theorem should be the right generalization to non-unitary representations.

5.2. *An algebraic reduction of the conjecture*

Vogan further reduced the claim of his conjecture to an entirely algebraic statement in the algebra $U(\mathfrak{g}) \otimes C(\mathfrak{p})$. Let us first recall that in 2.2 we described a diagonal copy $\mathfrak{k}_\Delta$ of $\mathfrak{k}$ inside $U(\mathfrak{g}) \otimes C(\mathfrak{p})$. $U(\mathfrak{k}_\Delta)$ and $Z(\mathfrak{k}_\Delta)$ denote the corresponding universal enveloping algebra and its center. It is easy to see that they are also embedded into $U(\mathfrak{g}) \otimes C(\mathfrak{p})$; namely, if $u \in U(\mathfrak{k})$ is a PBW monomial, then its image in $U(\mathfrak{g}) \otimes C(\mathfrak{p})$ is the sum of $u \otimes 1$ and terms of the form $w \otimes a$, with w having smaller degree than u.

We can now state Vogan's algebraic conjecture that implies the theorem in 5.1.

Theorem: *Let $z \in Z(\mathfrak{g})$. Then there is a unique $\zeta(z)$ in the center $Z(\mathfrak{k}_\Delta)$ of $U(\mathfrak{k}_\Delta)$, and there are K-invariant elements $a, b \in U(\mathfrak{g}) \otimes C(\mathfrak{p})$, such that*

$$z \otimes 1 = \zeta(z) + Da + bD.$$

To see that this theorem implies the theorem in 5.1, let $\tilde{x} \in (X \otimes S)(\gamma)$ be non-zero, such that $D\tilde{x} = 0$ and $\tilde{x} \notin \operatorname{Im} D$. Note that both $z \otimes 1$ and $\zeta(z)$ act as scalars on $\tilde{x}$. The first of these scalars is the infinitesimal character Λ of X applied to z, and the second is the $\mathfrak{k}$-infinitesimal character of γ applied to $\zeta(z)$, that is, $(\gamma + \rho_c)(\zeta(z))$.

On the other hand, since $(z \otimes 1 - \zeta(z))\tilde{x} = Da\tilde{x}$, and $\tilde{x} \notin \operatorname{Im} D$, it follows that $(z \otimes 1 - \zeta(z))\tilde{x} = 0$. Thus the above two scalars are the same, i.e., $\Lambda(z) = (\gamma + \rho_c)(\zeta(z))$.

In 5.6 we will show that under identifications $Z(\mathfrak{g}) \cong S(\mathfrak{h})^W \cong P(\mathfrak{h}^*)^W$ and $Z(\mathfrak{k}_\Delta) \cong Z(\mathfrak{k}) \cong S(\mathfrak{t})^{W_K} \cong P(\mathfrak{t}^*)^{W_K}$ the homomorphism ζ corresponds to the restriction of polynomials on $\mathfrak{h}^*$ to $\mathfrak{t}^*$. Here the already mentioned inclusion of $\mathfrak{t}^*$ into $\mathfrak{h}^*$ is given by extending functionals from $\mathfrak{t}$ to $\mathfrak{h}$, letting them act by 0 on $\mathfrak{a} = \mathfrak{p}^{\mathfrak{t}}$. It follows that $\Lambda = \gamma + \rho_c$, as claimed.

5.3. *A differential complex induced by Dirac operator*

Let us first note that the Clifford algebra $C(\mathfrak{p})$ has a natural $\mathbb{Z}_2$-gradation into even and odd parts:

$$C(\mathfrak{p}) = C^0(\mathfrak{p}) \oplus C^1(\mathfrak{p}).$$

This gradation induces a $\mathbb{Z}_2$-gradation on $U(\mathfrak{g}) \otimes C(\mathfrak{p})$ in an obvious way.

We define a map d from $U(\mathfrak{g}) \otimes C(\mathfrak{p})$ into itself, as $d = d^0 \oplus d^1$, where

$$d^0 : U(\mathfrak{g}) \otimes C^0(\mathfrak{p}) \to U(\mathfrak{g}) \otimes C^1(\mathfrak{p})$$

is given by

$$d^0(a) = Da - aD, \tag{2}$$

and

$$d^1 : U(\mathfrak{g}) \otimes C^1(\mathfrak{p}) \to U(\mathfrak{g}) \otimes C^0(\mathfrak{p})$$

is given by

$$d^1(a) = Da + aD. \tag{3}$$

In other words, if ϵ_a denotes the sign of a, that is, 1 for even a and -1 for odd a, then $d(a) = Da - \epsilon_a aD$ (for homogeneous a, i.e., those a which have sign).

We will use the formula for D^2 from Lemma 2.2, namely

$$D^2 = -\Omega_{\mathfrak{g}} \otimes 1 + \Omega_{\mathfrak{k}_\Delta} + C$$

to prove that our d induces a differential on the K-invariants in $U(\mathfrak{g}) \otimes C(\mathfrak{p})$.

Proposition: *Let d be the map defined in (2) and (3). Then*

(i) *d is K-equivariant, hence induces a map from $(U(\mathfrak{g}) \otimes C(\mathfrak{p}))^K$ into itself.*

(ii) *$d^2 = 0$ on $(U(\mathfrak{g}) \otimes C(\mathfrak{p}))^K$.*

Proof: (i) is trivial, since D is K-invariant.

Let $a \in (U(\mathfrak{g}) \otimes C(\mathfrak{p}))^K$ be even or odd. Then

$$d^2(a) = d(Da - \epsilon_a aD) = D^2 a - \epsilon_{Da} DaD - \epsilon_a (DaD - \epsilon_{aD} aD^2) = D^2 a - aD^2,$$

since obviously $\epsilon_{aD} = \epsilon_{Da} = -\epsilon_a$. From the formula for D^2 (Lemma 2.2), we see that a will commute with D^2 if and only if it commutes with $\Omega_{\mathfrak{k}_\Delta}$. If a is K-invariant, then this clearly holds, as a then commutes with all of $U(\mathfrak{k}_\Delta)$. □

Thus we see that d is a differential on $(U(\mathfrak{g})\otimes C(\mathfrak{p}))^K$, of degree 1 with respect to the above defined $\mathbb{Z}_2$-gradation. Note that we do not have a $\mathbb{Z}$-gradation on $(U(\mathfrak{g})\otimes C(\mathfrak{p}))^K$ so that d is of degree 1, i.e., this is not a complex in the usual sense.

5.4. *Determination of cohomology of the complex*

We want to calculate the cohomology of d. Before we state the result, let us note the following:

Proposition: *$Z(\mathfrak{k}_\Delta)$ is in the kernel of d.*

Proof: Since D is K-invariant, it commutes with $\mathfrak{k}_\Delta$, and thus with $U(\mathfrak{k}_\Delta)$ and in particular with $Z(\mathfrak{k}_\Delta)$. Since $Z(\mathfrak{k}_\Delta)\subset(U(\mathfrak{g})\otimes C^0(\mathfrak{p}))^K$, the claim follows. □

We now state the following theorem which implies Theorem 5.2.

Theorem: *Let d be the differential on $(U(\mathfrak{g})\otimes C(\mathfrak{p}))^K$ constructed above. Then* $\operatorname{Ker} d = Z(\mathfrak{k}_\Delta)\oplus\operatorname{Im} d$. *In particular, the cohomology of d is isomorphic to $Z(\mathfrak{k}_\Delta)$.*

The proof uses the standard method of filtering the algebra (the filtration comes from the usual filtration on $U(\mathfrak{g})$), and then passing to the graded algebra. This graded algebra is of course $S(\mathfrak{g})\otimes C(\mathfrak{p})$. The analogue of our theorem in the graded setting is easy; the complex we get is closely related to the standard Koszul complex associated to the vector space $\mathfrak{p}$. Namely, the operator $\bar{d}$ induced by d on $S(\mathfrak{g})\otimes C(\mathfrak{p})$ is given by supercommuting with

$$\bar{D}=\sum_i Z_i\otimes Z_i,$$

and one easily calculates that

$$\bar{d}(u\otimes Z_{i_1}\dots Z_{i_k})=-2\sum_{j=1}^{k}uZ_{i_j}\otimes Z_{i_1}\dots\hat{Z}_{i_j}\dots Z_{i_k}.$$

Upon identifying $C(\mathfrak{p})$ and $\bigwedge(\mathfrak{p})$ as vector spaces and writing

$$S(\mathfrak{g})\otimes C(\mathfrak{p})=S(\mathfrak{k})\otimes(S(\mathfrak{p})\otimes\bigwedge(\mathfrak{p})),$$

we see that

$$\bar{d}=(-2)id\otimes d_{\mathfrak{p}},$$

where d_p is the Koszul differential for the vector space $\mathfrak{p}$. In particular, $\bar{d}$ is a differential with cohomolgy $S(\mathfrak{k}) \otimes \mathbb{C}$, which embeds into $S(\mathfrak{g}) \otimes C(\mathfrak{p})$ by embedding $\mathbb{C}$ into $S(\mathfrak{p}) \otimes \bigwedge(\mathfrak{p})$ as the constants. Passing to K-invariants, we see that

$$\operatorname{Ker} \bar{d} = S(\mathfrak{k})^K \otimes 1 \oplus \operatorname{Im} \bar{d}$$

on $(S(\mathfrak{g}) \otimes C(\mathfrak{p}))^K$. This is the graded version of our theorem.

One can now go back to the original setting by an easy induction on the degree of the filtration. The main point is that one can reconstruct an element of $Z(\mathfrak{k}_\Delta)$ from its top term.

We refer the reader to our paper [12] for the details of the above proof. Let us note a consequence, which immediately proves Vogan's conjecture, just put $b = a$.

Corollary: *Let $z \in Z(\mathfrak{g})$. Then there is a unique $\zeta(z) \in Z(\mathfrak{k}_\Delta)$, and there is an $a \in (U(\mathfrak{g}) \otimes C^1(\mathfrak{p}))^K$, such that*

$$z \otimes 1 = \zeta(z) + Da + aD.$$

Proof: This follows at once from Theorem 5.4, if we just notice that $z \otimes 1$ commutes with D (indeed, it is in the center of $U(\mathfrak{g}) \otimes C(\mathfrak{p})$), and being even, it is thus in $\operatorname{Ker} d$. Hence, it is of the form $\zeta(z)+d(a) = \zeta(z)+Da+aD$. □

5.5. *Dirac inequality and unitary representations with nonzero Dirac cohomology*

We first indicate how to check if a unitarizable X has non-zero Dirac cohomology.

Proposition: *Let X be an irreducible unitarizable $(\mathfrak{g}, K)$-module with infinitesimal character Λ. Assume that $X \otimes S$ contains a $\tilde{K}$-type γ, i.e., $(X \otimes S)(\gamma) \neq 0$. Assume further that $||\Lambda|| = ||\gamma + \rho_c||$. Then the Dirac cohomology of X,* Ker D, *contains $(X \otimes S)(\gamma)$.*

Proof: This again uses the formula for D^2 from Lemma 2.2. The formula implies that D^2 acts on $(X \otimes S)(\gamma)$ by the scalar

$$-(||\Lambda||^2 - ||\rho||^2) + (||\gamma + \rho_c||^2 - ||\rho_c||^2) + (||\rho_c||^2 - ||\rho||^2) = 0.$$

It follows from self-adjointness of D that $D = 0$ on $(X \otimes S)(\gamma)$. □

Note that Corollary 5.1 implies the converse of the above proposition: if X is irreducible unitarizable, with Dirac cohomology containing $(X \otimes S)(\gamma)$, then the infinitesimal character of X is $\Lambda = \gamma + \rho_c$. Hence $||\Lambda|| = ||\gamma + \rho_c||$. We note that all irreducible unitary representations with nonzero Dirac cohomology and strongly regular infinitesimal characters were described in [12]. They are all $A_{\mathfrak{q}}(\lambda)$-modules. See Proposition 5.6 where this is explained in a special case.

Finally, combining the above proposition with Corollary 5.1, we sharpen the Parthasarathy's Dirac inequality:

Theorem: (Extended Dirac Inequality) *Let X be an irreducible unitarizable $(\mathfrak{g}, K)$-module with infinitesimal character Λ. If $(X \otimes S)(\gamma) \neq 0$, then*

$$||\Lambda|| \leq ||\gamma + \rho_c||.$$

The equality holds if and only if some W conjugate of Λ is equal to $\gamma + \rho_c$.

5.6. *Fundamental series and determination of ζ*

Proposition: *Let X be an $A_{\mathfrak{b}}(\lambda)$-module (as in Theorem 4.4) with $\mathfrak{b}$ a θ-stable Borel subalgebra, i.e., X is a fundamental series representation. Assume that $\lambda|_{\mathfrak{a}} = 0$. Then the Dirac cohomology of X contains a $\tilde{K}$-type E_γ of highest weight $\gamma = \lambda + \rho_n$.*

Proof: The lowest K-type $\mu(\mathfrak{b}, \lambda)$ has highest weight $\lambda + 2\rho_n$. Since $-\rho_n$ is a weight of S, E_γ occurs in $\mu(\mathfrak{b}, \lambda) \otimes S$, hence in $X \otimes S$. Since the infinitesimal character of X is $\lambda + \rho = \gamma + \rho_c$, it follows that E_γ is in the kernel of the Dirac operator D, i.e., in the Dirac cohomology of X. □

Now we can describe the homomorphism ζ explicitly.

Theorem: *The homomorphism ζ satisfies the following commutative diagram:*

$$\begin{array}{ccc} Z(\mathfrak{g}) & \xrightarrow{\zeta} & Z(\mathfrak{k}) \\ \downarrow & & \downarrow \\ S(\mathfrak{h})^W & \xrightarrow{\text{Res}} & S(\mathfrak{t})^{W_K} \end{array}$$

Here the vertical arrows are the Harish-Chandra homomorphisms, and the map Res *corresponds to the restriction of polynomials on $\mathfrak{h}^*$ to $\mathfrak{t}^*$ under the identifications $S(\mathfrak{h})^W = P(\mathfrak{h}^*)^W$ and $S(\mathfrak{t})^{W_K} = P(\mathfrak{t}^*)^{W_K}$. As before, we can view $\mathfrak{t}^*$ as a subspace of $\mathfrak{h}^*$ by extending functionals from $\mathfrak{t}$ to $\mathfrak{h}$, letting them act by 0 on $\mathfrak{a}$.*

Proof: Let $\bar{\zeta} : P(\mathfrak{h}^*)^W \to P(\mathfrak{t}^*)^{W_K}$ be the homomorphism induced by ζ under the identifications via Harish-Chandra homomorphisms. Furthermore, let $\tilde{\bar{\zeta}} : \mathfrak{t}^*/W_K \to \mathfrak{h}^*/W$ be the morphism of algebraic varieties inducing the homomorphism $\bar{\zeta}$. We have to show that $\bar{\zeta}$ is the restriction map, or alternatively that $\tilde{\bar{\zeta}}$ is given by the inclusion map.

We know from the above proposition that the fundamental series representation $A_{\mathfrak{b}}(\lambda)$ has the lowest K-type

$$\mu(\mathfrak{b}, \lambda) = \lambda + 2\rho_n,$$

and infinitesimal character

$$\Lambda = \lambda + \rho.$$

On the other hand, it follows from Proposition 5.5 that if $\lambda|_{\mathfrak{a}} = 0$, then the Dirac cohomology of $A_{\mathfrak{b}}(\lambda)$ contains the $\tilde{K}$-type of highest weight $\gamma = \lambda + \rho_n$.

When proving that Theorem 5.2 implies Theorem 5.1, we saw that $\Lambda(z) = (\gamma + \rho_c)(\zeta(z))$, for all $z \in Z(\mathfrak{g})$. In our present situation we however have

$$\Lambda = \lambda + \rho = (\lambda + \rho_n) + \rho_c = \gamma + \rho_c,$$

so it follows that $\Lambda(\zeta(z)) = \Lambda(z)$ for all $z \in Z(\mathfrak{g})$. This means that $\tilde{\bar{\zeta}}(\Lambda) = \Lambda$, for all infinitesimal characters Λ of the above fundamental series representations.

It is clear that when λ ranges over all admissible weights in $\mathfrak{h}^*$ such that $\lambda|_{\mathfrak{a}} = 0$, then $\Lambda = \lambda + \rho$ form an algebraically dense subset of $\mathfrak{t}^*$. To see this, it is enough to note that such λ span a lattice in $\mathfrak{t}^*$. Hence $\tilde{\bar{\zeta}}$ is indeed the inclusion map. □

5.7. *Remark on finite-dimensional representations*

Both Vogan and Kostant pointed out to us that Theorem 5.6 can also be proved by considering finite-dimensional representations with non-zero Dirac cohomology. These are the representations V_λ of highest weight $\lambda \in \mathfrak{t}^* \subset \mathfrak{h}^*$. See [12], Remark 5.6.

6. Generalized Bott-Borel-Weil Theorem and construction of discrete series

6.1. *Kostant cubic Dirac operator*

Let G be a compact semisimple Lie group and R be a closed subgroup. Let $\mathfrak{g}$ and $\mathfrak{r}$ be the complexifications of the corresponding Lie algebras. Let

$\mathfrak{g} = \mathfrak{r} \oplus \mathfrak{p}$ be the orthogonal decomposition with respect to the Killing form. Choose an orthonormal basis $Z_1, \ldots, Z_n$ of $\mathfrak{p}$ with respect to the Killing form $\langle\ ,\ \rangle$. Kostant [18] defines his cubic Dirac operator to be the element

$$D = \sum_{i=1}^{n} Z_i \otimes Z_i + 1 \otimes v \in U(\mathfrak{g}) \otimes C(\mathfrak{p}),$$

where $v \in C(\mathfrak{p})$ is the image of the fundamental 3-form $\omega \in \bigwedge^3(\mathfrak{p}^*)$,

$$\omega(X,Y,Z) = -\frac{1}{2}\langle X, [Y,Z]\rangle$$

under the Chevalley identification $\bigwedge(\mathfrak{p}^*) \to C(\mathfrak{p})$. Kostant's cubic Dirac operator reduces to the ordinary Dirac operator when $(\mathfrak{g}, \mathfrak{r})$ is a symmetric pair, since $\omega = 0$ for the symmetric pair. Kostant ([18], Theorem 2.16) shows that

$$D^2 = \Omega_{\mathfrak{g}} \otimes 1 - \Omega_{\mathfrak{r}_\Delta} + C, \tag{4}$$

where C is the constant $||\rho||^2 - ||\rho_{\mathfrak{r}}||^2$. This is the generalization of Lemma 2.2. The sign change comes from the fact that Kostant uses a slightly different definition of $C(\mathfrak{p})$, requiring Z_i^2 to be 1 and not -1. Over $\mathbb{C}$, there is no substantial difference between the two conventions.

Now we can define the cohomology of the complex $(U(\mathfrak{g}) \otimes C(\mathfrak{p}))^R$ using Kostant's cubic Dirac operator exactly as in Section 5, i.e., by $d(a) = Da - \epsilon_a a D$. As before, $d^2 = 0$ on $(U(\mathfrak{g}) \otimes C(\mathfrak{p}))^R$. Since the degree of the cubic term is zero in the filtration of $U(\mathfrak{g}) \otimes C(\mathfrak{p})$ used in Section 5, the proof Theorem 5.4 goes through without change and we get

Theorem: *Let d be the differential on $(U(\mathfrak{g}) \otimes C(\mathfrak{p}))^R$ defined by Kostant's cubic Dirac operator as above. Then* $\operatorname{Ker} d = \operatorname{Im} d \oplus Z(\mathfrak{r}_\Delta)$. *In particular, the cohomology of d is isomorphic to $Z(\mathfrak{r}_\Delta)$.*

As a consequence we get an analogous homomorphism $\zeta : Z(\mathfrak{g}) \to Z(\mathfrak{r})$ for a reductive subalgebra $\mathfrak{r}$ in a semisimple Lie algebra $\mathfrak{g}$ and a more general version of Vogan's conjecture.

6.2. *Kostant's theorem on cohomology of homogeneous spaces*

If we fix a Cartan subalgebra $\mathfrak{t}$ of $\mathfrak{r}$ and extend $\mathfrak{t}$ to a Cartan subalgebra $\mathfrak{h}$ of $\mathfrak{g}$, then ζ is induced by the Harish-Chandra homomorphism exactly as in Theorem 5.6. This was proved in [19], by constructing a sufficiently large family of highest weight modules with known infinitesimal characters and nonzero Dirac cohomology.

Moreover, the homomorphism ζ induces the structure of a $Z(\mathfrak{g})$-module on $Z(\mathfrak{r})$, which has topological significance. Namely, Kostant has shown that from a well-known theorem of H. Cartan [4], which is by far the most comprehensive result on the real (or complex) cohomology of a homogeneous space, one has

Theorem: [19] *There exists an isomorphism*

$$H^*(G/R, \mathbb{C}) \cong \mathrm{Tor}_*^{Z(\mathfrak{g})}(\mathbb{C}, Z(\mathfrak{r})).$$

6.3. *A generalized Weyl character formula*

We assume that $\operatorname{rank} R = \operatorname{rank} G$. Let $W^1 \subset W_{\mathfrak{g}}$ be the subset of Weyl group elements that map the positive Weyl chamber for $\mathfrak{g}$ into the positive Weyl chamber for $\mathfrak{r}$. If λ is a dominant weight for $\mathfrak{g}$, then $\lambda + \rho(\mathfrak{g})$ lies in the interior of the Weyl chamber for $\mathfrak{g}$. It follows that $w(\lambda + \rho(\mathfrak{g}))$ lies in the interior of the Weyl chamber for $\mathfrak{r}$ for any $w \in W^1$. Thus, $w \centerdot \lambda = w(\lambda + \rho(\mathfrak{g})) - \rho(\mathfrak{r})$ is a dominant weight for $\mathfrak{r}$. Let V_λ denote the irreducible (finite dimensional) representation of $\mathfrak{g}$ with highest weight λ, and let $U_{w.\lambda}$ denote the irreducible (finite dimensional) representation of $\mathfrak{r}$ with highest weight $w \centerdot \lambda$.

Theorem: [8]

$$V_\lambda \otimes S^+ - V_\lambda \otimes S^- = \sum_{w \in W^1} (-1)^{l(w)} U_{w.\lambda}.$$

It follows that

$$ch(V_\lambda) = \frac{\sum_{w \in W^1} (-1)^{l(w)} ch(U_{w.\lambda})}{\sum_{w \in W^1} (-1)^{l(w)} ch(U_{w.0})}.$$

Note that the above formula reduces to the Weyl character formula when R is a maximal torus T.

6.4. *A generalized Bott-Borel-Weil Theorem*

We assume that U_μ is an irreducible representation of R (or $\tilde{R}$) so that $S \otimes U_\mu$ is a representation of R. The Dirac operator acts on the smooth and L^2-sections on the twisted spinor bundles over G/R, if we let $Z_i \in \mathfrak{g}$ act by differentiating from the right. So we have

$$D\colon L^2(G) \otimes_R (S \otimes U_\mu) \to L^2(G) \otimes_R (S \otimes U_\mu).$$

We write this action in another form:

$$D\colon \operatorname{Hom}_{\tilde{R}}(U_\mu^*, L^2(G) \otimes S) \to \operatorname{Hom}_{\tilde{R}}(U_\mu^*, L^2(G) \otimes S).$$

Then D is formally self-adjoint. By Peter-Weyl theorem, one has $L^2(G) \cong \oplus_{\lambda \in \widehat{G}} V_\lambda \otimes V_\lambda^*$. It follows that

$$\operatorname{Ker} D = \bigoplus_{\lambda \in \widehat{G}} V_\lambda \otimes \operatorname{Ker}\{D\colon \operatorname{Hom}_{\tilde{R}}(U_\mu^*, V_\lambda^* \otimes S) \circlearrowleft\}.$$

The proved Vogan's conjecture implies $\operatorname{Ker} D \neq 0$ if and only if there is some $\lambda \in \widehat{G}$ such that $\lambda + \rho(\mathfrak{g})$ is conjugate to $\mu + \rho(\mathfrak{r})$ by the Weyl group. Further consideration of the multiplicity results in

Theorem: *One has* $\operatorname{Ker} D = V_{w(\mu+\rho(\mathfrak{r}))-\rho(\mathfrak{g})}$ *if there exists a* $w \in W_{\mathfrak{g}}$ *so that* $w(\mu + \rho(\mathfrak{r})) - \rho(\mathfrak{g})$ *is dominant, and* $\operatorname{Ker} D$ *is zero if no such* w *exists.*

In the case $R = T$, a maximal torus, this is a version of Borel-Weil theorem.

Corollary: *Consider*

$$D^+\colon L^2(G) \otimes_R (S^+ \otimes U_\mu) \to L^2(G) \otimes_R (S^- \otimes U_\mu)$$

and the adjoint

$$D^-\colon L^2(G) \otimes_R (S^- \otimes U_\mu) \to L^2(G) \otimes_R (S^+ \otimes U_\mu).$$

One has

$$\operatorname{Index} D = \dim \ker D^+ - \dim \operatorname{Ker} D^- = (-1)^{l(w)} \dim V_{w(\mu+\rho(\mathfrak{r}))-\rho(\mathfrak{g})}$$

if there exists a $w \in W_{\mathfrak{g}}$ *so that* $w(\mu + \rho(\mathfrak{r})) - \rho(\mathfrak{g})$ *is dominant and it is zero if no such* w *exists.*

6.5. *Geometric construction of discrete series*

Let G be a linear semisimple noncompact Lie group. Let K be a maximal compact subgroup of G. Assume that $\operatorname{rank} G = \operatorname{rank} K$. Let $\mathfrak{g}_0 = \mathfrak{k}_0 + \mathfrak{p}_0$ be the Cartan decomposition of the Lie algebra of G. Then $\mathfrak{u}_0 = \mathfrak{k}_0 + i\mathfrak{p}_0$ is a compact real form of $\mathfrak{g} = \mathfrak{g}_0 \otimes_{\mathbb{R}} \mathbb{C}$. Let U be the compact analytic subgroup in the complexification $G_{\mathbb{C}}$ of G with Lie algebra $\mathfrak{u}_0$.

Borel showed that there exists a torsion free discrete subgroup Γ of G so that $\Gamma\backslash G$ and $X = \Gamma\backslash G/K$ are compact smooth manifolds. For any $\mu \in \widehat{K}$, the Dirac operator D acts on the smooth sections of the twisted spin bundle in a similar way described in 6.4.

$$D\colon C^\infty(G/K, S \otimes E_\mu) \to C^\infty(G/K, S \otimes E_\mu).$$

Note that the above action of D commutes with the left action of G. So we can consider the elliptic operator

$$D_\mu^+(X)\colon C^\infty(\Gamma\backslash G/K, S^+ \otimes E_\mu) \to C^\infty(\Gamma\backslash G/K, S^- \otimes E_\mu).$$

The index of $D_\mu^+(X)$ can be computed by Atiyah-Singer Index Theorem

$$\operatorname{Index} D_\mu^+(X) = \int_X f(\Theta, \Phi),$$

where Θ is the curvature of X and Φ is the curvature of the twisted spinor bundle over G/K. By the homogeneity, $f(\Theta, \Phi)$ is a multiple of the volume form depending only on μ, i.e., $f(\Theta, \Phi) = c(\mu)dx$. Thus

$$\operatorname{Index} D_\mu^+(X) = c(\mu)vol(\Gamma\backslash G/K).$$

Let $Y = U/K$ be the compact homogeneous space. By Hirzebruch proportionality principle, the index of

$$D_\mu^+(Y)\colon C^\infty(U/K, S^+ \otimes E_\mu) \to C^\infty(U/K, S^- \otimes E_\mu)$$

can be computed in the same way and

$$\operatorname{Index} D_\mu^+(Y) = (-1)^q c(\mu)vol(U/K),$$

where $q = \dim G/K = \dim U/K$. It follows that

$$\operatorname{Index} D_\mu^+(X) = (-1)^q \frac{vol(\Gamma\backslash G/K)}{vol(U/K)} \operatorname{Index} D_\mu^+(Y).$$

If we normalize the Haar measure so that $vol(U) = 1$, then

$$\operatorname{Index} D_\mu^+(X) = (-1)^q vol(\Gamma\backslash G) \operatorname{Index} D_\mu^+(Y).$$

Let $L^2(G) \cong \int_{\widehat{G}} H_j \otimes H_j^* d\mu(j)$ be the abstract Plancherel decomposition. Then the L^2 sections of the twisted spinor bundles are decomposed as

$$L^2(G/K, S \otimes E_\mu) \cong \int_{\widehat{G}} H_j \otimes \operatorname{Hom}_K(H_j, S \otimes E_\mu) d\mu(j).$$

It follows that the L^2-sections of $\operatorname{Ker} D$ are decomposed as

$$\operatorname{Ker} D \cong \int_{\widehat{G}} H_j \otimes \operatorname{Ker}\{D\colon \operatorname{Hom}_{\tilde{K}}(E^*_{\mu'}, H^*_j \otimes S) \circlearrowleft\} d\mu(j).$$

By the proved conjecture of Vogan, if H_j occurs in the decomposition of $\operatorname{Ker} D$ then it has infinitesimal character $\mu + \rho_c$. There are at most finitely many representations with a fixed infinitesimal character. Thus, if $\operatorname{Ker} D$ is nonzero, the corresponding H_j must be in the discrete spectrum, i.e., a discrete series representation. Namely, the finitely many points must be of nonzero Plancherel measure in order to contribute nontrivially to the direct integral. It follows from Corollary 6.4 and the above discussion of the indices that $w(\mu + \rho_c) - \rho$ is dominant for some $w \in W_{\mathfrak{g}}$. Write λ for $w(\mu+\rho_c)-\rho$. Then the infinitesimal character $\lambda+\rho$ of H_j is strongly regular and therefore H_j is an $A_{\mathfrak{q}}(\lambda)$-module [24]. In case λ is regular, this $A_{\mathfrak{q}}(\lambda)$-module is isomorphic to $A_{\mathfrak{b}}(\lambda)$ with the lowest K-type $\lambda + 2\rho_n$. Therefore, we obtain the following theorem, which is a part of the main results in [1].

Theorem: (Atiyah-Schmid [1]) *Let G be a linear group. Assume that there exists a $w \in W_{\mathfrak{g}}$ so that $\lambda = w(\mu + \rho_c) - \rho$ is dominant and regular. Then the kernel of the Dirac operator D acting on the L^2-sections of the twisted spinor bundle corresponding to E_μ is the discrete series representation $A_{\mathfrak{b}}(\lambda)$ with the lowest K-type $\lambda + 2\rho_n$.*

Combining this with some of the original arguments in [1] and with results of Harish-Chandra and Hecht-Schmid which say that the discrete series are exactly the $A_{\mathfrak{b}}(\lambda)$ with λ dominant, one can obtain a geometric representation of all discrete series representations. We note that Atiyah and Schmid also extended this geometric construction of discrete series to nonlinear groups.

7. Lie algebra cohomology

7.1. *Definition*

Let $\mathfrak{g}$ be a complex Lie algebra and V a $\mathfrak{g}$-module. The space of invariants of V with respect to $\mathfrak{g}$ is the vector space

$$V^{\mathfrak{g}} = \{v \in V \big| Xv = 0, \forall X \in \mathfrak{g}\}.$$

Note that this definition is consistent with the definition of invariant (fixed) vectors under a group action. Upon differentiation, the condition of being fixed becomes the condition of being annihilated.

One also considers invariants not with respect to whole $\mathfrak{g}$ but with respect to some subalgebra. Let us describe an important example. Let $\mathfrak{g}$ be semisimple, $\mathfrak{h}$ a Cartan subalgebra, Δ the set of roots for $\mathfrak{g}$ with respect to $\mathfrak{h}$, Δ^+ a set of positive roots, and $\mathfrak{n} = \mathfrak{n}^+$ and $\mathfrak{n}^-$ the corresponding subalgebras like in Section 1.2.

Then one is interested in the space $V^{\mathfrak{n}}$ of $\mathfrak{n}$-invariants in V, which is now not only a vector space but an $\mathfrak{h}$-module. We already know why this space is interesting; if V is finite dimensional irreducible, then $V^{\mathfrak{n}}$ is one dimensional (the highest weight subspace), and the $\mathfrak{h}$-action on $V^{\mathfrak{n}}$, i.e., the highest weight of V, determines V. If V is still finite dimensional but reducible, $V = \oplus_i V_i$, then each V_i contributes a weight vector to $V^{\mathfrak{n}}$ and $V^{\mathfrak{n}}$ thus encodes the information about this decomposition.

We consider the functor $V \mapsto V^{\mathfrak{n}}$ from the category $\mathcal{M}(\mathfrak{g})$ of $\mathfrak{g}$-modules into the category $Vect_{\mathbb{C}}$ of complex vector spaces. (Analogously, $V \mapsto V^{\mathfrak{n}}$ would be a functor from $\mathcal{M}(\mathfrak{g})$ into $\mathcal{M}(\mathfrak{h})$.) This functor is in general left exact, but not exact. This means that if

$$0 \to U \to V \to W \to 0$$

is a short exact sequence of $\mathfrak{g}$-modules (which is a slightly more precise way to write $W = V/U$), then

$$0 \to U^{\mathfrak{g}} \to V^{\mathfrak{g}} \to W^{\mathfrak{g}}$$

is exact, but $V^{\mathfrak{g}} \to W^{\mathfrak{g}}$ is not surjective in general. Thus we define the $\mathfrak{g}$-cohomology functors to be the right derived functors of the functor $V \mapsto V^{\mathfrak{g}}$.

To motivate the use of derived functors, let us consider the following simple example. Knowing how one can benefit from the usual duality operation for vector spaces, $V^* = \mathrm{Hom}_{\mathbb{C}}(V, \mathbb{C})$, one would like to have something similar for modules, say over $\mathbb{Z}$ (for simplicity). If one tries to consider $V^* = \mathrm{Hom}_{\mathbb{Z}}(V, \mathbb{Z})$, then one notices that for free modules of finite rank it works fine (including double dual giving back the same module), but for say $\mathbb{Z}_2$, one has

$$\mathrm{Hom}_{\mathbb{Z}}(\mathbb{Z}_2, \mathbb{Z}) = 0.$$

Namely, $\mathbb{Z}_2$ is not free, and its generator 1 can not be mapped anywhere but to 0. We can however resolve $\mathbb{Z}_2$ by free modules:

$$0 \to \mathbb{Z} \xrightarrow{2\cdot} \mathbb{Z} \to \mathbb{Z}_2 \to 0.$$

Thus the complex $0 \to \mathbb{Z} \xrightarrow{2\cdot} \mathbb{Z} \to 0$, concentrated in degrees -1 and 0, should replace the module $\mathbb{Z}_2$. If we take this resolution and plug it in the functor $\mathrm{Hom}_{\mathbb{Z}}(-,\mathbb{Z})$ instead of the module $\mathbb{Z}_2$ it resolves, we get the complex

$$0 \to \mathbb{Z} \xrightarrow{2\cdot} \mathbb{Z} \to 0$$

which is concentrated in degrees 0 and 1 (by contravariance, the arrows changed direction). Its zeroth cohomology is 0, corresponding to the observed fact $\mathrm{Hom}_{\mathbb{Z}}(\mathbb{Z}_2,\mathbb{Z}) = 0$. However, its first cohomology is $\mathbb{Z}_2$ (this is $\mathrm{Ext}^1_{\mathbb{Z}}(\mathbb{Z}_2,\mathbb{Z})$, the first derived functor of Hom). If we now apply duality again, we get back to our resolution of $\mathbb{Z}_2$, that is $\mathbb{Z}_2$ in the zeroth cohomology.

To make the above completely precise, one would need to pass to derived category, which is a way of making modules equal (isomorphic) to their resolutions. However, the point we wanted to make is visible: the functor itself maybe sometimes gives nothing, but considering also the derived functors may give more.

We now get back to our covariant, left exact functor $V \mapsto V^{\mathfrak{g}}$. In principle, the right derived functors are defined using injective resolutions $0 \to V \to I^{\cdot}$; so the i-th $\mathfrak{g}$-cohomology space of V would be

$$H^i(\mathfrak{g}; V) = H^i((I^{\cdot})^{\mathfrak{g}}).$$

Namely, as above, we replace V by its resolution, plug the resolution in the functor, get a complex with induced differential, and then take cohomology. However, this is not very explicit, as injective resolutions are rather complicated to write down. Fortunately we have a better possibility: note that

$$V^{\mathfrak{g}} = \mathrm{Hom}_{\mathfrak{g}}(\mathbb{C}, V),$$

where $\mathbb{C}$ is the trivial $\mathfrak{g}$-module; simply identify every map $\phi : \mathbb{C} \to V$ with $\phi(1) \in V^{\mathfrak{g}}$. This means that our $\mathfrak{g}$-cohomology $H^i(\mathfrak{g}; V)$ is equal to $\mathrm{Ext}^i_{\mathfrak{g}}(\mathbb{C}, V)$, and to define it (calculate it) we can resolve $\mathbb{C}$ by free modules (or projectives) instead of resolving V by injectives. This is done by the so called standard, or Koszul complex

$$U(\mathfrak{g}) \otimes \textstyle\bigwedge^{\cdot} \mathfrak{g} \xrightarrow{\epsilon} \mathbb{C} \to 0,$$

where the $\mathfrak{g}$-action is given by left multiplication in the $U(\mathfrak{g})$-factor, the differential is the deRham differential

$$d(u\otimes X_1\wedge\cdots\wedge X_k)=\sum_i(-1)^{i-1}uX_i\otimes X_1\wedge\dots\widehat{X}_i\cdots\wedge X_k+$$
$$\sum_{i<j}(-1)^{i+j}u\otimes[X_i,X_j]\wedge X_1\wedge\dots\widehat{X}_i\dots\widehat{X}_j\cdots\wedge X_k,$$

and ϵ is the augmentation map, given by $1\mapsto 1$ and $\mathfrak{g}U(\mathfrak{g})\mapsto 0$.

To see that this is indeed a resolution, one considers the graded version, $S(\mathfrak{g})\otimes\bigwedge\mathfrak{g}$ and is thus lead to the analogous question of resolving the trivial module $\mathbb{C}$ over a polynomial algebra. Now for polynomials in one variable, it is obvious how to do this:

$$0\to\mathbb{C}[X]\xrightarrow{X\cdot}\mathbb{C}[X]\to\mathbb{C}\to 0$$

is clearly the required resolution. To increase the number of variables, this is tensored with itself several times; the introduction of signs which leads to the exterior algebra is forced by the requirement $d^2=0$. The reader is invited to try to construct a resolution for two variables from scratch and see how the exterior algebra appears quite naturally.

Now $H^i(\mathfrak{g};V)$ is the i-th cohomology of the complex

$$\mathrm{Hom}^{\cdot}_{U(\mathfrak{g})}(U(\mathfrak{g})\otimes\textstyle\bigwedge^{\cdot}\mathfrak{g},V)=\mathrm{Hom}^{\cdot}_{\mathbb{C}}(\bigwedge^{\cdot}\mathfrak{g},V)$$

with the induced differential

$$d\alpha(X_1\wedge\cdots\wedge X_k)=\sum_i X_i\alpha(X_1\wedge\dots\widehat{X}_i\cdots\wedge X_k)+\sum_{i<j}(-1)^{i+j}\alpha([X_i,X_j]\wedge\dots),$$

for $\alpha\in\mathrm{Hom}^{k-1}_{\mathbb{C}}(\bigwedge^{\cdot}\mathfrak{g},V)$. Of course, $H^0(\mathfrak{g};V)=V^{\mathfrak{g}}$.

7.2. *Properties and basic applications*

Here are some properties of these functors:

(1) Any short exact sequence

$$0\to U\to V\to W\to 0$$

of $\mathfrak{g}$-modules gives rise to a long exact sequence of cohomology

$$0\to U^{\mathfrak{g}}\to V^{\mathfrak{g}}\to W^{\mathfrak{g}}$$
$$\to H^1(\mathfrak{g};U)\to H^1(\mathfrak{g};V)\to H^1(\mathfrak{g};W)\to H^2(\mathfrak{g};U)\to\dots$$

This is a completely general property of derived functors. It can be used to calculate cohomology in some case, or for various proofs, like the one below.

(2) If $\mathfrak{g}$ is semisimple and if V has nontrivial infinitesimal character, then $H^i(\mathfrak{g}; V) = 0$ for all i. This can be proved by noting that since any $Z \in Z(\mathfrak{g})$ with no constant term vanishes on the trivial module $\mathbb{C}$, then by a standard homological argument it follows that the action of Z on the standard complex which is a resolution of $\mathbb{C}$ must be homotopic to 0.

From this statement one can obtain Weyl's theorem mentioned in 1.2: any finite dimensional module over a semisimple Lie algebra is a direct sum of irreducibles. This statement is equivalent to the statement that any short exact sequence

$$0 \to U \xrightarrow{i} V \xrightarrow{p} W \to 0$$

of finite dimensional $\mathfrak{g}$ modules is split, i.e., there is an $s : W \to V$ such that $p \circ s$ is the identity on W. Indeed, such a splitting exhibits V as a direct sum of U and W and we can then decompose modules completely by induction on the dimension.

Now to get a splitting, it suffices to know that the sequence

$$\operatorname{Hom}_{\mathfrak{g}}(W, V) \xrightarrow{p\circ} \operatorname{Hom}_{\mathfrak{g}}(W, W) \to 0$$

is exact; then we get our splitting s as a preimage of id_W. But the functor $\operatorname{Hom}_{\mathfrak{g}}(W, -)$ is a composition of two exact functors: $\operatorname{Hom}_{\mathbb{C}}(W, -)$, which is clearly exact, and $(-)^{\mathfrak{g}}$ which is also exact on finite dimensional modules by the long exact sequence and the fact that $H^1(\mathfrak{g}; -)$ vanishes on all finite dimensional modules. The vanishing of H^1 has to be checked separately for the trivial module $\mathbb{C}$; this is an easy calculation.

(3) Similarly, Levi's theorem that any Lie algebra is a semidirect product of the radical and a semisimple subalgebra can be obtained from the fact that the second cohomology of finite dimensional modules over a semisimple Lie algebra vanishes. Namely, by definition the radical $\mathfrak{r}$ of $\mathfrak{g}$ is the biggest solvable ideal, so there is an exact sequence

$$0 \to \mathfrak{r} \to \mathfrak{g} \to \mathfrak{l} \to 0$$

of Lie algebras, with $\mathfrak{l}$ semisimple. Now the nontrivial extensions of $\mathfrak{l}$ by $\mathfrak{r}$ can be seen to correspond to $H^2(\mathfrak{l}; \mathfrak{r})$. As the last space is 0, every extension is trivial and hence the above sequence splits.

7.3. *Theorems of Casselman-Osborne and Kostant*

We next consider a parabolic subalgebra $\mathfrak{q}$ of $\mathfrak{g}$, with a Levi decomposition

$$\mathfrak{q} = \mathfrak{l} \oplus \mathfrak{u};$$

this is a slightly different Levi decomposition from the one mentioned above, with $\mathfrak{u}$ the nilradical (the largest nilpotent ideal) and $\mathfrak{l}$ a reductive subalgebra. A special case arises in the situation already studied; for a Cartan subalgebra $\mathfrak{h}$ and $\mathfrak{n}$ coming from a choice of positive roots, define

$$\mathfrak{b} = \mathfrak{h} \oplus \mathfrak{n}.$$

This is a maximal solvable subalgebra of $\mathfrak{g}$ and such are called Borel subalgebras. A parabolic subalgebra of $\mathfrak{g}$ can be defined as any subalgebra containing a Borel subalgebra. One should think of Borel subalgebras as algebras of upper triangular matrices, and of parabolic subalgebras as algebras of block upper triangular matrices with some fixed shape of blocks. Then the Levi factor $\mathfrak{l}$ is the algebra of corresponding block diagonal matrices, and the nilradical $\mathfrak{u}$ is the algebra of all upper triangular matrices that are zero on blocks. This is for example exactly true (in some complex basis) if $\mathfrak{g} = \mathfrak{sl}(n, \mathbb{C})$.

We consider the $\mathfrak{u}$-cohomology spaces $H^i(\mathfrak{u}; V)$ for a $\mathfrak{g}$-module V. These spaces are actually $\mathfrak{l}$-modules in a natural way; namely, $\mathfrak{l}$ acts on the complex defining $H^i(\mathfrak{u}; V)$ by acting both on $\bigwedge \mathfrak{u}$ and on V. This action commutes with the differential, hence descends to cohomology.

Now $Z(\mathfrak{g})$, the center of the enveloping algebra of $\mathfrak{g}$, acts on the complex $\mathrm{Hom}^{\cdot}(\bigwedge \mathfrak{u}, V)$ and its cohomology $H^i(\mathfrak{u}; V)$ in two ways. One action is through the Harish-Chandra homomorphism $Z(\mathfrak{g}) \to Z(\mathfrak{l})$ and the $\mathfrak{l}$ (i.e., $U(\mathfrak{l})$) action described above, and the other is acting on V only. This second action is well defined only for the center and not for the rest of $U(\mathfrak{g})$.

Theorem: (Casselman-Osborne) *The above two actions of $Z(\mathfrak{g})$ agree on cohomology $H^i(\mathfrak{u}; V)$.*

This is a similar statement to Property (2) of 7.2, about the action of the center on the standard complex which was used to get vanishing of $\mathfrak{g}$-cohomology. A proof can be found in [28], Theorem 3.1.5, or [16], Theorem 4.149. We will also see how to get this in a different way, using Clifford algebra actions and Dirac type operators.

Theorem: (Kostant) *Let $\mathfrak{h}$ be a Cartan subalgebra of $\mathfrak{g}$ and let $\mathfrak{b} = \mathfrak{h} \oplus \mathfrak{n}$ be a Borel subalgebra corresponding to a choice of positive roots. Let V_λ be* *the irreducible finite dimensional $\mathfrak{g}$-module with highest weight λ. Then*

$$H^i(\mathfrak{n}; V_\lambda) = \bigoplus_{w \in W, l(w)=i} \mathbb{C}_{w(\lambda+\rho)-\rho}$$

as $\mathfrak{h}$-modules. Here $\mathbb{C}_\mu$ is the one dimensional $\mathfrak{h}$-module with weight μ, W is the Weyl group, and $l(w)$, the length of $w \in W$, is the smallest number of simple root reflections needed to write w as their product.

A proof of this statement and also its generalization to the case of $\mathfrak{u}$-cohomology can be found e.g. in [16], theorems 4.135 and 4.139. The original proof is greatly simplified by using the (more recent) Casselman-Osborne theorem to conclude that only the weights $w(\lambda + \rho) - \rho$ can appear.

Kostant interpreted this theorem as an algebraic version of the Borel-Weil-Bott theorem, which realizes finite dimensional $\mathfrak{g}$-modules as global sections or cohomology of line bundles on the flag variety $\mathcal{B}$ of $\mathfrak{g}$. (The flag variety consists of all Borel, i.e., maximal solvable subalgebras of $\mathfrak{g}$). He further used it to obtain an algebraic proof of the Weyl character formula, which explicitly calculates the (global) character of irreducible representations of compact groups.

7.4. $\mathfrak{n}$-*homology and Casselman subrepresentation theorem*

Lie algebra homology is defined in a similar way as cohomology. One starts by defining coinvariants

$$V_\mathfrak{g} = V/\mathfrak{g}V = V \otimes_{U(\mathfrak{g})} \mathbb{C}$$

of a $\mathfrak{g}$-module V; this is the largest quotient of V on which $\mathfrak{g}$ acts trivially. This is a right exact functor, and Lie algebra homology functors are the left derived functors

$$H_i(\mathfrak{g}; V) = \mathrm{Tor}_i^\mathfrak{g}(\mathbb{C}, V) = H_i(U(\mathfrak{g}) \otimes \textstyle\bigwedge^{\cdot} \mathfrak{g} \otimes_{U(\mathfrak{g})} V) = H_i(\textstyle\bigwedge^{\cdot} \mathfrak{g} \otimes V).$$

The differential is again induced by the deRham differential of the standard complex, which is used to resolve the variable $\mathbb{C}$ and thus define the derived functors.

To illustrate the importance of ($\mathfrak{n}$-)homology, we mention a very standard construction of representations, namely the real parabolic induction. Start with the Cartan decomposition

$$\mathfrak{g}_0 = \mathfrak{k}_0 \oplus \mathfrak{p}_0$$

of a real semisimple Lie algebra $\mathfrak{g}_0$. Let $\mathfrak{a}_0 \subset \mathfrak{p}_0$ be a maximal abelian subalgebra (subspace). Then one can consider the $(\mathfrak{g}_0, \mathfrak{a}_0)$-roots, by the same principle as for $\mathfrak{h}$ in $\mathfrak{g}$ (the adjoint action of $\mathfrak{a}_0$ on $\mathfrak{g}_0$ diagonalizes etc.) Let $\mathfrak{n}_0$ be the sum of positive root spaces (for a choice of positive roots). Then one shows that

$$\mathfrak{g}_0 = \mathfrak{k}_0 \oplus \mathfrak{a}_0 \oplus \mathfrak{n}_0;$$

this is called the Iwasawa decomposition of $\mathfrak{g}_0$. There is also a group version of this fact: multiplication defines a diffeomorphism

$$K \times A \times N \to KAN = G$$

where A and N are the connected subgroups of G corresponding to $\mathfrak{a}_0$ and $\mathfrak{n}_0$. We denote by L the centralizer of A in G; then $L = MA$, where $M = L \cap K$ is the centralizer of A in K. The subgroup

$$P = MAN$$

is called a minimal (real) parabolic subgroup of G. Note a slight problem with notation: the Lie algebra of P is not $\mathfrak{p}$ from Cartan decomposition (that one is not a Lie algebra at all). Nevertheless, this is common notation.

Here is what is meant by induction from P to G. Let V be a (finite dimensional) representation of P. Consider the principal bundle $G \to G/P$ and construct the associated G-equivariant vector bundle

$$\begin{array}{c} G \times_P V \\ \downarrow \\ G/P \end{array}$$

The space $G \times_P V$ consists of classes of the equivalence relation on $G \times V$ defined by setting

$$(gp, v) \sim (g, p \cdot v), \quad g \in G, p \in P, v \in V.$$

The continuous sections of this bundle form a representation $\mathrm{Ind}_c(V)$ of G called the continuously induced representation. Here G acts on the sections by left translation. (One can also consider smooth sections, or L^2 sections of this bundle.) The sections can also be interpreted as functions from G into V with the appropriate transformation property for P; in this way one can avoid the language of bundles. We will denote by $\mathrm{Ind}(V)$ the $(\mathfrak{g}, K)$-module of K-finite vectors in $\mathrm{Ind}_c(V)$.

Let us consider the case

$$V = \sigma \otimes \lambda \otimes 1,$$

where σ is an irreducible finite dimensional representation of M, λ is a character of A (which is the same as a character of $\mathfrak{a}$), and 1 denotes the trivial representation of N. This is basically the only case one needs to study. In this case, $\mathrm{Ind}(V)$ is called the principal series representation.

Casselman proved the following version of Frobenius reciprocity: for any $(\mathfrak{g}, K)$-module W,

$$\mathrm{Hom}_{(\mathfrak{g},K)}(W, \mathrm{Ind}(V)) \xrightarrow{\cong} \mathrm{Hom}_{(\mathfrak{p},M)}(W, V).$$

Here $\mathfrak{p}$ is the Lie algebra of P; note that M is a maximal compact subgroup of P. Since the action of $\mathfrak{n}$ on V is trivial, the second Hom-space is further equal to

$$\mathrm{Hom}_{(\mathfrak{p},M)}(W/\mathfrak{n}W, V).$$

If $W/\mathfrak{n}W$ is not equal to zero, then one can choose V so that the last space is nonzero. It follows that W maps nontrivially into $\mathrm{Ind}(V)$, and if W is irreducible this map has to be an embedding. So we get Casselman's subrepresentation theorem. Namely, $W/\mathfrak{n}W$ is not equal to zero; in fact, this space contains leading exponents of asymptotic expansions of matrix coefficients of W. See [5].

Let us finish this section by mentioning briefly two more facts. First, there is a version of Poincaré duality for Lie algebra cohomology. Second, there is a strong relationship between $\mathfrak{n}$-cohomology and Beilinson-Bernstein localization theory. In this theory, irreducible $(\mathfrak{g}, K)$-modules are realized as global sections (or cohomology) of certain sheaves on the flag variety $\mathcal{B}$ of $\mathfrak{g}$. The relationship is the following: the geometric fiber of the sheaf corresponding to a $(\mathfrak{g}, K)$-module V at a point $\mathfrak{b} \in X$ is exactly the $\mathfrak{n}$-cohomology of V, where $\mathfrak{n} = [\mathfrak{b}, \mathfrak{b}]$ is the nilradical of $\mathfrak{b}$.

8. $(\mathfrak{g}, K)$-cohomology

8.1. *Definition*

One can study relative Lie algebra cohomology for the pair $(\mathfrak{g}, \mathfrak{k})$. It has however become more usual to study pairs $(\mathfrak{g}, K)$. The main case we are interested in is the one suggested by our choice of notation: $\mathfrak{g}$ is a semisimple complex Lie algebra and K is a maximal compact subgroup of a Lie group G with complexified Lie algebra $\mathfrak{g}$. This setting can however be generalized; K could be another compact subgroup of G, or a complex reductive subgroup of a complex group with Lie algebra $\mathfrak{g}$. In principle K does not even have to be reductive, but then it is not as easy as below to write down resolutions. One can also consider similar pairs $(\mathcal{A}, K)$ where $\mathcal{A}$ is an associative algebra. $\mathcal{A}$ could be $U(\mathfrak{g})$, or a quotient U_θ corresponding to an infinitesimal character (θ is a Weyl group orbit of an element of $\mathfrak{h}^*$).

For the purpose of this section, let us concentrate on the usual (and the most interesting) case of $(\mathfrak{g}, K)$, the first one mentioned above.

Formally, $(\mathfrak{g}, K)$-cohomology is analogous to $\mathfrak{g}$-cohomology. Namely, one can consider the functor

$$V \mapsto V^{\mathfrak{g},K} = \{v \in V \big| Xv = 0, kv = v, \text{ for all } X \in \mathfrak{g}, k \in K\}$$

of taking $(\mathfrak{g}, K)$-invariants. It is a functor from the category $\mathcal{M}(\mathfrak{g}, K)$ of $(\mathfrak{g}, K)$-modules into the category of complex vector spaces, which is left exact. The $(\mathfrak{g}, K)$-cohomology functors $V \mapsto H^i(\mathfrak{g}, K; V)$ are the right derived functors of $V \mapsto V^{\mathfrak{g},K}$.

As before, one can write

$$V^{\mathfrak{g},K} = \mathrm{Hom}_{(\mathfrak{g},K)}(\mathbb{C}, V),$$

and thus

$$H^i(\mathfrak{g}, K; V) = \mathrm{Ext}^i_{(\mathfrak{g},K)}(\mathbb{C}, V).$$

As before, rather than resolving V by injectives, we use a projective resolution of the trivial module $\mathbb{C}$. This is the relative standard complex

$$U(\mathfrak{g}) \otimes_{U(\mathfrak{k})} \textstyle\bigwedge^{\cdot}(\mathfrak{g}/\mathfrak{k}) \xrightarrow{\epsilon} \mathbb{C} \to 0;$$

Since we are considering compact K, we can replace $\mathfrak{g}/\mathfrak{k}$ by the K-invariant direct complement $\mathfrak{p}$. The differential d of the above complex and the map ϵ are similar as before:

$$d(u \otimes X_1 \wedge \cdots \wedge X_k) = \sum_i (-1)^{i-1} uX_i \otimes X_1 \wedge \ldots \widehat{X}_i \cdots \wedge X_k +$$
$$\sum_{i<j} (-1)^{i+j} u \otimes [X_i, X_j]_{\mathfrak{p}} \wedge X_1 \wedge \ldots \widehat{X}_i \ldots \widehat{X}_j \cdots \wedge X_k,$$

for any compact K; here $[X_i, X_j]_{\mathfrak{p}}$ denotes the projection of $[X_i, X_j]$ to $\mathfrak{p}$ along $\mathfrak{k}$. If K is a symmetric subgroup, like in our main case when K is the maximal compact subgroup, then this projection is always zero, so the second sum actually vanishes. ϵ is as before the augmentation map, given by $1 \otimes 1 \mapsto 1$ and $\mathfrak{g}U(\mathfrak{g}) \otimes 1 \mapsto 0$.

The relative standard complex is obtained from the standard complex for $\mathfrak{g}$, by taking coinvariants with respect to the $\mathfrak{k}$-action given by right multiplication on $U(\mathfrak{g})$ and adjoint action on $\bigwedge(\mathfrak{g})$, and also with respect to the action of $\bigwedge(\mathfrak{k})$ on $\bigwedge(\mathfrak{g})$ by (exterior) multiplication. Exactness is proved using this and exactness of the standard complex for $\mathfrak{g}$. The fact that this is a projective resolution follows from general homological principles: any finite K-module is projective, and the functor $W \mapsto U(\mathfrak{g}) \otimes_{U(\mathfrak{k})} W$ from finite K-modules to $(\mathfrak{g}, K)$-modules preserves projectives, as it is left adjoint to the exact functor of forgetting the $\mathfrak{g}$-action.

Using the above resolution, we can now identify $H^i(\mathfrak{g}, K; V)$ with the ith cohomology of the complex

$$\mathrm{Hom}^{\cdot}_{(\mathfrak{g},K)}(U(\mathfrak{g}) \otimes_{U(\mathfrak{k})} \textstyle\bigwedge^{\cdot}(\mathfrak{p}), V) = \mathrm{Hom}^{\cdot}_K(\bigwedge^{\cdot}(\mathfrak{p}), V),$$

with differential

$$df(X_1 \wedge \cdots \wedge X_k) = \sum_i (-1)^{i-1} X_i \cdot f(X_1 \wedge \ldots \widehat{X}_i \cdots \wedge X_k)$$

if K is symmetric, and for other K there is another sum, over $i < j$.

There is also a less often mentioned theory of $(\mathfrak{g}, K)$-homology. It is constructed by deriving the functor of $(\mathfrak{g}, K)$-coinvariants; so

$$H_i(\mathfrak{g}, K; V) = \mathrm{Tor}_i^{(\mathfrak{g},K)}(\mathbb{C}, V).$$

which is calculated using the same resolution of $\mathbb{C}$ as above.

8.2. *Applications*

A good reference for learning about various applications of $(\mathfrak{g}, K)$-cohomology to the theory of automorphic forms is a recent survey article [22]. Let us mention just one very classical application, the Matsushima formula. Let $\Gamma \subset G$ be a cocompact lattice and let E be a finite dimensional representation of G. Then the group cohomology of Γ with coefficients in E, which is also equal to the cohomology of the space $\Gamma\backslash G/K$ with coefficients in E, can be expressed as

$$H^*(\Gamma, E) \cong \bigoplus_{\pi \in \widehat{G}} m(\pi, \Gamma) H^*(\mathfrak{g}, K; H_\pi \otimes E),$$

where $m(\pi, \Gamma)$ is the multiplicity of the unitary representation (π, H_π) of G in $L^2(\Gamma\backslash G)$.

Another application is a construction of derived Zuckerman functors. Namely, let $(\mathfrak{g}, K)$ be a pair as above, with K complex algebraic (e.g., the complexification of a maximal compact subgroup). Let $T \subset K$ be a closed reductive subgroup. Let $R(K)$ be the algebra of regular functions on K. Then one can express the derived Zuckerman modules of a $(\mathfrak{g}, T)$-module V as

$$\Gamma^i_{K,T}(V) = H^i(\mathfrak{k}, T; R(K) \otimes V).$$

Here the $(\mathfrak{k}, T)$ cohomology is taken with respect to the tensor product of the regular action with the given action on V. K acts by right translation on $R(K)$, and the $\mathfrak{g}$-action is obtained by twisting the given action π_V on V: if we regard an element of $R(K) \otimes V$ as a regular function $F : K \to V$, then for $X \in \mathfrak{g}$ the function $\pi(X)F$ is given by

$$(\pi(X)F)(k) = \pi_V(\mathrm{Ad}(k)X)(F(k)).$$

There are several versions of this construction, due to Wallach ([33], Chapter 6), Duflo-Vergne, and Miličić-Pandžić.

8.3. *The Vogan-Zuckerman classification*

The central result about $(\mathfrak{g}, K)$-cohomology (for K the maximal compact subgroup of G) is the classification of irreducible unitary $(\mathfrak{g}, K)$-modules with nonzero $(\mathfrak{g}, K)$-cohomology obtained by Vogan and Zuckerman in [32]. This result can be stated as follows.

Let V be irreducible unitary, of the same infinitesimal character as a finite dimensional representation F. Note that there is only one possible F for any given V. Then $V \otimes F^*$ has nonzero $(\mathfrak{g}, K)$-cohomology if and only if V is an $A_{\mathfrak{q}}(\lambda)$ module, as described in previous sections. (We also saw that in this case the lowest K-type of V gives rise to Dirac cohomology.)

Note that if V and F do not have the same infinitesimal character then $H^*(\mathfrak{g}, K; V \otimes F^*) = 0$. In case there is cohomology, it is equal to

$$\mathrm{Hom}_{L\cap K}(\textstyle\bigwedge^{i-\dim(\mathfrak{u}\cap\mathfrak{p})}(\mathfrak{l}\cap\mathfrak{p}), \mathbb{C}),$$

where L is the Levi subgroup involved in the definition of $A_{\mathfrak{q}}(\lambda)$, $\mathfrak{l}$ is the (complexified) Lie algebra of L and $\mathfrak{u}$ is the nilradical of $\mathfrak{q}$.

For more details, see [22] or [32].

9. Relationship of Dirac cohomology to other kinds of cohomology

In this section we briefly sketch some results from [13], which grew out of the ideas of [31].

9.1. *Kostant's cubic Dirac operator*

In [18], Kostant has constructed a cubic Dirac operator D corresponding to any decomposition

$$\mathfrak{g} = \mathfrak{r} \oplus \mathfrak{s}$$

of a semisimple Lie algebra $\mathfrak{g}$ with $\mathfrak{r}$ a reductive subalgebra such that the Killing form is non-degenerate on $\mathfrak{r}$. The definition is as follows: let Z_i be an orthonormal basis for $\mathfrak{s}$ and let $v \in \bigwedge^3 \mathfrak{s}$ correspond to the alternating trilinear form $(X, Y, Z) \mapsto B([X, Y], Z)$ under the isomorphism $(\bigwedge^3 \mathfrak{s})^* \cong \bigwedge^3 \mathfrak{s}$ induced by the Killing form. Then

$$D = \sum_i Z_i \otimes Z_i + 1 \otimes v \ \in U(\mathfrak{g}) \otimes C(\mathfrak{s}).$$

It is easy to see that D is independent of the choice of basis Z_i and $\mathfrak{r}$-invariant for the adjoint action. Following Kostant, we now use defining relations $Z_i^2 = 1$ instead of $Z_i^2 = -1$ for $C(\mathfrak{s})$.

Kostant has shown that the formula we had for the Dirac operator corresponding to the Cartan decomposition, i.e.,

$$D^2 = \Omega_{\mathfrak{g}} \otimes 1 - \Omega_{\mathfrak{r}_\Delta} + (||\rho||^2 - ||\rho_{\mathfrak{r}}||^2)$$

still holds in this more general situation. The ingredients of this formula are defined analogously as before.

9.2. *Applications to* $\mathfrak{u}$*-cohomology*

Let us consider a special case when $\mathfrak{r} = \mathfrak{l}$ and $\mathfrak{s} = \mathfrak{u} \oplus \bar{\mathfrak{u}}$ correspond to a θ-stable parabolic subalgebra

$$\mathfrak{q} = \mathfrak{l} \oplus \mathfrak{u}.$$

Note that $\mathfrak{u}$ and $\bar{\mathfrak{u}}$ are both isotropic for B, that B identifies $\bar{\mathfrak{u}}$ with $\mathfrak{u}^*$, and that $\bar{\mathfrak{u}}$ is complex conjugate to $\mathfrak{u}$ with respect to the real form $\mathfrak{g}_0$ implicit in the above definitions.

Let $S = \bigwedge^{\cdot} \mathfrak{u}$ be a space of spinors for $C(\mathfrak{s})$. Since $\mathfrak{s}$ is even-dimensional, S is unique up to isomorphism. There are two $\mathfrak{l}$-actions on S; one is given by the adjoint action of $\mathfrak{l}$ on $\mathfrak{u}$, and the other is the spin action, coming from $\mathfrak{l} \to \mathfrak{so}(\mathfrak{s}) \to C(\mathfrak{s})$. These two actions are related by a twist by $\rho_{\bar{\mathfrak{u}}}$; see [20].

Let X be a $(\mathfrak{g}, K)$-module. Then the space

$$X \otimes S = \bigwedge^{\cdot} \mathfrak{u} \otimes X$$

has an action of D and of the $\mathfrak{u}$-homology operator ∂. Moreover, since we can identify

$$\bigwedge^{\cdot} \mathfrak{u} \otimes X \cong \mathrm{Hom}(\bigwedge^{\cdot} \bar{\mathfrak{u}}, X)$$

using $(\bigwedge^{\cdot} \mathfrak{u})^* \cong \bigwedge^{\cdot} \mathfrak{u}^* \cong \bigwedge^{\cdot} \bar{\mathfrak{u}}$, we also get an action of the $\bar{\mathfrak{u}}$-cohomology operator d on $X \otimes S$.

One checks that

$$D = d + 2\partial$$

on $X \otimes S$. To see this, take a basis u_i for $\mathfrak{u}$ and a dual basis u_i^* for $\bar{\mathfrak{u}}$; so $B(u_i, u_j^*) = \delta_{ij}$. Now write

$$D = \sum_i u_i^* \otimes u_i + \sum_i u_i \otimes u_i^* - \frac{1}{4}\sum_{i,j} 1 \otimes [u_i^*, u_j^*]u_i u_j - \frac{1}{4}\sum_{i,j} 1 \otimes [u_i, u_j]u_i^* u_j^*.$$

Denote by C (respectively C^-) the element of $U(\mathfrak{g}) \otimes C(\mathfrak{s})$ composed of the first and the third term (respectively the second and the fourth term) in the above expression for D. Then show that the action of C on $X \otimes S$ induces the differential d, while the action of C^- induces 2∂.

The "half Diracs" C and C^- are independent of the choice of basis u_i and $\mathfrak{l}$-invariant. They however do depend on the choice of $\mathfrak{u}$ inside $\mathfrak{s}$, while D does not. Furthermore, both C and C^- square to zero, and their supercommutator $CC^- + C^-C$ is equal to D^2.

It is convenient to introduce another element,

$$E = -\frac{1}{2}\sum_i 1 \otimes u_i^* u_i.$$

This operator acts as a degree operator on $X \otimes S$. It satisfies the following commuting relations:

$$[E, C] = C; \quad [E, C^-] = -C^-; \quad [E, D^2] = 0.$$

It follows that E, C, C^- and D^2 span a four dimensional superalgebra inside $(U(\mathfrak{g}) \otimes C(\mathfrak{s}))^{\mathfrak{l}}$. This algebra is actually $\mathbb{Z}$-graded, if we set C^- to be of degree -1, D^2 and E of degree 0 and C of degree 1. This grading is compatible with the obvious grading of $X \otimes S$ coming from the standard grading of $\bigwedge^{\cdot} \mathfrak{u}$.

This superalgebra was used by physicists under the name supersymmetric algebra. It is denoted by $l(1,1)$ in Kac's classification [14]. It is completely solvable and its representation theory is easy but not trivial.

As an application of the above facts, let us mention that one can easily prove that the main result of [12] holds for C and C^- and in this way get the Casselman-Osborne theorem (see 7.3) as a corollary. Thus the formal analogy between the two results becomes more concrete.

There are two questions that arise from the above considerations. The first one, asked by Vogan in [31], is to relate the Dirac cohomology, $\bar{\mathfrak{u}}$-cohomology and $\mathfrak{u}$-homology of a $(\mathfrak{g}, K)$-module X. This could be useful for example to pass between different choices for $\mathfrak{u}$ within the same $\mathfrak{s}$. The second question was implicitly posed by Kostant in [20]. His remark was that $X \otimes S$ can be formed not just in the "Levi factor case", i.e., for $\mathfrak{r} = \mathfrak{l}$, but also for a wider class of $\mathfrak{r}$ described above. On $X \otimes S$ there is always a cubic Dirac operator, and its square is the Laplacian. So it looks natural to try and study this more general setting and see how to make use of that.

At this moment, we have some results regarding the first question. Namely, in certain special cases all three (co)homology modules are equal up to appropriate twists. The twist is coming from the already mentioned identification of spin and adjoint actions on $\bigwedge(\mathfrak{u})$, and it is given by $\rho_{\bar{\mathfrak{u}}}$.

9.3. *Hodge decomposition*

Let us take X to be unitary, so it carries an invariant positive definite hermitian form. We need to put a similar form on S. There is a natural, $\mathfrak{l}$-invariant one, given by the Killing form B on $\bigwedge^{\cdot} \mathfrak{u}$. This form is however indefinite if $\mathfrak{u}$ intersects both $\mathfrak{k}$ and $\mathfrak{p}$ as will usually be the case. The operators d and 2∂ can be shown to be adjoint with respect to this form, but one can not in general obtain a "Hodge decomposition".

A good case where one can get the "Hodge theory approach" to work is when $\mathfrak{l} = \mathfrak{k}$ so that $\mathfrak{u} \subset \mathfrak{p}$ and B is positive definite on $\mathfrak{u}$. This is possible only in the Hermitian symmetric case (i.e., when $\mathfrak{k}$ has a center). In this case the Dirac operator D is "ordinary" (i.e., has no cubic term) - it is the same D we studied in Section 5. We know that in this situation D^2 is a scalar on each $\tilde{K}$-type, so in particular, D^2 acts (locally) finitely on $X \otimes S$. Note also that D^2 is negative semidefinite in the present situation. Namely, the relations in the Clifford algebra are now $Z_i^2 = 1$, not -1, hence D is skew-symmetric and not symmetric as before. So all eigenvalues of D^2 are non-positive.

Because of these facts, one can use the easy variant of Hodge theory for finite dimensional spaces (see [33], Scholium 9.4.4), and conclude that

$$\operatorname{Ker} D = \operatorname{Ker} D^2 = \operatorname{Ker} d \cap \operatorname{Ker} \partial;$$

$$\operatorname{Ker} d = \operatorname{Ker} D \oplus \operatorname{Im} d; \quad \operatorname{Ker} \partial = \operatorname{Ker} D \oplus \operatorname{Im} \partial.$$

In particular, the cohomology of both d and ∂ is equal to the Dirac cohomology of D, $\operatorname{Ker} D$, as vector spaces. To compare them as $\mathfrak{l} = \mathfrak{k}$-modules involves the above mentioned twist.

The same argument proves an analogous result for X finite dimensional; here one uses the "admissible form" on X, i.e., the one invariant for the compact form of $\mathfrak{g}$. This case was known to Vogan and it is also implicit in [20].

9.4. *A counterexample*

Here is a simple example which shows that the equality of the three cohomology modules does not hold for general $(\mathfrak{sl}(2,\mathbb{C}), SO(2))$-modules. Consider the module V which is a nontrivial extension of the discrete series representation of highest weight -2 by the trivial module $\mathbb{C}$:

$$0 \to \mathbb{C} \to V \to W \to 0.$$

V is a submodule of the module $V_{-1,0}$ from the end of Section 1. The weights of V (for the basis element $\begin{pmatrix} 0 & -i \\ i & 0 \end{pmatrix}$ of $\mathfrak{k}$) are $\cdots -4, -2, 0$. We are considering the case $\mathfrak{l} = \mathfrak{k}$, $\mathfrak{u}$ is spanned by $u = X = \frac{1}{2}\begin{pmatrix} 1 & i \\ i & -1 \end{pmatrix}$ and $\bar{\mathfrak{u}}$ is spanned by $u^* = Y = \frac{1}{2}\begin{pmatrix} 1 & -i \\ -i & -1 \end{pmatrix}$. Now

$$V \otimes S = V \otimes 1 \oplus V \otimes u,$$

with $d : V \otimes 1 \to V \otimes u$ given by $d(v \otimes 1) = u^* \cdot v \otimes u$, and $\partial : V \otimes u \to V \otimes 1$ given by $\partial(v \otimes u) = u \cdot v \otimes 1$.

By an easy direct calculation one sees that the $\mathfrak{u}$-homology of V is given by

$$H_0(\partial) = 0; \quad H_1(\partial) = \mathbb{C}v_0 \otimes u,$$

the $\bar{\mathfrak{u}}$-cohomology of V is given by

$$H^0(d) = \mathbb{C}v_0 \otimes 1; \quad H^1(d) = \mathbb{C}v_0 \otimes u \oplus \mathbb{C}v_{-2} \otimes u,$$

and the Dirac cohomology of V is given by

$$H_D(V) = \operatorname{Ker} D = \mathbb{C}v_0 \otimes u.$$

So we see

$$H_D(V) = H_{\cdot}(\partial) \neq H^{\cdot}(d).$$

9.5. *Dirac cohomology and* $(\mathfrak{g}, K)$*-cohomology*

In the rest of this section we make a few comments on the relationship of Dirac cohomology and $(\mathfrak{g}, K)$-cohomology. It was proved in [12] that if X is unitary and has $(\mathfrak{g}, K)$-cohomology, i.e.,

$$H^*(\mathfrak{g}, K; X \otimes F^*) = H^*(\mathrm{Hom}_K^{\cdot}(\textstyle\bigwedge^{\cdot}\mathfrak{p}, X \otimes F^*)) \neq 0$$

for a finite dimensional F (which then necessarily has the same infinitesimal character as X), then X also has Dirac cohomology.

In the following we assume that $\dim \mathfrak{p}$ is even. Then we can write $\mathfrak{p}$ as a direct sum of isotropic vector spaces $\mathfrak{u}$ and $\bar{\mathfrak{u}} \cong \mathfrak{u}^*$. One considers the spinor spaces $S = \bigwedge^{\cdot}\mathfrak{u}$ and $S^* = \bigwedge^{\cdot}\bar{\mathfrak{u}}$; then

$$S \otimes S^* \cong \textstyle\bigwedge^{\cdot}(\mathfrak{u} \oplus \bar{\mathfrak{u}}) = \bigwedge^{\cdot}\mathfrak{p}.$$

It follows that we can identify the $(\mathfrak{g}, K)$-cohomology of $X \otimes F^*$ with

$$H^*(\mathrm{Hom}_{\tilde{K}}^{\cdot}(F \otimes S, X \otimes S)).$$

There are several possible actions of the Dirac operator D on the above complex; similarly as before, they can be related to the coboundary operator d and the boundary operator ∂ for $(\mathfrak{g}, K)$-homology, which also acts on the same complex after appropriate identifications.

Now if X is unitary, Wallach has proved that $d = 0$ (see [33], Proposition 9.4.3, or [2]). Using similar arguments one can analize the above mentioned Dirac actions and the actions of the corresponding "half-Diracs". In particular, it follows that

$$H^*(\mathfrak{g}, K; X \otimes F^*) = \mathrm{Hom}_{\tilde{K}}^{\cdot}(H_D(F), H_D(X)).$$

This can be concluded from the fact that the eigenvalues of D^2 are of opposite signs on $F \otimes S$ and $X \otimes S$; see [33], 9.4.6. One may hope to generalize some of these facts, either with respect to X, or with respect to F.

We finish by a remark that for nice and natural proofs it would be useful to formalize some constructions in the category of modules over the Clifford algebra, although this may not seem necessary as the category is extremely simple. For example, one can construct a coproduct of $C(\mathfrak{p})$ by defining

$$c(Z) = \frac{1}{\sqrt{2}}(Z \otimes 1 + 1 \otimes Z)$$

for $Z \in \mathfrak{p}$. This coproduct is an algebra morphism in the graded sense. It is not coassociative, but it is cocommutative. There is no counit, but there is an antiautomorphism, the already mentioned α (see 2.4) which can be used instead of an antipode. It seems worthwhile to try to use this structure to clarify notions like tensor products and dual modules.

10. Multiplicities of automorphic forms

In this section we prove a formula for multiplicities of automorphic forms which sharpens the result of Langlands and Hotta-Parthasarathy. Let G be a linear semisimple noncompact Lie group. Let K be a maximal compact subgroup of G. Assume that $\operatorname{rank} G = \operatorname{rank} K$. Let $\mathfrak{g}_0 = \mathfrak{k}_0 + \mathfrak{p}_0$ be the Cartan decomposition of the Lie algebra of G. Then $\mathfrak{u}_0 = \mathfrak{k}_0 + i\mathfrak{p}_0$ is a compact real form of $\mathfrak{g} = \mathfrak{g}_0 \otimes_{\mathbb{R}} \mathbb{C}$. Let U be the compact analytic subgroup in the complexification $G_{\mathbb{C}}$ of G with Lie algebra $\mathfrak{u}_0$.

10.1. *Hirzebruch proportionality principle*

Let Γ be a torsion free discrete subgroup of G so that $\Gamma\backslash G$ and $X = \Gamma\backslash G/K$ are compact smooth manifolds. Borel showed that such a Γ always exists. Then the regular representation on $\Gamma\backslash G$ is decomposed discretely with finite multiplicities:

$$L^2(\Gamma\backslash G) \cong \bigoplus_{\pi\in\widehat{G}} m(\Gamma,\pi)\mathcal{H}_\pi.$$

Let X_π be the Harish-Chandra module of $\mathcal{H}_\pi$. For any $\mu \in \widehat{K}$, we define the Dirac operators D, $D^+_\mu(X)$ and $D^+_\mu(Y)$ as in 6.5. As we saw in 6.5, if we normalize the Haar measure so that $vol(U) = 1$, then

$$\operatorname{Index} D^+_\mu(X) = (-1)^q vol(\Gamma\backslash G) \operatorname{Index} D^+_\mu(Y).$$

On the other hand,

$$\operatorname{Index} D^+_\mu(X) = \sum_{\pi\in\widehat{G}} m(\Gamma,\pi) \operatorname{Index} D^+_\mu(X_\pi),$$

where $D^+_\mu(X_\pi) : \operatorname{Hom}_{\tilde{K}}(E^*_\mu, X_\pi \otimes S^+) \to \operatorname{Hom}_{\tilde{K}}(E^*_\mu, X_\pi \otimes S^-)$ is the linear map defined by $\phi \mapsto D\circ\phi$ for any $\phi \in \operatorname{Hom}_{\tilde{K}}(E^*_\mu, X_\pi \otimes S^+)$.

10.2. *Dimension of automorphic forms*

If $\operatorname{Index} D^+_\mu(X_\pi) \neq 0$, then the Dirac cohomology $H_D(X_\pi)$ contains E_{μ^*}. It follows from the proved Vogan's conjecture that the infinitesimal character of X_π is given by $\mu^* + \rho_c$. If we assume that $\lambda = w(\mu^* + \rho_c) - \rho$ is dominant for some $w \in W$, then X_π is isomorphic to $A_{\mathfrak{q}}(\lambda)$ for some θ-stable parabolic subalgebra $\mathfrak{q}$. If in addition we assume that λ is regular with respect to the noncompact roots $\Delta^+(\mathfrak{p})$, then X_π is uniquely determined as a discrete series $A_{\mathfrak{b}}(\lambda)$. Since $\text{Index}\, D^+_\mu(A_{\mathfrak{b}}(\lambda)) = \dim D^+_\mu(A_{\mathfrak{b}}(\lambda)) - \text{codim} D^+_\mu(A_{\mathfrak{b}}(\lambda)) = (-1)^q$ and $\text{Index}\, D^+_\mu(Y)$ has been calculated in Corollary 6.4., we obtain the following theorem.

Theorem: *Let $\pi = A_{\mathfrak{b}}(\lambda)$ be a discrete series representation with λ regular with respect to all noncompact roots. Assume that λ is dominant and can be written as $\lambda = \mu - \rho_n$ for some highest weight $\mu \in \widehat{K}$. Then*

$$m(\Gamma, \pi) = vol(\Gamma\backslash G) d_\pi,$$

where d_π is the formal degree of π:

$$d_\pi = \frac{\Pi_{\alpha\in\Delta^+(\mathfrak{g},\mathfrak{t})}(\lambda + \rho, \alpha)}{\Pi_{\alpha\in\Delta^+(\mathfrak{g},\mathfrak{t})}(\rho, \alpha)}.$$

This sharpens the result of Langlands [21] and Hotta-Parthasarathy [11], who proved the above formula for discrete series representations whose K-finite matrix coefficients are in $L^1(G)$. Trombi-Varadarajan [27] proved that if the K-finite matrix coefficients of the discrete series $A_{\mathfrak{b}}(\lambda)$ are in $L^1(G)$, then for all $\alpha \in \Delta^+(\mathfrak{p})$ and all $w \in W_{\mathfrak{g}}$

$$\langle \lambda + \rho, \alpha \rangle > |\langle w\rho, \alpha \rangle|.$$

Hecht-Schmid [9] proved this is also a sufficient condition. Our assumption on the regularity of λ with respect to the noncompact roots amounts to the condition that for all $\alpha \in \Delta^+(\mathfrak{p})$

$$\langle \lambda + \rho, \alpha \rangle > |\langle \rho, \alpha \rangle|.$$

Therefore, our condition is weaker than that assumed by Langlands and Hotta-Parthasarathy.

10.3. *A final remark*

Borel and Wallach [2] proved that for any finite-dimensional representation of G,

$$H^*(\Gamma, F) = \bigoplus_{\pi\in\widehat{G}} m(\Gamma, \pi) H^*(\mathfrak{g}, K, X_\pi \otimes F).$$

We still assume that $\text{rank}\, G = \text{rank}\, K$. If the highest weight of F is regular, then it follows from a similar argument as in 10.2. that X_π is uniquely determined as a discrete series $A_{\mathfrak{b}}(\lambda)$, and therefore,

$$\dim H^*(\Gamma, F) = vol(\Gamma\backslash G)d_\pi \dim H^*(\mathfrak{g}, K, X_\pi \otimes F).$$

Acknowledgements

The research of the first author was partially supported by RGC-CERG grants of Hong Kong SAR and National Nature Science Foundation of China. The research of the second author was partially supported by a grant from the Ministry of Science and Technology of Republic of Croatia. Part of these notes was written during authors' visit to CNRS, University of Paris VII and IMS at the National University of Singapore. The authors thank these institutions for their generous support and hospitality.

References

1. M. Atiyah and W. Schmid, "A geometric construction of the discrete series for semisimple Lie groups", *Invent. Math.*, **42** (1977), 1–62,
2. A. Borel and N. Wallach, *Continuous cohomology, discrete subgroups, and representations of reductive groups*, Second edition, Mathematical Surveys and Monographs 67, American Mathematical Society, Providence, RI, 2000.
3. R. Cahn, P. Gilkey and J. Wolf, "Heat equation, proportionality principle, and volume of fundamental domains", 43-54, Differential Geometry and Relativity, Mathematical Phys. and Appl. Math., Vol. 3, Reidel, Dordrecht, 1976.
4. H. Cartan, "La transgression dans un groupe de Lie et dans un espace fibré principal", *Colloque de Topologie algébrique*, C.B.R.M. Bruxelles, (1950), 57–71.
5. W. Casselman and D. Miličić, "Asymptotic behavior of matrix coefficients of admissible representations", *Duke Math. Jour.* **49** (1982), 869–930.
6. W. Casselman and M. S. Osborne, "The $\mathfrak{n}$-cohomology of representations with an infinitesimal character", *Compositio Math.*, **31** (1975), 219–227.
7. C. Chevalley, *The algebraic theory of spinors*, Columbia University Press, 1954.
8. B. Gross, B. Kostant, P. Ramond and S. Sternberg, "The Weyl character formula, the half-spin representations, and equal rank subgroups", *Proc. Nat. Acad. Sci. U.S.A.*, **95** (1998), 8441–8442.
9. H. Hecht and W. Schmid, "On integrable representations of a semisimple Lie group", *Math. Ann.*, **220** (1976), 147–149.
10. R. Hotta, "On a realization of the discrete series for semisimple Lie groups", *Jour. Math. Soc. Japan*, **23** (1971), 384–407.

11. R. Hotta and R. Parthasarathy, "A geometric meaning of the multiplicities of integrable discrete classes in $L^2(\Gamma\backslash G)$", *Osaka Jour. Math.*, **10** (1973), 211–234.
12. J.-S. Huang and P. Pandžić, "Dirac cohomology, unitary representations and a proof of a conjecture of Vogan", *J. Amer. Math. Soc.*, **15** (2002), 185–202.
13. , J.-S. Huang, P. Pandžić and D. Renard, "Dirac operators and $\mathfrak{n}$-cohomology", in preparation.
14. V. Kac, "Lie superalgebras", *Adv. in Math.*, **26** (1977), 8–96.
15. A. W. Knapp, *Representation theory of semisimple groups: an overview based on examples*, Princeton University Press, 1986.
16. A. W. Knapp and D. A. Vogan, Jr., *Cohomological induction and unitary representations*, Princeton University Press, 1995.
17. B. Kostant, "Clifford algebra analogue of the Hopf-Koszul-Samelson theorem, the ρ-decomposition $C(\mathfrak{g}) = \operatorname{End} V_\rho \otimes C(P)$, and the $\mathfrak{g}$-module structure of $\wedge\mathfrak{g}$", *Adv. in Math.*, **125** (1997), 275–350.
18. B. Kostant, "A cubic Dirac operator and the emergence of Euler number multiplets of representations for equal rank subgroups", *Duke Math. Jour.*, **100** (1999), 447–501.
19. B. Kostant, "Dirac cohomology for the cubic Dirac operator", *Studies in memory of I. Schur*, 69–93, Progress in Math. vol. 210, 2003.
20. B. Kostant, "A generalization of the Bott-Borel-Weil theorem and Euler number multiplets of representations", *Lett. Math. Phys.*, **52** (2000), 61–78.
21. R. Langlands, "The dimension of spaces of automorphic forms", *Amer. Jour. Math.*, **85** (1963), 99–125.
22. J.-S. Li and J. Schwermer, "Automorphic representations and cohomology of arithmetic groups", 102–137, *Challenges for the 21st century*, Proceedings of International Conference on Fundamental Sciences: Mathematics and Theoretical Physics (Singapore, 2000), World Scientific Publishing.
23. R. Parthasarathy, "Dirac operator and the discrete series", *Ann. of Math.*, **96** (1972), 1–30.
24. S. A. Salamanca-Riba, "On the unitary dual of real reductive Lie groups and the $A_{\mathfrak{q}}(\lambda)$ modules: the strongly regular case", *Duke Math. Jour.*, **96** (1998), 521–546.
25. W. Schmid, "Homogeneous complex manifolds and representations of semisimple Lie groups", Dissertation, UC Berkeley, 1967, 223–286, in *Representation Theory and Harmonic Analysis on Semisimple Lie Groups*, Edited by P. Sally and D. Vogan, American Mathematical Society, 1989.
26. W. Schmid, "L^2-cohomology and discrete series", *Ann. of Math.*, **103** (1976), 375–394.
27. P. C. Trombi and V. S. Varadarajan, "Asymptotic behaviour of eigenfunctions on a semisimple Lie group: the discrete spectrum", *Acta. Math.*, **129** (1972), 237–280.
28. D. A. Vogan, Jr., *Representations of real reductive Lie groups*, Birkhäuser, Boston-Basel-Stuttgart, 1981.
29. D. A. Vogan, Jr., "Unitarizability of certain series of representations", *Ann.*

of Math., **120** (1984), 141–187.
30. D. A. Vogan, Jr., "Dirac operators and unitary representations", 3 talks at MIT Lie groups seminar, Fall of 1997.
31. D. A. Vogan, Jr., "$\mathfrak{n}$-cohomology in representation theory", a talk at "Functional Analysis VII", Dubrovnik, Croatia, September 2001.
32. D. A. Vogan, Jr. and G. J. Zuckerman, "Unitary representations with non-zero cohomology", *Compositio Math.*, **53** (1984), 51–90.
33. N. R. Wallach, *Real Reductive Groups*, Volume I, Academic Press, 1988.

On Multiplicity Free Actions

Chal Benson and Gail Ratcliff

Department of Mathematics
East Carolina University
Greenville, NC 27858, U.S.A.
Email: bensonf@mail.ecu.edu, ratcliffg@mail.ecu.edu

Contents

1 Preliminaries 223
1.1 Algebraic groups 223
1.2 Regular functions 223
1.3 Algebraic groups as Lie groups 224
1.4 Structure theory 224
1.5 Rational representations 226
1.6 Highest weight theory 228
1.7 The contragredient representation 229
1.8 Decompositions and multiplicities 229
1.9 Group actions 230
1.10 Section 1 notes 231
2 Multiplicity free actions 231
2.1 Borel orbits 232
2.2 Quasi-regular representations 233
2.3 Maximal unipotent subgroups 235
2.4 S-varieties 236
2.5 Spherical pairs 237
2.6 Section 2 notes 238
3 Linear multiplicity free actions 239
3.1 Connectivity of G 239
3.2 Borel orbits 241

3.3 Fundamental highest weights for a multiplicity free action 245
3.4 Section 3 notes 247
4 Examples of multiplicity free decompositions 248
4.1 $\mathbf{GL(n)} \otimes \mathbf{GL(m)}$ 248
4.2 $\mathbf{S^2(GL(n))}$ 252
4.3 $\mathbf{\Lambda^2(GL(n))}$ 254
4.4 $\mathbf{SO(n)} \times \mathbb{C}^\times$ 257
4.5 $\mathbf{GL(n)} \oplus_{\mathbf{GL(n)}} \mathbf{\Lambda^2(GL(n))}$ 258
4.6 Section 4 notes 260
5 A recursive criterion for multiplicity free actions 262
5.1 $\mathbf{GL(n)}$ 263
5.2 $\mathbf{GL(n)} \otimes \mathbf{GL(n)}$ 264
5.3 $\mathbf{GL(n)} \oplus_{\mathbf{GL(n)}} \mathbf{\Lambda^2(GL(n))}$ 264
6 The classification of linear multiplicity free actions 266
6.1 Irreducible multiplicity free actions 266
6.2 Decomposable actions 268
6.3 Saturated indecomposable multiplicity free actions 268
6.4 Non-saturated indecomposable multiplicity free actions 270
6.5 Completing the classification 270
6.6 Proof outline 272
6.7 Section 6 notes 273
7 Invariant polynomials and differential operators 273
7.1 Polynomial coefficient differential operators 273
7.2 Invariants in $\mathcal{PD}(V)$ 276
7.3 A canonical basis for the invariants 278
7.4 The fundamental invariants 279
7.5 The algebra $\mathbb{C}[V_{\mathbb{R}}]^K$ 281
7.6 Section 7 notes 283
8 Generalized binomial coefficients 284
8.1 The polynomials q_λ 284
8.2 The generalized binomial coefficients 285
8.3 Eigenvalues for operators in $\mathcal{PD}(V)^G$ 288
8.4 Examples 289
8.5 Section 8 notes 293
9 Eigenvalues for operators in $\mathcal{PD}(V)^G$ 294
9.1 Eigenvalue polynomials 294
9.2 A Harish-Chandra homomorphism for multiplicity free actions 296
9.3 Characterizing the eigenvalue polynomials 298
9.4 $\mathbf{GL(n)} \otimes \mathbf{GL(n)}$ yet again 299
9.5 Section 9 notes 301
References 301

1. Preliminaries

Much of the literature on multiplicity free actions is set in the framework of algebraic groups. We begin by summarizing the basic definitions and results we require concerning such groups and their representations.

1.1. *Algebraic groups*

The general linear group $GL(n, \mathbb{C})$ can be viewed as an algebraic group. Letting $gl(n, \mathbb{C})$ denote the space of $n \times n$ complex matrices, the group $GL(n, \mathbb{C})$ can be identified with the zero set for the polynomial function $p(A, w) = det(A)w - 1$ on $gl(n, \mathbb{C}) \times \mathbb{C}$. This determines the structure of $GL(n, \mathbb{C})$ as an affine variety. One calls G a *reductive complex (linear) algebraic group* when

- (*linear*) G is an algebraic subgroup of $GL(n, \mathbb{C})$, and
- (*reductive*) $\mathbb{C}^n$ is a direct sum of G-irreducible subspaces.

The classical examples are

$$GL(n.\mathbb{C}), SL(n, \mathbb{C}), O(n, \mathbb{C}), SO(n, \mathbb{C}), Sp(2n, \mathbb{C})$$

and direct products of these groups. The torus $(\mathbb{C}^\times)^n$ is a direct product of copies of $GL(1, \mathbb{C}) = \mathbb{C}^\times$. A reductive complex algebraic group is connected (in the Zariski topology) if and only if it is irreducible as an algebraic variety. The classical examples are all connected except for $O(n, \mathbb{C})$, which has two components. *We will assume that our algebraic groups G are connected unless noted otherwise.*

1.2. *Regular functions*

$\mathbb{C}[G]$ denotes the ring of *regular functions* on G. This is the coordinate ring of G as an affine variety. More concretely

$\mathbb{C}[G]$ is the algebra generated by the matrix entries of G and det^{-1}.

A function $f : G \to \mathbb{C}$ is regular if and only if f is the restriction of a regular function on $GL(n, \mathbb{C})$. So

$$\mathbb{C}[G] \cong \mathbb{C}[GL(n, \mathbb{C})]/I(G), \text{ where } I(G) = \{f \in \mathbb{C}[GL(n, \mathbb{C})] \ : \ f(G) = 0\}.$$

Examples 1.2.1: For $g = [a_{ij}] \in GL(n, \mathbb{C})$ let $z_{ij}(g) = a_{ij}$. Then

- $\mathbb{C}[GL(n, \mathbb{C})] = \mathbb{C}[z_{ij}, det^{-1}]$,
- $\mathbb{C}[SL(n, \mathbb{C})] = \mathbb{C}[z_{ij}]$,
- $\mathbb{C}[(\mathbb{C}^\times)^n] = \mathbb{C}[z_{11}, z_{11}^{-1}, \ldots, z_{nn}, z_{nn}^{-1}]$.

1.3. *Algebraic groups as Lie groups*

As an algebraic group, G carries the Zariski topology. As a set of $n \times n$ complex matrices, G also has a subspace topology from $gl(n, \mathbb{C})$. In fact, G is a smooth submanifold of $gl(n, \mathbb{C})$, viewed as a real vector space of dimension $(2n)^2$. In this way G is seen as a (real) Lie group with Lie algebra

$$\mathfrak{g} = \{A \in gl(n, \mathbb{C}) \ : \ e^{tA} \in G \text{ for all } t \in \mathbb{R}\}.$$

Moreover $\mathfrak{g}$ is closed under multiplication by i and hence is a *complex* Lie-subalgebra of $gl(n, \mathbb{C})$. Alternatively one can define the (complex) Lie algebra $\mathfrak{g}$ for G as

$$\mathfrak{g} = \{A \in gl(n, \mathbb{C}) \ : \ f \in I(G) \implies Af \in I(G)\}$$

where

$$Af(g) = \left.\frac{d}{dt}\right|_{t=0} f(g(I + tA)).$$

When G is reductive, $\mathfrak{g}$ is a complex reductive Lie algebra. This implies that

$$\mathfrak{g} = \mathfrak{z}(\mathfrak{g}) \oplus \mathfrak{g}'$$

where $\mathfrak{z}(\mathfrak{g})$ denotes the center of $\mathfrak{g}$ and the derived subalgebra $\mathfrak{g}' = [\mathfrak{g}, \mathfrak{g}]$ is semi-simple, a direct sum of simple ideals. $\mathfrak{g}'$ is the Lie algebra of the commutator subgroup $G' = (G, G)$.

1.4. *Structure theory*

A maximal connected solvable algebraic subgroup B of G is called a *Borel subgroup*. The following facts concerning such subgroups are well known:

- Any two Borel subgroups are conjugate in G.
- Given any Borel subgroup B, there is an *opposite* Borel subgroup B^- with the property that B^-B is Zariski dense in G and contains an open neighborhood of I.
- B is the semidirect product $B = HN$ of its commutator subgroup $N = (B, B)$ with a maximal torus H in G. The group N is a maximal unipotent subgroup of G.

The Lie algebra $\mathfrak{h}$ of H is a maximal abelian subalgebra of $\mathfrak{g}$. For $\alpha \in \mathfrak{h}^*$ let

$$\mathfrak{g}_\alpha = \{Y \in \mathfrak{g} \ : \ [X, Y] = \alpha(X)Y \text{ for all } X \in \mathfrak{h}\}.$$

Then

$$\Delta = \Delta(\mathfrak{g}, \mathfrak{h}) = \{\alpha \in \mathfrak{h}^* \ : \ \mathfrak{g}_\alpha \neq 0\}$$

is the set of *roots* for $\mathfrak{g}$ (relative to $\mathfrak{h}$). Each root space $\mathfrak{g}_\alpha$ is one dimensional and

$$\mathfrak{g} = \mathfrak{h} \oplus \bigoplus_{\alpha \in \Delta} \mathfrak{g}_\alpha.$$

There is a subset Δ^+ of Δ, called the *positive roots*, such that N has Lie algebra

$$\mathfrak{n} = \bigoplus_{\alpha \in \Delta^+} \mathfrak{g}_\alpha.$$

Now

- $\Delta = \Delta^+ \cup (-\Delta^+)$,
- $\mathfrak{b} = \mathfrak{h} \oplus \mathfrak{n}$ is the Lie algebra of B, and
- N^- has Lie algebra $\mathfrak{n}^- = \bigoplus_{\alpha \in \Delta^+} \mathfrak{g}_{-\alpha}$.

For each $\alpha \in \Delta^+$ there are elements $X_\alpha \in \mathfrak{g}_\alpha$, $X_{-\alpha} \in \mathfrak{g}_{-\alpha}$, $H_\alpha \in \mathfrak{h}$ which form an $sl(2)$-triple:

$$[H_\alpha, X_\alpha] = 2X_\alpha, \quad [H_\alpha, X_{-\alpha}] = -2X_{-\alpha}, \quad [X_\alpha, X_{-\alpha}] = H_\alpha.$$

For each $\alpha \in \Delta$ we have a *root reflection*

$$s_\alpha : \mathfrak{h}^* \to \mathfrak{h}^*, \quad s_\alpha(\lambda) = \lambda - \langle \lambda, \alpha \rangle \alpha$$

where

$$\langle \lambda, \alpha \rangle = \lambda(H_\alpha).$$

The *Weyl group* $W = W(\mathfrak{g}, \mathfrak{h})$ is the subgroup of $GL(\mathfrak{h}^*)$ generated by $\{s_\alpha \ : \ \alpha \in \Delta\}$. It is a finite reflection group that acts by permutations on the set Δ.

Example 1.4.1: The *standard Borel subgroup* in $G = GL(n, \mathbb{C})$ is

$$B_n = \left\{ \begin{bmatrix} z_1 & & \\ & \ddots & \star \\ \mathbf{O} & & z_n \end{bmatrix} \ : \ z_j \in \mathbb{C}^\times \right\},$$

the group of invertible upper triangular matrices. We have

$$B_n = H_n N_n$$

where H_n denotes the diagonal matrices in $GL(n,\mathbb{C})$ and N_n denotes the unipotent upper triangular matrices. The opposite Borel subgroup for B_n is

$$B_n^- = H_n N_n^-$$

where B_n^- and N_n^- are the invertible and unipotent lower triangular matrices respectively.

The Lie algebra $\mathfrak{h}_n$ of H_n is the set of all diagonal matrices. Letting $\varepsilon_i \in \mathfrak{h}^*$ denote the functional

$$\varepsilon_i(diag(z_1,\ldots,z_n)) = z_i$$

one has

$$\text{roots } \Delta = \{\varepsilon_i - \varepsilon_j \ : \ i \neq j\} \text{ and positive roots } \Delta^+ = \{\varepsilon_i - \varepsilon_j \ : \ i < j\}.$$

For $\alpha = \varepsilon_i - \varepsilon_j \in \Delta^+$ we have

$$X_\alpha = E_{ij}, \quad X_{-\alpha} = E_{ji}, \quad H_\alpha = E_{ii} - E_{jj}.$$

The root reflection s_α satisfies $s_\alpha(\varepsilon_k) = \varepsilon_{\tau(k)}$ where $\tau \in S_n$ is the transposition that interchanges i with j. Thus the Weyl group is isomorphic to S_n.

1.5. *Rational representations*

Definition 1.5.1: Let (σ, V) be a representation of G.

(1) (σ, V) is said to be *rational* (or regular) if it is finite dimensional and its matrix coefficients

$$g \mapsto \xi(\sigma(g)v), \quad \xi \in V^*, v \in V$$

all belong to $\mathbb{C}[G]$.

(2) (σ, V) is *locally rational* (or locally regular) if $dim(V) = \infty$ and for any finite dimensional subspace F of V there is a $\sigma(G)$-invariant subspace W with $F \subset W \subset V$ for which $\sigma|W$ is rational.

We let $\widehat{G}$ denote the set of equivalence classes of irreducible rational representations of G and sometimes write V_σ for the representation space of $\sigma \in \widehat{G}$.

Note that subrepresentations of rational (or locally rational) representations are rational (resp. locally rational).

Example 1.5.2: As $G = \mathbb{C}^\times$ is abelian, its irreducible representations are given by characters $\rho : \mathbb{C}^\times \to \mathbb{C}^\times$. For such a character to be rational we require $\rho \in \mathbb{C}[G] = \mathbb{C}[z, 1/z]$. So ρ is holomorphic on $\mathbb{C}^\times$ and hence determined by its restriction to the unit circle $\mathbb{T}$. One concludes that

$$(\mathbb{C}^\times)\widehat{\ } = \{\rho_n \ : \ n \in \mathbb{Z}\}$$

where $\rho_n(z) = z^n$. The character

$$\rho(z) = \frac{z}{|z|}$$

gives a representation of $\mathbb{C}^\times$ which is not rational.

This example illustrates the:

Weyl Unitarian Trick: Rational representations of G are determined by holomorphic extension from a maximal compact connected subgroup K of G (viewed as a real Lie group).

It now follows from the representation theory for compact Lie groups that rational representations are completely reducible. Moreover, the Unitarian Trick establishes a bijection between $\widehat{G}$ and the set $\widehat{K}$ of unitary equivalence classes of irreducible unitary representations of K. So one can work entirely in the compact group setting, should one so prefer. Maximal compact subgroups for the classical groups $GL(n.\mathbb{C})$, $SL(n, \mathbb{C})$, $O(n, \mathbb{C})$, $SO(n, \mathbb{C})$, $Sp(2n, \mathbb{C})$ are

$$U(n), SU(n), O(n, \mathbb{R}), SO(n, \mathbb{R}), Sp(2n) = U(2n) \cap Sp(2n, \mathbb{C}).$$

Each rational representation (σ, V) of G is smooth. That is, $t \mapsto \sigma(e^{tX})v$ is a smooth map $\mathbb{R} \to V$ for each $v \in V$, $X \in \mathfrak{g}$. Thus we obtain a derived representation

$$\sigma(X)v = \frac{d}{dt}\bigg|_{t=0} \sigma(e^{tX})v$$

of the Lie algebra $\mathfrak{g}$ on V. (Note that we are using the same notation for the representation σ on both the Lie group and the Lie algebra.)

When there is no ambiguity, we will denote the action of G (or $\mathfrak{g}$) on V by

$$\sigma(g)v = g \cdot v \text{ (or } \sigma(X)v = X \cdot v).$$

1.6. *Highest weight theory*

Let $B = HN$ be a Borel subgroup in G and (σ, V) be a rational representation. By the Lie-Kolchin Theorem, there are non-zero $\sigma(B)$-eigenvectors. That is, there are vectors $v \neq 0$ in V such that

$$\sigma(b)v = \psi(b)v \quad \text{for all } b \in B,$$

where $\psi : B \to \mathbb{C}^\times$ is a regular character. As $N = (B, B)$, we have $\psi|_N = 1$ and hence

- $\{v \in V \ : \ v \text{ is a } B\text{-eigenvector}\} = V^N$, the N-fixed vectors in V, and
- ψ is determined by $\psi|_H \in \widehat{H}$.

Highest weight theory asserts that

$$\sigma \text{ is irreducible} \iff dim(V^N) = 1.$$

For each $\sigma \in \widehat{G}$, there is a non-zero B-eigenvector $v_\sigma \in V_\sigma$, unique up to scalar multiples. This is the *highest weight vector* for σ.

The corresponding character $\psi : B \to \mathbb{C}^\times$ can be differentiated to give a functional λ on the Lie algebra $\mathfrak{b}$, with $\lambda(\mathfrak{n}) = 0$. We have $X \cdot v_\sigma = \lambda(X)v_\sigma$ for all $X \in \mathfrak{b}$. As noted above, λ is determined by its value on the Lie algebra $\mathfrak{h}$ of H.

The functional λ in $\mathfrak{h}^*$ is the *highest weight* for σ. We can extend λ to $\mathfrak{b}$ (or $\mathfrak{b}^-$) by taking $\lambda(\mathfrak{n}) = 0$ (resp. $\lambda(\mathfrak{n}^-) = 0$). Thus v_σ is, up to scalars, the unique vector in V with

$$X \cdot v_\sigma = \lambda(X)v_\sigma \text{ for all } X \in \mathfrak{b}.$$

Highest weight theory asserts, moreover, that

$$\sigma \text{ is determined up to equivalence by its highest weight.}$$

Given a representation of G with highest weight λ, we will denote the corresponding representation as V_λ. (Keep in mind that this all depends on the initial choice of a Borel subgroup.) For an element b in the subgroup B, we will denote the corresponding character by $b \mapsto b^\lambda$.

The highest weights for $GL(n, \mathbb{C})$ (with respect to the standard Borel subgroup) are

$$\{diag(h_1, \ldots, h_n) \mapsto \sum_{i=1}^{n} d_i h_i \ : \ d_1, \ldots, d_n \in \mathbb{Z} \text{ with } d_1 \geq d_2 \geq \cdots \geq d_n\}.$$

1.7. *The contragredient representation*

Given a representation of G on V, we define the *contragredient representation* of G on V^* by:

$$g \cdot \xi(v) = \xi(g^{-1} \cdot v) \text{ for } \xi \in V^*,\ v \in V.$$

If V has highest weight vector v with highest weight λ with respect to some Borel subgroup B, then V^* has a highest weight vector v^* with weight $-\lambda$ with respect to the *opposite* Borel subgroup B^-.

1.8. *Decompositions and multiplicities*

Let (ρ, W) be a rational representation of G and $\sigma \in \widehat{G}$. One has a direct sum decomposition

$$W = \bigoplus_{\sigma \in \widehat{G}} W^\sigma$$

of W into *σ-isotypic components*

$$\begin{aligned} W^\sigma &= \sum \{V \ :\ V \text{ is a } \rho(G)\text{-invariant subspace of } W \text{ with } \rho|V \simeq \sigma\} \\ &= \sum \{T(V_\sigma) \ :\ T : V_\sigma \to W \text{ intertwines } \sigma \text{ with } \rho\}. \end{aligned}$$

Then

$$W^\sigma \simeq m(\sigma,\rho) V_\sigma = \underbrace{V_\sigma \oplus \cdots \oplus V_\sigma}_{m(\sigma,\rho)}$$

as G-modules where the *multiplicity* $m(\sigma, \rho)$ of σ in ρ is given by

$$m(\sigma,\rho) = dim(W^\sigma)/dim(V_\sigma) = dim\big(Hom_G(V_\sigma, W)\big).$$

($Hom_G(V_\sigma, W)$ is the space of linear maps $V_\sigma \to W$ intertwining σ with ρ.)

The multiplicity $m(\sigma, \rho)$ can also be characterized using highest weight theory. Let B be a Borel subgroup of G and λ be the highest weight for σ. Then

$$m(\sigma,\rho) = dim\left(W^{B,\lambda}\right)$$

where

$$W^{B,\lambda} = \{w \in W \ :\ b \cdot w = b^\lambda w\}$$

is the space of weight vectors in W with weight λ.

1.9. *Group actions*

We use the notation

$$G : X$$

to indicate that there is a rational action of G on a variety X,

$$G \times X \to X, \quad (g, x) \mapsto g \cdot x.$$

This means $G \times X \to X$ is a morphism of algebraic varieties satisfying

$$(gh) \cdot x = g \cdot (h \cdot x) \quad \text{and} \quad e \cdot x = x.$$

The variety X may be affine or, more generally, quasi-projective. That is, the intersection of a closed with an open set in $\mathbb{CP}^n$.

It follows that if $f \in \mathbb{C}[X]$, the regular functions on X, then

$$x \mapsto f(g^{-1} \cdot x)$$

is a regular function. Thus we obtain a representation ρ of G on $\mathbb{C}[X]$,

$$(\rho(g)f)(x) = (g \cdot f)(x) = f(g^{-1} \cdot x).$$

Lemma 1.9.1: *The representation ρ of G on $\mathbb{C}[X]$ is locally rational.*

Remark 1.9.2: *Linear* actions $G : X$ will be our main concern. That is, X will usually be a vector space on which G acts by a rational representation. Then $\mathbb{C}[X]$ is the algebra of polynomials on the vector space X and one can give an easy proof of Lemma 1.9.1. Let $\mathcal{P}_k(X)$ denote the space of polynomials on X homogeneous of degree k. For any finite dimensional subspace F of $\mathbb{C}[X]$, we have $F \subset W$ where

$$W = \sum_{k=0}^{n} \mathcal{P}_k(X)$$

for n sufficiently large. Now W is a finite dimensional, G-invariant subspace and we have a rational representation.

Now for $\sigma \in \widehat{G}$ define the σ-isotypic component $\mathbb{C}[X]^\sigma$ as:

$$\mathbb{C}[X]^\sigma = \sum \{T(V_\sigma) \ : \ T \in Hom_G(V_\sigma, \mathbb{C}[X])\}.$$

Lemma 1.9.3: *We have an algebraic direct sum*

$$\mathbb{C}[X] = \bigoplus_{\sigma \in \widehat{G}} \mathbb{C}[X]^\sigma.$$

Thus

$$\mathbb{C}[X] \cong \bigoplus_{\sigma \in \widehat{G}} m(\sigma, \mathbb{C}[X]) V_\sigma$$

as G-modules, where the (possibly infinite) multiplicity $m(\sigma, \mathbb{C}[X])$ *of* σ *in* $\mathbb{C}[X]$ *is*

$$m(\sigma, \mathbb{C}[X]) = dim(Hom_G(V_\sigma, \mathbb{C}[X])) = dim\left(\mathbb{C}[X]^{B,\lambda}\right),$$

where λ *is the highest weight for the representation* σ.

We can now introduce our principal objects of study.

Definition 1.9.4: $G : X$ is a *multiplicity free action* if $m(\sigma, \mathbb{C}[X]) \leq 1$ for each $\sigma \in \widehat{G}$.

Given an action $G : X$ and an algebraic subgroup H of G, one has

$$H : X \text{ multiplicity free} \implies G : X \text{ multiplicity free.} \tag{1}$$

Indeed, if $G : X$ fails to be multiplicity free, then some representation $\sigma \in \widehat{G}$ occurs in $\mathbb{C}[X]$ with multiplicity greater than one. Thus all irreducible constituents of $\sigma|_H$ occur with multiplicity greater than one. So $H : X$ also fails to be multiplicity free.

1.10. *Section 1 notes*

The material in this section is standard and there are many excellent references. One is the book [17] by Goodman and Wallach. See Chapter 1 and Appendix D in [17] for further details on foundational material concerning algebraic groups and Lie groups. See Section 11.3 of [17] for proofs of the assertions concerning Borel subgroups. Our treatment of the isotypic decomposition for $\mathbb{C}[X]$ follows Section 12.1 in [17].

2. Multiplicity free actions

This work is mainly devoted to the study of linear multiplicity free actions. In this section, however, we consider actions $G : X$ in the more general context of algebraic varieties. Our main purpose is to describe some noteworthy non-linear examples.

2.1. *Borel orbits*

There is a simple criterion for multiplicity free actions.

Theorem 2.1.1: *If a Borel subgroup B in G has a (Zariski) dense orbit in X then $G : X$ is multiplicity free.*

Proof: Suppose that $B \cdot x_\circ$ is dense in X. Let $\sigma \in \widehat{G}$ occur in $\mathbb{C}[X]$ (that is $m(\sigma, \mathbb{C}[X]) > 0$) and let λ be the highest weight for σ. Let $f_1, f_2 \in \mathbb{C}[X]$ be two B-highest weight vectors with weight λ. One has

$$f_j(b^{-1} \cdot x_\circ) = b^\lambda f_j(x_\circ).$$

As f_j is regular and $B \cdot x_\circ$ dense in X, we see that f_j is completely determined by the value $f_j(x_\circ)$. In particular, $f_j(x_\circ) \neq 0$ (as $f_j \neq 0$) and we can write

$$f_2 = \frac{f_2(x_\circ)}{f_1(x_\circ)} f_1.$$

So the space of λ-highest weight vectors in $\mathbb{C}[X]$ is one dimensional and hence $m(\sigma, \mathbb{C}[X]) = 1$. □

Since any two Borel subgroups are conjugate in G, the criterion in Theorem 2.1.1 does not depend on the choice of Borel subgroup B.

Suppose that $B \cdot x_\circ$ is dense in X. Since B is connected, it follows that X must be an *irreducible* variety and that $B \cdot x_\circ$ is a Zariski open set. Thus $X \backslash (B \cdot x_\circ)$ is a closed set which contains no open subsets in X. We conclude that there is *only one* dense open B-orbit in X.

Conversely, if $G : X$ is a multiplicity free action with X an *irreducible affine* variety then X contains an open (hence dense) B-orbit. We will prove this result for linear multiplicity free actions in the following section. (See Theorem 3.2.8 below.) One special case of the converse admits, however, a direct proof. This is the case where G is an algebraic torus $G = A \cong (\mathbb{C}^\times)^n$. In this case the Borel subgroup is A itself and one has the following.

Proposition 2.1.2: *Let X be an irreducible affine variety and A be a torus. If $A : X$ is multiplicity free then there is an open (hence dense) A-orbit in X.*

Proof: One can choose weight vectors $f_1, \ldots, f_r$ in $\mathbb{C}[X]$ that generate $\mathbb{C}[X]$ as an algebra. Since $A : X$ is multiplicity free the weights $\{\lambda_1, \ldots, \lambda_r\}$

for $f_1, \dots, f_r$ must be linearly independent. Choose a point $x_\circ \in X$ for which $f_j(x_\circ) \neq 0$ for $1 \le j \le r$. Define a map $\rho : \mathbb{C}[X] \to \mathbb{C}[A]$ by

$$(\rho f)(a) = (a \cdot f)(x_\circ) = f(a^{-1} \cdot x_\circ).$$

We claim that ρ is injective. Indeed, for $f = f^m = f_1^{m_1} \cdots f_r^{m_r}$ one has

$$(\rho f)(a) = a^{m_1\lambda_1} \cdots a^{m_r\lambda_r} f(x_\circ) = a^{m\lambda} f(x_\circ),$$

and hence

$$(\rho f)(a) = \sum_m c_m a^{m\lambda} f^m(x_\circ)$$

for $f = \sum_m c_m f^m \in \mathbb{C}[X]$. If $(\rho f)(a) = 0$ for all $a \in A$, then linear independence of the λ_j's implies that $c_m = 0$ for all m.

Thus every regular function on X is determined by its restriction to $A \cdot x_\circ$. It follows that $A \cdot x_\circ$ is open and dense in X. □

2.2. *Quasi-regular representations*

We continue to let G denote a reductive complex linear algebraic group. The left and right actions of G on $X = G$

- $g \cdot x = gx$, and
- $g \cdot x = xg^{-1}$

give rise to the *left and right regular representations*

- $L(g)f(x) = f(g^{-1}x)$, and
- $R(g)f(x) = f(xg)$ respectively

of G on $\mathbb{C}[G]$. For any algebraic subgroup H of G we let

$$\mathbb{C}[G/H] = \mathbb{C}[G]^{R(H)}, \quad \mathbb{C}[H\backslash G] = \mathbb{C}[G]^{L(H)}$$

and define the *left (resp. right) quasi-regular representation* of G as the restriction of L to $\mathbb{C}[G/H]$ (resp. R to $\mathbb{C}[H\backslash G]$). The representations L and R of G on $\mathbb{C}[G/H]$ and $\mathbb{C}[H\backslash G]$ are equivalent via the intertwining operator

$$T : \mathbb{C}[G/H] \to \mathbb{C}[H\backslash G], \quad T(f) = \check{f}, \quad (\check{f}(x) = f(x^{-1})).$$

In fact the homogeneous space G/H is a smooth quasi-projective variety with coordinate ring $\mathbb{C}[G/H]$. If H is a reductive or normal subgroup then G/H is an affine variety. We remark that one can have $\mathbb{C}[G/H] = \mathbb{C}$ even when $dim(G/H) > 0$. This situation occurs whenever G/H is a projective variety, in particular when H is a Borel subgroup of G. In any case, the left action $G : (G/H)$ is rational and gives rise to the left quasi-regular representation. Similar remarks apply for $H\backslash G$ and the right quasi-regular representation.

The isotypic decomposition for the quasi-regular representations is given by *Frobenius Reciprocity*:

Theorem 2.2.1: *As a G-module we have*

$$\mathbb{C}[G/H] \cong \bigoplus_{\sigma \in \widehat{G}} dim(V_{\sigma^*}^H) V_\sigma.$$

In particular, $\mathbb{C}[G] \cong \oplus_\sigma dim(V_\sigma) V_\sigma$.

Proof: Lemma 1.9.3 applies here since $G : (G/H)$ is a rational action. It suffices to show that $Hom_G(V_\sigma, \mathbb{C}[G/H]) \cong V_{\sigma^*}^H$. For this, one verifies that

$$\Phi : V_{\sigma^*}^H \to Hom_G(V_\sigma, \mathbb{C}[G/H]), \quad \Phi(\xi)(v)(g) = \xi(\sigma(g^{-1})v)$$

is an isomorphism with inverse

$$\Lambda : Hom_G(V_\sigma, \mathbb{C}[G/H]) \to V_{\sigma^*}^H, \quad \Lambda(T)(v) = (T(v))(e). \qquad \square$$

Corollary 2.2.2: *The action of G on G/H (or on $H\backslash G$) is multiplicity free if and only if $dim(V_\sigma^H) \leq 1$ for all $\sigma \in \widehat{G}$.*

Note that if H_1 and H_2 are algebraic subgroups of G with $H_1 \subset H_2$ then $V_\sigma^{H_2} \subset V_\sigma^{H_1}$. Thus if $G : (G/H_1)$ is a multiplicity free action then so is $G : (G/H_2)$.

The proof of Theorem 2.2.1 shows that the σ-isotypic component in $\mathbb{C}[G]$ for the left regular representation is

$$\begin{aligned} \mathbb{C}[G]^{\sigma,L} &= \{T(v) \ : \ v \in V_\sigma, T \in Hom_G(V_\sigma, \mathbb{C}[G])\} \\ &= \{\Phi(\xi)(v) \ : \ v \in V_\sigma, \xi \in V_\sigma^*\} \\ &= \{m\check{}_{\xi,v} \ : \ v \in V_\sigma, \xi \in V_\sigma^*\}, \end{aligned}$$

where $m_{\xi,v}(g) = \xi(\sigma(g)v)$ is a matrix coefficient and $m\check{}_{\xi,v}(g) = m_{\xi,v}(g^{-1})$. The right regular representation preserves $\mathbb{C}[G]^{\sigma,L}$, since L and R commute. As a module for $G \times G$, $\mathbb{C}[G]^{\sigma,L}$ is isomorphic to $V_\sigma \otimes V_{\sigma^*} = V_\sigma \otimes V_\sigma^*$. Indeed, the map

$$\Psi : V_\sigma \otimes V_\sigma^* \to \mathbb{C}[G]^{\sigma,L}, \quad \Psi(v \otimes \xi) = m\check{}_{\xi,v}$$

intertwines $\sigma \otimes \sigma^*$ with the left-right regular representation

$$(LR)(g_1, g_2)f(x) = f(g_1^{-1} x g_2)$$

of $G \times G$ on $\mathbb{C}[G]$. This gives the *Peter-Weyl Theorem:*

Theorem 2.2.3: *The left-right action of $G \times G$ on G is multiplicity free. As a $G \times G$-module we have*

$$\mathbb{C}[G] \cong \bigoplus_{\sigma \in \widehat{G}} \sigma \otimes \sigma^*.$$

2.3. *Maximal unipotent subgroups*

Let $B = HN$ be a Borel subgroup in G. For each $\sigma \in \widehat{G}$ we have $dim(V_\sigma^N) = 1$, by highest weight theory. Theorem 2.2.1 now shows:

Theorem 2.3.1: *The action of G on G/N is multiplicity free. Moreover, each irreducible representation $\sigma \in \widehat{G}$ occurs exactly once in $\mathbb{C}[G/N]$.*

The *Borel-Weil Theorem* provides an explicit model for the irreducible representation with specified highest weight $\lambda \in \mathfrak{h}^*$ inside $\mathbb{C}[G/N^-]$.

Theorem 2.3.2: *The irreducible representation with highest weight $\lambda \in \mathfrak{h}^*$ is given by the left regular representation of G on*

$$R_\lambda = \{f \in \mathbb{C}[G] \ : \ f(gb) = b^{-\lambda} f(g) \ \textit{for all}\ b \in B^-\}.$$

Moreover, a highest weight vector in R_λ is given on the dense subset NHN^- of G by

$$f_\lambda(nhn^-) = h^{-\lambda}.$$

Proof: First observe that R_λ is a $L(G)$-invariant subspace of $\mathbb{C}[G/N^-]$. If $f \in R_\lambda$ is a B-highest weight vector then

$$f(nhn^-) = f(h) = (h^{-1} \cdot f)(e)$$

and also $f(h) = h^{-\lambda} f(e)$. Thus f has weight λ and $f = f(e) f_\lambda$. So R_λ is $L(G)$-irreducible with highest weight λ and highest weight vector f_λ. □

Corollary 2.3.3: *The span of $R_\lambda R_\mu$ is $R_{\lambda+\mu}$.*

Proof: The span of $R_\lambda R_\mu$ is a G-invariant subspace of $R_{\lambda+\mu}$. □

Remark 2.3.4: It follows from Theorem 2.1.1 that $G : (G/B)$ is also multiplicity free. But $V_\sigma^B = \{0\}$ unless $\sigma \in \widehat{G}$ is the trivial representation. So $\mathbb{C}[G/B] = \mathbb{C}$ and this is an uninteresting example. Alternatively, one can note that G/B is a flag manifold, hence projective, hence compact. So every regular function on G/B is constant.

2.4. *S-varieties*

Let $B = HN$ be a Borel subgroup in G with opposite Borel subgroup $B^- = HN^- = N^-H$. For $i = 1, \ldots k$, let $\sigma_i \in \widehat{G}$ act on the space V_i, and let $v_i \in V_i$ be a B^--highest weight vector with weight $-\lambda_i$. Let $v = v_1 + \cdots + v_k \in V_1 \oplus \cdots \oplus V_k$. Then

$$X = \overline{G \cdot v},$$

the Zariski-closure of the orbit of v in $V_1 \oplus \cdots \oplus V_k$, is a G-invariant subvariety, called an *S-variety.*

Since $N^- \cdot v = \{v\}$, we see that

$$B \cdot v = NHN^- \cdot v$$

is dense in $G \cdot v$ and hence also in X. It follows from Theorem 2.1.1 that $G : X$ is a multiplicity free action. In the present context, however, we will show directly that $G : X$ is multiplicity free by exhibiting the decomposition of $\mathbb{C}[X]$.

Theorem 2.4.1: *The multiplicity free decomposition of* $\mathbb{C}[X]$ *is*

$$\mathbb{C}[X] \cong \bigoplus_{\lambda \in \Lambda} R_\lambda,$$

where $\Lambda = \{a_1\lambda_1 + \cdots + a_k\lambda_k \ : \ a_j \in \mathbb{N}\}$.

(Throughout, $\mathbb{N}$ denotes the non-negative integers, including zero.)

Proof: We can lift an element ξ of V_i^* to V, and define a function f_ξ on G by $f_\xi(g) = \xi(g \cdot v) = \xi(g \cdot v_i)$. Then for $b \in B^-$, $f_\xi(gb) = \xi(gb \cdot v_i) = b^{-\lambda_i}\xi(g \cdot v_i) = b^{-\lambda_i} f_\xi(v)$. Thus f_ξ is in the G-irreducible space R_{λ_i} defined above. Hence $\{\xi|_X : \xi \in V_i^*\}$ is equivalent to the G-irreducible R_{λ_i}.

Note that $\mathbb{C}[X]$ is generated by the restriction of elements of each V_i^* to X. So by Corollary 2.3.3, we conclude that the irreducible components are products of the subspaces R_{λ_i}. □

There is another characterization of S-varieties:

Theorem 2.4.2: *Let X be an irreducible affine G-variety with a G-open orbit, such that the isotropy subgroup of any point in the open orbit contains a maximal unipotent subgroup. Then X is an S-variety.*

Proof: One can find a (rational) representation σ of G and a G-equivariant embedding of X into V_σ. (See [44].)

So assume that $X \subset V_\sigma$ and let $v \in X$ be any point in the open G-orbit. Choose a Borel subgroup B with $N \subset B$ such that $N \subset Stab_G(v)$. Then we can write $v = v_1 + \cdots + v_k$, where each v_i is a weight vector with weight λ_i, and the $\lambda_i' s$ are distinct. Since v is stabilized by N, each v_i is a highest weight vector.

Thus our variety X is the closure (in V) of $G \cdot v$, where $v = v_1 + \cdots + v_k$. □

2.5. *Spherical pairs*

Suppose that H is a *reductive* algebraic subgroup of G. We say that (G, H) is a *spherical pair* if $dim(V_\sigma^H) \leq 1$ for all $\sigma \in \widehat{G}$. Equivalently, the actions of G on G/H and $H \backslash G$ are multiplicity free. In this section we will summarize some results concerning spherical pairs, without proof.

Let U and K denote maximal compact connected subgroups of G and H. These are compact real Lie groups. Recall that the irreducible rational representations of G and H correspond to irreducible unitary representations of U and K via the Unitarian Trick. So

$$(G, H) \text{ is spherical } \iff dim(V_\sigma^K) \leq 1 \text{ for all } \sigma \in \widehat{U}.$$

We say that (U, K) is a *compact Gelfand pair* in this case. It is known that (U, K) is a Gelfand pair if and only if the algebra $L^1(U//K)$ of integrable K-bi-invariant functions on U is abelian with respect to convolution.

Table 1: The classical compact Riemannian symmetric spaces

	U	K	G	H
1	$SO(n, \mathbb{R})$	$\{I\}$	$SO(n, \mathbb{C})$	$\{I\}$
2	$SU(n)$	$\{I\}$	$SL(n, \mathbb{C})$	$\{I\}$
3	$Sp(2n)$	$\{I\}$	$Sp(2n, \mathbb{C})$	$\{I\}$
4	$SU(n)$	$SO(n, \mathbb{R})$	$SL(n, \mathbb{C})$	$SO(n, \mathbb{C})$
5	$SU(2n)$	$Sp(2n)$	$SL(2n, \mathbb{C})$	$Sp(2n, \mathbb{C})$
6	$SU(p+q)$	$S(U(p) \times U(q))$	$SL(p+q, \mathbb{C})$	$S(GL(p, \mathbb{C}) \times GL(q, \mathbb{C}))$
7	$SO(p+q, \mathbb{R})$	$SO(p, \mathbb{R}) \times SO(q, \mathbb{R})$	$SO(p+q, \mathbb{C})$	$SO(p, \mathbb{C}) \times SO(q, \mathbb{C})$
8	$SO(2n, \mathbb{R})$	$U(n)$	$SO(2n, \mathbb{C})$	$GL(n, \mathbb{C})$
9	$Sp(2n)$	$U(n)$	$Sp(2n, \mathbb{C})$	$GL(n, \mathbb{C})$
10	$Sp(p+q)$	$Sp(p) \times Sp(q)$	$Sp(p+q, \mathbb{C})$	$Sp(p, \mathbb{C}) \times Sp(q, \mathbb{C})$

The preceding remarks show that the spherical pairs (G, H) are precisely the complexifications of the compact Gelfand pairs (U, K). It is well known that (U, K) is a Gelfand pair whenever U/K is a Riemannian symmetric space. The classification of irreducible symmetric spaces of compact type produces ten families of examples, listed in Table 1, together with seventeen exceptional cases. In entries 8 and 9, $U(n)$ is embedded in $SO(2n, \mathbb{R})$ and $Sp(2n)$ via

$$A + iB \mapsto \left[\begin{array}{c|c} A & B \\ \hline -B & A \end{array}\right].$$

In each entry of Table 1, G is a *simple* complex algebraic group (and U is a simple compact Lie group). This means that the adjoint representation for G on its Lie algebra $\mathfrak{g}$ is irreducible. Hence $dim(\mathfrak{g}^H) \leq 1$ for spherical pairs (G, H) with G simple. In particular, the center of H is at most one dimensional.

The spherical pairs (G, H) with G simple were classified by Krämer. In addition to the examples that arise from symmetric spaces, there are six further families and six exceptional cases. These are listed in Table 2. The compact form (U, K) for each entry in Table 2 is a compact Gelfand pair for which U/K is not a symmetric space.

Table 2: Spherical pairs with G simple, not arising from symmetric spaces

	G	H
1	$SL(p+q, \mathbb{C})$ $(p \neq q)$	$SL(p, \mathbb{C}) \times SL(q, \mathbb{C})$
2	$SL(2n+1, \mathbb{C})$	$\mathbb{C}^\times \times Sp(2n, \mathbb{C})$
3	$SL(2n+1, \mathbb{C})$	$Sp(2n, \mathbb{C})$
4	$Sp(2n, \mathbb{C})$	$Sp(2n-2, \mathbb{C}) \times SO(2, \mathbb{C})$
5	$SO(2n+1, \mathbb{C})$	$GL(n, \mathbb{C})$
6	$SO(4n+2, \mathbb{C})$	$SL(2n+1, \mathbb{C})$
7	$SO(10, \mathbb{C})$	$Spin(7, \mathbb{C}) \times SO(2, \mathbb{C})$
8	$SO(9, \mathbb{C})$	$Spin(7, \mathbb{C})$
9	$SO(8, \mathbb{C})$	$G_2^{\mathbb{C}}$
10	$SO(7, \mathbb{C})$	$G_2^{\mathbb{C}}$
11	$E_6^{\mathbb{C}}$	$Spin(10, \mathbb{C})$
12	$G_2^{\mathbb{C}}$	$SL(3, \mathbb{C})$

2.6. *Section 2 notes*

Theorem 2.1.1 is due to Servedio, [49]. The converse was proved by Vinberg in [52].

The structure of G/H as a quasi-projective algebraic variety is discussed in Section 11.2.1 of [17].

The classical Peter-Weyl Theorem concerns the decomposition of $L^2(K)$ under the action of $K \times K$ for compact Lie groups K. Let $\mathcal{E}_\sigma$ denote the space of matrix coefficients $m_{\xi,v}(k) = \xi(\sigma(k)v)$ for $\sigma \in \widehat{K}$. One has a Hilbert space direct sum

$$L^2(K) = \bigoplus_{\sigma \in \widehat{K}} \mathcal{E}_\sigma$$

with $K \times K$ acting on $\mathcal{E}_\sigma$ by a copy of $\sigma^* \otimes \sigma$. Theorem 2.2.3 yields the algebraic content of this theorem, via the Unitarian Trick. A complete proof also requires two analytic facts:

- Schur Orthogonality: $\mathcal{E}_\sigma \perp \mathcal{E}'_\sigma$ in $L^2(K)$ for $\sigma \not\cong \sigma'$; and
- $\bigcup_{\sigma \in \widehat{K}} \mathcal{E}_\sigma$ is dense in $L^2(K)$.

See, for example, Section 5.2 in [15] for proofs.

We refer the reader to Chapter IV in [21] or Chapter 1 in [16] for information concerning Gelfand pairs (U, K). In particular, for the fact that Riemannnian symmetric spaces yield Gelfand pairs. The classification of Riemannian symmetric spaces, including the exceptional cases, can be found in Chapter X of [20]. In [17], Section 12.3, it is shown that symmetric spaces yield spherical pairs (G, H) by showing that there is an open Borel orbit in G/H. This approach involves a complex version of the Iwasawa decomposition.

Krämer's classification of spherical pairs (G, H) with G simple appeared in [36]. This classification was pushed further by Brion [8] and by Mikityuk [40] to encompass all spherical pairs (G, H) with G *semi*-simple. This results in eight additional families of examples. The classification of spherical pairs can also be found in Vinberg's recent survey article [53]. S-varieties were introduced by Vinberg and Popov [54].

3. Linear multiplicity free actions

We now restrict our attention to *linear* actions $G : V$. That is, G is a reductive algebraic group acting on a complex vector space V by some rational representation.

3.1. *Connectivity of G*

We generally assume that our groups G are connected unless stated otherwise. Proposition 3.1.3 below shows that this entails no great loss of generality. For Lemma 3.1.1 and Proposition 3.1.2 we assume G is connected and B denotes a Borel subgroup.

Lemma 3.1.1: *Let $h \in \mathbb{C}[V]$ be a highest weight vector for B with prime decomposition*

$$h = p_1^{m_1} \cdots p_k^{m_k}.$$

Then each irreducible factor p_j is a B-highest weight vector.

Proof: Let $\psi : B \to \mathbb{C}^\times$ be the weight for h. Then for $b \in B$,

$$\psi(b)h = b \cdot h = (b \cdot p_1)^{m_1} \cdots (b \cdot p_k)^{m_k}$$

and each $b \cdot p_j$ is an irreducible polynomial. By uniqueness of the prime decomposition for h we conclude that, for each j, $b \cdot p_j$ is a non-zero multiple of one of the prime factors. That is, $b \cdot p_j \in \cup_{l=1}^k \mathbb{C}^\times p_l$. But B acts continuously on this space, and B is connected, so B must act by a scalar on each p_j. □

Proposition 3.1.2: *If $G : V$ is* not *a multiplicity free action then the multiplicities $\{m(\sigma, \mathbb{C}[V]) : \sigma \in \widehat{G}\}$ are unbounded.*

Proof: As $G : V$ is not multiplicity free we can find a pair of linearly independent B-highest weight vectors h_1, h_2 in $\mathbb{C}[V]$ with common weight $\psi : B \to \mathbb{C}^\times$. In view of Lemma 3.1.1 we can remove any common irreducible factors and assume h_1, h_2 are relatively prime. For each N and $0 \le k \le N$, $h_1^k h_2^{N-k}$ has weight ψ^N. Moreover, $\{h_1^k h_2^{N-k} : 0 \le k \le N\}$ is a linearly independent set in $\mathbb{C}[V]$. For otherwise we could express some $h_1^{k_\circ} h_2^{N-k_\circ}$ as a linear combination

$$h_1^{k_\circ} h_2^{N-k_\circ} = \sum_{k=k_\circ+1}^{N} c_k h_1^k h_2^{N-k}$$

and conclude that h_1 divides $h_2^{N-k_\circ}$. This contradicts the fact that h_1, h_2 are relatively prime. Thus the irreducible representation with highest weight ψ^N occurs in $\mathbb{C}[V]$ with multiplicity at least N. □

Proposition 3.1.3: *Let G be multiply-connected with identity component $G_\circ$. Then $G : V$ is a multiplicity free action if and only if $G_\circ : V$ is a multiplicity free action.*

Proof: If $G_\circ : V$ is multiplicity free then so is $G : V$, in view of (1). Conversely, suppose that $G : V$ is a multiplicity free action but that $G_\circ : V$ is not multiplicity free. Let

$$\mathbb{C}[V] = \bigoplus_{\lambda} P_\lambda$$

denote the decomposition of $\mathbb{C}[V]$ into distinct G-irreducible subspaces and $\{g_1, \ldots, g_\ell\}$ be a complete set of coset representatives for $G_\circ$ in G. Proposition 3.1.2 ensures that for any N there is some $\sigma \in \widehat{G_\circ}$ with $m(\sigma, \mathbb{C}[V]) \geq N\ell 2^\ell$. As the P_λ's are $G_\circ$-invariant, each copy of σ in $\mathbb{C}[V]$ is contained in some P_λ. Let W denote a subspace of some P_λ on which $G_\circ$ acts by a copy of σ. Since $G_\circ$ is normal in G, $g_i \cdot W \subset P_\lambda$ is $G_\circ$-invariant with $G_\circ$ acting via a copy of $\sigma_i \in \widehat{G_\circ}$, where

$$\sigma_i(g) = \sigma(g_i g g_i^{-1}).$$

As P_λ is G-irreducible, we must have

$$P_\lambda = \bigoplus_{j \in J} (g_j \cdot W) \tag{2}$$

for some subset J of $\{1, \ldots, \ell\}$. Thus P_λ is equivalent to $\sum_{j\in J} \sigma_j$ as a $G_\circ$-module.

Equation 2 contains at most ℓ factors, so at least $N2^\ell$ distinct P_λ's must contain copies of σ. As there are only 2^ℓ possibilities for J, at least N of these P_λ's must be equivalent as $G_\circ$-modules. Thus we have shown that for each N one can find N distinct irreducible G-modules that are equivalent as $G_\circ$-modules. This is impossible since $G/G_\circ$ is a finite group. □

3.2. *Borel orbits*

We will prove the converse of Theorem 2.1.1 in the context of linear actions $G : V$. We use the notation introduced in Section 1.4: Choose a Borel subgroup $B = HN$ in G and let $\Delta^+ \subset \mathfrak{h}^*$ be the associated set of positive roots.

Now suppose that $h \in \mathbb{C}[V]$ is a B-highest weight vector with weight $\lambda \in \mathfrak{h}^*$. Let

$$P = P_h = \{g \in G \ : \ g \cdot h \in \mathbb{C}^\times h\}. \tag{3}$$

This is a parabolic subgroup of G that contains B. We have $X \cdot h = \lambda(X)h$ for all $X \in \mathfrak{p}$, the Lie algebra of P.

- Let L and U be the Levi component and unipotent part of P, so $P = LU$.
- On the Lie algebra level we can write

$$\mathfrak{p} = \mathfrak{h} \oplus \sum_{\alpha \in \Delta_P} \mathfrak{g}_\alpha, \quad \text{where } \Delta_P = \{\alpha \in \Delta \mid \langle \lambda, \alpha \rangle \geq 0\}.$$

Now setting

$$\Delta_L = \{\alpha \in \Delta \mid \langle \lambda, \alpha \rangle = 0\},$$
$$\Delta_U = \{\alpha \in \Delta \mid \langle \lambda, \alpha \rangle > 0\} = \{\alpha \in \Delta \mid -\alpha \notin \Delta_P\},$$

one has

$$\mathfrak{l} = \mathfrak{h} \oplus \sum_{\alpha \in \Delta_L} \mathfrak{g}_\alpha, \quad \mathfrak{u} = \sum_{\alpha \in \Delta_U} \mathfrak{g}_\alpha.$$

- Letting $\mathfrak{u}^- = \sum_{\alpha \in \Delta_U} \mathfrak{g}_{-\alpha}$ we have $\mathfrak{u} \subset \mathfrak{n}$, $\mathfrak{u}^- \subset \mathfrak{n}^-$ and

$$\mathfrak{g} = \mathfrak{p} \oplus \mathfrak{u}^- = \mathfrak{l} \oplus \mathfrak{u} \oplus \mathfrak{u}^-.$$

We will make extensive use of the following map (which depends on h). Let

$$V_\circ = \{z \in V \ : \ h(z) \neq 0\},$$

a principal open set in V, and define

$$\chi = \chi_h : V_\circ \to \mathfrak{g}^*$$

via

$$\chi(z)(X) = \frac{(X \cdot h)(z)}{h(z)}.$$

Lemma 3.2.1: *χ is P-equivariant.*

Proof: Note that the action of P on V preserves $V_\circ$. Also P acts on $\mathfrak{g}^*$ via the coadjoint action

$$(g \cdot f)(X) = f(Ad(g^{-1})X).$$

For $g \in P$ one has

$$\chi(g^{-1} \cdot z)(X) = \frac{(X \cdot h)(g^{-1} \cdot z)}{h(g^{-1} \cdot z)} = \frac{(g \cdot X)(g \cdot h)(z)}{(g \cdot h)(z)}$$
$$= \frac{\lambda(g)(g \cdot X)(h)(z)}{\lambda(g)h(z)} = \chi(z)(g \cdot X).$$

Thus $\chi(g^{-1} \cdot z)(X) = (g^{-1} \cdot \chi(z))(X)$ so χ is P-equivariant. □

Lemma 3.2.2: *The stabilizer of λ in P is $Stab_P(\lambda) = L$. Moreover, $Stab_U(\lambda) = \{e\}$.*

Proof: Here $\lambda \in \mathfrak{h}^*$ is regarded as a functional on all of $\mathfrak{g}$ with $\lambda(\mathfrak{n}) = \{0\} = \lambda(\mathfrak{n}^-)$. For $X \in \mathfrak{p}$ one has $X \cdot \lambda = 0$ if and only if $\lambda[X, \mathfrak{g}] = \{0\}$. Clearly $\mathfrak{h}$ stabilizes λ because $[\mathfrak{h}, \mathfrak{g}] \subset \mathfrak{n} + \mathfrak{n}^-$. Also for any $\alpha \in \Delta_L$ and any $Y = H + \sum_{\beta \in \Delta} c_\beta X_\beta \in \mathfrak{g}$,

$$\lambda([X_\alpha, Y]) = c_{-\alpha}\lambda(H_\alpha) = c_{-\alpha}\langle\lambda, \alpha\rangle = 0.$$

Thus X_α stabilizes λ for each $\alpha \in \Delta_L$. So $\mathfrak{l} = \mathfrak{h} + \sum_{\alpha \in \Delta_L} \mathfrak{g}_\alpha$ stabilizes λ.

We have shown that $L \subset Stab_P(\lambda)$. For the reverse inclusion, let

$$X = H + \sum_{\alpha \in \Delta_L} a_\alpha X_\alpha + \sum_{\alpha \in \Delta_U} b_\alpha X_\alpha \in \mathfrak{p}$$

stabilize λ. Then for $\beta \in \Delta_U$,

$$\lambda[X, X_{-\beta}] = b_\beta \lambda(H_\beta) = b_\beta \langle\lambda, \beta\rangle.$$

But $\langle\lambda, \beta\rangle > 0$, and hence $b_\beta = 0$. Thus $Stab_P(\lambda) \subset L$ and the stabilizer of λ in U is trivial. □

Proposition 3.2.3: *The image of $\chi : V_\circ \to \mathfrak{g}^*$ is $\lambda + \mathfrak{p}^\perp$, a single P-orbit in $\mathfrak{g}^*$.*

Proof: For all $z \in V_\circ$ and $X \in \mathfrak{p}$,

$$\chi(z)(X) = \frac{(X \cdot h)(z)}{h(z)} = \frac{\lambda(X)h(z)}{h(z)} = \lambda(X).$$

Thus $\chi(z) - \lambda$ annihilates $\mathfrak{p}$ and $\chi(V_\circ) \subset \lambda + \mathfrak{p}^\perp$. Note that $dim(\lambda + \mathfrak{p}^\perp) = dim(\mathfrak{u})$. As U is unipotent and acts without stabilizer on λ we conclude that $U \cdot \lambda$ is both open and closed in $\lambda + \mathfrak{p}^\perp$ and hence $U \cdot \lambda = \lambda + \mathfrak{p}^\perp$. As $P = LU$ and L stabilizes λ we also have $P \cdot \lambda = \lambda + \mathfrak{p}^\perp$. As χ is P-equivariant we must have $P \cdot \lambda = \chi(V_\circ)$. □

Corollary 3.2.4: *$\lambda = \chi(z_\circ)$ for some $z_\circ \in V_\circ$.*

Let

$$\Sigma = \Sigma_h = \chi^{-1}(\lambda) = \{z \in V_\circ \ : \ \chi(z) = \lambda\}. \tag{4}$$

The group L acts on Σ because χ is P-equivariant and L stabilizes λ.

Lemma 3.2.5: *$U \times \Sigma \cong V_\circ$ via $(g, z) \mapsto g \cdot z$.*

Proof: Given $z \in V_\circ$ one has $\chi(z) = g \cdot \lambda$ for some $g \in U$ in view of Proposition 3.2.3. Thus $\lambda = g^{-1} \cdot \chi(z) = \chi(g^{-1} \cdot z)$, so $g^{-1} \cdot z \in \Sigma$. Now $(g, g^{-1} \cdot z) \in U \times \Sigma$ maps to z.

To see that $(g, z) \mapsto g \cdot z$ is injective, suppose that $g \cdot z = g' \cdot z'$ for some $g, g' \in U$, $z, z' \in \Sigma$. Applying χ gives

$$g \cdot \lambda = g \cdot \chi(z) = g' \cdot \chi(z') = g' \cdot \lambda$$

and thus $g^{-1}g'$ stabilizes λ. As U acts without stabilizer on λ it follows that $g = g'$ and thus also $z = z'$. □

Lemma 3.2.6: *There is a unique parabolic subgroup $P = P_h$ of lowest possible dimension. Moreover, $P \subset P_{h'}$ for all B-highest weight vectors $h' \in \mathbb{C}[V]$.*

Proof: Suppose that $h \in \mathbb{C}[V]$ is a B-highest weight vector with prime decomposition

$$h = p_1^{m_1} \cdots p_r^{m_r}.$$

The proof of Lemma 3.1.1 shows that P_h acts by a character on each irreducible factor p_j. Thus

$$P_h = P_{p_1} \cap \cdots \cap P_{p_r}.$$

Now assume that $h, h' \in \mathbb{C}[V]$ are two highest weight vectors and that P_h has minimal dimension. Let $p_1, \ldots, p_r$ be the prime factors of h and $q_1, \ldots, q_s$ the prime factors of h'. Letting $h'' = p_1 \cdots p_r q_1 \cdots q_s$ one has

$$P_{h''} = P_{p_1} \cap \cdots \cap P_{p_r} \cap P_{q_1} \cap \cdots \cap P_{q_s} = P_h \cap P_{h'}.$$

If $P_h \not\subset P_{h'}$ then $P_{h''}$ is a proper subset of P_h and hence $dim(P_{h''}) < dim(P_h)$, a contradiction. Thus $P_h \subset P_{h'}$ as claimed. Moreover if $P_{h'}$ also has minimal dimension then $P_{h'} \subset P_h$ and now $P_h = P_{h'}$. This shows uniqueness. □

Lemma 3.2.7: *Let $P = P_h = LU$ be the unique parabolic subgroup of minimal dimension. Then (L, L) acts trivially on $\Sigma = \Sigma_h$.*

Proof: First note that as P acts on h by a character, $(L, L) \subset (P, P)$ acts trivially on h. Suppose, however, that (L, L) does not act trivially on Σ. It follows that there is some highest weight vector $h' \in \mathbb{C}[\Sigma]$ for the action $L : \Sigma$ that is not fixed by (L, L). Recall that $B = HN$ with $N \subset U$ and that $V_\circ = U \cdot \Sigma$. Thus we can extend h' to a B-semi-invariant function on $V_\circ$.

For N sufficiently large, $h^N h'$ is a regular function on V. Now $h^N h' \in \mathbb{C}[V]$ is a B-highest weight vector and $P \subset P_{h^N h'}$ by Lemma 3.2.6. Thus P acts by a character on $h^N h'$ and hence (L, L) acts trivially on $h^N h'$. As (L, L) fixes both h and $h^N h'$, it must fix h', a contradiction. □

Lemma 3.2.7 implies that for the minimal parabolic $P = P_h = LU$, the action of L on $\Sigma = \Sigma_h$ is a torus action. Lemma 3.2.5 implies that the variety Σ is affine and irreducible.

Theorem 3.2.8: *If $G : V$ is multiplicity free then there is an open B-orbit in V.*

Proof: Suppose that there is no open B-orbit in V. Let $P = P_h = LU$ be the parabolic subgroup of minimal dimension and $\Sigma = \Sigma_h$.

We claim that $H : \Sigma$ is not multiplicity free. Indeed, if $H : \Sigma$ were multiplicity free then there would be an open H-orbit in Σ, $H \cdot v_\circ$ say, by Proposition 2.1.2. Now as $U \cdot \Sigma = V_\circ$, one has $UH \cdot v_\circ$ open in $V_\circ$ and hence open in V. As $U \subset N$, then $B \cdot v_\circ = HN \cdot v_\circ$ is open in V, a contradiction.

As $H : \Sigma$ is not multiplicity free we can find a pair of linearly independent highest weight vectors $h_1, h_2 \in \mathbb{C}[\Sigma]$ with a common weight for the action $L : \Sigma$. As in the proof of Lemma 3.2.7, one can extend h_1 and h_2 to B-semi-invariant functions on $V_\circ$. Now for N sufficiently large, $h^N h_1$ and $h^N h_2$ are regular on V and share a common weight. Hence the action $G : V$ fails to be multiplicity free. □

Note that $B \cdot v_\circ$ is open in V if and only if $dim(\mathfrak{b} \cdot v_\circ) = dim(B \cdot v_\circ) = dim(V)$. Thus $B \cdot v_\circ$ is open if and only if $\mathfrak{b} \cdot v_\circ = V$. We obtain an infinitesimal version of Theorems 2.1.1 and 3.2.8:

Corollary 3.2.9: *$G : V$ is multiplicity free if and only if $\mathfrak{b} \cdot v_\circ = V$ for some point $v_\circ \in V$.*

We will apply Corollary 3.2.9 to the study of examples in Section 4.

3.3. *Fundamental highest weights for a multiplicity free action*

Let $G : V$ be a linear multiplicity free action, $B = HN$ a Borel subgroup of G and $v_\circ$ a point in V with $B \cdot v_\circ$ open in V. Let

$$\Lambda \subset \mathfrak{h}^*$$

denote the set of highest weights λ for representations $\sigma_\lambda \in \widehat{G}$ that occur in $\mathbb{C}[V]$. The character $B \to \mathbb{C}^\times$ given by $\lambda \in \Lambda$ will be denoted $b \mapsto b^\lambda$.

For $\lambda \in \Lambda$, let $P_\lambda \subset \mathbb{C}[V]$ be the irreducible subspace of $\mathbb{C}[V]$ on which G acts by a copy of σ_λ. Then

$$\mathbb{C}[V] = \bigoplus_{\lambda \in \Lambda} P_\lambda \tag{5}$$

is the multiplicity free decomposition of $\mathbb{C}[V]$ under the action of G. As G preserves the subspaces $\mathcal{P}_m(V)$ of polynomials homogeneous of degree m, each P_λ is contained in some $\mathcal{P}_m(V)$.

Each P_λ contains a B-highest weight vector h_λ, which is unique up to scalar multiples. As shown in the proof of Theorem 2.1.1, we must have $h_\lambda(v_\circ) \neq 0$ and we can normalize our choice of h_λ by the condition

$$h_\lambda(v_\circ) = 1.$$

For $\lambda, \mu \in \Lambda$,

$$b \cdot (h_\lambda h_\mu) = (b \cdot h_\lambda)(b \cdot h_\mu) = (b^\lambda h_\lambda)(b^\mu h_\mu) = b^{\lambda+\mu} h_\lambda h_\mu.$$

Hence

$$\lambda + \mu \in \Lambda \quad \text{and} \quad h_{\lambda+\mu} = h_\lambda h_\mu.$$

In particular, Λ is an additive semigroup in $\mathfrak{h}^*$.

Next suppose $\lambda \in \Lambda$ and let

$$h_\lambda = p_1^{m_1} \cdots p_k^{m_k}$$

be the prime decomposition for h_λ. Lemma 3.1.1 shows that each p_j is a B-highest weight vector. Suitably normalizing the p_j's, we can now say that $p_j = h_{\lambda_j}$ for some $\lambda_j \in \Lambda$ and $\lambda = m_1\lambda_1 + \cdots + m_k\lambda_k$.

Now let

$$\Lambda' = \{\lambda \in \Lambda \ : \ h_\lambda \text{ is an irreducible polynomial}\}.$$

We know that $\Lambda' \neq \emptyset$ and that Λ' generates the semigroup Λ. We will show that Λ' is a $\mathbb{Q}$-linearly independent subset of $\mathfrak{h}^*$. Indeed, suppose that $\lambda_1, \ldots, \lambda_k \in \Lambda'$ satisfy a non-trivial linear dependence relation over $\mathbb{Q}$. Clearing denominators we obtain a non-trivial linear dependence

$$a_1\lambda_1 + \cdots + a_k\lambda_k = 0$$

with integer coefficients a_j. Let

$$L = \{j \ : \ a_j \geq 0\}$$

and set $m_j = |a_j|$ for all $1 \leq j \leq k$. Then

$$\sum_{j \in L} m_j \lambda_j = \sum_{j \notin L} m_j \lambda_j$$

and hence

$$\prod_{j \in L} h_{\lambda_j}^{m_j} = \prod_{j \notin L} h_{\lambda_j}^{m_j}$$

since these are both highest weight vectors with a common weight and take value 1 at the point $v_\circ$. But this is impossible, since the h_{λ_j}'s are distinct irreducibles and $\mathbb{C}[V]$ is a unique factorization domain.

It now follows that Λ' is a finite set with at most $dim(\mathfrak{h})$ elements.

In summary, we have proved the following.

Proposition 3.3.1: *$\Lambda' = \{\lambda \in \Lambda \; : \; h_\lambda$ is irreducible$\}$ is a $\mathbb{Q}$-linearly independent subset of $\mathfrak{h}^*$ with at most $dim(H)$ elements. The additive semigroup Λ is freely generated by Λ'. Writing*

$$\Lambda' = \{\lambda_1, \ldots, \lambda_r\},$$

the decomposition for $\mathbb{C}[V]$ can be written

$$\mathbb{C}[V] = \bigoplus_{\lambda} P_\lambda$$

where the sum is taken over all $\mathbb{N}$-linear combinations $\lambda = m_1\lambda_1 + \cdots m_r\lambda_r$. The highest weight vector in the irreducible subspace P_λ is $h_\lambda = h_{\lambda_1}^{m_1} \cdots h_{\lambda_r}^{m_r}$.

Definition 3.3.2: The number r is the *rank* of the multiplicity free action $G : V$, $\{\lambda_1, \ldots, \lambda_r\}$ are the *fundamental highest weights* and $\{h_1 = h_{\lambda_1}, \ldots, h_r = h_{\lambda_r}\}$ are the *fundamental highest weight vectors.*

3.4. *Section 3 notes*

Many results in this section are due to Roger Howe. Proposition 3.3.1 is from [23]. Proposition 3.1.3 appeared in [3] but the proof was shown to the authors by Howe.

The proof given for Theorem 3.2.8 is due to Friedrich Knop [31] and based on ideas from [10]. The more standard proof (see [52]) is shorter but requires use of a hard result of Rosenlicht. For background on parabolic subgroups see Section V.7 in [29].

A linear action $G : V$ is called a *prehomogeneous vector space* when there is an open G-orbit in V. We refer the reader to [28] for a survey of this subject, including classification and applications to analysis. Theorem 3.2.8 implies that each linear multiplicity free action is, in particular, a prehomogeneous vector space.

Proposition 3.1.2 shows that linear actions $G : V$ are of three types:

(1) $G : V$ is multiplicity free.
(2) Some $\sigma \in \widehat{G}$ occurs in $\mathbb{C}[V]$ with infinite multiplicity.
(3) Each $\sigma \in \widehat{G}$ has $m(\sigma, \mathbb{C}[V]) < \infty$ but these multiplicities are unbounded.

It is known that (2) is equivalent to the existence of a non-constant G-invariant in $\mathbb{C}[V]$. (See Theorem 5.5 in [3].) In this case, *every* $\sigma \in \widehat{G}$ which occurs in $\mathbb{C}[V]$ does so with infinite multiplicity.

4. Examples of multiplicity free decompositions

4.1. GL(n) ⊗ GL(m)

Here we consider the action of $G = GL(n, \mathbb{C}) \times GL(m, \mathbb{C})$ on $\mathbb{C}^n \otimes \mathbb{C}^m$ via the the outer tensor product of the defining representations for the two factors. Identifying $\mathbb{C}^n \otimes \mathbb{C}^m$ with the space $M_{n,m}(\mathbb{C})$ of complex $n \times m$-matrices (via $e_i \otimes e_j \leftrightarrow E_{i,j}$) one has

$$(g, g') \cdot v = gv(g')^t \tag{6}$$

In fact it will prove convenient to *twist* the action by composing with the automorphisms $g \mapsto (g^{-1})^t$ on both factors. This gives

$$(g, g') \cdot v = (g^{-1})^t v (g')^{-1} \tag{7}$$

and the associated representation on $\mathbb{C}[M_{n,m}(\mathbb{C})]$ becomes

$$(g, g') \cdot p(v) = p(g^t v g').$$

Of course the decompositions for $\mathbb{C}[M_{n,m}(\mathbb{C})]$ under the two actions (6) and (7) are the same. Twisting by $g \mapsto (g^{-1})^t$ has the effect of interchanging representations with their contragredients. So decomposing $\mathbb{C}[M_{n,m}(\mathbb{C})]$ with respect to (7) amounts to decomposing the symmetric algebra $S(M_{n,m}(\mathbb{C})) \cong \mathbb{C}[M_{n,m}(\mathbb{C})^*]$ with respect to (6).

Recall that the upper triangular matrices in $GL(k, \mathbb{C})$ give the standard Borel subgroup B_k with Lie algebra $\mathfrak{b}_k$. The diagonal matrices H_k in $GL(k, \mathbb{C})$ give the maximal torus with Lie algebra $\mathfrak{h}_k \cong \mathbb{C}^k$. We let

$$B = B_n \times B_m, \quad \mathfrak{b} = \mathfrak{b}_n \times \mathfrak{b}_m, \quad H = H_n \times H_m, \quad \mathfrak{h} = \mathfrak{h}_n \times \mathfrak{h}_m,$$

$\varepsilon_j \in \mathfrak{h}_n^*$ be $\varepsilon_j(diag(h_1, \ldots, h_n)) = h_j$ and likewise for $\varepsilon_j' \in \mathfrak{h}_m^*$. We sometimes use the shorthand

$$\lambda = (\mu, \nu) = (\mu_1, \ldots, \mu_n; \nu_1, \ldots, \nu_m)$$

for the weight $\lambda = \mu_1\varepsilon_1 + \cdots + \mu_n\varepsilon_n + \nu_1\varepsilon_1' + \cdots + \nu_m\varepsilon_m'$ in $\mathfrak{h}^*$. The dominant weights have $\mu \in \mathbb{Z}^n$, $\nu \in \mathbb{Z}^m$ with $\mu_1 \geq \cdots \geq \mu_n$ and $\nu_1 \geq \cdots \geq \nu_m$. We say that λ is *non-negative* and write $\lambda \geq 0$ when $\mu_j \geq 0$ and $\nu_j \geq 0$ for all j.

For any $(h, h') = (diag(h_1, \ldots, h_n), diag(h_1' \ldots, h_m')) \in H$ one has

$$(h, h') \cdot z_{i,j} = h_i h_j' z_{i,j}$$

where $z_{ij} : M_{n,m}(\mathbb{C}) \to \mathbb{C}$ is the (i, j)-th entry function. So $z_{i,j}$ is a weight vector with weight $\varepsilon_i + \varepsilon_j'$. Hence also $z^\alpha = \prod_{i,j} z_{ij}^{\alpha_{ij}}$ is a weight vector with weight

$$\sum_{i,j} \alpha_{ij}(\varepsilon_i + \varepsilon_j') \geq 0.$$

As the z^α's form a basis for $\mathbb{C}[M_{n,m}(\mathbb{C})]$, all weight vectors in $\mathbb{C}[M_{n,m}(\mathbb{C})]$ are non-negative. Thus the highest weights $\lambda = (\mu, \nu)$ that occur in $\mathbb{C}[M_{n,m}(\mathbb{C})]$ all have

$$\mu_1 \geq \cdots \geq \mu_n \geq 0, \quad \nu_1 \geq \cdots \geq \nu_m \geq 0.$$

This explains why we are using the action (7) in place of (6).

Let us assume that $n \geq m$ and set

$$v_\circ = \left[\frac{I_m}{O}\right] \in M_{n,m}(\mathbb{C}).$$

A typical element X in the Lie algebra $\mathfrak{b}$ for B has the form

$$X = \left(\left[\begin{array}{c|c} A & B \\ \hline O & C \end{array}\right], D\right) \tag{8}$$

where A, D are $m \times m$ upper triangular, C is $(n-m) \times (n-m)$ upper triangular and B is an arbitrary $m \times (n-m)$-matrix. The derived action for (7) gives

$$X \cdot v_\circ = -\left[\begin{array}{c|c} A & B \\ \hline O & C \end{array}\right]^t \left[\frac{I_m}{O}\right] - \left[\frac{I_m}{O}\right] D = -\left[\frac{A^t + D}{B^t}\right]. \tag{9}$$

Here A^t, D are arbitrary $m \times m$ lower and upper triangular matrices respectively and B^t is an arbitrary matrix of size $(n-m) \times m$. We conclude that

$\mathfrak{b} \cdot v_\circ = M_{n,m}(\mathbb{C})$. Hence $B \cdot v_\circ$ is open in $M_{n,m}(\mathbb{C})$, so (7) is a multiplicity free action.

The element X in Equation 8 belongs to $\mathfrak{h}$ when $B = O$ and A, C, D are diagonal,

$$A = diag(a_1, \ldots, a_m), \quad C = diag(c_1, \ldots, c_{n-m}), \quad D = diag(d_1, \ldots, d_m)$$

say. For such X, Equation 9 shows $X \cdot v_\circ = 0$ if and only if $D = -A$. So the stabilizer $\mathfrak{h}_\circ$ of $v_\circ$ in $\mathfrak{h}$ is

$$\mathfrak{h}_\circ = \left\{ \left(\begin{bmatrix} a_1 & & & & & \\ & \ddots & & & & \\ & & a_m & & & \\ & & & c_1 & & \\ & & & & \ddots & \\ & & & & & c_{n-m} \end{bmatrix}, \begin{bmatrix} -a_1 & & \\ & \ddots & \\ & & -a_m \end{bmatrix} \right) \right\}.$$

If $\lambda = (\mu, \nu)$ is the highest weight for a representation that occurs in $\mathbb{C}[M_{n,m}(\mathbb{C})]$ then we must have $\lambda(\mathfrak{h}_\circ) = \{0\}$. That is,

$$(\mu_1 - \nu_1)a_1 + \cdots + (\mu_m - \nu_m)a_m + \mu_{m+1}c_1 + \cdots + \mu_n c_{n-m} = 0$$

for all a_j, c_j. Thus we must have

$$\mu_j = \nu_j \text{ for } 1 \leq j \leq m \text{ and } \mu_j = 0 \text{ for } j > m.$$

We have now shown that the only candidates for highest weights of representations occurring in $\mathbb{C}[M_{n,m}(\mathbb{C})]$ have the form

$$\lambda = (\mu, \mu) \text{ where } \mu_1 \geq \cdots \geq \mu_m \geq 0.$$

Note that a representation with highest weight $\lambda = (\mu, \mu)$ is equivalent to $\sigma_n^\mu \otimes \sigma_m^\mu$, the outer tensor product of the irreducible representations of $GL(n, \mathbb{C})$ and $GL(m, \mathbb{C})$ with highest weight μ. We will show that all such weights λ *do* occur in $\mathbb{C}[M_{n,m}(\mathbb{C})]$ by exhibiting a λ-highest weight vector in $\mathbb{C}[M_{n,m}]$. For this, let

$$\xi_k(z) = \begin{vmatrix} z_{11} & \cdots & z_{1k} \\ \vdots & & \vdots \\ z_{k1} & \cdots & z_{kk} \end{vmatrix},$$

the leading minor determinant for $z \in M_{n,m}(\mathbb{C})$, $1 \leq k \leq m$. For $(h, h') = (diag(h_1, \ldots, h_n), diag(h'_1 \ldots, h'_m)) \in H$ one computes

$$(h, h') \cdot \xi_k(z) = \xi_k(hzh') = \xi_k([h_i h'_j z_{ij}]) = (h_1 \cdots h_k)(h'_1 \cdots h'_k)\xi_k(z).$$

Hence ξ_k is a highest weight vector with weight $(\varepsilon_1+\cdots+\varepsilon_k, \varepsilon'_1+\cdots+\varepsilon'_k) = (1^k, 1^k)$. For any $\mu_1 \geq \cdots \geq \mu_m \geq 0$

$$\xi_\mu = \xi_1^{\mu_1-\mu_2}\xi_2^{\mu_2-\mu_3}\cdots\xi_{m-1}^{\mu_{m-1}-\mu_m}\xi_m^{\mu_m} \tag{10}$$

has weight $\lambda = (\mu, \mu)$ and degree

$$|\mu| = \mu_1 + \cdots + \mu_m.$$

In summary, we have proved:

Theorem 4.1.1: *The action of $G = GL(n,\mathbb{C}) \times GL(m,\mathbb{C})$ on $\mathbb{C}^n \otimes \mathbb{C}^m \cong M_{n,m}(\mathbb{C})$ is multiplicity free. For $n \geq m$ one has the decomposition*

$$\mathbb{C}[M_{n,m}(\mathbb{C})] = \bigoplus_\mu P_\mu$$

where the sum is over all $\mu \in \mathbb{N}^m$ with $\mu_1 \geq \cdots \geq \mu_m \geq 0$. The polynomial ξ_μ given by (10) is a highest weight vector in P_μ with weight (μ,μ) and $P_\mu \simeq \sigma_n^\mu \otimes \sigma_m^\mu$ as a G-module. Moreover

$$\mathcal{P}_k(M_{n,m}(\mathbb{C})) = \bigoplus_{|\mu|=k} P_\mu.$$

We observe the following points regarding this example.

- The action has rank $m = min(n,m)$. Here $\{(1^k, 1^k) \;:\; 1 \leq k \leq m\}$ are the fundamental highest weights and $\xi_1, \ldots, \xi_m$ are the fundamental highest weight vectors. Indeed, the determinant of a matrix is an irreducible polynomial in the matrix entries.
- For the more conventional (untwisted) action (6), the decomposition of $\mathbb{C}[M_{n,m}(\mathbb{C})]$ is as in Theorem 4.1.1, but now ξ_μ is a highest weight vector for the *opposite* Borel with weight $(-\mu, -\mu)$.
- The group $GL(n,\mathbb{C}) \times GL(m,\mathbb{C})$ can be replaced by $SL(n,\mathbb{C}) \times GL(m,\mathbb{C})$ or $SL(n,\mathbb{C}) \times SL(m,\mathbb{C}) \times \mathbb{C}^\times$.
- If we restrict to $SL(n,\mathbb{C}) \times SL(m,\mathbb{C})$ then the action remains multiplicity free *provided* $n \neq m$. The Borel subalgebra $\mathfrak{b}'$ consists of matrices X as in (8) where $tr(A) + tr(C) = 0 = tr(D)$. Equation 9 shows $\mathfrak{b}' \cdot v_\circ = M_{n,m}(\mathbb{C})$ when $n > m$.
- When $n = m$, the action $\mathbf{SL(n)} \otimes \mathbf{SL(n)}$ fails to be multiplicity free. Indeed, $det : M_{n,n}(\mathbb{C}) \to \mathbb{C}$ is a non-constant $(SL(n,\mathbb{C}) \times SL(n,\mathbb{C}))$-invariant polynomial. So $\mathbb{C}[M_{n,n}(\mathbb{C})]$ contains two copies of the trivial representation of $SL(n,\mathbb{C}) \times SL(n,\mathbb{C})$.

- When $m = 1$ this example reduces to the action of $GL(n, \mathbb{C})$ on $\mathbb{C}^n$ by the defining representation (or its twisted form, the contragredient representation on $(\mathbb{C}^n)^*$). In this case the decomposition in Theorem 4.1.1 reduces to

$$\mathbb{C}[z_1, \ldots, z_n] = \bigoplus_{k=0}^{\infty} \mathcal{P}_k(\mathbb{C}^n).$$

The action has rank one with fundamental highest weight (1^1) and z_1^k is a highest weight vector in $\mathcal{P}_k(\mathbb{C}^n)$.

4.2. $\mathbf{S^2(GL(n))}$

Next we consider the action of $GL(n, \mathbb{C})$ on the symmetric 2-tensors $S^2(\mathbb{C}^2)$ via the symmetric square of the defining representation. Identifying $S^2(\mathbb{C}^n)$ with the complex $n \times n$-symmetric matrices

$$Sym(n, \mathbb{C}) = \{A \in M_{n,n}(\mathbb{C}) \ : \ A^t = A\}$$

our action reads

$$g \cdot v = gvg^t. \tag{11}$$

As in the preceding example we prefer to twist the action by $g \mapsto (g^{-1})^t$. This gives

$$g{\cdot}v = (g^{-1})^t v g^{-1}, \quad g{\cdot}p(v) = p(g^t v g) \quad \text{for } v \in Sym(n, \mathbb{C}),\ p \in \mathbb{C}[Sym(n, \mathbb{C})]. \tag{12}$$

As for $\mathbf{GL(n) \otimes GL(m)}$, twisting ensures that all weights $\lambda = (\lambda_1, \ldots, \lambda_n)$ that occur in $\mathbb{C}[Sym(n, \mathbb{C})]$ are non-negative. Again we let

$$\xi_\lambda = \xi_1^{\lambda_1 - \lambda_2} \cdots \xi_{n-1}^{\lambda_{n-1} - \lambda_n} \xi_n^{\lambda_n}$$

for dominant weights $\lambda_1 \geq \cdots \geq \lambda_n \geq 0$, where ξ_k is a leading minor determinant restricted to $Sym(n, \mathbb{C}) \subset M_{n,n}(\mathbb{C})$.

Theorem 4.2.1: *The action of $GL(n, \mathbb{C})$ on $Sym(n, \mathbb{C})$ is multiplicity free. We have the decomposition*

$$\mathbb{C}[Sym(n, \mathbb{C})] = \bigoplus_{\lambda} P_\lambda$$

where the sum is over all $\lambda \in \mathbb{N}^n$ with $\lambda_1 \geq \cdots \geq \lambda_n \geq 0$. The polynomial ξ_λ is a highest weight vector in P_λ with weight 2λ. Moreover,

$$\mathcal{P}_k(Sym(n, \mathbb{C})) = \bigoplus_{|\lambda| = k} P_\lambda.$$

Proof: The proof parallels that for Theorem 4.1.1. First we note that for the derived action of $gl(n,\mathbb{C})$ one has

$$X \cdot I = -(X^t + X).$$

We obtain all $n\times n$-symmetric matrices as X ranges over all upper triangular matrices. Thus $\mathfrak{b}_n \cdot I = Sym(n,\mathbb{C})$ and hence $v_\circ = I$ has an open Borel orbit. This proves that our action is multiplicity free.

Suppose that $f \in \mathbb{C}[Sym(n,\mathbb{C})]$ is a highest weight vector with weight μ. For $h = diag(h_1,\ldots,h_n) \in H_n$,

$$f(h^2) = (h \cdot f)(I) = h^\mu f(I) = h_1^{\mu_1} \cdots h_n^{\mu_n} f(I).$$

Now for $h = diag(\pm 1,\ldots,\pm 1)$ we have $h^2 = I$ and hence $f(I) = h^\mu f(I)$ for all such h. As $B_n \cdot I$ is open in $Sym(n,\mathbb{C})$, f is determined by the value $f(I)$ and we must have $f(I) \neq 0$. We conclude that $h^\mu = 1$ for all $h = diag(\pm 1,\ldots,\pm 1)$ and thus each μ_j must be even. (Note that $H_\circ = \{diag(\pm 1,\ldots,\pm 1)\}$ is the stabilizer of $v_\circ = I$ in H_n.)

Since all weights that occur in $\mathbb{C}[Sym(n,\mathbb{C})]$ are non-negative we conclude that all dominant weights μ that occur have the form

$$\mu = 2\lambda = (2\lambda_1,\ldots,2\lambda_n)$$

for some $\lambda_1 \geq \cdots \geq \lambda_n \geq 0$. To complete the proof, one checks that the polynomial ξ_λ is a weight vector with weight 2λ. Indeed for $h \in H_n$ and $1 \leq k \leq n$,

$$h \cdot \xi_k(z) = \xi_k\big([h_i h_j z_{ij}]\big) = h_1^2 \cdots h_k^2 \xi(z),$$

so ξ_k is a (2^k)-weight vector. □

As regards this example we note:

- The action has rank n with fundamental highest weights (2^k) and fundamental highest weight vectors ξ_k, $1 \leq k \leq n$. It is known that the determinant of a symmetric matrix is irreducible as a polynomial in the entries z_{ij} with $i \leq j$.
- For the untwisted action (11), the decomposition of $\mathbb{C}[Sym(n,\mathbb{C})]$ is as in Theorem 4.2.1 but now ξ_λ is a highest weight vector for the opposite Borel with weight -2λ.
- $\mathbf{S^2(SL(n))}$ is not a multiplicity free action because $det \in \mathbb{C}[Sym(n,\mathbb{C})]$ is a non-constant $SL(n,\mathbb{C})$-invariant.

4.3. $\Lambda^2(GL(n))$

We now turn to the action of $GL(n,\mathbb{C})$ on $\Lambda^2(\mathbb{C}^n)$, the skew-symmetric 2-tensors. Identifying $\Lambda^2(\mathbb{C}^n)$ with

$$Skew(n,\mathbb{C}) = \{A \in M_{n,n}(\mathbb{C}) \ : \ A^t = -A\},$$

the action is given by Equation 11. As in the preceding examples, we will employ the twisted version, as in Equation 12.

The leading minor determinant ξ_j vanishes on $Skew(n,\mathbb{C})$ when j is odd and is a perfect square when j is even. We let $\nu_j \in \mathbb{C}[Skew(n,\mathbb{C})]$ denote the Pfaffian polynomial:

$$\nu_j(z) = Pf \begin{bmatrix} z_{1,1} & \cdots & z_{1,2j} \\ \vdots & & \vdots \\ z_{2j,1} & \cdots & z_{2j,2j} \end{bmatrix},$$

the square root of the principal $2j \times 2j$ subdeterminant. So $deg(\nu_j) = j$ and $\nu_j^2 = \xi_{2j}$ for $j = 1, \ldots, \lfloor n/2 \rfloor$. The polynomials ν_j are given explicitly by the formula

$$\nu_j(z) = \frac{1}{j!2^j} \sum_{\sigma \in S_{2j}} sign(\sigma) \prod_{\ell=1}^{j} z_{\sigma(2\ell-1)\sigma(2\ell)}.$$

Let $m = \lfloor n/2 \rfloor$ and write

$$\nu_\lambda = \nu_1^{\lambda_1-\lambda_2} \cdots \nu_{m-1}^{\lambda_{m-1}-\lambda_m} \nu_m^{\lambda_m}$$

for $\lambda_1 \geq \cdots \geq \lambda_m \geq 0$.

Theorem 4.3.1: *The action of $GL(n,\mathbb{C})$ on $Skew(n,\mathbb{C})$ is multiplicity free. We have the decomposition*

$$\mathbb{C}[Skew(n,\mathbb{C})] = \bigoplus_{\lambda} P_\lambda$$

where the sum is over all $\lambda \in \mathbb{N}^m$ with $\lambda_1 \geq \cdots \geq \lambda_m \geq 0$. The polynomial ν_λ is a highest weight vector in P_λ with weight

$$\widetilde{\lambda} = (\lambda_1, \lambda_1, \lambda_2, \lambda_2, \ldots, \lambda_m, \lambda_m).$$

Moreover,

$$\mathcal{P}_k(Skew(n,\mathbb{C})) = \bigoplus_{|\lambda|=k} P_\lambda.$$

Proof: Let $J = \begin{bmatrix} 0 & 1 \\ -1 & 0 \end{bmatrix}$ and let $\widetilde{J}$ be the $n \times n$-matrix

$$\widetilde{J} = \underbrace{J \oplus \cdots \oplus J}_{m\text{-times}} \text{ for } n = 2m \text{ even and}$$

$$\widetilde{J} = \underbrace{J \oplus \cdots \oplus J}_{m\text{-times}} \oplus 0 \text{ for } n = 2m+1 \text{ odd.}$$

We claim that $\widetilde{J}$ has an open B_n-orbit in $Skew(n, \mathbb{C})$. Hence our action is multiplicity free. In fact, for $X \in gl(n, \mathbb{C})$,

$$X \cdot \widetilde{J} = -(X^t \widetilde{J} + \widetilde{J} X) = -(\widetilde{J} X - (\widetilde{J} X)^t).$$

First suppose that $n = 2m$ and let $X \in \mathfrak{b}_n$. Decompose X into 2×2-blocks X_{ij}, $1 \leq i, j \leq m$. Now $\widetilde{J} X$ has 2×2-blocks $J X_{ij}$, which are arbitrary for $i < j$. Taking $X_{ii} = \begin{bmatrix} 0 & 0 \\ 0 & c_i \end{bmatrix}$ gives $J X_{ii} = \begin{bmatrix} 0 & c_i \\ 0 & 0 \end{bmatrix}$. This shows that an arbitrary strictly upper triangular matrix can be written in the form $\widetilde{J} X$ for some $X \in \mathfrak{b}_n$. So $\mathfrak{b}_n \cdot \widetilde{J} = Skew(n, \mathbb{C})$ when n is even. When $n = 2m+1$ the same argument shows that for suitable entries in the first $n-1$ rows and columns of $X \in \mathfrak{b}_n$, $\widetilde{J} X$ contains any desired strictly upper triangular matrix in its first $n-1$ rows and columns. As the last column of $\widetilde{J} X$ is

$$\begin{bmatrix} x_{2,n} \\ -x_{1,n} \\ \vdots \\ x_{2m,n} \\ -x_{2m-1,n} \\ 0 \end{bmatrix}$$

we conclude that $\mathfrak{b}_n \cdot \widetilde{J} = Skew(n, \mathbb{C})$ when n is odd.

Suppose that μ is a highest weight that occurs in $\mathbb{C}[Skew(n, \mathbb{C})]$ and that f is a μ-weight vector. Since all weights in $\mathbb{C}[Skew(n, \mathbb{C})]$ are non-negative, we have $\mu_1 \geq \cdots \geq \mu_n \geq 0$. Moreover $f(\widetilde{J}) \neq 0$ as $B_n \cdot \widetilde{J}$ is open. For $h \in H_n$ we have

$$h^\mu f(\widetilde{J}) = (h \cdot f)(\widetilde{J}) = f(h \widetilde{J} h).$$

But

$$h \widetilde{J} h = \begin{bmatrix} h_1 h_2 J & & \\ & h_3 h_4 J & \\ & & \ddots \end{bmatrix}$$

so that $h\widetilde{J}h = \widetilde{J}$ for elements $h = diag(h_1, \ldots, h_n)$ of the form

$$h = \begin{cases} diag(h_1, h_1^{-1}, h_2, h_2^{-1}, \ldots, h_m, h_m) & \text{for } n = 2m \\ diag(h_1, h_1^{-1}, h_2, h_2^{-1}, \ldots, h_m, h_m, h_{2m+1}) & \text{for } n = 2m+1 \end{cases}.$$

It follows that $h^\mu = 1$ for all such h and thus

$$\mu_1 = \mu_2, \mu_3 = \mu_4, \ldots, \mu_{2m-1} = \mu_{2m} \text{ and } \mu_n = 0 \text{ when } n \text{ is odd.}$$

Thus all highest weights occurring in $\mathbb{C}[Skew(n, \mathbb{C})]$ have the form

$$\widetilde{\lambda} = (\lambda_1, \lambda_1, \lambda_2, \lambda_2, \ldots, \lambda_m, \lambda_m)$$

for some $\lambda_1 \geq \cdots \geq \lambda_m \geq 0$. To complete the proof one just needs to check that ν_λ has weight $\widetilde{\lambda}$. This reduces to the calculation

$$\begin{aligned}(h \cdot \nu_j)(z) &= Pf\left(\begin{bmatrix} h_1 & & \\ & \ddots & \\ & & h_{2j}\end{bmatrix}\begin{bmatrix} z_{1,1} & \cdots & z_{1,2j} \\ \vdots & & \vdots \\ z_{2j,1} & \cdots & z_{2j,2j}\end{bmatrix}\begin{bmatrix} h_1 & & \\ & \ddots & \\ & & h_{2j}\end{bmatrix}\right) \\ &= det\begin{bmatrix} h_1 & & \\ & \ddots & \\ & & h_{2j}\end{bmatrix} Pf\begin{bmatrix} z_{1,1} & \cdots & z_{1,2j} \\ \vdots & & \vdots \\ z_{2j,1} & \cdots & z_{2j,2j}\end{bmatrix} \\ &= h_1 \cdots h_{2j} Pf\begin{bmatrix} z_{1,1} & \cdots & z_{1,2j} \\ \vdots & & \vdots \\ z_{2j,1} & \cdots & z_{2j,2j}\end{bmatrix},\end{aligned}$$

which shows that ν_j is a weight vector with weight (1^{2j}). □

Regarding the action $\mathbf{\Lambda^2(GL(n))}$ we note:

- The Pfaffian of a skew symmetric matrix is irreducible as a polynomial in the entries z_{ij} with $i < j$. Thus $\mathbf{\Lambda^2(GL(n))}$ has rank $m = \lfloor n/2 \rfloor$ with fundamental highest weights (1^{2j}) and fundamental highest weight vectors ν_j.
- The untwisted action has the same decomposition but now ν_λ is a lowest weight vector with weight $-\widetilde{\lambda}$.
- For the trace zero upper triangular matrices $\mathfrak{b}'_n$, one sees, as in the proof of Theorem 4.3.1, that $\mathfrak{b}'_n \cdot \widetilde{J} = Skew(n, \mathbb{C})$ when n is odd. So $\mathbf{\Lambda^2(SL(2m+1))}$ is a multiplicity free action. On the other hand, this fails for $n = 2m$ even. Indeed, $det \in \mathbb{C}[Skew(2m, \mathbb{C})]$ is a non-constant $SL(2m, \mathbb{C})$-invariant so $\mathbf{\Lambda^2(SL(2m))}$ is not multiplicity free.

4.4. SO(n) × $\mathbb{C}^\times$

The group $G = SO(n,\mathbb{C}) \times \mathbb{C}^\times$ acts on $V = \mathbb{C}^n$ in the usual way:

$$(g,t) \cdot v = tgv.$$

The decomposition for $\mathbb{C}[V] = \mathbb{C}[z_1, \ldots, z_n]$ is given by the classical theory of spherical harmonics. Let

$$\varepsilon(z) = z_1^2 + \cdots + z_n^2$$

and

$$\Delta = \varepsilon(\partial) = \partial_1^2 + \cdots + \partial_n^2,$$

the constant coefficient differential operator with Wick symbol $p(z,w) = \varepsilon(w)$. Since ε is an $SO(n,\mathbb{C})$-invariant polynomial, Δ is an $SO(n,\mathbb{C})$-invariant operator. It follows that the space of *harmonic polynomials*

$$\mathcal{H} = Ker(\Delta) = \{p \in \mathbb{C}[V] \ : \ \Delta p = 0\}$$

is $SO(n,\mathbb{C})$-invariant, as is

$$\mathcal{H}_m = \mathcal{H} \cap \mathcal{P}_m(V),$$

the harmonic polynomials homogeneous of degree m. It is well known that $\mathcal{H}_m$ is an irreducible module for $SO(n,\mathbb{C})$ and

$$\mathcal{P}_m(V) = \mathcal{H}_m \oplus \varepsilon\mathcal{P}_{m-2}(V). \tag{13}$$

This gives, in particular,

$$dim(\mathcal{H}_m) = \dim(\mathcal{P}_m) - dim(\mathcal{P}_{m-2}) = \binom{m+n-1}{m} - \binom{m+n-3}{m-2},$$

for $m \geq 2$. Now Equation 13 leads inductively to the decomposition

$$\mathcal{P}_m(V) = \bigoplus_{k+2\ell=m} P_{k,\ell} \qquad (P_{k,\ell} = \mathcal{H}_k\varepsilon^\ell) \tag{14}$$

of $\mathcal{P}_m(V)$ into irreducible $SO(n,\mathbb{C})$-modules. The modules $\{P_{k,\ell} \ : \ k+2\ell = m\}$ appearing in (14) are clearly inequivalent because their dimensions are distinct.

Equation 14 now gives a decomposition

$$\mathbb{C}[V] = \bigoplus_{k,\ell \geq 0} P_{k,\ell} \tag{15}$$

of $\mathbb{C}[V]$ into irreducibles for $G = SO(n,\mathbb{C})$. This is *not*, however, multiplicity free because $\mathcal{H}_{k,\ell} \simeq \mathcal{H}_{k,\ell'}$ as $SO(n,\mathbb{C})$-modules for $\ell \neq \ell'$. We use the scalars $\mathbb{C}^\times$ to repair this defect.

The action of $\mathbb{C}^\times$ on $\mathbb{C}[V]$ by scalars commutes with Δ and hence preserves the $\mathcal{H}_m$'s and $\mathcal{P}_{k,\ell}$'s. Now $\mathbb{C}^\times$ acts on $P_{k,\ell}$ by the character $t \mapsto t^{-(k+2\ell)}$. Hence $P_{k,\ell}$ and $P_{k,\ell'}$ are inequivalent as $\mathbb{C}^\times$-modules when $\ell \neq \ell'$. Thus (15) is multiplicity free as a decomposition for the group $G = SO(n,\mathbb{C}) \times \mathbb{C}^\times$.

A highest weight vector in $P_{k,\ell}$ is given by $(z_1+iz_2)^k\varepsilon(z)^\ell$, for a suitably chosen Borel subgroup in $G = SO(n,\mathbb{C}) \times \mathbb{C}^\times$. The multiplicity free action $G : \mathbb{C}^n$ has rank 2 with fundamental highest weight vectors $z_1 + iz_2$ and $\varepsilon(z)$.

4.5. $\mathbf{GL(n) \oplus_{GL(n)} \Lambda^2(GL(n))}$

The group $G = GL(n,\mathbb{C})$ acts diagonally on

$$V = V_1 \oplus V_2 = \mathbb{C}^n \oplus \Lambda^2(\mathbb{C}^n) \cong \mathbb{C}^n \oplus Skew(n,\mathbb{C}).$$

For consistency with Sections 4.1 and 4.3 we twist the action of G on both V_1 and V_2. We have seen how to decompose $\mathbb{C}[V_1]$ and $\mathbb{C}[V_2]$ under the action of G. Writing $\sigma^\mu = \sigma_n^\mu$ for the irreducible representation of G with highest weight $\mu \geq 0$ we have

$$\mathbb{C}[V_1] \simeq \bigoplus_k \sigma^{(k)}, \qquad \mathbb{C}[V_2] = \bigoplus_\lambda \sigma^{\widetilde{\lambda}}$$

as G-modules. Here the second sum is over all $\lambda = (\lambda_1 \geq \cdots \geq \lambda_m \geq 0)$ $(m = \lfloor n/2 \rfloor)$ and $\widetilde{\lambda} = (\lambda_1, \lambda_1, \ldots, \lambda_m, \lambda_m)$. Thus we can write

$$\mathbb{C}[V] = \mathbb{C}[V_1] \otimes \mathbb{C}[V_2] \simeq \sum_{k,\lambda} \sigma^{(k)} \otimes \sigma^{\widetilde{\lambda}}.$$

Tensor product representations of $GL(n,\mathbb{C})$ can be decomposed using the *Littlewood-Richardson rules.* To apply this technique, one identifies each highest weight $\mu \geq 0$ with a Young's diagram consisting of μ_j boxes on row j. The Littlewood-Richardson rules ensure that $\sigma^{(k)} \otimes \sigma^{\widetilde{\lambda}}$ has a multiplicity free decomposition. Moreover, the representations σ^μ that occur in the decomposition are given by diagrams that can be obtained from that for $\widetilde{\lambda}$ by adding k boxes, no two of which fall in the same column. For example, when $\lambda = (3,2)$, and $k = 3$, we must add three boxes to the Young's diagram for $\widetilde{\lambda} = (3,3,2,2)$. For $n \geq 5$ this produces five diagrams:

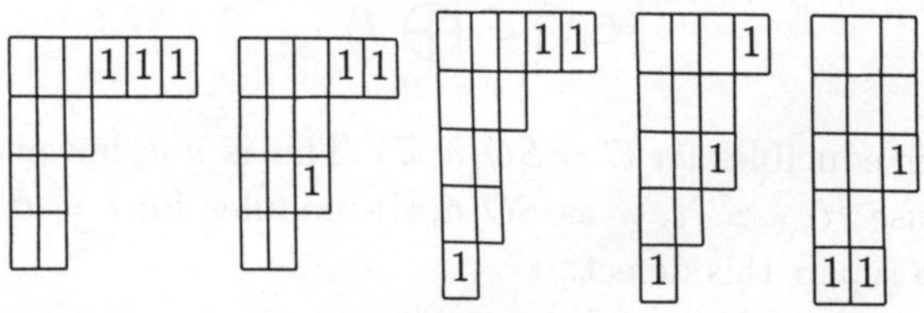

The three boxes added to the diagram for $(3,3,2,2)$ have been marked with 1's in each case. This exercise with the Littlewood-Richardson rules shows

$$\sigma^{(3)} \otimes \sigma^{(3,3,2,2)} \simeq \sigma^{(6,3,2,2)} \oplus \sigma^{(5,3,3,2)} \oplus \sigma^{(5,3,2,2,1)} \oplus \sigma^{(4,3,3,2,1)} \oplus \sigma^{(3,3,3,2,2)}.$$

In general, the Littlewood-Richardson rules yield

$$\sigma^{(k)} \otimes \sigma^{\widetilde{\lambda}} = \sum_{\mu} \sigma^{\mu}$$

where the sum is over all highest weights $\mu \geq 0$ of the form

$$\mu = (\lambda_1 + j_1, \lambda_1, \lambda_2 + j_2, \lambda_2, \dots), \quad j_1 + \cdots + j_m = k, \quad j_i \leq \lambda_{i-1} - \lambda_i.$$

Note that λ can be recovered from μ by extracting every other row. This shows that the representations σ^{μ} that occur in $\sigma^{(k)} \otimes \sigma^{\widetilde{\lambda}}$ and in $\sigma^{(k')} \otimes \sigma^{\widetilde{\lambda'}}$ are distinct for $\lambda' \neq \lambda$. Moreover, the irreducibles σ^{μ} in $\sigma^{(k)} \otimes \sigma^{\widetilde{\lambda}}$ and $\sigma^{(k')} \otimes \sigma^{\widetilde{\lambda}}$ are clearly distinct when $k \neq k'$ as these satisfy $|\mu| = 2|\lambda| + k$ and $|\mu| = 2|\lambda| + k'$ respectively. This shows that $\mathbb{C}[V]$ has a multiplicity free decomposition under the action of G. Note that an arbitrary highest weight $\mu \geq 0$ can be written in the form "$\mu = (\lambda_1 + j_1, \lambda_1, \lambda_2 + j_2, \lambda_2, \dots)$" by letting $j_\ell = \mu_{2\ell-1} - \mu_{2\ell}$. We have proved the following.

Theorem 4.5.1: *The diagonal action of $GL(n,\mathbb{C})$ on $\mathbb{C}^n \oplus \Lambda^2(\mathbb{C}^n)$ is multiplicity free. Moreover*

$$\mathbb{C}[\mathbb{C}^n \oplus \Lambda^2(\mathbb{C}^n)] \simeq \bigoplus_{\mu \geq 0} \sigma^{\mu}$$

as a $GL(n,\mathbb{C})$-module. That is, every non-negative highest weight occurs in $\mathbb{C}[\mathbb{C}^n \oplus \Lambda^2(\mathbb{C}^n)]$ with multiplicity one.

There is another viewpoint on this example. One can identify $V = V_1 \oplus V_2 = \mathbb{C}^n \oplus \Lambda^2(\mathbb{C}^n)$ with $Skew(n+1,\mathbb{C})$ by regarding the first row (or column) of an $(n+1) \times (n+1)$ skew symmetric matrix as an element of $\mathbb{C}^n$, and the remaining entries as an element of $\Lambda^2(\mathbb{C}^n) \cong Skew(n,\mathbb{C})$. For $z \in Skew(n+1,\mathbb{C})$, we write z' for the element of $Skew(n,\mathbb{C})$ obtained by removing the first row and column of z. Under this identification, the diagonal action of $GL(n,\mathbb{C})$ on $V_1 \oplus V_2$ is realized on $Skew(n+1,\mathbb{C})$ by restricting the action of $GL(n+1,\mathbb{C})$ to the subgroup $GL(n,\mathbb{C}) \subset GL(n+1,\mathbb{C})$, embedded as

$$\left\{ \left[\begin{array}{c|c} 1 & 0 \\ \hline 0 & A \end{array} \right] \ : \ A \in GL(n,\mathbb{C}) \right\}.$$

The fundamental highest weight vectors arising from the separate actions of $GL(n,\mathbb{C})$ on V_1 and V_2 are z_{12} and $\nu'_k(z)$ for $k=1,\ldots,\lfloor n/2\rfloor$, where $\nu'_k(z)$ is the Pfaffian of the first $2k$ rows and columns of the $n\times n$ matrix z'. There are additional fundamental highest weight vectors, $\nu_k(z)$ for $k=1,\ldots,\lfloor(n+1)/2\rfloor$. These are the Pfaffians of the first $2k$ rows and columns of the $(n+1)\times(n+1)$ matrix z. Note that $\nu_1(z)=z_{12}$, so our fundamental highest weight vectors are the ν_k's together with the ν'_k's. For a highest weight $\mu\geq 0$, a μ-highest weight vector in $\mathbb{C}[V]$ is

$$\nu_1^{\mu_1-\mu_2}(\nu'_1)^{\mu_2-\mu_3}\nu_2^{\mu_3-\mu_4}\cdots\nu_m^{\mu_{n-1}-\mu_n}(\nu'_m)^{\mu_n}$$

when $n=2m$ is even, and

$$\nu_1^{\mu_1-\mu_2}(\nu'_1)^{\mu_2-\mu_3}\nu_2^{\mu_3-\mu_4}\cdots\nu_m^{\mu_{n-2}-\mu_{n-1}}(\nu'_m)^{\mu_{n-1}-\mu_n}(\nu_{m+1})^{\mu_n}$$

when $n=2m+1$ is odd. In particular, this is a rank n multiplicity free action with fundamental highest weights $\{(1^k)\ :\ 1\leq k\leq n\}$.

4.6. *Section 4 notes*

The decomposition for action $\mathbf{GL(n)}\otimes\mathbf{GL(m)}$ is called $GL(n)-GL(m)$ *duality*. This decomposition and those for the actions $\mathbf{S^2(GL(n))}$ and $\mathbf{\Lambda^2(GL(n))}$ were popularized by Howe in [22]. Reference [23] explains how these results were rediscovered independently by various mathematicians and are implicit in earlier work of Weyl and Schur. $GL(n)-GL(m)$ duality is equivalent to *Schur duality*, which gives the decomposition for $(\mathbb{C}^n)^{\otimes m}$ under the action of $GL(n,\mathbb{C})\times S_m$. In fact, as explained in [23], both dualities can be derived from the *First Fundamental Theorem* (FFT) of Invariant Theory.

We refer the reader to Section B.2.6 in [17] for the theory of Pfaffian polynomials. The irreducibility of determinants and Pfaffians on $Sym(n,\mathbb{C})$ and $Skew(n,\mathbb{C})$ is proved in Section B.2.7 of [17]. The theory of spherical harmonics can be found in Section 5.2.3 in [17] as well as many other sources. The Littlewood-Richardson rules are discussed in Section I.9 of [39].

Compact forms for the actions $\mathbf{Gl(n)}\otimes\mathbf{GL(m)}$, $\mathbf{S^2(GL(n))}$, $\mathbf{\Lambda^2(GL(n))}$ and $\mathbf{SO(n)}\times\mathbb{C}^\times$ arise in connection with Hermitian symmetric spaces. In fact, let G/K be an irreducible Hermitian symmetric space of non-compact type. Here G is a semi-simple real Lie group and K is a compact Lie subgroup. The symmetric space structure gives a Cartan decomposition $\mathfrak{g}=\mathfrak{k}+\mathfrak{p}$ for the Lie algebra. The complex structure on

$T_e(G/K)$ is viewed as an $\mathbb{R}$-linear map $J : \mathfrak{p} \to \mathfrak{p}$ with $J^2 = -I$. Now

$$\mathfrak{p}_{\mathbb{C}} = \mathfrak{p}_+ \oplus \mathfrak{p}_-$$

where $\mathfrak{p}_\pm$ are the $(\pm i)$-eigenspaces for J on the complexification of $\mathfrak{p}$. The complexified adjoint action of K on $\mathfrak{g}_{\mathbb{C}}$ preserves $\mathfrak{p}_\pm$. For the classical irreducible Hermitian symmetric spaces this construction leads to the following actions. (See [20].)

- Type A III: $G/K = SU(n,m)/S(U(n) \times U(m))$. The action of $K = S(U(n) \times U(m))$ on $\mathfrak{p}_+$ can be identified with the action of K on $M_{n,m}(\mathbb{C})$ via $(k,k') \cdot v = kv(k')^*$. The group K has a one dimensional center and we essentially have a compact form of the multiplicity free action $\mathbf{SL(n)} \otimes \mathbf{GL(m)}$. More precisely, the action here agrees with action (6) twisted by $k' \mapsto ((k')^t)^{-1} = \overline{k'}$ on the second factor.
- Type C I: $G/K = Sp(2n,\mathbb{R})/U(n)$. The action of $K = U(n)$ on $\mathfrak{p}_+$ can be identified with that of K on $Sym(n,\mathbb{C})$ via $k \cdot v = kvk^t$. So this is a compact form of $\mathbf{S^2(GL(n))}$.
- Type D III: $G/K = SO^*(2n)/U(n)$. The action of $K = U(n)$ on $\mathfrak{p}_+$ can be identified with that of K on $Skew(n,\mathbb{C})$ via $k \cdot v = kvk^t$. This is a compact form of the multiplicity free action $\mathbf{\Lambda^2(GL(n))}$.
- Type BD I: $G/K = SO_\circ(n,2)/SO(n) \times SO(2)$. The action of $K = SO(n,\mathbb{R}) \times \mathbb{T}$ on $\mathfrak{p}_+$ can be identified with the action of K on $\mathbb{C}^n$ via $(k,t) \cdot z = tkz$. This is a compact form of $\mathbf{SO(n)} \times \mathbb{C}^\times$.

The classification of irreducible Hermitian symmetric spaces includes two exceptional cases, in addition to the four families described above.

- Type E III: In this case $K = Spin(10) \times \mathbb{T}$ and G has Lie algebra $\mathfrak{e}_{6(-14)}$, a certain real form for the complex Lie algebra $\mathfrak{e}_6$. One can identify $\mathfrak{p}_+$ with $\Lambda^{\text{even}}(\mathbb{C}^5) \approxeq \mathbb{C}^{16}$ and $Spin(10)$ acts by the positive half-spin representation. It is known that (with the scalars included) this action in multiplicity free.
- Type E VII: Here $K = E_6 \times \mathbb{T}$ and G has Lie algebra $\mathfrak{e}_{7(-25)}$, a real form of $\mathfrak{e}_7$. In this case $\mathfrak{p}_+$ can be identified with an exceptional Jordan algebra $\mathcal{J}$ of dimension 27. The representation of E_6 on $\mathcal{J}$ is described in [11]. This action (with the scalars included) is multiplicity free.

Thus one has:

Theorem 4.6.1: *([25]) The linear action $K : \mathfrak{p}_+$ associated to any irreducible Hermitian symmetric space of non-compact type is multiplicity free.*

5. A recursive criterion for multiplicity free actions

In this section we present a recursive criterion for multiplicity free actions due to Knop [31]. Given a linear action $G : V$ we let

$$\Psi = \Psi(V) \subset \mathfrak{h}^*$$

be the set of all weights for the representation of H on V, listed with multiplicities. As usual, $B = HN$ is a Borel subgroup and $\Delta = \Delta^+ \cup (-\Delta^+)$ are the roots for G, as in Section 1.4.

For a *highest* weight $\lambda \in \Psi$ let

$$S_\lambda = \{\alpha \in \Delta^+ \ : \ \langle \lambda, \alpha \rangle > 0\}.$$

If $z \in V$ is a λ-highest weight vector then $\mathfrak{g} \cdot z = \mathbb{C}z$ if and only if $S_\lambda = \emptyset$.

If $S_\lambda = \emptyset$ for all highest weights $\lambda \in \Psi$, then G acts by scalars on every weight space in V. In this case, we essentially have a torus action. In particular, $G : V$ is multiplicity free if and only if $H : V$ is multiplicity free. This happens if and only if the set Ψ of weights for $H : V$ is linearly independent.

Suppose that $\lambda \in \Psi$ is a highest weight with $S_\lambda \neq \emptyset$. Let $z_\circ \in V$ be a λ-highest weight vector and let $f_\circ \in V^*$ be the corresponding $(-\lambda)$-lowest weight vector normalized so that $f_\circ(z_\circ) = 1$. Let $P = P_{f_\circ}$ and $\Sigma = \Sigma_{f_\circ}$ be as in Equations 3 and 4. Now $P^- = LU^-$ is the opposite parabolic subgroup to $P = LU$ and $P^- = P_{z_\circ}$.

From the definition of Σ one has that $z \in \Sigma$ if and only if $f_\circ(z) \neq 0$ and $(X \cdot f_\circ)(z) = -\lambda(X) f_\circ(z)$ for all $X \in \mathfrak{g} = \mathfrak{p} + \mathfrak{u}^-$. But $X \cdot f_\circ = -\lambda(X) f_\circ$ for $X \in \mathfrak{p}$, so

$$\Sigma = \{z \in V \ : \ f_\circ(z) \neq 0 \text{ and } (\mathfrak{u}^- \cdot f_\circ)(z) = 0\}.$$

Hence Σ is a open set in the subspace

$$W = (\mathfrak{u}^- \cdot f_\circ)^\perp$$

of V. Recall that Σ is invariant under the action of the Levi component L of P. As $U \cdot \Sigma = V_\circ = \{z \in V \ : \ f_\circ(z) \neq 0\}$ by Lemma 3.2.5, we see that there is an open B-orbit in V if and only if there is an open $L \cap B$-orbit in Σ. Equivalently, $G : V$ is multiplicity free if and only if $L : W$ is multiplicity free.

We know, moreover, that $\mathfrak{l} = \mathfrak{h} + \sum_{\alpha \in \Delta(L)} \mathfrak{g}_\alpha$ where

$$\Delta(L) = \{\alpha \in \Delta \ : \ \langle \lambda, \alpha \rangle = 0\} = \Delta \backslash S_\lambda.$$

The positive roots for L are $\Delta^+(L) = \Delta^+ \backslash S_\lambda$. Each of the root vectors $\{X_{-\alpha} \ : \ \alpha \in S_\lambda\} \subset \mathfrak{u}^-$ acts non-trivially on $f_\circ$, so the set of weights in

$\mathfrak{u}^- \cdot f_\circ$ is $\{-\lambda + \alpha \ : \ \alpha \in S_\lambda\}$. Thus the set of weights in $W = (\mathfrak{u}^- \cdot f_\circ)^\perp$ is $\Psi \backslash (\lambda - S_\lambda)$.

In summary, we have a recursive algorithm that begins with the pair

$$(\Delta_0^+ = \Delta^+, \Psi_0 = \Psi)$$

where

- Δ^+ is the set of positive roots for G, and
- $\Psi \subset \mathfrak{h}^*$ is the set of all weights for the representation of G on V, listed with multiplicity.

Given a pair (Δ_n^+, Ψ_n) do the following:

- For each highest weight $\lambda \in \Psi_n$ let $S_\lambda = \{\alpha \in \Delta_n^+ \ : \ \langle \lambda, \alpha \rangle > 0\}$;
- If $S_\lambda = \emptyset$ for all highest weights $\lambda \in \Psi_n$, then $G : V$ is a multiplicity free action if and only if Ψ_n is linearly independent;
- Otherwise, choose a highest weight $\lambda \in \Psi_n$ with $S_\lambda \neq \emptyset$ and apply the above steps to the pair $(\Delta_{n+1}^+, \Psi_{n+1}) = (\Delta_n^+ \backslash S_\lambda, \Psi_n \backslash (\lambda - S_\lambda))$.

Eventually all the S_λ's are empty and the algorithm terminates at the second step above.

To illustrate this method, we revisit some of the examples described in Section 4.

5.1. GL(n)

Here $G = GL(n, \mathbb{C})$ acts on $V = \mathbb{C}^n$ as usual. One has

$$\Delta_0^+ = \Delta^+ = \{\varepsilon_i - \varepsilon_j \ : \ i < j\}, \quad \Psi_0 = \Psi = \{\varepsilon_1, \ldots, \varepsilon_n\}.$$

We know $e_1 = (1, 0, \ldots, 0)$ is a highest weight vector in V with weight ε_1, and so

$$S_{\varepsilon_1} = \{\varepsilon_1 - \varepsilon_2, \ldots, \varepsilon_1 - \varepsilon_n\}.$$

The coordinate vector $z_1 \in V^*$ dual to e_1 has parabolic subgroup $P = P_{z_1}$ with Lie algebra $\mathfrak{p}$ spanned by $\{E_{11}\} \cup \{E_{ij} \ : \ i > 2\}$. The Levi component is $L = GL(1, \mathbb{C}) \times GL(n-1, \mathbb{C})$. The nilpotent $\mathfrak{u}^- = Span\{E_{1j} \ : \ j \geq 2\}$ gives $\mathfrak{u}^- z_1 = Span\{z_2, \ldots, z_n\}$ and hence $W = (\mathfrak{u}^- z_1)^\perp = \mathbb{C} e_1$. Thus we have

$$\Delta_1^+ = \Delta_0^+ \backslash S_{\varepsilon_1} = \{\varepsilon_i - \varepsilon_j \ : \ 2 \leq i < j\}, \quad \Psi_1 = \Psi_0 \backslash \{\varepsilon_2, \ldots, \varepsilon_n\} = \{\varepsilon_1\}.$$

Since $S_{\varepsilon_1} = \emptyset$, the process terminates. One concludes that $G : V$ is multiplicity free since Ψ_1 is a linearly independent set.

In practice, it is not necessary to identify L and W at each stage in the induction. We illustrate this in the next example.

5.2. GL(n) ⊗ GL(n)

Here $G = GL(n,\mathbb{C}) \times GL(n,\mathbb{C})$ acts on $V = \mathbb{C}^n \otimes \mathbb{C}^n$. Now

$$\Delta_0^+ = \Delta^+ = \{\varepsilon_i - \varepsilon_j \ : \ 1 \le i < j \le n\} \cup \{\varepsilon_i' - \varepsilon_j' \ : \ 1 \le i < j \le n\}$$

and

$$\Psi_0 = \Psi = \{\varepsilon_i + \varepsilon_j' \ : \ 1 \le i \le n, 1 \le j \le n\}.$$

$\lambda = \varepsilon_1 + \varepsilon_1'$ is the only highest weight in Ψ_0. One has

$$S_{\varepsilon_1+\varepsilon_1'} = \{\varepsilon_1 - \varepsilon_2, \ldots, \varepsilon_1 - \varepsilon_n\} \cup \{\varepsilon_1' - \varepsilon_2', \ldots, \varepsilon_1' - \varepsilon_n'\}.$$

Thus

$$\Delta_1^+ = \{\varepsilon_i - \varepsilon_j \ : \ 2 \le i < j \le n\} \cup \{\varepsilon_i' - \varepsilon_j' \ : \ 2 \le i < j \le n\}$$

and

$$\Psi_1 = \{\varepsilon_1 + \varepsilon_1'\} \cup \{\varepsilon_i + \varepsilon_j' \ : \ 2 \le i \le n, 2 \le j \le n\}.$$

From these roots and weights we can see that $L : W$ is equivalent to the action of $\mathbb{C}^\times \times GL(n-1,\mathbb{C}) \times \mathbb{C}^\times \times GL(n-1,\mathbb{C})$ on $\mathbb{C} \oplus (\mathbb{C}^{n-1} \otimes \mathbb{C}^{n-1})$.

At the next stage of the algorithm one takes $\lambda = \varepsilon_2 + \varepsilon_2'$ and obtains

$$\Delta_2^+ = \{\varepsilon_i - \varepsilon_j \ : \ 3 \le i < j\} \cup \{\varepsilon_i' - \varepsilon_j' \ : \ 3 \le i < j\},$$

$$\Psi_2 = \{\varepsilon_1 + \varepsilon_1', \varepsilon_2 + \varepsilon_2'\} \cup \{\varepsilon_i + \varepsilon_j' \ : \ 3 \le i, j\}.$$

After n steps, the process terminates with

$$\Delta_n^+ = \emptyset, \quad \Psi_n = \{\varepsilon_1 + \varepsilon_1', \ldots, \varepsilon_n + \varepsilon_n'\}.$$

At this point $L \cong (\mathbb{C}^\times)^n \times (\mathbb{C}^\times)^n$ and $S_\lambda = \emptyset$ for all $\lambda \in \Psi_n$. As Ψ_n is linearly independent, $G : V$ is multiplicity free.

5.3. GL(n) $\oplus_{\mathbf{GL(n)}}$ Λ²(GL(n))

For the diagonal action of $G = GL(n,\mathbb{C})$ on $V = \mathbb{C}^2 \oplus \Lambda^2(\mathbb{C}^n)$ one has

$$\Delta_0^+ = \Delta^+ = \{\varepsilon_i - \varepsilon_j \ : \ i < j\}, \quad \Psi_0 = \Psi = \{\varepsilon_1, \ldots, \varepsilon_n\} \cup \{\varepsilon_i + \varepsilon_j \ : \ i < j\}.$$

Choose $\lambda = \varepsilon_1$, so that $S_\lambda = \{\varepsilon_1 - \varepsilon_j \ : \ 2 \le j\}$ and

$$\Delta_1^+ = \{\varepsilon_i - \varepsilon_j \ : \ 2 \le i < j\}, \quad \Psi_1 = \{\varepsilon_1\} \cup \{\varepsilon_i + \varepsilon_j \ : \ i < j\}.$$

Next choose $\lambda = \varepsilon_1 + \varepsilon_2$ to obtain $S_\lambda = \{\varepsilon_2 - \varepsilon_j \ : \ 3 \le j\}$ and

$$\Delta_2^+ = \{\varepsilon_i - \varepsilon_j \ : \ 3 \le i < j\}, \quad \Psi_2 = \{\varepsilon_1, \varepsilon_1 + \varepsilon_2\} \cup \{\varepsilon_i + \varepsilon_j \ : \ 2 \le i < j\}.$$

After n steps, the process terminates with $\Delta_n^+ = \emptyset$ and

$$\Psi_n = \{\varepsilon_1, \varepsilon_1 + \varepsilon_2, \varepsilon_2 + \varepsilon_3, \ldots, \varepsilon_{n-1} + \varepsilon_n\}.$$

As this is a linearly independent set, one concludes $G : V$ is a multiplicity free action.

Alternatively, one could begin with $\lambda = \varepsilon_1 + \varepsilon_2$ in the first step. This gives $S_\lambda = \{\varepsilon_1 - \varepsilon_j \ : \ 3 \le j\} \cup \{\varepsilon_2 - \varepsilon_j \ : \ 3 \le j\}$ and

$$\Delta_1^+ = \{\varepsilon_1 - \varepsilon_2\} \cup \{\varepsilon_i - \varepsilon_j \ : \ 3 \le i < j\},$$

$$\Psi_1 = \{\varepsilon_1, \ldots, \varepsilon_n\} \cup \{\varepsilon_1 + \varepsilon_2\} \cup \{\varepsilon_i + \varepsilon_j \ : \ 3 \le i < j\}$$

Next one could choose a highest weight in Ψ_1 that was not available in V, namely $\lambda = \varepsilon_3 + \varepsilon_4$. This gives $S_\lambda = \{\varepsilon_3 - \varepsilon_j, \varepsilon_4 - \varepsilon_j \ : \ 5 \le j\}$ and

$$\Delta_2^+ = \{\varepsilon_1 - \varepsilon_2, \varepsilon_3 - \varepsilon_4\} \cup \{\varepsilon_i - \varepsilon_j \ : \ 5 \le i < j\},$$

$$\Psi_2 = \{\varepsilon_1 \ldots, \varepsilon_n\} \cup \{\varepsilon_1 + \varepsilon_2, \varepsilon_3 + \varepsilon_4\} \cup \{\varepsilon_i + \varepsilon_j \ : \ 5 \le i < j\}.$$

We could then continue in this manner (choosing $\lambda = \varepsilon_5 + \varepsilon_6$) or take $\lambda = \varepsilon_1$. The latter choice gives $S_\lambda = \{\varepsilon_1 - \varepsilon_2\}$,

$$\Delta_3^+ = \{\varepsilon_3 - \varepsilon_4\} \cup \{\varepsilon_i - \varepsilon_j \ : \ 5 \le i < j\}, \text{ and}$$

$$\Psi_3 = \{\varepsilon_1, \varepsilon_3, \ldots, \varepsilon_n\} \cup \{\varepsilon_1 + \varepsilon_2, \varepsilon_3 + \varepsilon_4\} \cup \{\varepsilon_i + \varepsilon_j \ : \ 5 \le i < j\}.$$

For n even, the algorithm could terminate with

$$\Psi_n = \{\varepsilon_1, \varepsilon_3, \varepsilon_5, \ldots, \varepsilon_{n-1}\} \cup \{\varepsilon_1 + \varepsilon_2, \varepsilon_3 + \varepsilon_4, \ldots, \varepsilon_{n-1} + \varepsilon_n\}.$$

This example illustrates how different choices for a highest weight $\lambda \in \Psi_k$ can yield different paths through the algorithm and result in different terminal sets.

6. The classification of linear multiplicity free actions

Let (π, V) be a rational representation of some reductive algebraic group G and consider the question of whether or not this gives a multiplicity free action. The answer depends only on the algebraic subgroup $\pi(G)$ of $GL(V)$. One says that two rational representations (π_1, V_1), (π_2, V_2) of groups G_1, G_2 are *geometrically equivalent* if $\pi_1(G_1)$ coincides with $\pi_2(G_2)$ under some isomorphism $V_1 \to V_2$. For example, (π, V) is geometrically equivalent to $(\pi \circ \varphi, V)$ for any $\varphi \in Aut(G)$. Taking $\varphi(g) = (g^t)^{-1}$ on $G = GL(n, \mathbb{C})$, we see that any rational representation of $GL(n, \mathbb{C})$ is geometrically equivalent to its contragredient.

Linear multiplicity free actions have been completely classified up to geometric equivalence. In this section we present the results of this classification.

6.1. *Irreducible multiplicity free actions*

The multiplicity free actions $G : V$ where G acts irreducibly on V were classified by Victor Kac in [26], building on earlier work including that of Sato and Kimura [47]. The open Borel orbit criterion provided the main technique used to achieve this classification.

First suppose that the image $\mathbf{G}$ of G in $GL(V)$ contains a copy of the scalars. In this case, G is reductive but not semisimple. The group $\mathbf{G}$ coincides with the image of $G' \times \mathbb{C}^\times$, where $G' = (G, G)$ is the commutator subgroup in G, a semisimple group with finite center. The possibilites for $\mathbf{G}'$, the image of G' in $GL(V)$, are listed in Table 3, up to geometric equivalence. That is, for each such $\mathbf{G}'$, the joint action of $\mathbf{G}'$ and $\mathbb{C}^\times$ on V is multiplicity free.

Bold faced type is used in Table 3 to indicate a subgroup of $GL(V)$, the notation indicating the representation involved. The first three entries denote the defining representations of $SL(n, \mathbb{C})$, $SO(n, \mathbb{C})$ and $Sp(2n, \mathbb{C})$ on $\mathbb{C}^n$, $\mathbb{C}^n$ and $\mathbb{C}^{2n}$ respectively. $\mathbf{S^2(SL(n))}$ and $\mathbf{\Lambda^2(SL(n))}$ denote the images in $GL(S^2(\mathbb{C}^n))$ and $GL(\Lambda^2(\mathbb{C}^n))$ of the symmetric and skew-symmetric squares of the defining representation of $SL(n)$. $\mathbf{SL(n) \otimes SL(m)}$ denotes the image of the representation of $SL(n) \times SL(m)$ on $\mathbb{C}^n \otimes \mathbb{C}^m$ (outer tensor product of the two defining representations) and similarly for $\mathbf{SL(n) \otimes Sp(2m)}$. $\mathbf{Spin(7)}$, $\mathbf{Spin(9)}$ denote the image of the spin representations of $Spin(7, \mathbb{C})$, $Spin(9, \mathbb{C})$ on $\mathbb{C}^8$ and $\mathbb{C}^{16}$ respectively. $\mathbf{Spin(10)}$ indicates the positive spin representation of $Spin(10, \mathbb{C})$ on $\mathbb{C}^{16}$. $\mathbf{G_2}$ and $\mathbf{E_6}$ denote actions on $\mathbb{C}^7$ and $\mathbb{C}^{27}$ respectively. The conditions on n and m

Table 3: Irreducible multiplicity free actions $\mathbf{G}' \times \mathbb{C}^\times : V$, $\mathbf{G}'$ semisimple

Semisimple Group $\mathbf{G}'$ ($\mathbf{G}' \times \mathbb{C}^\times : V$ is multiplicity free)	Degrees of fundamental highest weight vectors	Rank
$\mathbf{SL(n)}$ $(n \geq 1)$	1	1
$\mathbf{SO(n)}$ $(n \geq 3)$	$1, 2$	2
$\mathbf{Sp(2n)}$ $(n \geq 2)$	1	1
$\mathbf{S^2(SL(n))}$ $(n \geq 2)$	$1, 2, \ldots, n$	n
$\mathbf{\Lambda^2(SL(n))}$ $(n \geq 4)$	$1, 2, \ldots, \lfloor n/2 \rfloor$	$\lfloor n/2 \rfloor$
$\mathbf{SL(n) \otimes SL(m)}$ $(n, m \geq 2)$	$1, 2, \ldots, \min(n, m)$	$\min(n, m)$
$\mathbf{Sp(2n) \otimes SL(2)}$ $(n \geq 2)$	$1, 2, 2$	3
$\mathbf{Sp(2n) \otimes SL(3)}$ $(n \geq 2)$	$1, 2, 2, 3, 3, 4$	6
$\mathbf{Sp(4) \otimes SL(n)}$ $(n \geq 4)$	$1, 2, 2, 3, 4, 4$	6
$\mathbf{Spin(7)}$	$1, 2$	2
$\mathbf{Spin(9)}$	$1, 2, 2$	3
$\mathbf{Spin(10)}$	$1, 2$	2
$\mathbf{G_2}$	$1, 2$	2
$\mathbf{E_6}$	$1, 2, 3$	3

in Table 3 are imposed to eliminate redundancies caused by isomorphisms in low dimensions. Note that the actions $\mathbf{SO(n)}$, $\mathbf{S^2(SL(n))}$, $\mathbf{\Lambda^2(SL(n))}$ and $\mathbf{SL(n) \otimes SL(m)}$ were discussed in Section 4.

Table 4: Irreducible multiplicity free actions $\mathbf{G}' : V$, $\mathbf{G}'$ semisimple

Group
$\mathbf{SL(n)}$ $(n \geq 2)$
$\mathbf{Sp(2n)}$ $(n \geq 2)$
$\mathbf{\Lambda^2(SL(2m+1))}$ $(m \geq 2)$
$\mathbf{SL(n) \otimes SL(m)}$ $(n, m \geq 2, n \neq m)$
$\mathbf{Sp(4) \otimes SL(n)}$ $(n \geq 5)$

The scalars $\mathbb{C}^\times$ act on $\mathcal{P}_m(V)$ by the character $t \mapsto t^{-m}$. Thus $G' \times \mathbb{C}^\times : V$ is a multiplicity free action if and only if the representations of G' on each $\mathcal{P}_m(V)$ are multiplicity free. If we remove the scalars, then multiplicities can arise across different degrees of homogeneity in $\mathbb{C}[V]$. This happens, for example, when $G = SO(n, \mathbb{C}) \times \mathbb{C}^\times$, as discussed in Section 4.4. In fact, most of the actions from Table 3 fail to be multiplicity free when the scalars are removed. Table 4 lists the irreducible multiplicity free actions for semisimple groups up to geometric equivalence.

6.2. *Decomposable actions*

Given linear actions $G_1 : V_1$, $G_2 : V_2$, one can form the product action $G_1 \times G_2 : V_1 \oplus V_2$. If $G_1 : V_1$, $G_2 : V_2$ are both multiplicity free with decompositions $\mathbb{C}[V_j] = \bigoplus_{\lambda \in \Lambda_j} P_\lambda$, then $\mathbb{C}[V_1 \oplus V_2] \cong \mathbb{C}[V_1] \otimes \mathbb{C}[V_2]$ decomposes in a multiplicity free fashion

$$\mathbb{C}[V_1 \oplus V_2] = \bigoplus_{\lambda \in \Lambda_1, \lambda' \in \Lambda_2} P_\lambda \otimes P_{\lambda'}$$

under the action of $G_1 \times G_2$. Thus products of multiplicity free actions are multiplicity free.

One says that an action $G : V$ is *decomposable* if it is geometrically equivalent to a product action $G_1 \times G_2 : V_1 \oplus V_2$ with non-zero V_j. Otherwise, we say $G : V$ is *indecomposable.* Each irreducible multiplicity free action is indecomposable, but the converse is far from true. The action $\mathbf{GL(n)} \oplus_{\mathbf{GL(n)}} \mathbf{\Lambda^2(GL(n))}$ described in Section 4.5 provides one example of an indecomposable multiplicity free action that is not irreducible.

6.3. *Saturated indecomposable multiplicity free actions*

The classification of linear multiplicity free actions was completed independently by the authors [5] and Andrew Leahy [38]. Given a non-irreducible linear action $G : V$, one decomposes V as a direct sum of G-irreducible subspaces

$$V = V_1 \oplus \cdots \oplus V_m.$$

$G : V$ is said to be *saturated* if the image $\mathbf{G}$ of G in $GL(V)$ contains a full torus $(\mathbb{C}^\times)^m$. That is, the dimension of the center in $\mathbf{G}$ equals the number m of irreducible summands. Given $G : V$ one can always form a saturated action $G' \times (\mathbb{C}^\times)^m : V$. This action is multiplicity free if and only if the representation of G' on

$$\mathcal{P}_{k_1}(V_1) \otimes \cdots \otimes \mathcal{P}_{k_m}(V_m)$$

is multiplicity free for each $(k_1, \ldots, k_m)$.

The saturated indecomposable multiplicity free actions consist of the irreducible actions in Table 3 together with the actions listed in Table 5. Each entry in Table 5 denotes the image $\mathbf{G}'$ of the semisimple part G' of G

Table 5: Indecomposable non-irreducible saturated multiplicity free actions

Semisimple Group $\mathbf{G'}$ ($\mathbf{G'} \times (\mathbb{C}^\times)^2 : V$ is multiplicity free)	Degrees of fundamental highest weight vectors (rank)
$\mathbf{SL(n)} \oplus_{\mathbf{SL(n)}} \mathbf{SL(n)}$ $(n \geq 2)$	$1, 1, 2$ (3)
$\mathbf{SL(n)} \oplus_{\mathbf{SL(n)}} \mathbf{SL(n)}^*$ $(n \geq 3)$	$1, 1, 2$ (3)
$\mathbf{SL(n)} \oplus_{\mathbf{SL(n)}} \mathbf{\Lambda^2(SL(n))}$ $(n \geq 4)$	$1, 2, \ldots, \lfloor n/2 \rfloor$, $1, 2, \ldots, \lfloor (n+1)/2 \rfloor$ (n)
$\mathbf{SL(n)}^* \oplus_{\mathbf{SL(n)}} \mathbf{\Lambda^2(SL(n))}$ $(n \geq 4)$	$1, 2, \ldots, \lfloor n/2 \rfloor$, $1, 2, \ldots, \lfloor (n-1)/2 \rfloor$ $(n \dot{-} 1)$
$\mathbf{SL(n)} \oplus_{\mathbf{SL(n)}} (\mathbf{SL(n)} \otimes \mathbf{SL(m)})$ $(n, m \geq 2)$	$1, 2, \ldots, \min(n, m)$, $1, 2, \ldots, \min(n, m+1)$ $(\min(n, m) + \min(n, m+1))$
$\mathbf{SL(n)}^* \oplus_{\mathbf{SL(n)}} (\mathbf{SL(n)} \otimes \mathbf{SL(m)})$ $(n \geq 3, m \geq 2)$	$1, 2, \ldots, \min(n, m)$, $1, 2, \ldots, \min(n, m+1)$ $(\min(n, m) + \min(n, m+1))$
$\mathbf{SL(2)} \oplus_{\mathbf{SL(2)}} (\mathbf{SL(2)} \otimes \mathbf{Sp(2n)})$ $(n \geq 2)$	$1, 1, 2, 2, 2$ (5)
$(\mathbf{SL(n)} \otimes \mathbf{SL(2)}) \oplus_{\mathbf{SL(2)}} (\mathbf{SL(2)} \otimes \mathbf{SL(m)})$ $(n, m \geq 2)$	$1, 1, 2, 2, 2$ (5)
$(\mathbf{SL(n)} \otimes \mathbf{SL(2)}) \oplus_{\mathbf{SL(2)}} (\mathbf{SL(2)} \otimes \mathbf{Sp(2m)})$ $(n, m \geq 2)$	$1, 1, 2, 2, 2, 2$ (6)
$(\mathbf{Sp(2n)} \otimes \mathbf{SL(2)}) \oplus_{\mathbf{SL(2)}} (\mathbf{SL(2)} \otimes \mathbf{Sp(2m)})$ $(n, m \geq 2)$	$1, 1, 2, 2, 2, 2, 2$ (7)
$\mathbf{Sp(2n)} \oplus_{\mathbf{Sp(2n)}} \mathbf{Sp(2n)}$ $(n \geq 2)$	$1, 1, 2, 2$ (4)
$\mathbf{Spin(8)} \oplus_{\mathbf{Spin(8)}} \mathbf{SO(8)}$	$1, 1, 2, 2, 2$ (5)

in $GL(V)$. In each case, V has two irreducible summands, $V = V_1 \oplus V_2$, and one simple factor in $\mathbf{G'}$ is acting diagonally. Thus, for example, $\mathbf{SL(n)} \oplus_{\mathbf{SL(n)}} (\mathbf{SL(n)} \otimes \mathbf{SL(m)})$ denotes the image of $SL(n, \mathbb{C}) \times SL(m, \mathbb{C})$ under the representation on $V = V_1 \oplus V_2 = (\mathbb{C}^n) \oplus (\mathbb{C}^n \otimes \mathbb{C}^m)$ where $SL(n, \mathbb{C})$ acts diagonally on V_1 and V_2. $\mathbf{Spin(8)} \oplus_{\mathbf{Spin(8)}} \mathbf{SO(8)}$ denotes the image of the action of $Spin(8, \mathbb{C})$ on $\mathbb{C}^8 \oplus \mathbb{C}^8$ via the direct sum of the positive spin representation with the natural representation (via $SO(8, \mathbb{C})$). The notation $\mathbf{SL(n)}^*$ denotes the contragredient to the defining representation. For each group $\mathbf{G'}$ in Table 5 the saturated action $\mathbf{G'} \times (\mathbb{C}^\times)^2 : V$ is multiplicity free. Together, Tables 3 and 5 classify all saturated indecomposable multiplicity free actions up to geometric equivalence. In particular, the classification shows that a saturated indecomposable multiplicity free action can have at most two irreducible summands.

6.4. *Non-saturated indecomposable multiplicity free actions*

Only one entry in Table 5 remains multiplicity free when the torus $(\mathbb{C}^\times)^2$ is removed, namely

$$(\mathbf{SL(n)} \otimes \mathbf{SL(2)}) \oplus_{\mathbf{SL(2)}} (\mathbf{SL(2)} \otimes \mathbf{SL(m)}) \quad \text{for } n, m \geq 3$$

In addition, for each group $\mathbf{G}'$ in Table 5 one can consider the joint action of $\mathbf{G}' \times \mathbb{C}^\times$ where $\mathbb{C}^\times$ acts on $V = V_1 \oplus V_2$ via

$$t \cdot (v_1, v_2) = (t^a v_1, t^b v_2)$$

for some integers a, b. Such actions are multiplicity free in the cases listed in Table 6.

Table 6: Non-saturated multiplicity free actions

$(\mathbf{SL(n)} \otimes \mathbf{SL(2)}) \oplus_{\mathbf{SL(2)}} (\mathbf{SL(2)} \otimes \mathbf{SL(m)})$	$n, m \geq 3$
$\mathbf{SL(n)} \oplus_{\mathbf{SL(n)}} \mathbf{SL(n)}$	$n \geq 3, a \neq b$
$\mathbf{SL(n)}^* \oplus_{\mathbf{SL(n)}} \mathbf{SL(n)}$	$n \geq 3, a \neq -b$
$\mathbf{SL(2m)} \oplus_{\mathbf{SL(2m)}} \mathbf{\Lambda^2(SL(2m))}$	$m \geq 2, b \neq 0$
$\mathbf{SL(2m+1)} \oplus_{\mathbf{SL(2m+1)}} \mathbf{\Lambda^2(SL(2m+1))}$	$m \geq 2, a \neq -mb$
$\mathbf{SL(2m)}^* \oplus_{\mathbf{SL(2m)}} \mathbf{\Lambda^2(SL(2m))}$	$m \geq 2, b \neq 0$
$\mathbf{SL(2m+1)}^* \oplus_{\mathbf{SL(2m+1)}} \mathbf{\Lambda^2(SL(2m+1))}$	$m \geq 2, a \neq mb$
$\mathbf{SL(n)} \oplus_{\mathbf{SL(n)}} (\mathbf{SL(n)} \otimes \mathbf{SL(m)})$	$2 \leq n < m, a \neq 0$
$\mathbf{SL(n)} \oplus_{\mathbf{SL(n)}} (\mathbf{SL(n)} \otimes \mathbf{SL(m)})$	$m \geq 2, n \geq m+2, a \neq b$
$\mathbf{SL(n)}^* \oplus_{\mathbf{SL(n)}} (\mathbf{SL(n)} \otimes \mathbf{SL(m)})$	$2 \leq n < m, a \neq 0$
$\mathbf{SL(n)}^* \oplus_{\mathbf{SL(n)}} (\mathbf{SL(n)} \otimes \mathbf{SL(m)})$	$m \geq 2, n \geq m+2, a \neq -b$
$(\mathbf{SL(2)} \otimes \mathbf{SL(2)}) \oplus_{\mathbf{SL(2)}} (\mathbf{SL(2)} \otimes \mathbf{SL(m)})$	$m \geq 3, a \neq 0$
$(\mathbf{SL(n)} \otimes \mathbf{SL(2)}) \oplus_{\mathbf{SL(2)}} (\mathbf{SL(2)} \otimes \mathbf{Sp(2m)})$	$n \geq 3, m \geq 2, b \neq 0$

6.5. *Completing the classification*

Let G be a connected complex algebraic *reductive* group acting on V via some rational representation π. The commutator subgroup G' of G is semi-simple and, by lifting to a finite covering if necessary, we can suppose that

$G = G' \times A$ where A is some algebraic torus. Write V as a direct sum

$$V = W_1 \oplus W_2 \oplus \cdots \oplus W_r$$

of $\pi(G)$-invariant subspaces W_j which are indecomposable under the action of G'. Letting π_j denote the action of G on W_j, we have $\pi(G') = \mathbf{G_1}' \times \mathbf{G_2}' \times \cdots \times \mathbf{G_r}'$ acting on V via the product action, where $\mathbf{G_j}' = \pi_j(G')$.

Theorem 6.5.1: *([5]) $G : V$ is a multiplicity free action if and only if A contains a direct product of the form $A_1 \times A_2 \times \cdots \times A_r$ where each A_j is a torus of dimension at most 2 and the actions $\mathbf{G_j}' \times A_j : W_j$ are multiplicity free for $j = 1, \ldots, r$.*

One can choose the A_j's in Theorem 6.5.1 to be minimal, in the sense that the action of $\mathbf{G_j}' \times B$ on W_j fails to be multiplicity free for all proper subgroups B of A_j. Some factors A_j can be trivial and $A_1 \times \cdots \times A_r$ need not act on $W_1 \oplus \cdots \oplus W_r$ via a product action. That is, the A_j's can act diagonally on the indecomposable G'-summands W_j.

Example 6.5.2: As an example, consider $G' = SL(n) \times SL(n)$ for $n \geq 2$ acting on $V = W_1 \oplus W_2 = (\mathbb{C}^n \oplus \mathbb{C}^n) \oplus (\mathbb{C}^n \oplus \mathbb{C}^n)$ via

$$(g, h) \cdot (x_1, y_1, x_2, y_2) = (gx_1, gy_1, hx_2, hy_2)$$

for $g, h \in SL(n)$ and $x_j, y_j \in \mathbb{C}^n$. Here $\mathbf{G_1}' = \mathbf{SL(n)} \oplus_{\mathbf{SL(n)}} \mathbf{SL(n)} = \mathbf{G_2}'$ appear in Table 5. Let $A = (\mathbb{C}^\times)^3$ act on V via

$$(t_1, t_2, t_3) \cdot (x_1, y_1, x_2, y_2) = (t_1 t_2 x_1, t_1^2 t_2 y_1, t_1 t_2 x_2, t_1 t_2^2 y_2) \qquad \text{for } t_j \in \mathbb{C}^\times.$$

The joint action of $G' \times A$ on V is multiplicity free. Indeed, if we let $A_1 = \mathbb{C}^\times \times \{1\} \times \{1\}$ and $A_2 = \{1\} \times \mathbb{C}^\times \times \{1\}$ then we see that the actions $\mathbf{G_j}' \times A_j : W_j$ appear in Table 6. Here, however, one can't find subtori A_j as in Theorem 6.5.1 that act independently on W_1 and W_2.

Theorem 6.5.1 completes the classification of multiplicity free actions because we have exhibited all of the possibilities for the groups $\mathbf{G_j}' \subset GL(W_j)$ and for the actions of the A_j's on the W_j's. More precisely, if $G : V$ is multiplicity free then for each $1 \leq j \leq r$ we must have either:

(1) W_j is $\mathbf{G_j}'$ irreducible and

  (a) $\pi_j(A_j) = \mathbb{C}^\times$ and $\mathbf{G_j}'$ appears in Table 3, or

  (b) $A_j = \{1\}$ and $\mathbf{G_j}'$ appears in Table 4.

(2) W_j is a sum of two $\mathbf{G_j}'$-irreducible subspaces and

  (a) $\pi_j(A_j) = (\mathbb{C}^\times)^2$ and $\mathbf{G_j}'$ appears in Table 5, or

(b) $\mathbf{G_j}'$ appears together with $\pi_j(A_j) \cong \mathbb{C}^\times$ in Table 6, or
(c) $A_j = \{1\}$ and $\mathbf{G_j}' = (\mathbf{SL(n)} \otimes \mathbf{SL(2)}) \oplus_{\mathbf{SL(2)}} (\mathbf{SL(2)} \otimes \mathbf{SL(m)})$ with $n, m \geq 3$.

6.6. *Proof outline*

The saturated indecomposable multiplicity free actions in Table 5 are obtained from the irreducible actions in Table 3 by extensive case-by-case analysis. Suppose that $G = G'$ is semisimple and that (π, V) is an indecomposable action with $V = V_1 \oplus V_2$, a direct sum of two G-irreducible subspaces. Let $\pi_j = \pi|V_j$ and $\mathbf{G_j} = \pi_j(G) \subset GL(V_j)$ for $j = 1, 2$. If $G \times (\mathbb{C}^\times)^2 : V$ is multiplicity free then it is clear that both $G \times \mathbb{C}^\times : V_1$ and $G \times \mathbb{C}^\times : V_2$ must be multiplicity free. Thus both $\mathbf{G_1}$ and $\mathbf{G_2}$ appear in Table 3.

Define normal subgroups $\mathbf{K_1}$ and $\mathbf{K_2}$ of $\mathbf{G_1}$ and $\mathbf{G_2}$ by

$$\mathbf{K_1} := \pi_1(\mathrm{Ker}(\pi_2)) \quad \text{and} \quad \mathbf{K_2} := \pi_2(\mathrm{Ker}(\pi_1)). \tag{16}$$

The map

$$F : \mathbf{G_1}/\mathbf{K_1} \to \mathbf{G_2}/\mathbf{K_2} \tag{17}$$

given by $F(\pi_1(g)\mathbf{K_1}) := \pi_2(g)\mathbf{K_2}$ is a well-defined group isomorphism.

If $\mathbf{K_1} = \mathbf{G_1}$ then it follows that $\mathbf{K_2} = \mathbf{G_2}$ and we have $\pi(G) = \mathbf{G_1} \times \mathbf{G_2}$. In this case, our action decomposes as a direct product of the multiplicity free actions $\mathbf{G_j} \times (\mathbb{C}^\times)$.

Next suppose that the $\mathbf{K_j}$'s are proper subgroups of the $\mathbf{G_j}$'s. Note that $\mathbf{K_j}$ need not be connected. We write $\mathbf{K_j^\circ}$ for the identity component in $\mathbf{K_j}$. Since $\mathbf{K_j}$ is a normal subgroup of $\mathbf{G_j}$, so is $\mathbf{K_j^\circ}$. As $\mathbf{G_j}$ appears in Table 3, $\mathbf{G_j}$ is either a simple group or a product of two simple factors. Thus we can write $\mathbf{G_j} = \mathbf{K_j^\circ H_j}$, where either $\mathbf{H_j} = \mathbf{G_j}$ (when $\mathbf{K_j^\circ} = \{e\}$) or $\mathbf{H_j}$ is one of two simple factors in $\mathbf{G_j}$. As $\mathbf{H_j} \cong \mathbf{G_j}/\mathbf{K_j^\circ}$ covers $\mathbf{G_j}/\mathbf{K_j}$, the derivative of F yields an isomorphism between the Lie algebras of $\mathbf{H_1}$ and $\mathbf{H_2}$. In fact, Table 3 shows that $\mathbf{H_j}$ is simply connected except when $\mathbf{H_j} = \mathbf{SO(n)}$. Thus, we can realize F as a group isomorphism $\mathbf{H_1} \cong \mathbf{H_2}$ or as a covering of one of the $\mathbf{H_j}$'s by the other. We write $F : \mathbf{H_1} \to \mathbf{H_2}$ for this map after interchanging the roles of $\mathbf{G_1}$ and $\mathbf{G_2}$ when $\mathbf{H_2}$ covers $\mathbf{H_1}$. If we define a new group L by $L = \mathbf{K_1^\circ} \times \mathbf{H_1} \times \mathbf{K_2^\circ}$ and a representation σ of L on $V = V_1 \oplus V_2$ by

$$\sigma(k_1, h_1, k_2) := (k_1 h_1, F(h_1) k_2) \in GL(V_1) \times GL(V_2)$$

then we see that $\pi(G) = \sigma(L) = \mathbf{G_1} \oplus_{\mathbf{H_1}} \mathbf{G_2}$. That is, $\mathbf{H_1}$ acts diagonally on V_1 and V_2 via F.

In summary, we have shown that any saturated indecomposable multiplicity free action with two irreducible summands is obtained from two entries in Table 3 via diagonalization along simple factors. Combining pairs of entries in Table 3 which share a common simple factor yields a large number of indecomposable actions. It is necessary to examine each of these in turn to determine which are multiplicity free. Various techniques can be applied in each case. One can look for an open Borel orbit. (Many candidates can be eliminated easily because the underlying vector space has dimension greater than that of a Borel subgroup.) One can apply the Littlewood-Richardson rules (and variants for the classical groups $SO(n, \mathbb{C})$, $Sp(2n, \mathbb{C})$) or use the recursive criterion from Section 5. Table 5 is the end result of this analysis.

If $V = V_1 \oplus \cdots \oplus V_r$ is a sum of r irreducible subspaces, then the above reasoning applies to each pair of groups $\mathbf{G_i} = \pi_i(G)$, $\mathbf{G_j} = \pi_j(G)$. For the action to be indecomposable, at least one simple factor in each $\mathbf{G_i}$ must act diagonally on at least one V_j with $j \neq i$. Diagonalization from Tables 3 and 5 produces indecomposable actions with more than two irreducible summands. The analysis shows, however, that no such actions are multiplicity free.

6.7. *Section 6 notes*

The articles [7] and [31] contain formulas for the fundamental highest weight vectors and further detailed information for each action in the classification. For the irreducible actions, much of this is due to Howe and Umeda [24].

7. Invariant polynomials and differential operators

7.1. *Polynomial coefficient differential operators*

For each $f \in \mathbb{C}[V]$ we have the multiplication operator $M_f \in End(\mathbb{C}[V])$,

$$M_f(h) = fh.$$

Let $\mathcal{P}(V) = \{M_f \; : \; f \in \mathbb{C}[V]\}$, so $\mathcal{P}(V) \cong \mathbb{C}[V]$. For each $v \in V$ we have the directional derivative $\partial_v \in End(\mathbb{C}[V])$,

$$(\partial_v h)(w) = \lim_{t \to 0} \frac{f(w + tv) - f(w)}{t}.$$

The algebra $\mathcal{D}(V)$ generated by $\{\partial_v \; : \; v \in V\}$ is the algebra of constant coefficient differential operators. The embedding $V \hookrightarrow \mathcal{D}(V)$, $v \mapsto \partial_v$ extends to an isomorphism from the symmetric algebra $S(V)$ to $\mathcal{D}(V)$. Thus

$$\mathcal{D}(V) \cong S(V) \cong \mathbb{C}[V^*]$$

as algebras.

We let $\mathcal{PD}(V)$ be the subalgebra of $End(\mathbb{C}[V])$ generated by $\mathcal{P}(V)$ and $\mathcal{D}(V)$. This is the algebra of *polynomial coefficient differential operators.* The product rule for derivatives shows that the map

$$\mathcal{P}(V) \otimes \mathcal{D}(V) \to \mathcal{PD}(V), \quad p \otimes L \mapsto pL$$

given by multiplication is a vector space isomorphism. Composing this with the algebra isomorphisms

$$\mathbb{C}[V \oplus V^*] \cong \mathbb{C}[V] \otimes \mathbb{C}[V^*] \cong \mathcal{P}(V) \otimes \mathcal{D}(V)$$

produces a vector space isomorphism

$$\delta : \mathbb{C}[V \oplus V^*] \to \mathcal{PD}(V), \quad p \mapsto p(z, \partial).$$

We say that $p \in \mathbb{C}[V \oplus V^*]$ is the *Wick symbol* for the operator $p(z, \partial)$. The inverse map

$$\sigma : \mathcal{PD}(V) \to \mathbb{C}[V \oplus V^*]$$

of δ is called the *polarized symbol map.*

The above discussion can be made more concrete by introducing coordinates. Let $(z_1, \ldots, z_n)$ be coordinates on V with respect to some basis $\{e_1, \ldots, e_n\}$ and $(w_1, \ldots, w_n)$ be coordinates on V^* with respect to the dual basis $\{e_1^*, \ldots, e_n^*\}$. We have

$$\mathbb{C}[V] = \mathbb{C}[z_1, \ldots, z_n], \quad \mathbb{C}[V \oplus V^*] = \mathbb{C}[z_1, \ldots, z_n, w_1, \ldots, w_n].$$

The monomials

$$z^\alpha = z_1^{\alpha_1} \cdots z_n^{\alpha_n}$$

form a basis for $\mathbb{C}[V]$ and the differential operators

$$\partial^\alpha = \partial_1^{\alpha_1} \cdots \partial_n^{\alpha_n}, \quad \left(\partial_j = \frac{\partial}{\partial z_j}\right)$$

form a basis for $\mathcal{D}(V)$ as $\alpha = (\alpha_1, \ldots, \alpha_n)$ ranges over all multi-indices $\alpha \in \mathbb{N}^n$. The operator $p(z, \partial)$ with Wick symbol

$$p(z, w) = \sum_{\alpha, \beta} c_{\alpha,\beta} z^\alpha w^\beta.$$

is

$$p(z, \partial) = \sum_{\alpha,\beta} c_{\alpha,\beta} z^\alpha \partial^\beta.$$

The symbol mapping is not an algebra isomorphism because $\mathbb{C}[V \oplus V^*]$ is abelian whereas $\mathcal{PD}(V)$ is not. In fact, one has the Heisenberg commutation relations

$$[\partial_i, z_j] = \delta_{i,j}$$

in $\mathcal{PD}(V)$. To obtain an algebra isomorphism, one can pass to the associated graded algebras. The algebra $\mathbb{C}[V \oplus V^*]$ is filtered by

$$\mathbb{C}^{(k)}[V \oplus V^*] = \sum_{a+b \leq k} \mathcal{P}_a(V) \otimes \mathcal{P}_b(V^*).$$

Then

$$\mathcal{PD}^{(k)}(V) = \delta(\mathbb{C}^{(k)}[V \oplus V^*])$$

gives a filtration of the algebra $\mathcal{PD}(V)$. That is

$$\mathbb{C} = \mathcal{PD}^{(0)} \subset \mathcal{PD}^{(1)} \subset \cdots \subset \mathcal{PD}^{(k)}(V) \subset \cdots, \qquad \bigcup_{k=0}^{\infty} \mathcal{PD}^{(k)}(V) = \mathcal{PD}(V),$$

and

$$\mathcal{PD}^{(k)}(V)\mathcal{PD}^{(\ell)}(V) \subset \mathcal{PD}^{(k+\ell)}(V)$$

in view of the commutation relations. In terms of our basis,

$$\mathcal{PD}^{(k)}(V) = Span\{z^\alpha \partial^\beta \ : \ |\alpha| + |\beta| \leq k\}$$

where $|\alpha| = \alpha_1 + \cdots + \alpha_n$. The map $\delta : \mathbb{C}[V \oplus V^*] \to \mathcal{PD}(V)$ induces an algebra isomorphism from $gr(\mathbb{C}[V \oplus V^*]) \cong \mathbb{C}[V \oplus V^*]$ to

$$gr(\mathcal{PD}(V)) = \sum_{k=0}^{\infty} \mathcal{PD}^{(k+1)}(V)/\mathcal{PD}^{(k)}(V),$$

an abelian algebra that is canonically isomorphic to $\mathcal{PD}(V)$ as a vector space.

Lemma 7.1.1: *$\mathcal{PD}(V)$ is strongly dense in $End(\mathbb{C}[V])$. That is,*

$$\mathcal{PD}(V)|\mathcal{X} = Hom(\mathcal{X}, \mathbb{C}[V])$$

for any finite dimensional subspace $\mathcal{X}$ of $\mathbb{C}[V]$.

Proof: Let $\mathcal{X} \subset \mathbb{C}[V] = \mathbb{C}[z_1, \ldots, z_n]$ be a finite dimensional subspace and $T \in Hom(\mathcal{X}, \mathbb{C}[V])$. Let $\{p_1, \ldots, p_m\}$ be a basis for $\mathcal{X}$, ordered so that $deg(p_1) \leq \cdots \leq deg(p_m)$. By taking suitable linear combinations, one can ensure that p_k contains a monomial m_k which

- has degree $deg(p_k)$ and
- $p_1, \ldots, p_{k-1}$ contain no non-zero multiples of m_k.

Let $f_j = T(p_j)$ for $1 \leq j \leq m$, and let $m_1 = z^\alpha$. Then the operator

$$D_1 = f_1 \frac{1}{\alpha!} \partial^\alpha \quad (\alpha! = \alpha_1! \cdots \alpha_n!)$$

has $D_1(p_1) = f_1$.

Assume inductively that there is some $D_k \in \mathcal{PD}(V)$ with $D_k(p_j) = f_j$ for $1 \leq j \leq k$. Let $m_{k+1} = z^\beta$. Then $\partial^\beta p_1 = \cdots = \partial^\beta p_k = 0$, and $\partial^\beta p_{k+1} = \beta!$ Now

$$D_{k+1} = \left(f_{k+1} - D_k p_{k+1}\right) \frac{1}{\beta!} \partial^\beta + D_k$$

satisfies $D_{k+1}(p_j) = f_j$ for $1 \leq j \leq k+1$. By induction we conclude that there is an operator $D \in \mathcal{PD}(V)$ with $D|\mathcal{X} = T$. □

7.2. *Invariants in $\mathcal{PD}(V)$*

Now suppose that a reductive algebraic group G is acting linearly on V. For the moment, we do not assume that $G : V$ is a multiplicity free action. The group G acts on $\mathcal{PD}(V)$ via conjugation:

$$(g \cdot D)(f) = g \cdot D(g^{-1} \cdot f).$$

In terms of coordinates one calculates:

$$g \cdot M_{z_j} = \sum_\ell (g^{-1})_{j\ell} M_{z_\ell}, \quad g \cdot \partial_j = \sum_\ell (g^t)_{j\ell} \partial_\ell.$$

It follows that the polarized symbol mapping intertwines the action of G on $\mathcal{PD}(V)$ with its action on $\mathbb{C}[V \oplus V^*] = \mathbb{C}[z_1, \ldots, z_n, w_1, \ldots, w_n]$ via

$$g \cdot p(z, w) = p(g^{-1} z, g^t w).$$

This formula agrees with that for the representation of G on $\mathbb{C}[V \oplus V^*]$ arising from $G : V$ and its contragredient $G : V^*$,

$$g \cdot (v, \xi) = (g \cdot v, g \cdot \xi) = (g \cdot v, \xi \circ g^{-1}).$$

Thus we have shown:

Lemma 7.2.1: *The G-invariants in* $\mathcal{PD}(V)$ *are*

$$\mathcal{PD}(V)^G = \mathcal{PD}(V) \cap End_G(\mathbb{C}[V]) = \{p(z,\partial) \;:\; p \in \mathbb{C}[V \oplus V^*]^G\}.$$

Let K be a maximal compact connected subgroup in G. The Unitarian Trick ensures that an operator $D \in \mathcal{PD}(V)$ is G-invariant if and only if it is K-invariant. Now we define the K-average of D via

$$D^\natural = \int_K (k \cdot D) dk$$

where dk denotes normalized Haar measure on K. The operator D belongs to a finite dimensional subspace $\mathcal{PD}^{(m)}(V)$ for some m, and the integral converges in $\mathcal{PD}^{(m)}(V)$. So $D^\natural$ is a polynomial coefficient differential operator. A change of variables and unimodularity of K shows that $k_\circ \cdot D^\natural = D^\natural$ for each $k_\circ \in K$. Thus $D^\natural \in \mathcal{PD}(V)^G$.

Lemma 7.2.2: $\mathcal{PD}(V)^G$ *acts irreducibly on* $\mathbb{C}[V]^{B,\lambda}$ *for all dominant weights* λ.

Proof: For $D \in \mathcal{PD}(V)^G$, $b \in B$ and h a λ-highest weight vector, we have

$$b \cdot (Dh) = D(b \cdot h) = D(b^\lambda h) = b^\lambda Dh.$$

Thus $\mathcal{PD}(V)^G$ preserves the space $\mathbb{C}[V]^{B,\lambda}$ of λ-highest weight vectors.

Next let h_1, h_2 be two λ-highest weight vectors in $\mathbb{C}[V]$ and consider the finite dimensional G-invariant space

$$\mathcal{X} = Span(G \cdot h_1) + Span(G \cdot h_2).$$

Since $Span(G \cdot h_1)$ and $Span(G \cdot h_2)$ are equivalent as G-modules, there is some $T \in End_G(\mathcal{X})$ with $T(h_1) = h_2$. By Lemma 7.1.1 there is an operator $D \in \mathcal{PD}(V)$ with $D|\mathcal{X} = T$. Now $D^\natural \in \mathcal{PD}(V)^G$ and

$$\begin{aligned} D^\natural h_1 &= \int_K k \cdot D(k^{-1} \cdot h_1) dk \\ &= \int_K k \cdot T(k^{-1} \cdot h_1) dk \quad (\text{as } k^{-1} \cdot h_1 \in \mathcal{X}) \\ &= \int_K T(h_1) dk \quad (\text{as } T \text{ is } G\text{-invariant}) \\ &= T(h_1) \\ &= h_2. \end{aligned}$$

This shows that $\mathcal{PD}(V)^G$ acts irreducibly on $\mathbb{C}[V]^{B,\lambda}$. □

Theorem 7.2.3: *$G : V$ is a multiplicity free action if and only if $\mathcal{PD}(V)^G$ is abelian.*

Proof: Let $G : V$ be a multiplicity free action and

$$\mathbb{C}[V] = \bigoplus_{\lambda \in \Lambda} P_\lambda$$

be the decomposition of $\mathbb{C}[V]$ into pair-wise inequivalent G-irreducibles. Schur's Lemma ensures that any operator $D \in \mathcal{PD}(V)^G$ must preserve each P_λ and acts on P_λ as multiplication by some scalar. It follows that any two operators in $\mathcal{PD}(V)^G$ commute.

Conversely, suppose that $\mathcal{PD}(V)^G$ is abelian. As $\mathcal{PD}(V)^G$ acts irreducibly on $\mathbb{C}[V]^{B,\lambda}$ we must have $dim(\mathbb{C}[V]^{B,\lambda}) \leq 1$ for all dominant weights λ. Hence $G : V$ is multiplicity free. □

Thus when $G : V$ is multiplicity free, both algebras $\mathbb{C}[V \oplus V^*]^G$ and $\mathcal{PD}(V)^G$ are abelian. Although δ and σ are not algebra maps, they induce algebra isomorphisms

$$\mathbb{C}[V \oplus V^*]^G \cong gr\big(\mathbb{C}[V \oplus V^*]^G\big) \cong gr\big(\mathcal{PD}(V)^G\big)$$

between the associated graded algebras. Concretely, this means that although one generally has $p(z,\partial)q(z,\partial) \neq (pq)(z,\partial)$, the operators $p(z,\partial)q(z,\partial)$ and $(pq)(z,\partial)$ have the same top degree terms. Here "degree" in $\mathcal{PD}(V)$ is defined using the filtration $\mathcal{PD}^{(k)}(V)$ from Section 7.1. In particular, $z^\alpha \partial^\beta$ has degree $|\alpha| + |\beta|$.

7.3. *A canonical basis for the invariants*

Suppose that $G : V$ is a (linear) multiplicity free action. The trivial representation of G occurs in $\mathbb{C}[V]$ on $\mathcal{P}_0(V) = \mathbb{C}$, the constant polynomials. As the representation of G on $\mathbb{C}[V]$ is multiplicity free, it follows that $\mathbb{C}[V]^G = \mathbb{C}$. That is, there are no non-constant G-invariants in $\mathbb{C}[V]$. Because of the connection with G-invariant differential operators it is, however, of interest to study the G-invariants in $\mathbb{C}[V \oplus V^*]$.

Let

$$\mathbb{C}[V] = \bigoplus_{\lambda \in \Lambda} P_\lambda$$

denote the multiplicity free decomposition of $\mathbb{C}[V]$ under the action of G. Then

$$\mathbb{C}[V \oplus V^*] = \mathbb{C}[V] \otimes \mathbb{C}[V]^* = \bigoplus_{\lambda,\lambda' \in \Lambda} P_\lambda \otimes P_{\lambda'}^*,$$

where the subspaces $P_\lambda \otimes P^*_{\lambda'}$ are G-invariant in $\mathbb{C}[V \oplus V^*]$. Thus

$$\mathbb{C}[V \oplus V^*]^G = \bigoplus_{\lambda,\lambda' \in \Lambda} \left(P_\lambda \otimes P^*_{\lambda'}\right)^G.$$

But

$$\left(P_\lambda \otimes P^*_{\lambda'}\right)^G \cong \left(Hom(P_{\lambda'}, P_\lambda)\right)^G = Hom_G(P_{\lambda'}, P_\lambda) = \begin{cases} 0 \text{ if } \lambda \neq \lambda' \\ \mathbb{C} \text{ if } \lambda = \lambda' \end{cases}$$

by Schur's Lemma. The first isomporphism is given by $f \otimes \xi \mapsto (p \mapsto \xi(p)f)$. The element in $P_\lambda \otimes P^*_\lambda$ that corresponds to I_{P_λ} under the isomorphism $P_\lambda \otimes P^*_\lambda \cong Hom(P_\lambda, P_\lambda)$ is

$$\widetilde{p}_\lambda = \sum_{j=1}^{d_\lambda} f_j \otimes f_j^* \tag{18}$$

where $d_\lambda = dim(P_\lambda)$ and $\{f_j \ : \ 1 \leq j \leq d_\lambda\}$ is any basis for P_λ with dual basis $\{f_j^*\}$. Thus we have shown that

$$\mathbb{C}[V \oplus V^*]^G = \bigoplus_{\lambda \in \Lambda} \left(P_\lambda \otimes P^*_\lambda\right)^G = \bigoplus_{\lambda \in \Lambda} \mathbb{C}\widetilde{p}_\lambda.$$

So $\{\widetilde{p}_\lambda \mid \lambda \in \Lambda\}$ is a basis for $\mathbb{C}[V \oplus V^*]^G$. As Equation 18 does not depend on the choice of basis $\{f_j\}$ for P_λ, the basis $\{\widetilde{p}_\lambda \mid \lambda \in \Lambda\}$ for $\mathbb{C}[V \oplus V^*]^G$ is canonical. Applying Wick quantization we obtain a canonical basis for $\mathcal{PD}(V)^G$.

We call the polynomials $\widetilde{p}_\lambda$ the *unnormalized canonical invariants*. To achieve some simplification in formulae to be derived below, we also introduce the (*normalized*) *canonical invariants*

$$p_\lambda = \frac{1}{d_\lambda}\widetilde{p}_\lambda, \quad (d_\lambda = dim(P_\lambda)).$$

In summary we have proved the following.

Proposition 7.3.1: *$\{p_\lambda \ : \ \lambda \in \Lambda\}$ and $\{p_\lambda(z, \partial) \ : \ \lambda \in \Lambda\}$ are canonical vector space bases for $\mathbb{C}[V \oplus V^*]^G$ and $\mathcal{PD}(V)^G$ respectively.*

7.4. *The fundamental invariants*

Now let r be the rank of the multiplicity free action $G : V$ and

$$\Lambda' = \{\lambda_1, \ldots, \lambda_r\}$$

be the set of fundamental highest weights. Recall that $\Lambda = \{m_1\lambda_1 + \cdots + m_r\lambda_r \mid m \in \mathbb{N}^r\}$. (See Proposition 3.3.1.)

Definition 7.4.1: The *fundamental invariants* for $G : V$ are $\{\gamma_1, \ldots, \gamma_r\}$ where

$$\gamma_j = p_{\lambda_j}.$$

For $\lambda \in \Lambda$ let $|\lambda| \in \mathbb{N}$ denote the degree of homogeneity of P_λ. That is, $P_\lambda \subset \mathcal{P}_{|\lambda|}(V)$. Then the canonical invariant p_λ is homogeneous of degree $2|\lambda|$.

For any weights $\mu, \nu \in \mathfrak{h}^*$ we will write

$$\mu \prec \nu$$

when $\nu - \mu$ is a sum of positive roots.

Lemma 7.4.2: *For any $\lambda, \mu \in \Lambda$ there are values $c_\nu = c_{\lambda,\mu,\nu}$ for which*

$$p_\lambda p_\mu = \sum_\nu c_\nu p_\nu,$$

where the sum is over all $\nu \in \Lambda$ with $|\nu| = |\lambda| + |\mu|$ and $\nu \preceq \lambda + \mu$. Moreover, $c_{\lambda+\mu} \neq 0$.

Proof: The product $p_\lambda p_\mu$ is G-invariant and belongs to $\mathcal{P}_{|\lambda|+|\mu|}(V) \otimes \mathcal{P}_{|\lambda|+|\mu|}(V^*)$. As the p_ν's form a homogeneous basis for $\mathbb{C}[V \oplus V^*]^G$ we conclude that

$$p_\lambda p_\mu = \sum_{|\nu|=|\lambda|+|\mu|} c_\nu p_\nu$$

for some values c_ν. Let $\{f_j\}_{j=1}^{d_\lambda}$ and $\{h_j\}_{j=1}^{d_\mu}$ be bases of weight vectors for P_λ and P_μ so that f_1, h_1 are highest weight vectors. We know that all other weights in an irreducible representation space precede the highest weight in the partial ordering defined above.

We have

$$p_\lambda p_\mu = \frac{1}{d_\lambda d_\mu} \sum_{i,j} f_i h_j \otimes f_i^* h_j^*.$$

The $\mathbb{C}[V]$-components $f_i h_j$ in this sum are weight vectors with weights $\lambda_i + \mu_j \preceq \lambda + \mu$. It follows that $c_\nu = 0$ unless $\nu \prec \lambda + \mu$. Moreover, the term $f_1 h_1 \otimes f_1^* h_1^*$ contains the $(\lambda + \mu)$-highest weight vector $f_1 h_1$. We conclude that $c_{\lambda+\mu} \neq 0$. □

Theorem 7.4.3: $\mathbb{C}[V \oplus V^*]^G = \mathbb{C}[\gamma_1, \ldots, \gamma_r]$. *That is, $\mathbb{C}[V \oplus V^*]^G$ is a polynomial ring freely generated by the fundamental invariants.*

Proof: Given $m \in \mathbb{N}^r$, let $\lambda = m_1\lambda_1 + \cdots + m_r\lambda_r \in \Lambda$. Lemma 7.4.2 shows that

$$\gamma^m = \gamma_1^{m_1} \cdots \gamma_r^{m_r} = p_{\lambda_1}^{m_1} \cdots p_{\lambda_r}^{m_r} = a_\lambda p_\lambda + \sum_{\nu \prec \lambda} a_\nu p_\nu$$

for some coefficients a_ν with $a_\lambda \neq 0$. As $\{p_\lambda\}$ is a basis for $\mathbb{C}[V \oplus V^*]^G$, we conclude that $\{\gamma^m \mid m \in \mathbb{N}^r\}$ is also a basis for $\mathbb{C}[V \oplus V^*]^G$. □

Corollary 7.4.4: *$\mathcal{PD}(V)^G$ is a polynomial ring freely generated by* $\{D_j = \gamma_j(z, \partial) \; : \; 1 \leq j \leq r\}$.

Proof: From Theorem 7.4.3 we see that $\{\gamma^m(z, \partial) \; : \; m \in \mathbb{N}^r\}$ is a basis for the vector space $\mathcal{PD}(V)^G$. Also, given $m = (m_1, \ldots, m_r)$,

$$D^m = D_1^{m_1} \cdots D_r^{m_r} = \gamma_1(z, \partial)^{m_1} \cdots \gamma_r(z, \partial)^{m_r}$$

differs from $\gamma^m(z, \partial)$ by an element of $\mathcal{PD}^{(2|\lambda|-1)}(V)$, where $\lambda = m_1\lambda_1 + \cdots m_r\lambda_r$. By induction on degree in $\mathcal{PD}(V)$ we conclude that $\{D^m \; : \; m \in \mathbb{N}^r\}$ is a vector space basis for $\mathcal{PD}(V)^G$. Thus $\mathcal{PD}(V)^G = \mathbb{C}[D_1, \ldots, D_r]$.□

7.5. *The algebra* $\mathbb{C}[V_\mathbb{R}]^K$

An alternative viewpoint on $\mathbb{C}[V \oplus V^*]^G$ will prove useful. Let K denote a maximal compact connected Lie subgroup of G and $\langle \cdot, \cdot \rangle$ be any K-invariant positive definite Hermitian inner product on V. The conjugate-linear vector space isomorphism

$$V \to V^*, \quad v \mapsto v^* = \langle \cdot, v \rangle$$

is K-equivariant (but not G-equivariant). In view of the Unitarian Trick we obtain an algebra isomorphism

$$\mathbb{C}[V_\mathbb{R}]^K = \mathbb{C}[V \oplus \overline{V}]^K \cong \mathbb{C}[V \oplus V^*]^K = \mathbb{C}[V \oplus V^*]^G. \tag{19}$$

Here $\overline{V}$ denotes V with the conjugate complex structure and $V_\mathbb{R}$ is the underlying real vector space for V.

Introducing coordinates $(z_1, \ldots, z_n)$ on V with respect to an *orthonormal* basis, one has $\mathbb{C}[V] = \mathbb{C}[z_1, \ldots, z_n]$. This polynomial ring also carries an inner product, namely

$$\langle p, q \rangle_\mathcal{F} = \big(p(\partial)\overline{q}\big)(0) = \big(\overline{q}(\partial)p\big)(0),$$

the so-called *Fischer inner product.* Here $p(\partial) = p(\partial_1, \ldots, \partial_n)$ for $p = p(z_1, \ldots, z_n)$ and for $q(z) = \sum_\alpha c_\alpha z^\alpha$, $\overline{q}(z) = \sum_\alpha \overline{c_\alpha} z^\alpha$.

Thus

$$\langle z^\alpha, z^\beta\rangle_{\mathcal{F}} = \delta_{\alpha,\beta}\alpha! = \delta_{\alpha,\beta}\alpha_1!\cdots\alpha_n!$$

for multi-indices $\alpha = (\alpha_1,\ldots,\alpha_n)$, $\beta = (\beta_1,\ldots,\beta_n)$.

Lemma 7.5.1: *The subspaces $\{P_\lambda : \lambda \in \Lambda\}$ in $\mathbb{C}[V]$ are pair-wise orthogonal with respect to the Fischer inner product.*

Proof: This follows from the fact that $K \subset U(V)$ and $U(V)$ preserves $\langle\cdot,\cdot\rangle_{\mathcal{F}}$. This can be seen using an alternative formula for the Fisher inner product:

$$\langle p,q\rangle_{\mathcal{F}} = \frac{1}{\pi^n}\int p(z)\overline{q(z)}e^{-|z|^2}dzd\overline{z}. \tag{20}$$

In (20), $n = dim_{\mathbb{C}}(V)$, $|z|^2 = \langle z,z\rangle$ and "$dzd\overline{z}$" denotes Lebesgue measure on $V_{\mathbb{R}}$ normalized using $\langle\cdot,\cdot\rangle$. We see that $\langle k\cdot p, k\cdot q\rangle_{\mathcal{F}} = \langle p,q\rangle_{\mathcal{F}}$ for $k \in U(V)$ via a change of variables in (20), since both $|z|^2$ and $dzd\overline{z}$ are $U(V)$-invariant.

To establish (20) it suffices to verify that

$$\int z^\alpha\overline{z}^{\alpha'}e^{-|z|^2}dzd\overline{z} = \pi^n\delta_{\alpha,\alpha'}\alpha!.$$

For this, use polar coordinates $z_j = r_je^{i\theta_j}$ to write

$$\int z^\alpha\overline{z}^{\alpha'}e^{-|z|^2}dzd\overline{z} = \prod_{j=1}^n\int_0^\infty\int_0^{2\pi} r_j^{\alpha_j+\alpha_j'+1}e^{i(\alpha_j-\alpha_j')\theta_j}e^{-r_j^2}d\theta_jdr_j.$$

The integral in θ_j is zero unless $\alpha_j = \alpha_j'$, in which case one has

$$2\pi\int_0^\infty r_j^{2\alpha_j+1}e^{-r_j^2}dr_j = 2\pi\int_0^\infty s^{\alpha_j}e^{-s}\frac{ds}{2} = \pi\alpha_j!. \qquad \square$$

Using isomorphism (19) we can regard the canonical invariants p_λ as elements of $\mathbb{C}[V_{\mathbb{R}}]^K = \mathbb{C}[z_1,\ldots,z_n,\overline{z}_1,\ldots,\overline{z}_n]^K$. We have

$$p_\lambda(z,\overline{z}) = \frac{1}{d_\lambda}\sum_{j=1}^{d_\lambda}e_j(z)\overline{e}_j(\overline{z}) = \frac{1}{d_\lambda}\sum_{j=1}^{d_\lambda}|e_j(z)|^2$$

where $\{e_j\}$ is any orthonormal basis for P_λ (with respect to $\langle\cdot,\cdot\rangle_{\mathcal{F}}$). Note that each $p_\lambda \in \mathbb{C}[V_{\mathbb{R}}]^K$ is real valued and non-negative.

We later require the formula

$$\sum_{|\lambda|=k}d_\lambda p_\lambda = \frac{|z|^{2k}}{k!}. \tag{21}$$

Indeed $\sum_{|\lambda|=k} d_\lambda p_\lambda(z,\overline{z}) = \sum_e |e(z)|^2$ where e ranges over an orthonormal basis for $\mathcal{P}_k(V)$ obtained by concatenation of orthonormal bases for $\{P_\lambda : |\lambda| = k\}$. The sum is, however, independent of the basis and we can use $\{z^\alpha/\sqrt{\alpha!} : |\alpha| = k\}$ to compute

$$\sum_{|\lambda|=k} d_\lambda p_\lambda = \sum_{|\alpha|=k} z^\alpha \overline{z}^\alpha/\alpha! = \frac{1}{k!}\sum_{|\alpha|=k} \frac{k!}{\alpha!} z^\alpha \overline{z}^\alpha = |z|^{2k}/k!$$

as stated.

On $\mathbb{C}[V_\mathbb{R}] = \mathbb{C}[z_1,\ldots,z_n,\overline{z}_1,\ldots,\overline{z}_n]$ we consider the Hermitian inner product

$$\langle p,q\rangle_* = \big(p(\partial,\overline{\partial})\overline{q}\big)(0) = \big(\overline{q}(\partial,\overline{\partial})p\big)(0).$$

This "doubled Fischer inner product" is determined by

$$\langle z^\alpha \overline{z}^\beta, z^{\alpha'}\overline{z}^{\beta'}\rangle_* = \delta_{\alpha,\alpha'}\delta_{\beta,\beta'}\alpha!\beta!.$$

Proposition 7.5.2: *$\{p_\lambda \mid \lambda \in \Lambda\}$ is an orthogonal basis for $\mathbb{C}[V_\mathbb{R}]^K$ with respect to the inner product $\langle\cdot,\cdot\rangle_*$. Moreover $\langle p_\lambda, p_\lambda\rangle_* = 1/d_\lambda$.*

Proof: Let $\{e_j\}$, $\{f_j\}$ be $\langle\cdot,\cdot\rangle_\mathcal{F}$-orthonormal bases for P_λ, P_μ. Using Lemma 7.5.1 we compute

$$\begin{aligned}
\langle p_\lambda, p_\mu\rangle_* &= \big(p_\lambda(\partial,\overline{\partial})\overline{p}_\mu(z,\overline{z})\big)(0)\\
&= \frac{1}{d_\lambda d_\mu}\sum_{i,j}\big(e_i(\partial)\overline{e}_i(\overline{\partial})\overline{f}_j(z)f_j(\overline{z})\big)(0)\\
&= \frac{1}{d_\lambda d_\mu}\sum_{i,j}\big(e_i(\partial)\overline{f}_j(z)\big)(0)\big(\overline{e}_i(\overline{\partial})f_j(\overline{z})\big)(0)\\
&= \frac{1}{d_\lambda d_\mu}\sum_{i,j}|\langle e_i, f_j\rangle_\mathcal{F}|^2\\
&= \frac{\delta_{\lambda,\mu}d_\lambda}{d_\lambda d_\mu}\\
&= \frac{\delta_{\lambda,\mu}}{d_\lambda}.
\end{aligned}$$

□

7.6. *Section 7 notes*

Theorems 7.2.3, 7.4.3 and Corollary 7.4.4 are from [24]. An action whose invariants form a polynomial ring is said to be *coregular*. The coregular actions for simple groups are classified in [27] and [48].

The Fischer inner product is also called the *Fock inner product*, especially in connection with Equation 20. Fock space $\mathcal{F}$ is the Hilbert space completion of $\mathbb{C}[V]$ with respect to $\langle\cdot,\cdot\rangle_{\mathcal{F}}$. This can be identified as the space of holomorphic functions on V square integrable with respect to the Gaussian measure $e^{-|z|^2}dzd\overline{z}$ [14]. The inner product $\langle\cdot,\cdot\rangle_*$ can be regarded as the restriction of $\langle\cdot,\cdot\rangle_{\mathcal{F}}\otimes\langle\cdot,\cdot\rangle_{\mathcal{F}^*}$ from $\mathcal{F}\otimes\mathcal{F}^*$ to $\mathbb{C}[V\oplus\overline{V}]\cong\mathbb{C}[V]\otimes\mathbb{C}[V]^*$. One can identify $\mathcal{F}\otimes\mathcal{F}^*$ with the space of Hilbert-Schmidt operators on $\mathcal{F}$. Now $\langle\cdot,\cdot\rangle_{\mathcal{F}}$ induces the Hilbert-Schmidt norm. There is also a connection with the Berezin star product, as explained in [1].

8. Generalized binomial coefficients

We continue to assume that $G : V$ is a multiplicity free action. As in the previous section, $\{p_\lambda \mid \lambda\in\Lambda\}$ are the canonical invariants. We view these as living in $\mathbb{C}[V_{\mathbb{R}}]^K$.

8.1. *The polynomials* q_λ

Let $\Delta = \partial\cdot\overline{\partial} = \partial_1\overline{\partial}_1+\cdots+\partial_n\overline{\partial}_n$ and consider the operator $T:\mathbb{C}[V_{\mathbb{R}}]\to\mathbb{C}[V_{\mathbb{R}}]$,

$$(Tp)(z,\overline{z}) = (e^{\Delta}p)(z,-\overline{z}) = e^{-\Delta}(p(z,-\overline{z})).$$

Note that T is an involutive automorphism. Indeed, writing $(Mp)(z,\overline{z}) = p(z,-\overline{z})$ one has

$$T = M\circ e^{\Delta} = e^{-\Delta}\circ M = T^{-1}. \tag{22}$$

Definition 8.1.1: Let $q_\lambda = T(p_\lambda) = (-1)^{|\lambda|}e^{-\Delta}p_\lambda$ for each $\lambda\in\Lambda$.

The two formulas in the definition for q_λ agree because $p_\lambda(z,-\overline{z}) = (-1)^{|\lambda|}p_\lambda(z,\overline{z})$.

Lemma 8.1.2: *$\{q_\lambda\ :\ \lambda\in\Lambda\}$ is a vector space basis for $\mathbb{C}[V_{\mathbb{R}}]^K$.*

Proof: First note that $q_\lambda\in\mathbb{C}[V_{\mathbb{R}}]$ is K-invariant because p_λ is K-invariant and Δ is a $U(V)$-invariant operator. Moreover

$$q_\lambda = (-1)^{|\lambda|}p_\lambda + r_\lambda$$

where $p_\lambda\in\mathcal{P}_{2|\lambda|}(V_{\mathbb{R}})$ and r_λ is of lower degree. As $\{p_\lambda\}$ is a basis for $\mathbb{C}[V_{\mathbb{R}}]^K$, so is $\{q_\lambda\}$. □

8.2. *The generalized binomial coefficients*

As q_λ belongs to $\mathbb{C}[V_\mathbb{R}]^K$, it can be written as a finite linear combination of the canonical invariants p_ν.

Definition 8.2.1: The *generalized binomial coefficients* $\left[\begin{smallmatrix}\lambda\\ \nu\end{smallmatrix}\right]$ are defined for $\lambda, \nu \in \Lambda$ via

$$q_\lambda = \sum_{\nu\in\Lambda} (-1)^{|\nu|} \begin{bmatrix}\lambda\\ \nu\end{bmatrix} p_\nu.$$

The proof of Lemma 8.1.2 shows that

$$\begin{bmatrix}\lambda\\ \lambda\end{bmatrix} = 1, \quad \begin{bmatrix}\lambda\\ \nu\end{bmatrix} = 0 \text{ when } |\lambda| = |\nu| \text{ but } \lambda \neq \nu \text{ and } \quad \begin{bmatrix}\lambda\\ \nu\end{bmatrix} = 0 \text{ for } |\nu| > |\lambda|.$$

So in fact

$$q_\lambda = \sum_{|\nu|\leq|\lambda|} (-1)^{|\nu|} \begin{bmatrix}\lambda\\ \nu\end{bmatrix} p_\nu = (-1)^{|\lambda|} p_\lambda + \sum_{|\nu|<|\lambda|} (-1)^{|\nu|} \begin{bmatrix}\lambda\\ \nu\end{bmatrix} p_\nu.$$

The terminology "generalized binomial coefficient" is motivated by Example 8.4.1 below. Our immediate goal is to develop some combinatorial properties of these coefficients and to relate them to eigenvalues for operators in $\mathcal{PD}(V)^G$. We begin with the following.

Proposition 8.2.2:

$$\sum_{\mu\in\Lambda} (-1)^{|\mu|} \begin{bmatrix}\lambda\\ \mu\end{bmatrix} \begin{bmatrix}\mu\\ \nu\end{bmatrix} = \begin{cases} (-1)^{|\lambda|} & \textit{for } \lambda = \nu \\ 0 & \textit{for } \lambda \neq \nu \end{cases}.$$

Proof: First note that as $q_\lambda = T(p_\lambda)$ and $T^2 = I$ one has

$$p_\lambda = T(q_\lambda) = \sum_\nu (-1)^{|\nu|} \begin{bmatrix}\lambda\\ \nu\end{bmatrix} T(p_\nu) = \sum_\nu (-1)^{|\nu|} \begin{bmatrix}\lambda\\ \nu\end{bmatrix} q_\nu, \tag{23}$$

an interesting formula in its own right. But now

$$\begin{aligned} q_\lambda &= \sum_\mu (-1)^{|\mu|} \begin{bmatrix}\lambda\\ \mu\end{bmatrix} p_\mu = \sum_\mu (-1)^{|\mu|} \begin{bmatrix}\lambda\\ \mu\end{bmatrix} \left(\sum_\nu (-1)^{|\nu|} \begin{bmatrix}\mu\\ \nu\end{bmatrix} q_\nu\right) \\ &= \sum_\nu (-1)^{|\nu|} \left(\sum_\mu (-1)^{|\mu|} \begin{bmatrix}\lambda\\ \mu\end{bmatrix} \begin{bmatrix}\mu\\ \nu\end{bmatrix}\right) q_\nu \end{aligned}$$

and the result follows from the linear independence of $\{q_\lambda \ : \ \lambda \in \Lambda\}$. □

Proposition 8.2.3: *For $\lambda \in \Lambda$ and $k \in \mathbb{N}$*

$$\frac{\Delta}{k!}^k p_\lambda = \sum_{|\nu|=|\lambda|-k} \begin{bmatrix} \lambda \\ \nu \end{bmatrix} p_\nu \tag{24}$$

and

$$\frac{\Delta}{k!}^k q_\lambda = (-1)^k \sum_{|\nu|=|\lambda|-k} \begin{bmatrix} \lambda \\ \nu \end{bmatrix} q_\nu \tag{25}$$

Proof: We have

$$\sum_\nu (-1)^{|\nu|} \begin{bmatrix} \lambda \\ \nu \end{bmatrix} p_\nu = q_\lambda = T(p_\lambda) = (-1)^{|\lambda|} e^{-\Delta} p_\lambda$$
$$= (-1)^{|\lambda|} \sum_k (-1)^k \frac{\Delta}{k!}^k p_\lambda.$$

Equating homogeneous components of degree $2(|\lambda| - k)$ on both sides of this equation yields (24). Applying the operator T to both sides of (24) yields (25) since

$$T(\Delta^k p_\lambda) = (-1)^{|\lambda|-k} e^{-\Delta} \Delta^k p_\lambda = (-1)^k \Delta^k \left((-1)^{|\lambda|} e^{-\Delta} p_\lambda\right) = (-1)^k \Delta^k q_\lambda.$$

□

The next theorem is the key to all subsequent results in this section.

Theorem 8.2.4: *(Yan's Pieri Formula) For $\nu \in \Lambda$, $k \in \mathbb{N}$,*

$$\frac{|z|}{k!}^{2k} d_\nu p_\nu = \sum_{|\lambda|=|\nu|+k} \begin{bmatrix} \lambda \\ \nu \end{bmatrix} d_\lambda p_\lambda.$$

Proof: First note that for polynomials $p, q \in \mathbb{C}[V_\mathbb{R}]$ one has

$$\langle z_j p, q \rangle_* = \langle p, \partial_j q \rangle_*, \quad \langle \overline{z}_j p, q \rangle_* = \langle p, \overline{\partial}_j q \rangle_*$$

and hence $\langle |z|^2 p, q\rangle_* = \langle p, \Delta q\rangle_*$. Now using Proposition 7.5.2 and Equation 24,

$$\begin{aligned}
\frac{|z|}{k!}^{2k} p_\nu &= \sum_{|\lambda|=|\nu|+k} \frac{\langle (|z|^{2k}/k!)p_\nu, p_\lambda\rangle_*}{\langle p_\lambda, p_\lambda\rangle_*} p_\lambda \\
&= \sum_{|\lambda|=|\nu|+k} d_\lambda \langle p_\nu, (\Delta^k/k!)p_\lambda\rangle_* p_\lambda \\
&= \sum_{|\lambda|=|\nu|+k} d_\lambda \begin{bmatrix}\lambda\\ \nu\end{bmatrix} \langle p_\nu, p_\nu\rangle_* p_\lambda \\
&= \sum_{|\lambda|=|\nu|+k} \frac{d_\lambda}{d_\nu} \begin{bmatrix}\lambda\\ \nu\end{bmatrix} p_\lambda.
\end{aligned}$$

□

Corollary 8.2.5: *For $|\lambda| = |\nu| + k$ one has*

$$\begin{bmatrix}\lambda\\ \nu\end{bmatrix} = \frac{1}{k!}\sum \begin{bmatrix}\varepsilon_1\\ \nu\end{bmatrix}\begin{bmatrix}\varepsilon_2\\ \varepsilon_1\end{bmatrix}\cdots\begin{bmatrix}\varepsilon_{k-1}\\ \varepsilon_{k-2}\end{bmatrix}\begin{bmatrix}\lambda\\ \varepsilon_{k-1}\end{bmatrix}$$

where the sum is over all $(\varepsilon_1, \ldots, \varepsilon_{k-1})$ with $|\varepsilon_j| = |\nu| + j$.

Proposition 8.2.6: *The generalized binomial coefficients are non-negative real numbers.*

Proof: Corollary 8.2.5 shows that it suffices to prove $\left[\begin{smallmatrix}\lambda\\ \nu\end{smallmatrix}\right] \geq 0$ for $|\lambda| = |\nu| + 1$. Let $\{f_1, \ldots, f_{d_\nu}\}$ be an orthonormal basis for P_ν with respect to the Fischer inner product. Since $z_i f_j \in \mathcal{P}_{|\nu|+1}(V) = \bigoplus_{|\lambda|=|\nu|+1} P_\lambda$, we can write

$$z_i f_j = \sum_{|\lambda|=|\nu|+1} f_\lambda(i,j)$$

where $f_\lambda(i,j) \in P_\lambda$. Hence also

$$|z|^2 d_\nu p_\nu = \sum_{i=1}^{n}\sum_{j=1}^{d_\nu} z_i f_j \overline{z_i f_j} = \sum_{|\lambda|=|\lambda'|=|\nu|+1} \sum_{i=1}^{n}\sum_{j=1}^{d_\nu} f_\lambda(i,j)\overline{f_{\lambda'}(i,j)}.$$

The sum $\sum_{|\lambda|=|\lambda'|=|\nu|+1} P_\lambda \otimes \overline{P}_{\lambda'}$ is direct in $\mathbb{C}[V_\mathbb{R}] = \mathbb{C}[V] \otimes \mathbb{C}[\overline{V}]$ and each $P_\lambda \otimes \overline{P}_{\lambda'}$ is a K-invariant subspace. Since $|z|^2 d_\nu p_\nu$ is a K-invariant polynomial, it follows that

$$\sum_{i=1}^{n}\sum_{j=1}^{d_\nu} f_\lambda(i,j)\overline{f_{\lambda'}(i,j)} \in P_\lambda \otimes \overline{P}_{\lambda'}$$

is K-invariant for each $|\lambda| = |\nu| + 1 = |\lambda'|$. But

$$(P_\lambda \otimes \overline{P}_{\lambda'})^K = \begin{cases} \{0\} & \text{for } \lambda \neq \lambda' \\ \mathbb{C}p_\lambda & \text{for } \lambda = \lambda' \end{cases}$$

via Schur's Lemma, as in Section 7.3. We conclude that $\sum_{i=1}^n \sum_{j=1}^{d_\nu} f_\lambda(i,j) \overline{f_{\lambda'}(i,j)} = 0$ for $\lambda \neq \lambda'$ and that $\sum_{i=1}^n \sum_{j=1}^{d_\nu} |f_\lambda(i,j)|^2 = c_\lambda p_\lambda$ for some $c_\lambda \in \mathbb{C}$. As p_λ and $\sum_{i=1}^n \sum_{j=1}^{d_\nu} |f_\lambda(i,j)|^2$ are both non-negative real valued polynomials on $V_\mathbb{R}$, we must have $c_\lambda \geq 0$. Hence

$$|z|^2 d_\nu p_\nu = \sum_{|\lambda|=|\nu|+1} c_\lambda p_\lambda$$

for some values $c_\lambda \geq 0$. Theorem 8.2.4 now implies $\left[{\lambda \atop \nu}\right] = c_\lambda / d_\lambda \geq 0$. □

8.3. *Eigenvalues for operators in $\mathcal{PD}(V)^G$*

The operators $p_\nu(z,\partial)$ in the canonical basis for $\mathcal{PD}(V)^G$ act by scalars on each subspace $P_\lambda \subset \mathbb{C}[V]$. This is a consequence of Schur's Lemma, G-invariance of $p_\nu(z,\partial)$ and the fact that the decomposition $\mathbb{C}[V] = \oplus P_\lambda$ is multiplicity free.

Definition 8.3.1: For $\nu, \lambda \in \Lambda$ let $\widehat{p}_\nu(\lambda) \in \mathbb{C}$ denote the eigenvalue of $p_\nu(z,\partial)$ on P_λ. That is, $p_\nu(z,\partial)|_{P_\lambda} = \widehat{p}_\nu(\lambda) I_{P_\lambda}$.

Proposition 8.3.2: $d_\nu \widehat{p}_\nu(\lambda) = \left[{\lambda \atop \nu}\right]$ *for all* $\lambda, \nu \in \Lambda$.

Proof: Note that

$$p_\nu(z,\partial) p_\lambda(z,\overline{z}) = \widehat{p}_\nu(\lambda) p_\lambda(z,\overline{z})$$

as $p_\lambda(z,\overline{z}) \in P_\lambda \otimes \overline{P}_\lambda$. Now using Theorem 8.2.4,

$$\begin{aligned} d_\nu p_\nu(z,\partial) e^{|z|^2} &= d_\nu p_\nu(z,\overline{z}) e^{|z|^2} \\ &= d_\nu p_\nu(z,\overline{z}) \sum_k \frac{|z|^{2k}}{k!} \\ &= \sum_k \sum_{|\lambda|=|\nu|+k} \begin{bmatrix} \lambda \\ \nu \end{bmatrix} d_\lambda p_\lambda(z,\overline{z}) \\ &= \sum_\lambda \begin{bmatrix} \lambda \\ \nu \end{bmatrix} d_\lambda p_\lambda(z,\overline{z}). \end{aligned}$$

But we can also use Equation 21 to write

$$e^{|z|^2} = \sum_k \frac{|z|^{2k}}{k!} = \sum_k \sum_{|\lambda|=k} d_\lambda p_\lambda(z,\overline{z}) = \sum_\lambda d_\lambda p_\lambda(z,\overline{z}).$$

Thus

$$d_\nu p_\nu(z,\partial)e^{|z|^2} = \sum_\lambda d_\nu d_\lambda \widehat{p}_\nu(\lambda) p_\lambda(z,\overline{z}).$$

Comparing these two expressions for $d_\nu p_\nu(z,\partial)e^{|z|^2}$ gives $d_\nu \widehat{p}_\nu(\lambda) = \left[{\lambda \atop \nu}\right]$ as claimed. □

Together Propositions 8.2.6 and 8.3.2 yield:

Corollary 8.3.3: *The eigenvalues $\widehat{p}_\nu(\lambda)$ are non-negative real numbers.*

8.4. *Examples*

Here are two examples to illustrate the circle of ideas developed above. Even the most basic example is of interest in this context.

Example 8.4.1: GL(n): Consider the usual action of $G = GL(n,\mathbb{C})$ on $V = \mathbb{C}^n$. We have maximal compact subgroup $K = U(n)$ and the standard inner product $\langle z, z'\rangle = z \cdot \overline{z}'$ is K-invariant. $\mathbb{C}[V]$ decomposes as

$$\mathbb{C}[V] = \mathbb{C}[z_1 \ldots, z_n] = \bigoplus_{m\in\mathbb{N}} \mathcal{P}_m(V)$$

under the actions of G and of K. Using the orthonormal basis $\{z^\alpha/\sqrt{\alpha!} \;:\; |\alpha| = m\}$ for $\mathcal{P}_m(V)$ (with respect to $\langle\cdot,\cdot\rangle_{\mathcal{F}}$) we compute the canonical invariant $p_m(z,\overline{z}) \in \mathbb{C}[V_\mathbb{R}]^K$:

$$p_m(z,\overline{z}) = \frac{1}{d_m}\sum_{|\alpha|=m} \frac{z^\alpha \overline{z}^\alpha}{\alpha!} = \frac{1}{d_m}\frac{1}{m!}\sum_{|\alpha|=m}\binom{m}{\alpha}|z|^{2\alpha} = \frac{1}{d_m}\frac{1}{m!}|z|^{2m}.$$

Substituting

$$d_m = dim(\mathcal{P}_m(V)) = \binom{m+n-1}{m}$$

one obtains

$$p_m(z,\overline{z}) = \frac{(n-1)!}{(m+n-1)!}|z|^{2m}.$$

The fundamental invariant is $\gamma = p_1 = |z|^2/n$ and the above computation shows that $\mathbb{C}[V_\mathbb{R}]^K = \mathbb{C}[\gamma]$, illustrating Theorem 7.4.3.

Next we compute

$$\frac{|z|^{2(m-k)}}{(m-k)!}d_k p_k = \frac{|z|^{2(m-k)}}{(m-k)!}\frac{|z|}{k!}^{2k} = \binom{m}{k}\frac{|z|}{m!}^{2m} = \binom{m}{k}d_m p_m.$$

Now Theorem 8.2.4 implies

$$\begin{bmatrix} m \\ k \end{bmatrix} = \binom{m}{k}$$

for $k, m \in \mathbb{N}$. This fact motivates the terminology "generalized binomial coefficient".

The polynomials $\{q_m \mid m \in \mathbb{N}\}$ are given by

$$\begin{aligned} q_m(z,\overline{z}) &= \sum_k (-1)^k \binom{m}{k} p_k(z,\overline{z}) \\ &= \sum_k (-1)^k \binom{m}{k} \frac{(n-1)!}{(k+n-1)!}|z|^{2k} \\ &= L_m^{(n-1)}(|z|^2) \end{aligned}$$

where

$$L_m^{(r)}(x) = r!\sum_{k=0}^{m}\binom{m}{k}\frac{(-x)^k}{(k+r)!}$$

is the *generalized Laguerre polynomial* of order r and degree m, normalized to have value 1 at $x = 0$.

Now according to Proposition 8.3.2 the operator $d_k p_k(z, \partial) \in \mathcal{PD}(V)^G$ has eigenvalue

$$d_k\widehat{p}_k(m) = \binom{m}{k}$$

on $\mathcal{P}_m(V)$. One can see this directly because

$$\begin{aligned} d_k p_k(z,\partial)(z_1^m) &= \sum_{|\alpha|=k} \frac{z^\alpha \partial^\alpha}{\alpha!}(z_1^m) = \frac{z_1^k \partial_1}{k!}^k (z_1^m) \\ &= \frac{z_1^k}{k!}\frac{m!}{(m-k)!}z_1^{m-k} = \binom{m}{k} z_1^m \end{aligned}$$

for $m \geq k$.

For this example Proposition 8.2.2 asserts that

$$\sum_{k=\ell}^{m}(-1)^k\binom{m}{k}\binom{k}{\ell} = \delta_{m,\ell}(-1)^m. \tag{26}$$

Equivalently, the lower triangular matrix A with entries $A_{m,k} = (-1)\binom{m}{k}$ ($0 \leq m, k \leq N$ say) is self inverting: $A^2 = I$. One can verify (26) directly as follows.

$$\begin{aligned} x^m = (1-(1-x))^m &= \sum_k (-1)^k \binom{m}{k} (1-x)^k \\ &= \sum_k (-1)^k \binom{m}{k} \sum_\ell (-1)^\ell \binom{k}{\ell} x^\ell \\ &= \sum_\ell (-1)^\ell \left[\sum_k (-1)^k \binom{m}{k} \binom{k}{\ell} \right] x^\ell. \end{aligned}$$

Example 8.4.2: GL(n) $\otimes$ GL(n): Recall the (twisted) action (7) of $G = GL(n,\mathbb{C}) \times GL(n,\mathbb{C})$ on $V = \mathbb{C}^n \otimes \mathbb{C}^n \cong M_{n,n}(\mathbb{C})$. Restricting to the maximal compact subgroup $K = U(n) \times U(n)$ we have

$$(k_1, k_2) \cdot z = \overline{k}_1 v k_2^*$$

where $k^* = \overline{k}^t = k^{-1}$. The inner product $\langle z, w\rangle = tr(zw^*)$ on V is K-invariant. The decomposition

$$\mathbb{C}[M_{n,n}(\mathbb{C})] = \bigoplus_{\lambda \in \Lambda} P_\lambda$$

under the action of K, given in Theorem 4.1.1, is indexed by partitions $\lambda = (\lambda_1 \geq \cdots \geq \lambda_n \geq 0)$.

Let $\Sigma \subset V$ denote the set of matrices of the form $h = diag(d_1, \ldots, d_n)$ with $d_j \in \mathbb{R}^+$.

Proposition 8.4.3: *([6]) For $\lambda \in \Lambda$, $p_\lambda \in \mathbb{C}[V_\mathbb{R}]^K$ is determined by its restriction to Σ via the formula*

$$p_\lambda(h) = c_\lambda s_\lambda(d_1^2, \ldots, d_n^2)$$

where s_λ is a Schur polynomial in n variables and c_λ is a positive constant.

The Schur polynomial s_λ arises as the character of the representation σ_n^λ of $GL(n,\mathbb{C})$ with highest weight λ,

$$s_\lambda(x_1, \ldots, x_n) = tr(\sigma_n^\lambda(diag(x_1, \ldots, x_n)),$$

and is given explicitly by the determinantal formula

$$s_\lambda(x_1, \ldots, x_n) = \frac{det\left[x_i^{\lambda_j + n - j}\right]}{det\left[x_i^{n-j}\right]}. \tag{27}$$

The Schur function $s_{(1^k)}$ is the k'th elementary symmetric function e_k. Thus the fundamental invariants $\gamma_k = p_{(1^k)} \in \mathbb{C}[V_\mathbb{R}]^K$ are determined, up to normalization, by

$$\gamma_k(h) = c_k e_k(d_1^2, \ldots, d_n^2)$$

on the cross-section Σ.

It is useful to identify partitions $\lambda \in \Lambda$ with their Young's diagrams. Given two diagrams $\lambda, \nu \in \Lambda$, we write $\nu \subset \lambda$ when ν is a sub-diagram of λ. That is, $\nu_j \le \lambda_j$ for all j. If $|\lambda| = |\nu| + k$ then $\mathcal{C}_{\lambda\nu}$ will denote the number of sequences $(\varepsilon_0, \ldots, \varepsilon_k)$ of Young's diagrams such that

- $\nu = \varepsilon_0 \subset \varepsilon_1 \subset \cdots \subset \varepsilon_{k-1} \subset \varepsilon_k = \lambda$, and
- $|\varepsilon_j| = |\nu| + j$ for $j = 1, \ldots, k$.

That is, ε_j is obtained from ε_{j-1} by adding a single box $\square$ to some row. Note that $\mathcal{C}_{\lambda\nu} = 0$ if $\nu \not\subset \lambda$. When $\nu \subset \lambda$, $\mathcal{C}_{\lambda\nu}$ is the number of *standard tableaux* of shape $\lambda - \nu$. That is, the number of ways to assign the values $1, 2, \ldots, k$ to the boxes of the skew-diagram $\lambda - \nu$ so that values increase as we move along rows from left to right and as we move down columns.

Proposition 8.4.4: *The generalized binomial coefficients can be expressed as*

$$\begin{bmatrix} \lambda \\ \nu \end{bmatrix} = \frac{1}{k!}\frac{d_\nu c_\nu}{d_\lambda c_\lambda}\mathcal{C}_{\lambda\nu}$$

for $|\lambda| = |\nu| + k$.

Proof: Note that the polynomial $\gamma(z) = |z|^2 = tr(zz^*)$ is given on Σ by

$$\gamma(h) = d_1^2 + \cdots + d_n^2 = s_{(1)}(d_1^2, \ldots, d_n^2).$$

Consider the case where $|\lambda| = |\nu| + 1$. Theorem 8.2.4 and Proposition 8.4.3 yield

$$d_\nu c_\nu s_{(1)} s_\nu = \sum_{|\lambda| = |\nu|+1} \begin{bmatrix} \lambda \\ \nu \end{bmatrix} d_\lambda c_\lambda s_\lambda.$$

On the other hand, the classical *Pieri formula* asserts:

$$s_{(1)} s_\nu = \sum_{\substack{|\lambda| = |\nu|+1 \\ \nu \subset \lambda}} s_\lambda.$$

Comparing these expressions gives

$$\begin{bmatrix} \lambda \\ \nu \end{bmatrix} = \left\{ \begin{array}{ll} \frac{d_\nu c_\nu}{d_\lambda c_\lambda} & \text{if } \nu \subset \lambda \\ 0 & \text{if } \nu \not\subset \lambda \end{array} \right\} = \frac{d_\nu c_\nu}{d_\lambda c_\lambda}\mathcal{C}_{\lambda\nu}.$$

An application of Corollary 8.2.5 now yields the result for $|\lambda| = |\nu| + k$. □

The dimensions $d_\mu = dim(P_\mu)$ are $d_\mu = dim(\sigma_n^\mu)^2$, in view of Theorem 4.1.1. A classical formula for these dimensions gives

$$d_\mu = \left[\prod_{i<j} \frac{\mu_i - \mu_j + j - i}{j - i} \right]^2 .$$

Remark 8.4.5: This example motivates the name "Yan's Pieri Formula" for Theorem 8.2.4. We will return to this example below in Section 9.4. As a byproduct of this subsequent analysis, one sees that the normalization constants c_λ appearing in Propositions 8.4.3 and 8.4.4 are given by $c_\lambda = 1/H(\lambda)$, where $H(\lambda)$ is as in Proposition 9.4.1.

8.5. *Section 8 notes*

The polynomials q_λ play a role in connection with analysis on the Heisenberg group. Let $H_V = V \times \mathbb{R}$ with product

$$(z,t)(z',t') = \big(z + z', t + t' - \frac{1}{2} Im\langle z, z'\rangle\big).$$

Any compact Lie subgroup K of $U(V)$ act by automorphisms on H_V via

$$k \cdot (z,t) = (kz, t).$$

It is known that $(K \ltimes H_V, K)$ is a Gelfand pair if and only if $K : V$ is a multiplicity free action. A generic set of spherical functions for such a Gelfand pair is completely determined by the q_λ-polynomials. In particular,

$$\phi(z,t) = q_\lambda(z) e^{-|z|^2/2} e^{it}$$

is one such spherical function. We refer the reader to [4] concerning this connection.

One can extend the inner product $\langle \cdot, \cdot \rangle_{\mathcal{F}}$ from $\mathbb{C}[V]$ to $\mathbb{C}[V_{\mathbb{R}}]$ using Equation 20. The q_λ's are then orthogonal with respect to $\langle \cdot, \cdot \rangle_{\mathcal{F}}$. In fact, they can be obtained via Gram-Schmidt orthogonalization using $\langle \cdot, \cdot \rangle_{\mathcal{F}}$ from the p_λ's [4].

Generalized binomial coefficients were first introduced in the setting of Hermitian symmetric spaces. See [12], [37] and [13]. Z. Yan subsequently defined generalized binomial coefficients in the more general context of multiplicity free actions in his unpublished manuscript [56]. This contains the first proofs of Theorem 8.2.4 and Proposition 8.3.2. Further combinatorial identities concerning the generalized binomial coefficients can be found in [6]. Both [56] and [6] use representation theory for the Heisenberg group and exploit the connection with spherical functions outlined above. The treatment given here achieves some simplification.

In [7] it is shown that the generalized binomial coefficients are (non-negative) *rational* numbers. Thus the same holds for the eigenvalues $\widehat{p}_\nu(\lambda)$. The proof involves extensive case-by-case analysis working from the classification for multiplicity free actions presented in Section 6.

For background on Schur functions, including the Pieri formula, we refer the reader to Chapter I in [39].

9. Eigenvalues for operators in $\mathcal{PD}(V)^G$

Recall that for a multiplicity free action $G : V$ each $D \in \mathcal{PD}(V)^G$ acts by scalars on the irreducible subspaces $\{P_\lambda \,:\, \lambda \in \Lambda\}$ in the decomposition of $\mathbb{C}[V]$. For $\nu \in \Lambda$, $\widehat{p}_\nu(\lambda)$ denotes the eigenvalue of $p_\nu(z, \partial)$ on P_λ. This is $d_\nu \widehat{p}_\nu(\lambda) = \left[{\lambda \atop \nu}\right]$. We will see that the map $\widehat{p}_\nu : \Lambda \to \mathbb{C}$ extends in a natural way to a polynomial function on the subspace $Span_{\mathbb{C}}(\Lambda)$ of $\mathfrak{h}^*$.

9.1. *Eigenvalue polynomials*

We say that an operator $D \in \mathcal{PD}(V)$ has *order m* if

$$D \in \delta\Big(\mathbb{C}[V] \otimes \sum_{j \leq m} \mathcal{P}_j(V^*)\Big).$$

Here $\delta : \mathbb{C}[V \oplus V^*] \to \mathcal{PD}(V)$ is the map given by Wick quantization, $p \mapsto p(z, \partial)$. In terms of coordinates, D has order m when $D \in Span\{z^\alpha \partial^\beta \,:\, |\beta| \leq m\}$.

Proposition 9.1.1: *Let* $D \in \mathcal{PD}(V)$ *and* $f = f_1^{a_1} \cdots f_r^{a_r} \in \mathbb{C}[V]$. *Then*

$$Df(z) = \beta_D(z; a_1, \ldots, a_r) f(z)$$

where $\beta_D \in \mathbb{C}[V][f_1^{-1}, \ldots, f_r^{-1}][a_1, \ldots, a_r]$.

Proof: The proof is by induction on the order of D. For operators of order 1, it is enough to consider $D = \partial_w$, a directional derivative. In this case

$$Df(z) = \sum_j a_j (\partial_w f_j) \left(\frac{f}{f_j}\right),$$

so $\beta_D = \sum_j a_j(\partial_w f_j)/f_j$ in this case.

For D of order greater than 1, write $D = g + \sum_i X_i E_i$ where $g \in \mathbb{C}[V]$ and the X_i's have order 1. Now $E_i f(z) = \beta_i(z; a_1, \ldots, a_r) f(z)$ for suitable β_i by the induction hypothesis. As

$$X_i E_i f(z) = X_i \beta_i f(z) + \beta_i X_i f(z) = \left[X_i \beta_i + \beta_i \sum_j a_j \frac{X_i(f_j)}{f_j} \right] f(z)$$

the result follows. □

Corollary 9.1.2:

(1) As a polynomial in $a = (a_1, \ldots, a_r)$, the degree of $\beta_D(z; a)$ is order(D).
(2) As polynomials in a, the homogeneous components of highest degree in $\beta_D(z; a)$ and $\sigma_D(z, \frac{df}{f}(z))$ agree. Here $\sigma_D \in \mathbb{C}[V \oplus V^*]$ denotes the Wick symbol of D and

$$\frac{df}{f} = a_1 \frac{df_1}{f_1} + \cdots + a_r \frac{df_r}{f_r}.$$

Proof: (1) follows by induction from the proof of Proposition 9.1.1. For (2), we first examine $Df(z)$. For $D = \partial_w$ one has

$$\sigma_D\left(z, \frac{df}{f}\right) = a_1 \frac{w^*(f_1)}{f_1} + \cdots + a_r \frac{w^*(f_r)}{f_r} = a_1 \frac{\partial_w f_1}{f_1} + \cdots + a_r \frac{\partial_w f_r}{f_r} = \beta_D(z; a).$$

Given any polynomial $b(a)$ in $a = (a_1 \ldots, a_r)$ we will write *top*(b) for the homogeneous component of highest degree. If we look closely at the induction step, with $D = g + \sum_i X_i E_i$, we see that

$$top(\beta_D) = \sum_i top(\beta_i)(z; a) \sum_j a_j \frac{X_i(f_j)}{f_j}.$$

By the induction hypothesis, $top(\beta_i) = top(\sigma_{E_i}(z, df/f))$, and we have just seen that $top(\sigma_{X_i})(z, df/f) = \sum_j a_j X_i(f_j)/f_j$. The result now follows since $top(\sigma_{X_i E_i}) = top(\sigma_{X_i} \sigma_{E_i})$. □

Now suppose that $G : V$ is a multiplicity free action with, as usual,

(1) $\Lambda \subset \mathfrak{h}^*$ the set of highest weights for the representations that occur in $\mathbb{C}[V]$,
(2) $\Lambda' = \{\lambda_1, \ldots, \lambda_r\}$ the set of fundamental highest weights, and
(3) $h_j = h_{\lambda_j}$ $(1 \leq j \leq r)$ the fundamental highest weight vectors.

Then $h = h_1^{a_1} \cdots h_r^{a_r}$ is a highest weight vector in P_λ for $\lambda = a_1\lambda_1 + \cdots + a_r\lambda_r \in \Lambda$. Thus for any $\nu \in \Lambda$,

$$\widehat{p_\nu}(\lambda)h(z) = p_\nu(z, \partial)h(z) = \beta_\nu(z; a_1, \ldots, a_r)h(z)$$

with $\beta_\nu = \beta_{p_\nu(z,\partial)}$ as in Proposition 9.1.1. It follows that $\beta_\nu(z; a) = \widehat{p_\nu}(\lambda)$ for all z. Thus $\beta_\nu = \beta_\nu(a)$ is a polynomial in the parameters a, independent of z. Since

$$\widehat{p_\nu}(a_1\lambda_1 + \cdots + a_r\lambda_r) = \beta_\nu(a_1, \ldots, a_r)$$

we see that $\widehat{p_\nu}$ extends in a natural way to a polynomial function on $Span_{\mathbb{C}}(\Lambda)$ in $\mathfrak{h}^*$. Moreover, by Corollary 9.1.2 we have

$$top(\widehat{p_\nu}(a_1\lambda_1 + \cdots + a_r\lambda_r)) = p_\nu\left(z, a_1\frac{dh_1}{h_1} + \cdots + a_r\frac{dh_r}{h_r}\right). \tag{28}$$

Note that although both arguments on the right hand side of Equation 28 depend on z, the result is independent of z, and gives a polynomial function of the parameters $a_1, \ldots, a_r$.

9.2. *A Harish-Chandra homomorphism for multiplicity free actions*

Let $B = HN$ be a Borel subgroup in G with $\Delta^+ \subset \mathfrak{h}^*$ the associated set of positive roots. Let W denote the Weyl group and

$$\rho = \frac{1}{2}\sum_{\alpha\in\Delta^+} \alpha$$

be half the sum of the positive roots. $\mathcal{U}(\mathfrak{g})$ is the universal enveloping algebra for $\mathfrak{g}$ with center $\mathcal{ZU}(\mathfrak{g})$. The Harish-Chandra homomorphism is an algebra isomorphism

$$H : \mathcal{ZU}(\mathfrak{g}) \to \mathbb{C}[\mathfrak{h}^*]^W$$

such that

$Z \in \mathcal{ZU}(\mathfrak{g})$ acts on V_λ by the scalar $H(Z)(\lambda + \rho)$ for all highest weights λ.

Let $G : V$ be a multiplicity free action. Knop constructs a map on $\mathcal{PD}(V)^G$ analogous to the Harish-Chandra homomorphism. Let

$$\mathfrak{a}^* = Span_{\mathbb{C}}(\Lambda) = Span_{\mathbb{C}}\{\lambda_1, \ldots, \lambda_r\} \subset \mathfrak{h}^*, \quad A = \mathfrak{a}^* + \rho$$

and define a map

$$h : \mathcal{PD}(V)^G \to \mathbb{C}[A], \quad h(D)(\lambda + \rho) = \beta_D(\lambda). \tag{29}$$

(As before, D acts on P_λ by the scalar $\beta_D(\lambda)$ when $\lambda \in \Lambda$.) Then the following diagram commutes:

$$\begin{array}{ccc} \mathcal{ZU}(\mathfrak{g}) & \xrightarrow{H} & \mathbb{C}[\mathfrak{h}^*]^W \\ \downarrow \pi & & \downarrow r \\ \mathcal{PD}(V)^G & \xrightarrow{h} & \mathbb{C}[A] \end{array} \tag{30}$$

The map π is induced by the representation of G on V, and r is the restriction map.

Theorem 9.2.1: *([31]) $h : \mathcal{PD}(V)^G \cong \mathbb{C}[A]^{W_\circ}$ as algebras where $W_\circ \subset Stab_W(\mathfrak{a}^* + \rho)$ is a finite reflection group.*

The proof of Theorem 9.2.1 is given in Lemmas 9.2.2 through 9.2.6 below:

Lemma 9.2.2: *Let $N = Stab_W(A)$. Then $Im(r) \subset \mathbb{C}[A]^N$ and these two rings have the same field of fractions.*

Proof: From (30) it is clear that $Im(r) \subset \mathbb{C}[A]^N$ and r factors through $Im(r)$:

$$\mathbb{C}[\mathfrak{h}^*]^W \twoheadrightarrow Im(r) \hookrightarrow \mathbb{C}[A]^N.$$

Letting $\mathfrak{h}^*//W$, $\mathcal{V}$ and $A//N$ denote the varieties with coordinate rings $\mathbb{C}[\mathfrak{h}^*]^W$, $Im(r)$ and $\mathbb{C}[A]^N$ respectively, we have

$$\mathfrak{h}^*//W \hookleftarrow \mathcal{V} \twoheadleftarrow A//N.$$

We see that $r^* : A//N \rightarrow \mathfrak{h}^*//W$ fails to be one-to-one whenever $\sigma \in W$, $\sigma \notin N$ but $\sigma(A) \cap A \neq \emptyset$. In this case, $\sigma(A) \neq A$, so $\sigma(A) \cap A$ is a lower-dimensional subvariety in A. Thus r^* fails to be one-to-one on $\cup_{\sigma \in W \setminus N}(\sigma(A) \cap A)$, whose complement is an open set in A. Thus the map $A//N \rightarrow \mathcal{V}$ is onto and generically one-to-one, so these varieties have the same field of rational functions. □

Lemma 9.2.3: *$\mathbb{C}[A]^N$ is the integral closure of $Im(r)$ in its field of fractions.*

Proof: Since $\mathbb{C}[A]$ is a polynomials ring, it is integrally closed in its quotient field $\mathbb{C}(A)$. Thus if $f \in \mathbb{C}(A)^N$ is integral over $\mathbb{C}[A]^N$ it is integral over $\mathbb{C}[A]$ and hence $f \in \mathbb{C}[A] \cap \mathbb{C}(A)^N = \mathbb{C}[A]^N$. This shows that $\mathbb{C}[A]^N$ is integrally closed in $\mathbb{C}(A)^N$.

It is known that $\mathbb{C}[\mathfrak{h}^*]$ is integral over $\mathbb{C}[\mathfrak{h}^*]^W$. (See Lemma 4.1.2 in [50].) Hence the homomorphic image $\mathbb{C}[A]$ of $\mathbb{C}[\mathfrak{h}^*]$ is integral over $Im(r)$ and in particular $\mathbb{C}[A]^N$ is integral over $Im(r)$. Using Lemma 9.2.2, we conclude that $\mathbb{C}[A]^N$ is the integral closure of $Im(r)$ in its quotient field. □

Lemma 9.2.4: $\mathbb{C}[A]^N \subset Im(h) \subset \mathbb{C}[A]$.

Proof: The map h is injective since each $D \in \mathcal{PD}(V)^G$ is completely determined by its eigenvalues $\{\beta_D(\lambda) : \lambda \in \Lambda\}$. As $\mathcal{PD}(V)^G$ is a polynomial ring (by Corollary 7.4.4), so is $Im(h)$. We have $Im(r) \subset Im(h) \subset \mathbb{C}[A]$. Given $f \in \mathbb{C}[A]^N$, we know that f is integral over $Im(r)$, hence over $Im(h)$. Thus $f \in Im(h)$. □

Lemma 9.2.5: $Im(h) = \mathbb{C}[A]^{W_\circ}$ *for some subgroup* $W_\circ \subset N$.

Proof: Apply Galois theory to the fraction fields of the rings $\mathbb{C}[A]^N \subset Im(h) \subset \mathbb{C}[A]$. □

Lemma 9.2.6: $W_\circ$ *is a finite reflection group.*

Proof: This follows from the fact that $\mathbb{C}[A]^{W_\circ} = Im(h)$ is a polynomial ring. (See Theorem 4.2.5 in [50].) □

9.3. *Characterizing the eigenvalue polynomials*

Recall that for $\nu \in \Lambda$, $\widehat{p}_\nu$ can be regarded as a polynomial function on $\mathfrak{a}^* = Span_{\mathbb{C}}(\Lambda)$. We now shift $d_\nu \widehat{p}_\nu$ to obtain a polynomial e_ν on $A = \mathfrak{a}^* + \rho$:

$$e_\nu(\lambda + \rho) = d_\nu \widehat{p}_\nu(\lambda)$$

for $\lambda \in \mathfrak{a}^*$. Note that

$$e_\nu = h\big(d_\nu p_\nu(z, \partial)\big)$$

where h is given by (29). Moreover, $e_\nu(\lambda + \rho) = \left[{\lambda \atop \nu}\right]$ when $\lambda \in \Lambda$, in view of Proposition 8.3.2.

These polynomials have the following remarkable property.

Theorem 9.3.1: *([31]) For $\nu \in \Lambda$ the polynomial e_ν is the unique polynomial such that:*

(1) e_ν is $W_\circ$-invariant.
(2) $deg(e_\nu) \leq |\nu|$.

(3) $e_\nu(\lambda+\rho) = 0$ *for all* $\lambda \in \Lambda$ *with* $|\lambda| \le |\nu|$, $\lambda \ne \nu$.
(4) $e_\nu(\nu+\rho) = 1$.

Proof: Property (1) holds because e_ν belongs to $Im(h) = \mathbb{C}[\mathfrak{a}^* + \rho]^{W_\circ}$. Property (2) holds by Corollary 9.1.2. Properties (3) and (4) are basic facts concerning the generalized binomial coefficients $\left[{\lambda \atop \nu}\right] = e_\nu(\lambda+\rho)$, as explained following Definition 8.2.1.

It remains to show uniqueness of e_ν. Let $d = |\nu|$ and set

$$\Lambda_d = \{\lambda \in \Lambda \ : \ |\lambda| \le d\}.$$

The map h restricts to yield an isomorphism

$$h : \mathcal{PD}^{(d)}(V)^G \to \mathbb{C}^{(d)}[\mathfrak{a}^* + \rho]^{W_\circ}$$

where $\mathcal{PD}^{(d)}(V)^G = \mathcal{PD}^{(d)}(V) \cap \mathcal{PD}(V)^G$ are the G-invariants in $\mathcal{PD}^{(d)}(V)$ and $\mathbb{C}^{(d)}[\mathfrak{a}^*+\rho]^{W_\circ} = (\sum_{m \le d} \mathcal{P}_m(\mathfrak{a}^*+\rho)) \cap \mathbb{C}[\mathfrak{a}^*+\rho]^{W_\circ}$. As $\{p_\lambda(z,\partial) \ : \ \lambda \in \Lambda_d\}$ is a basis for $\mathcal{PD}^{(d)}(V)^G$, we have that $\{e_\lambda \ : \ \lambda \in \Lambda_d\}$ is a basis for $\mathbb{C}^{(d)}[\mathfrak{a}^* + \rho]^{W_\circ}$. Now consider the linear map

$$\varepsilon : \mathbb{C}^{(d)}[\mathfrak{a}^* + \rho]^{W_\circ} \to \mathbb{C}^{\Lambda_d}, \quad \varepsilon(f) = \big(f(\lambda+\rho) \ : \ \lambda \in \Lambda_d\big).$$

In particular we have

$$\varepsilon(e_\nu) = \left(\begin{bmatrix} \lambda \\ \nu \end{bmatrix} : \ \lambda \in \Lambda_d \right).$$

As $\left[{\lambda \atop \lambda}\right] = 1$ and $\left[{\lambda \atop \nu}\right] = 0$ whenever $|\nu| > |\lambda|$ or $|\nu| = |\lambda|$ but $\nu \ne \lambda$, we see that $\{\varepsilon(e_\nu) \ : \ \nu \in \Lambda_d\}$ is a basis for $\mathbb{C}^{\Lambda_d}$. Thus ε is a vector space isomorphism. Now if $f \in \mathbb{C}[\mathfrak{a}^* + \rho]$ satisfies properties (1)-(4) then $f \in \mathbb{C}^{(d)}[\mathfrak{a}^*+\rho]^{W_\circ}$ and $\varepsilon(f) = \varepsilon(e_\nu)$. Hence $f = e_\nu$ as desired. □

9.4. GL(n) ⊗ GL(n) *yet again*

Recall the action of $G = GL(n,\mathbb{C}) \times GL(n,\mathbb{C})$ on $V = \mathbb{C}^n \otimes \mathbb{C}^n \cong M_{n,n}(\mathbb{C})$. Here $\mathfrak{h} = \mathfrak{h}_n \times \mathfrak{h}_n$ and the Weyl group W is isomorphic to $S_n \times S_n$. The highest weights that occur in $\mathbb{C}[V]$ have the form $(\lambda;\lambda)$ for $\lambda \in \mathfrak{h}_n^*$ non-negative and dominant. Thus we identify Λ with the set of all partitions $\lambda = (\lambda_1 \ge \cdots \ge \lambda_n \ge 0)$ and $\mathfrak{a}^* = Span_{\mathbb{C}}(\Lambda) = Span_{\mathbb{C}}\{(1^k;1^k) \ : \ 1 \le k \le n\}$ with $\mathfrak{h}^* \cong \mathbb{C}^n$. We can take

$$\rho = (n-1, n-2, \ldots, 1, 0)$$

under this identification. $W_\circ = Stab_W(\mathfrak{a}^* + \rho)$ is the diagonal subgroup in W which we identify with S_n acting as usual on $\mathbb{C}^n$:

$$\sigma \cdot (z_1, \ldots, z_n) = (z_{\sigma(1)}, \ldots, z_{\sigma(n)}).$$

Thus $\mathbb{C}[\mathfrak{a}^* + \rho]^{W_\circ}$ is identified with

$$\Lambda^*(n) = \{p \in \mathbb{C}[z_1, \ldots, z_n] \ : \ p(z+\rho) = p(\sigma \cdot z + \rho) \text{ for all } \sigma \in S_n\},$$

the algebra of *shifted symmetric polynomials* in n variables.

Now given $\nu \in \Lambda$ the *shifted Schur polynomial* $s^*_\nu(z)$ is defined by

$$s^*_\nu(z_1, \ldots, z_n) = \frac{det[(z_i + n - i \downharpoonright \nu_j + n - j)]}{det[(z_i + n - i \downharpoonright n - j)]}, \tag{31}$$

where $(y \downharpoonright k)$ is the *falling factorial*

$$(y \downharpoonright k) = y(y-1)\cdots(y-k+1).$$

It is shown in [42] that s^*_ν is a well defined polynomial function. The reader should compare Equation (31) with the determinantal formula (27) for Schur polynomials.

Recall that the canonical invariants p_ν for this example are given, up to normalization, by the Schur polynomials s_ν. (See Proposition 8.4.3.) The eigenvalue polynomials e_ν, which interpolate the generalized binomial coefficients, are, up to normalization, the shifted Schur polynomials.

Proposition 9.4.1: *The eigenvalue polynomial e_ν for partition ν is given by*

$$e_\nu(\lambda + \rho) = \frac{1}{H(\nu)} s^*_\nu(\lambda)$$

where

$$H(\nu) = \frac{\prod_i (\nu_i + n - i)!}{\prod_{i<j}(\nu_i - \nu_j - i + j)}$$

is the product of the hook-lengths for ν.

Proof: We apply the characterization Theorem 9.3.1. The definition of s^*_ν shows that s^*_ν is shifted symmetric with degree $|\nu|$.

Suppose that $\lambda \neq \nu$ is a partition with $|\lambda| \leq |\nu|$. We have $s^*_\nu(\lambda) = 0$ as required by property (3) in Theorem 9.3.1. Indeed, $\lambda_\ell < \nu_\ell$ for some ℓ. Thus for all $j \leq \ell \leq i$ one has $\lambda_i \leq \lambda_\ell < \nu_\ell \leq \nu_j$ and hence $(\lambda_i + n - i \downharpoonright \nu_j + n - j) = 0$. That is, the (i,j)'th entry in the determinant from the numerator in (31) vanishes for $j \leq \ell \leq i$. It follows that $s^*_\nu(\lambda) = 0$ as claimed.

Finally we check that $s^*_\nu(\nu) = H(\nu)$. First note that $(\nu_i + n - i \downharpoonright \nu_j + n - j) = 0$ for $i > j$. Hence

$$det[(\nu_i + n - i \downharpoonright \nu_j + n - j)] = \prod_i (\nu_i + n - i)!$$

The denominator in $s^*_\nu(\nu)$ is the Vandermonde determinant in the variables $\nu+\rho$. This gives

$$det[(\nu_i + n - i \downharpoonleft n - j)] = \prod_{i<j}(\nu_i - \nu_j - i + j),$$

so $s^*_\nu(\nu) = H(\nu)$ as claimed. □

9.5. *Section 9 notes*

The Harish-Chandra isomorphism first appeared in [19]. The reader can find a modern treatment in Section V.5 of [29]. One reference for facts concerning integrality and integral closures, used in the proof of Theorem 9.2.1, is the text [2] by Atiyah and Macdonald.

The results in this section are due to Knop. A considerably more general version Theorem 9.2.1 was proved in [30]. This asserts that the center of the ring of invariant differential operators for any smooth affine G-variety is a polynomial ring, canonically isomorphic to the ring of invariants for a finite reflection group.

Knop calls $W_\circ$ the *little Weyl group.* This group is given explicitly in [31] for each saturated indecomposable multiplicity free action in the classification from Section 6. In most cases, $W_\circ$ coincides with $N = Stab_W(A)$.

Theorem 9.3.1 was conjectured by Sahi, who proved a special case in [46]. Shifted Schur polynomials are due to Okounkov and Olshanski [42, 43]. The proof of vanishing is taken from [42], which includes many remarkable properties for these functions. Recent work of Knop yields the eigenvalue polynomials e_ν for many other multiplicity free actions. For the actions $GL(n,\mathbb{C}) : S^2(\mathbb{C}^n)$, $GL(n,\mathbb{C}) : \Lambda^2(\mathbb{C}^n)$, $SO(n,\mathbb{C}) \times \mathbb{C}^\times : \mathbb{C}^n$ and $E_6 \times \mathbb{C}^\times : \mathbb{C}^{27}$ this results in *shifted Jack polynomials* with various parameters. See [32, 33].

A multiplicity free action $G : V$ is said to be a *Capelli action* when the map $\pi : \mathcal{ZU}(\mathfrak{g}) \to \mathcal{PD}(V)^G$ in (30) is surjective. In [24] it is shown that the irreducible multiplicity free actions $\mathbf{G}' \times \mathbb{C}^\times : V$ in Table 3 are all Capelli actions except for

$$\mathbf{G}' = \mathbf{Sp(2n)} \otimes \mathbf{SL(3)}, \mathbf{Spin(9)}, \mathbf{E_6}.$$

References

1. D. Arnal, O. Boukary Baoua, C. Benson, and G. Ratcliff, *Invariant theory for the orthogonal group via star products*, J. Lie Theory **11** (2001), 441–458.
2. M. F. Atiyah and I. G. Macdonald, *Introduction to commutative algebra*, Addison-Wesley, Reading, Mass., 1969.

3. C. Benson, J. Jenkins, R. Lipsman, and G. Ratcliff, *A geometric criterion for Gelfand pairs associated with the Heisenberg group*, Pacific J. Math. **178** (1997), 1–36.
4. C. Benson, J. Jenkins, and G. Ratcliff, *Bounded K-spherical functions on Heisenberg groups*, J. Funct. Anal. **105** (1992), 409–443.
5. C. Benson and G. Ratcliff, *A classification for multiplicity free actions*, J. Algebra **181** (1996), 152–186.
6. ______, *Combinatorics and spherical functions on the Heisenberg group*, Representation Theory **2** (1998), 79–105.
7. ______, *Rationality of the generalized binomial coefficients for a multiplicity free action*, J. Austrl. Math. Soc. (Series A) **68** (2000), 387–410.
8. M. Brion, *Classification des espaces homogènes sphériques*, Compos. Math. **63** (1987), 189–208.
9. ______, *Spherical varieties: an introduction*, Prog. Math., vol. 80, pp. 11–26, Birkhäuser, Basel, 1989.
10. M. Brion, D. Luna, and T. Vust, *Espaces homogènes shpériques*, Invent. Math. **84** (1986), 565–619.
11. C. Chevalley and R. Shafer, *The exceptional Lie algebras F_4 and E_6*, Proc. Nat. Acad. Sci. Amer. **36** (1950), 137–141.
12. H. Dib, *Fonctions de Bessel sur une algèbre de Jordan*, J. Math. Pures Appl. **69** (1990), 403–448.
13. J. Faraut and A. Koranyi, *Analysis on symmetric cones*, Oxford University Press, New York, 1994.
14. G. Folland, *Harmonic analysis in phase space*, Princeton University Press, New Jersey, 1989.
15. ______, *A course in abstract harmonic analysis*, CRC Press, Boca Raton, 1995.
16. R. Gangolli and V.S. Varadarajan, *Harmonic analysis of spherical functions on real reductive groups*, Springer-Verlag, New York, 1988.
17. R. Goodman and N. Wallach, *Representations and invariants of the classical groups*, Encyclopedia of Mathematics and its Applications, vol. 68, Cambridge University Press, New York, 1998.
18. V. Guillemin and S. Sternberg, *Multiplicity free spaces*, J. Differential Geom. **19** (1984), 31–56.
19. Harish-Chandra, *On some applications of the enveloping algebra of a semi-simple Lie algebra*, Trans. Amer. Math. Soc. **70** (1951), 185–243.
20. S. Helgason, *Differential geometry, Lie groups, and symmetric spaces*, Academic Press, New York, 1978.
21. ______, *Groups and geometric analysis*, Academic Press, New York, 1984.
22. R. Howe, *Remarks on classical invariant theory*, Trans. Amer. Math. Soc. **313** (1989), 539–570.
23. ______, *Perspectives on invariant theory: Schur duality, multiplicity-free actions and beyond*, Israel Math. Conf. Proc., vol. 8, Bar-Ilan Univ., Ramat Gan, 1995.
24. R. Howe and T. Umeda, *The Capelli identity, the double commutant theorem and multiplicity-free actions*, Math. Ann. **290** (1991), 565–619.

25. K. Johnson, *On a ring of invariant polynomials on a hermitian symmetric space*, J. Algebra **62** (1980), 72–81.
26. V. Kac, *Some remarks on nilpotent orbits*, J. Algebra **64** (1980), 190–213.
27. V. Kac, V. L. Popov, and E. B. Vinberg, *Sur les groupes lin aires alg briques dont l'alg bre des invariants est libre*, C. R. Acad. Sci. Paris S r. A-B **283** (1976), A875–A878.
28. T. Kimura, *Introduction to prehomogeneous vector spaces*, Transl. Math. Mono., vol. 215, Amer. Math. Soc., Providence, Rhode Island, 2003.
29. A. Knapp, *Lie groups beyond an introduction*, Progress in Math., vol. 140, Birkhäuser, Boston, 1996.
30. F. Knop, *A Harish-Chandra homomorphism for reductive group actions*, Annals of Math. **140** (1994), 253–289.
31. ______, *Some remarks on multiplicity free spaces*, NATO Adv. Sci. Inst. Ser. C Math. Phys. Sci., vol. 514, pp. 301–317, Kluwer Acad. Publ., Dordrecht, 1998.
32. ______, *Construction of commuting difference operators for multiplicity free spaces*, Sel. Math., New ser. **6** (2000), 443–470.
33. ______, *Semisymmetric polynomials and the invariant theory of matrix vector pairs*, Representation Theory **5** (2001), 224–266.
34. F. Knop and S. Sahi, *Difference equations and symmetric polynomials defined by their zeroes*, International Math. Research Notes **10** (1996), 473–486.
35. B. Kostant and S. Sahi, *The Capelli identity, tube domains and the generalized Laplace transform*, Advances in Math. **87** (1991), 71–92.
36. M. Krämer, *Sphärische untergruppen in kompakten zusammenhängenden Liegruppen*, Compos. Math. **38** (1979), 129–153.
37. M. Lassalle, *Une formule de binôme généralisée pour les polynômes de Jack*, C. R. Acad. Sci. Paris, Série I **310** (1990), 253–256.
38. A. Leahy, *A classification of multiplicity free representations*, J. Lie Theory **8** (1998), 367–391.
39. I. G. Macdonald, *Symmetric functions and Hall polynomials, second edition*, Clarendon Press, Oxford, 1995.
40. I. V. Mikityuk, *On the integrability of invariant Hamiltonian systems with homogeneous configuration spaces*, Math USSR-Sb. **57** (1987), 527–546.
41. A. Okounkov and G. Olshanski, *Shifted Jack polynomials, binomial formula, and applications*, Math. Res. Letters **4** (1997), 69–78.
42. ______, *Shifted Schur functions*, St. Petersburg Math. **9** (1998), 239–300.
43. ______, *Shifted Schur functions II. The binomial formula for characters of classical groups and its applications*, Amer. Math. Soc. Transl. Ser 2 **181** (1998), 245–271.
44. V. L. Popov, *Stability of the action of an algebraic group on an algebraic variety*, Izv. Akad. Nauk SSSR Ser. Mat. **36** (1972), 367–379.
45. B. Ørsted and G. Zhang, *Weyl quantization and tensor products of Fock and Bergman spaces*, Indiana Math. Journal **43** (1994), 551–583.
46. S. Sahi, *The spectrum of certain invariant differential operators associated to Hermitian symmetric spaces*, Lie Theory and Geometry (J. L. Brylinski, ed.), Progress in Math., vol. 123, Birkhäuser, Boston, 1994, pp. 569–576.

47. M. Sato and T. Kimura, *A classification of irreducible prehomogeneous vector spaces and their relative invariants*, Nagoya Math. J. **65** (1977), 1–155.
48. G. Schwarz, *Representations of simple Lie groups with regular rings of invariants*, Invent. Math. **49** (1978), 167–191.
49. F. Servedio, *Prehomogeneous vector spaces and varieties*, Trans. Amer. Math. Soc. **176** (1973), 421–444.
50. T.A. Springer, *Invariant theory*, Lecture Notes in Math., vol. 585, Springer Verlag, New York, 1977.
51. R. Stanley, *Some combinatorial properties of Jack symmetric functions*, Advances in Math. **77** (1989), 76–115.
52. E. B. Vinberg, *Complexity of actions of reductive Lie groups*, Funct. Anal. and Appl. **20** (1986), 1–11.
53. ______, *Commutative homogeneous spaces and co-isotropic symplectic actions*, Russian Math. Surveys **56** (2001), 1–60.
54. E. B. Vinberg and V. L. Popov, *On a class of quasihomogeneous affine varieties*, Izv. Akad. Nauk SSSR Ser. Mat. **36** (1972), 749–764.
55. H. Weyl, *The classical groups, their invariants and representations*, Princeton University Press, Princeton, N.J., 1946.
56. Z. Yan, *Special functions associated with multiplicity-free representations*, unpublished preprint.

Multiplicity-Free Spaces and Schur–Weyl–Howe Duality

Roe Goodman

Rutgers University
Department of Mathematics
110 Frelinghuysen Rd
Piscataway NJ 08854-8019, USA
E-mail: goodman@math.rutgers.edu

Contents

Introduction 307
1. Representations and duality 307
   1.1. Representations of algebraic groups 307
   1.2. Examples 308
   1.3. Reductive groups and isotypic decompositions 309
   1.4. Multiplicities and duality 311
2. Proof of duality theorem and examples 313
   2.1. Density lemmas 313
   2.2. Proof of duality theorem 315
   2.3. Examples 316
3. Schur–Weyl duality 320
   3.1. Commutant of GL(n) action on tensors 320
   3.2. Highest weight theory 321
   3.3. Duality and N-fixed vectors 326
4. Commutant character formulas 329
   4.1. Characters 329
   4.2. Frobenius and determinant character formulas 329
   4.3. Proof of Frobenius character formula 330
   4.4. Proof of determinant character formula 331
5. Character formulas for Schur–Weyl duality 332
   5.1. Frobenius formula for $\mathfrak{S}_k$ characters 332
   5.2. Determinant formula for $\mathfrak{S}_k$ characters 334
   5.3. Schur–Weyl duality and GL(k)-GL(n) duality 336

6. Polynomial invariants and FFT 338
6.1. Invariant polynomials 338
6.2. Invariants of vectors and covectors 339
6.3. Polynomial FFT for GL(n) 341
6.4. Polynomial FFT for the orthogonal group 341
6.5. Polynomial FFT for the symplectic group 343
7. Tensor invariants and proof of FFT 344
7.1. Tensor invariants for GL(V) 344
7.2. Proof of polynomial FFT for GL(V) 345
7.3. Tensor invariants for orthogonal and symplectic groups 347
7.4. Proof of polynomial FFT for orthogonal and symplectic groups 350
8. Weyl algebra and Howe duality 351
8.1. Duality in the Weyl algebra 351
8.2. Howe duality for orthogonal/symplectic groups 355
8.3. Howe duality for GL(k) 357
9. Harmonic duality 358
9.1. Harmonic polynomials 358
9.2. Main theorem 360
10. Decomposition of harmonic polynomials 364
10.1. O(k) Harmonics (k *odd*) 365
10.2. O(k) Harmonics (k *even*) 371
10.3. Examples of harmonic decompositions 374
11. Symplectic group and oscillator representation 375
11.1. Real symplectic group 376
11.2. Holomorphic (coherent-state) model for oscillator representation 383
11.3. Bargmann–Segal transform 387
11.4. Real (oscillatory-wave) model for oscillator representation 389
11.5. Analytic vectors 391
12. Dual pair Sp(n, R)–O(k) 394
12.1. Decomposition of $\mathrm{H}^2(M_{n\times k})$ under $\mathrm{Mp}(n,\mathbb{R}) \times \mathrm{O}(k)$ 394
12.2. Square-integrable representations of $\mathrm{Sp}(n,\mathbb{R})$ 399
13. Brauer algebra and tensor harmonics 402
13.1. Centralizer algebra and Brauer diagrams 402
13.2. Generators for the centralizer algebra 405
13.3. Relations in the centralizer algebra 407
13.4. Harmonic tensors 409
13.5. Decomposition of harmonic tensors for Sp(V) 410
References 414

Dedicated to the memory of Irving E. Segal, who introduced me to the beauties and mysteries of representation theory.

Introduction

The unifying theme of these lectures is the *duality* between the irreducible representations occuring in a linear group action and irreducible representations of the commuting algebra relative to this action. This notion of duality in representation theory was introduced by Schur a century ago, and it has developed into an important tool with many applications. In keeping with the tutorial aspect, I have tried to tell the story starting from the beginning and including complete proofs of all the major results (at several points I refer to the lectures of Benson–Ratcliff in the present volume for details). Of course, this limits the scope of the lectures to the more classical parts of the theory: Schur–Weyl–Brauer duality for finite-dimensional representations, and Howe duality between finite-dimensional and infinite-dimensional highest-weight representations. Substantial parts of these lectures are based on joint work with Nolan Wallach and I would like to acknowledge his contributions to my understanding of representation theory. I would also like to thank Eng-Chye Tan and Chen-Bo Zhu for inviting me to give these lectures and for their wonderful hospitality.

1. Representations and duality

1.1. *Representations of algebraic groups*

Assume that $G \subset \mathrm{GL}(n, \mathbb{C})$ is an *algebraic group* (defined by a set of polynomial equations in the matrix entry functions). We denote by $\mathrm{Aff}(G)$ the commutative algebra of *regular functions* on G (the restrictions to G of polynomials in the matrix entry functions x_{ij} and $\det^{-1}$).

Let (ρ, L) be a representation of G on a complex vector space L. If L is finite-dimensional, then we say that ρ is *regular* (rational) if the representative functions $g \mapsto \mathrm{tr}(\rho(g)E)$, for $E \in \mathrm{End}(L)$, are regular. Every regular function on G arises as such a representative function. When L is infinite dimensional, we say that ρ is *locally regular* if for all $x \in L$ there is finite-dimensional G-invariant subspace M containing x so that $(\rho|_M, M)$ is a regular representation.

The most fundamental tool in representation theory is **Schur's Lemma:** *If E and F are irreducible, finite-dimensional representations of* *a group G, then*

$$\dim \operatorname{Hom}_G(E,F) = \begin{cases} 1 & \text{if } E \cong F \\ 0 & \text{if } E \ncong F \end{cases}$$

($\operatorname{Hom}_G(E,F)$ denotes the space of linear transformations $T : E \to F$ that intertwine the G actions on the two spaces). To prove Schur's Lemma, observe that the null space and range of T are G-invariant subspaces, so T must be either zero or bijective, with the first case holding if $E \ncong F$. When $E \cong F$ and S, T are two nonzero intertwining maps, take λ to be an eigenvalue of $S^{-1}T$. Since $S^{-1}T - \lambda I$ commutes with the action of G on E and has a nonzero null space, it must be zero.

1.2. *Examples*

(1) Let (π, V) be any regular (finite-dimensional) representation of G. We denote by $\mathcal{P}(V)$ the algebra of complex-valued polynomial functions on V. Define a representation of G on $\mathcal{P}(V)$ by

$$\rho(g)f(v) = f(\pi(g)^{-1}v) \quad \text{for } f \in \mathcal{P}(V) \text{ and } g \in G\,.$$

Since the G action is linear, it commutes with the $\mathbb{C}^\times$ action on V by scalar multiplication, and we have the direct-sum decomposition into finite-dimensional G-invariant subspaces

$$\mathcal{P}(V) = \bigoplus_{k \geq 0} \mathcal{P}^k(V)\,,$$

where $\mathcal{P}^k(V)$ is the space of homogeneous polynomials of degree k. The action of G on each of these spaces is regular, so the representation ρ is locally regular. Furthermore, the G action preserves the multiplication on $\mathcal{P}(V)$.

(2) With (π, V) as above, we can take the full tensor algebra

$$\mathcal{T}(V) = \bigoplus_{k \geq 0} V^{\otimes k}$$

with G action $\rho(g)(v_1 \otimes \cdots \otimes v_k) = \pi(g)v_1 \otimes \cdots \otimes \pi(g)v_k$. Since G leaves invariant each subspace $V^{\otimes k}$, the representation ρ is locally regular. As in the previous example, the G action preserves the (noncommutative) multiplication on $\mathcal{T}(V)$.

(3) Let $X \subset \mathbb{C}^m$ be an affine algebraic set (the zero set of a family of polynomials) and suppose that there is a regular G action on X

$$G \times X \longrightarrow X\,, \quad (g, x) \mapsto g \cdot x\,.$$

Set $L = \text{Aff}(X)$ (the restriction to X of the polynomial functions on $\mathbb{C}^m$). Let G act on L by $\rho(g)f(x) = f(g^{-1} \cdot x)$. We can prove that this representation is locally regular as follows.

Given $f \in \text{Aff}(X)$, set $V_f = \text{Span}\{\rho(g)f : g \in G\}$. The function $(g, x) \mapsto f(g^{-1} \cdot x)$ on $G \times X$ is regular, and $\text{Aff}(G \times X) = \text{Aff}(G) \otimes \text{Aff}(X)$, there are regular functions ϕ_i on G and ψ_i on X so that

$$f(g^{-1} \cdot x) = \sum_{k=1}^{p} \phi_k(g)\psi_k(x) .$$

In particular, $V_f \subset \text{Span}\{\psi_k\}$ is finite-dimensional, so we can choose $g_1, \ldots, g_r$ in G such that the functions $f_i = \rho(g_i)f$ give a basis for V_f. Now choose points $x_1, \ldots, x_k$ in X so that the evaluation functionals δ_{x_i} are a basis for V_f^*. Since

$$\langle \rho(g) f_i, \delta_{x_j} \rangle = \rho(gg_i) f(x_j) = \sum_{k=1}^{r} \phi_k(gg_i)\psi(x_j) ,$$

we see that the representation of G on V_f is regular. Thus (ρ, L) is locally regular.

1.3. *Reductive groups and isotypic decompositions*

A complex algebraic group G is called *reductive* if every finite dimensional regular representation decomposes as a direct sum of irreducible representations (this property is equivalent to every G-invariant subspace of a regular representation having a G-invariant complementary subspace). The *classical groups* are reductive:

- the general linear group $\text{GL}(n, \mathbb{C})$ of invertible $n \times n$ complex matrices
- the special linear group $\text{SL}(n, \mathbb{C})$ of $n \times n$ complex matrices of determinant one
- the orthogonal group $\text{O}(\mathbb{C}^n, \omega)$ of $n \times n$ matrices preserving a non-degenerate symmetric bilinear form $\omega(x, y) = x^t By$ on $\mathbb{C}^n$, where B is a symmetric invertible $n \times n$ matrix (defining equation $g^t Bg = B$)
- the special orthogonal group $\text{SO}(\mathbb{C}^n, \omega)$ of orthogonal matrices of determinant one
- the symplectic group $\text{Sp}(\mathbb{C}^{2n}, \omega)$ of $2n \times 2n$ matrices preserving a non-degenerate skew-symmetric bilinear form $\omega(x, y) = x^t Jy$ on $\mathbb{C}^{2n}$, where J is a skew-symmetric invertible $2n \times 2n$ matrix (defining equation $g^t Jg = J$)

Finite groups are shown to be reductive by the method of averaging over the group. The proof that classical groups are reductive can be carried out analytically by integrating over a compact real form (Weyl's *unitary trick* – see [16, Theorem 2.4.7]), or algebraically by using a *Casimir operator* (see [16, Theorem 2.4.5]). Direct products of reductive groups are reductive, and the quotient of a reductive group by a closed normal subgroup is reductive (this is obvious). An algebraic group is reductive if and only if its identity component is reductive.

Assume G is reductive, and let $\widehat{G}$ be the equivalence classes of irreducible finite-dimensional regular representations of G. For each $\lambda \in \widehat{G}$ fix a representation (π^λ, F^λ) in the class λ. Let λ^* be the equivalence class of the *contragredient* representation on the dual space $(F^\lambda)^*$.

Given a locally regular representation (ρ, L) of G, set

$$L_{(\lambda)} = \sum V \qquad (\text{sum of all } V \subset L \text{ such that } \rho|_V \cong F^\lambda).$$

Call $L_{(\lambda)}$ the *λ-isotypic component* of L. Define $\mathrm{Spec}(\rho) = \{\lambda \in \widehat{G} : L_{(\lambda)} \neq 0\}$ (the *G-spectrum* of (ρ, L)).

Proposition 1.1: $L = \bigoplus_{\lambda \in \mathrm{Spec}(L)} L_{(\lambda)}$ *(algebraic direct sum).*

Proof: We first verify that the sum is direct. Suppose, for the sake of contradiction, that $L_{(\lambda)} \cap L_{(\mu)} \neq 0$ for some $\lambda, \mu \in \widehat{G}$ with $\lambda \neq \mu$. Then there exists a G-invariant subspace $W \neq 0$ so that $W \subset L_{(\lambda)} \cap L_{(\mu)}$ and $\dim W < \infty$. Since G is reductive, $W = V_1 \oplus \cdots \oplus V_n$, where each V_j is an irreducible G-module. Hence $V_j \cong F^\lambda$ and also $V_j \cong F^\mu$, a contradiction.

To see that L is the sum of its isotypic components, set $L_0 = \bigoplus_\lambda L_{(\lambda)}$. If $L_0 \neq L$, then there exists a nonzero $x \in L \backslash L_0$. But x is contained in a finite dimensional G invariant subspace W that is the direct sum of irreducible G-invariant subspaces. Hence $W \subset L_0$, a contradiction. □

Corollary 1.2: *There is a linear projection $x \mapsto x^\natural$ from L onto the space L^G of G-fixed vectors.*

We now turn to the G-module structure of the isotypic components of a representation L. Denote by $\mathrm{Hom}_G(F^\lambda, L)$ the vector space of all linear maps $T : F^\lambda \to L$ that intertwine the G actions on these spaces. This is the space of *covariants of type* λ.

Theorem 1.3: *If (ρ, L) is a locally regular representation of a complex reductive algebraic group G, then*

$$L \cong \bigoplus_{\lambda \in \mathrm{Spec}(\rho)} E^\lambda \otimes F^\lambda,$$

where $E^\lambda = \mathrm{Hom}_G(F^\lambda, L)$ and G acts by $1 \otimes \rho$ on each summand. In particular, the multiplicity of λ in ρ is the dimension of the space of covariants of type λ.

Proof: Let $T \in \mathrm{Hom}_G(F^\lambda, L)$ be a nonzero intertwining operator. Then T is injective, by Schur's lemma. Conversely, if $W \subset L_{(\lambda)}$ is a G invariant irreducible subspace, then there is an intertwining map T so that $W = T(F^\lambda)$. This implies that the map $T \otimes v \mapsto Tv$ from $E^\lambda \otimes F^\lambda$ to $L_{(\lambda)}$ is surjective.

It remains to prove that the map $E^\lambda \otimes F^\lambda \to L_{(\lambda)}$ is injective. Suppose $v_j \in F^\lambda$ and $T_j \in E^\lambda$ satisfy $\sum_j T_j v_j = 0$. We may assume that $\{v_j\}$ is linearly independent. Fix a decomposition

$$L_{(\lambda)} = \bigoplus_i F_i \,, \quad F_i \cong F^\lambda \,.$$

This defines G-invariant projections $P_i : L_{(\lambda)} \to F^\lambda$, and by assumption

$$\sum_j P_i T_j v_j = 0 \quad \text{for all } i \,.$$

By Schur's lemma, $P_i T_j = c_{ij} I$ for some $c_{ij} \in \mathbb{C}$, so we conclude that $c_{ij} = 0$ for all i, j, by the linear independence of $\{v_j\}$. Hence $T_j = 0$ for all j. □

1.4. *Multiplicities and duality*

One says that L is *multiplicity-free* as a G module if $\dim E^\lambda = 1$ for all $\lambda \in \mathrm{Spec}(\rho)$. In this case L is uniquely determined as a G-module by its spectrum. For a detailed analysis of such representations when $L = \mathcal{P}(X)$ and X is a vector space or affine variety with regular G action see the lectures by Benson–Ratcliff in this volume.

In these lectures we will study representations (ρ, L) that are *not* multiplicity free. We want to determine

- The spectrum $\mathrm{Spec}(\rho) \subset \widehat{G}$,
- The multiplicities $m_\lambda = \dim E^\lambda$,
- Explicit models for the multiplicity spaces E^λ.

Let $\mathrm{End}_G(L)$ be the algebra of linear transformations on L that commute with the G action. There is a natural representation of this algebra on each multiplicity space E^λ. Indeed, if $A \in \mathrm{End}_G(L)$ and $T \in E^\lambda$, then the linear map $A \circ T : F^\lambda \to L$ also commutes with the G action on L, and hence is an element of E^λ. Following the ideas of I. Schur, H. Weyl and R. Howe, the unifying theme in our approach will be

Hidden Symmetry: Study the spaces E^λ as modules for *good* subalgebras of $\mathrm{End}_G(L)$.

The term *hidden symmetry* comes from applications of representation theory to quantum mechanics in cases where the *geometric symmetries* such as rotation invariance do not suffice to explain the multiplicities in the energy spectrum. In some cases, one can find a larger symmetry group containing G and extend the representation of G to a representation of this larger group on L that is multiplicity free. In other cases the hidden symmetries are given by a Lie algebra of differential operators commuting with the G action (see [30]).

When L is infinite-dimension (for example, when $L = \mathrm{Aff}(X)$ with X an affine G variety), then $\mathrm{End}(L)$ is too big to deal with purely algebraically. In the context of *unitary representations* on a Hilbert space, one uses the von Neumann algebra of *bounded* operators that commute with G. In our algebraic setting we shall assume that L is of countable dimension and that we have a subalgebra $\mathcal{R} \subset \mathrm{End}(L)$ that satisfies

(i) $\mathcal{R}$ acts irreducibly on L
(ii) $\mathcal{R}$ is invariant under G, relative to the action $\mathrm{Ad}(g)T = \rho(g)T\rho(g)^{-1}$, and the representation Ad of G on $\mathcal{R}$ is locally regular

In case $\dim L < \infty$ we take $\mathcal{R} = \mathrm{End}(L) \cong L \otimes L^*$ and these conditions are always satisfied. When $L = \mathcal{P}(X)$ with X a smooth affine G variety, we take $\mathcal{R} = \mathbb{D}(X)$, the *algebraic* differential operators on X (see Agricola [1]). In particular, if X is a vector space with linear G action, then $\mathbb{D}(X)$ is the *Weyl algebra* $\mathcal{PD}(X)$ of differential operators with polynomial coefficients, which we will examine in detail in Section 8.

Fix $\mathcal{R}$ satisfying the conditions (i) and (ii) and let

$$\mathcal{R}^G = \{T \in \mathcal{R} : \mathrm{Ad}(g)T = T \quad \text{for all } g \in G\}$$

(the commutant of $\rho(G)$ in $\mathcal{R}$).

Theorem 1.4: *Each multiplicity space E^λ is an irreducible $\mathcal{R}^G$ module. Furthermore, if $\lambda, \mu \in \mathrm{Spec}(\rho)$ and $E^\lambda \cong E^\mu$ as an $\mathcal{R}^G$ module, then $\lambda = \mu$.*

In the next section we will prove this theorem.[a] At this point we derive some consequences. The following corollary plays a fundamental role in our approach to Howe duality.

Corollary 1.5: *Let σ be the representation of $\mathcal{R}^G$ on L. Then (σ, L) is a semisimple $\mathcal{R}^G$ module, and each irreducible submodule E^λ occurs with finite multiplicity* $\dim F^\lambda$.

When L is finite-dimensional then $\mathcal{R} = \mathrm{End}(L)$, and from the inequivalence of the representations E^λ together with Schur's lemma we obtain the classical *Double Commutant Theorem.*

Corollary 1.6: *If* $\dim L < \infty$ *and* $\mathcal{B} = \mathrm{End}_G(L)$, *then* $\mathrm{Span}\{\rho(G)\}$ *consists of all linear transformations on L that commute with $\mathcal{B}$.*

Corollary 1.7 (Duality Correspondence): *Let* $\mathrm{Spec}(\sigma)$ *denote the set of equivalence classes of the irreducible representations of the algebra $\mathcal{R}^G$ that occur in L. Then the map $F^\lambda \to E^\lambda$ sets up a bijection between* $\mathrm{Spec}(\rho)$ *and* $\mathrm{Spec}(\sigma)$.

2. Proof of duality theorem and examples

2.1. *Density lemmas*

Lemma 2.1 (Dixmier–Schur): *Let L be a vector space over $\mathbb{C}$ of countable dimension. Let $\mathcal{R} \subset \mathrm{End}(L)$ be a subalgebra that acts irreducibly on L. Suppose $A \in \mathrm{End}(L)$ commutes with $\mathcal{R}$. Then $A = \lambda I$ for some $\lambda \in \mathbb{C}$.*

Proof: Suppose that A is not a multiple of the identity. Since $\mathcal{R}$ acts irreducibly, Schur's lemma implies that $A - \lambda I$ is invertible for all $\lambda \in \mathbb{C}$. Hence for every nonzero polynomial $p(x)$ in one variable the operator $p(A)$ is invertible (factor $p(x)$ into linear factors). Thus there is an algebra homomorphism from the field $\mathbb{C}(x)$ of rational functions in one variable into $\mathrm{End}(L)$ given by $p(x)/q(x) \mapsto p(A)q(A)^{-1}$. Fix a nonzero vector $v \in L$. Then the linear map $r(x) \mapsto r(A)v$ is injective from $\mathbb{C}(x)$ to L. But $\mathbb{C}(x)$ has uncountable dimension as a vector space over $\mathbb{C}$, since the functions $\{(x-\lambda)^{-1} : \lambda \in \mathbb{C}\}$ are linearly independent, a contradiction. □

Assume now that L has countable dimension as a complex vector space and that $\mathcal{R} \subset \mathrm{End}(L)$ is a subalgebra that acts irreducibly on L.

[a]See [16, Theorem 4.5.12] for the case that $\mathcal{R}$ is a graded algebra; the generalization presented here is due to Agricola [1].

Lemma 2.2 (Jacobson): *Let X be any finite-dimensional subspace of L. Then every $f \in \mathrm{Hom}(X, L)$ is of the form $r|_X$ for some $r \in \mathcal{R}$.*

Proof: Let $\{x_1, \ldots, x_n\}$ be a basis for X. Define

$$L^{(n)} = \underbrace{L \oplus \cdots \oplus L}_{n \text{ copies}}, \quad X^{(n)} = \underbrace{X \oplus \cdots \oplus X}_{n \text{ copies}}, \quad x^{(n)} = [x_1, \ldots, x_n] \in X^{(n)}.$$

Let $\mathcal{R}$ act on $L^{(n)}$ by $r \cdot [y_1, \ldots, y_n] = [ry_1, \ldots, ry_n]$ for $r \in \mathcal{R}$ and $y_i \in L$, and extend f to a linear map $f^{(n)} : X^{(n)} \to L^{(n)}$ by

$$f^{(n)}[y_1, \ldots, y_n] = [f(y_1), \ldots, f(y_n)].$$

Denote by $M = \mathcal{R} \cdot x^{(n)}$ the cyclic $\mathcal{R}$ submodule generated by $x^{(n)}$. Define $L_i \subset L^{(n)}$ to be the vectors that have arbitrary entries from L in the ith place and are zero in the other positions. Pick a maximal subset $I \subset \{1, \ldots, n\}$ with the property that the sum

$$N = M + \sum_{i \in I} L_i$$

is direct. Then N is an $\mathcal{R}$ submodule of $L^{(n)}$. Since $\mathcal{R}$ acts irreducibly on L, the $\mathcal{R}$ modules $N \cap L_j$ are either zero or L_j for each j. But if $N \cap L_j = 0$, then the sum $N + L_j$ would be direct, contradicting the choice of I. Hence $N = L^{(n)}$, proving that M has an $\mathcal{R}$-invariant complement. Thus there is a projection $P : L^{(n)} \to M$ that commutes with the action of $\mathcal{R}$. We can write

$$P[y_1, \ldots, y_n] = \left[\sum_{j=1}^{n} p_{1j} y_j, \ldots, \sum_{j=1}^{n} p_{nj} y_j \right]$$

where $p_{ij} \in \mathrm{End}(L)$. Since P commutes with $\mathcal{R}$ on $L^{(n)}$, the transformations p_{ij} all commute with $\mathcal{R}$ on L. Hence $p_{ij} \in \mathbb{C}I$ by Lemma 2.1 and $[p_{ij}]$ is a matrix of scalars.

Now calculate

$$f^{(n)} P[y_1, \ldots, y_n] = \left[\sum_{j=1}^{n} p_{1j} f(y_j), \ldots, \sum_{j=1}^{n} p_{nj} f(y_j) \right] = P f^{(n)} [y_1, \ldots, y_n].$$

Hence $f^{(n)}$ commutes with P. Since $x^{(n)} \in M$ we have

$$f^{(n)} x^{(n)} = f^{(n)} P x^{(n)} = P f^{(n)} x^{(n)} \in M.$$

Thus there exists $r \in \mathcal{R}$ so that $f^{(n)} x^{(n)} = r x^{(n)}$. Since $\{x_1, \ldots, x_n\}$ is a basis for X, this implies that $f = r|_X$. □

Corollary 2.3 (Burnside): *If* $\dim L < \infty$ *then* $\mathcal{R} = \mathrm{End}(L)$.

Now let (ρ, L) be a locally regular representation of G with $\dim L$ countable. Assume that $\mathcal{R} \subset \mathrm{End}(L)$ satisfies conditions (i) and (ii) stated before Theorem 1.4.

Lemma 2.4: *Let* $X \subset L$ *be a finite-dimensional* G *invariant subspace. Then* $\mathcal{R}^G|_X = \mathrm{Hom}_G(X, L)$.

Proof: Let $T \in \mathrm{Hom}_G(X, L)$. Then by Lemma 2.2 there exists $r \in \mathcal{R}$ such that $r|_X = T$. Since G is reductive, condition (ii) implies that there is a projection $r \mapsto r^\natural$ from $\mathcal{R} \to \mathcal{R}^G$. But the map $\mathcal{R} \to \mathrm{Hom}(X, L)$ given by $y \mapsto y|_X$ intertwines the G actions, since X is G-invariant. Hence $T = T^\natural = r^\natural|_X$. □

2.2. *Proof of duality theorem*

Take $\lambda \in \mathrm{Spec}(\rho)$ and let $Z_\lambda \subset L_{(\lambda)}$ be any irreducible G-submodule. Given $f \in L$, we denote by $U_f = \mathcal{R}^G f$ the cyclic $\mathcal{R}^G$ module generated by f. We write $\mathbb{C}[G]$ for the *group algebra* of G (the formal finite linear combinations of the elements of G).

(a) *If* $0 \neq M \subset L_{(\lambda)}$ *is an* $\mathcal{R}^G$*-module, then* $M \cap Z_\lambda \neq 0$.

To verify this, take $0 \neq m \in M$ and set $X = \mathrm{Span}\{\rho(G)m\}$. Then $\dim X < \infty$ and $X \subset L_{(\lambda)}$. Hence there exists $T \in \mathrm{Hom}_G(X, Z_\lambda)$ with $Tm \neq 0$. By Lemma 2.4 there exists $r \in \mathcal{R}^G$ with $r|_X = T$. Then $rm = Tm \in M \cap Z_\lambda$.

(b) *If* $0 \neq f \in Z_\lambda$ *then* $U_f \cap Z_\lambda = \mathbb{C}f$.

Take $u = rf \in U_f \cap Z_\lambda$. Since $Z_\lambda = \mathrm{Span}\{\rho(G)f\}$, we have

$$rZ_\lambda = \mathrm{Span}\{\rho(G)rf\} = \mathrm{Span}\{\rho(G)u\} \subset Z_\lambda$$

Thus $r|_{Z_\lambda} \in \mathrm{End}_G(Z_\lambda) = \mathbb{C}I$ by Schur's Lemma. So $u = r \cdot f \in \mathbb{C}f$, proving (b).

(c) *If* $f \in Z_\lambda$ *is nonzero, then* U_f *is an irreducible* $\mathcal{R}^G$*-module.*

Indeed, if $0 \neq M \subset U_f$ is an $\mathcal{R}^G$-submodule, then $0 \neq M \cap Z_\lambda \subset \mathbb{C}f$ by (a) and (b). Thus $f \in M$ and hence $M = U_f$, which proves (c).

(d) *Let* $f_1, \ldots, f_d$ *be a basis of* Z_λ. *Set* $M_i = U_{f_i}$. *Then the sum* $\sum_{i=1}^d M_i$ *is direct and* $M_i \cong M_j$ *as* $\mathcal{R}^G$ *modules.*

We have $\mathrm{Span}\{\rho(g)|_{Z_\lambda} : g \in G\} = \mathrm{End}(Z_\lambda)$ by Corollary 2.3. Thus for each i there exists an element $u_i \in \mathbb{C}[G]$ such that $\rho(u_i)f_j = \delta_{ij} f_j$. Suppose

$m_i \in M_i$ and $\sum_i m_i = 0$. There exist $r_i \in \mathcal{R}^G$ so that $m_i = r_i f_i$. Hence

$$0 = \rho(u_j)\Big(\sum_i m_i\Big) = \sum_i \rho(u_j) r_i f_i = \sum_i r_i \rho(u_j) f_i = r_j f_j = m_j$$

for $j = 1, \ldots, d$. This proves the first statement of (d). For the second, apply Corollary 2.3 again to obtain $u_{ji} \in \mathbb{C}[G]$ such that $\rho(u_{ji}) f_i = f_j$. Since M_i and M_j are irreducible by (c), the map $\rho(u_{ji}) : M_i \to M_j$ is an $\mathcal{R}^G$-module isomorphism, by Schur's Lemma.

(e) *Let* M_i *be as in* (d). *Then* $L_{(\lambda)} = \bigoplus_{i=1}^d M_i$.

Recall that $L_{(\lambda)}$ is the sum of all irreducible G-submodules of L that are in the class λ. Thus it is enough to show that if W_λ is such a submodule then

$$W_\lambda \subset \bigoplus_{i=1}^{d} M_i \,. \tag{1}$$

Take a G isomorphism $T : Z_\lambda \to W_\lambda$. Then Lemma 2.4 furnishes $r \in \mathcal{R}^G$ such that $r = T$ on Z_λ. Hence W_λ satisfies (1), which proves (e).

The first assertion of Theorem 1.4 now follows from (c), (d), and (e). To prove the second assertion, it suffices to prove the following.

(f) *Let* f_λ *and* f_μ *be nonzero vectors in irreducible* G *subspaces* Z_λ *and* Z_μ. *Suppose* $U_{f_\lambda} \cong U_{f_\mu}$ *as* $\mathcal{R}^G$*-modules. Then* $\lambda = \mu$.

Let $T : U_{f_\lambda} \to U_{f_\mu}$ be an $\mathcal{R}^G$-module isomorphism. Let X be a finite-dimensional G-invariant subspace containing f_λ and Tf_λ. There is a projection operator $P_\lambda : X \to L_{(\lambda)}$ onto the λ-isotypic component of L, and Lemma 2.4 furnishes $r \in \mathcal{R}^G$ such that $r|_X = P_\lambda$. Thus $r \cdot f_\lambda = f_\lambda$ so we have

$$Tf_\lambda = Trf_\lambda = rTf_\lambda = P_\lambda T f_\lambda \in L_{(\lambda)} \,.$$

Since T is an $\mathcal{R}^G$ module isomorphism, it follows that $U_{f_\mu} \subset L_{(\lambda)}$. Hence $f_\mu \in L_{(\lambda)}$, and so we conclude that $\mu = \lambda$. □

2.3. *Examples*

(1) (Product groups) Let H and K be reductive complex algebraic groups, and let $G = H \times K$ be the direct product algebraic group, where $\mathrm{Aff}(G) \cong \mathrm{Aff}(H) \otimes \mathrm{Aff}(K)$ under the natural pointwise multiplication map. We can use the duality theorem to prove that $\widehat{G} = \widehat{H} \times \widehat{K}$: *Every irreducible regular representation* (L, ρ) *of* G *is given by*

$$L = M \otimes N \,, \quad \rho(h,k) = \sigma(h) \otimes \tau(k) \quad \text{for } h \in H \text{ and } k \in K \tag{2}$$

where (σ, M) *is an irreducible representation of* H *and* (τ, N) *is an irreducible representation* K. To prove this, suppose first that (ρ, L) is defined by (2). Then Corollary 2.3 implies that $\mathrm{End}(L)$ is spanned by the transformations $\{\rho(h,k) : h \in H, k \in K\}$ and hence $\mathrm{End}_G(L) = \mathbb{C}I$, showing that L is irreducible.

Conversely, given an irreducible regular representation (ρ, L) of G, use Theorem 1.4 (with $\mathcal{R} = \mathrm{End}(L)$) to decompose L as a K-module:

$$L = \bigoplus_{\lambda \in \widehat{K}} E^\lambda \otimes F^\lambda \,. \tag{3}$$

Set $\sigma(h) = \rho(h,1)$ and $\tau(k) = \rho(1,k)$. Since $\sigma(h)$ commutes with $\tau(k)$, H acts on each E^λ by some representation μ^λ. We claim that E^λ is irreducible under H. To prove this, note that

$$\mathrm{End}_K(L) \cong \bigoplus_{\lambda \in \widehat{K}} \mathrm{End}(E^\lambda) \otimes I \,. \tag{4}$$

Given $T \in \mathrm{End}_K(L)$, we know by Corollary 2.3 that T is a linear combination of the transformations $\sigma(h)\tau(k)$. Under the isomorphism (4) the K-invariant transformations only act on E^λ. This proves that $\mathrm{End}_K(L)$ is spanned by $\{\sigma(h) : h \in H\}$, and hence E^λ is irreducible under H by Theorem 1.4. Thus each summand in (3) is an irreducible G module, by the earlier argument, so there can be only one summand.

(2) (Multiplicity-free representations of product groups) Suppose (ρ, L) is any locally regular representation of G that is multiplicity-free. By Example (1) the isotypic decomposition of L under $H \times K$ is of the form

$$L = \bigoplus_{(\alpha,\beta) \in \Lambda} E^\alpha \otimes F^\beta \tag{5}$$

where $\Lambda \subset \widehat{H} \times \widehat{K}$ and E^α is the irreducible H-module of type α, while F^β is the irreducible K-module of type β. Set $\sigma = \rho|_H$ and $\tau = \rho|_K$. Then $\mathrm{Spec}(\sigma)$ is the projection $\Lambda \to \widehat{H}$, whereas $\mathrm{Spec}(\tau)$ is the projection $\Lambda \to \widehat{K}$. In general Λ is not determined by these projections. If both of these projections are injective, we say that the representation ρ sets up a *duality correspondence* between $\mathrm{Spec}(\sigma)$ and $\mathrm{Spec}(\tau)$. Clearly such representations of G must be very special, and in these lectures they will play an important role. The next example is the most familiar of them.

(3) (Two-sided group action) Let K be any reductive complex algebraic group. Set $G = K \times K$ and $L = \mathrm{Aff}(K)$. Define the representation ρ of G on L by

$$\rho(x,y)f(k) = f(x^{-1}ky) \quad \text{for } k, x, y \in K \,.$$

From Example (1) we know that $\widehat{G} = \widehat{K} \times \widehat{K}$. Consider Aff$(K)$ as a K-module relative to the *right translation* action $\rho(1,k)$ and apply Theorem 1.3:

$$\mathrm{Aff}(K) = \bigoplus_{\lambda \in \widehat{K}} E^\lambda \otimes F^\lambda . \tag{6}$$

Here K acts on $E^\lambda = \mathrm{Hom}_K(F^\lambda, \mathrm{Aff}(K))$ by $\rho(k,1) \circ T$, where $T : F^\lambda \to \mathrm{Aff}(K)$ intertwines the action of K on F^λ with the right translation action of K on Aff(K).

We claim that $E^\lambda \cong F^{\lambda^*}$. To prove this, define a map $E^\lambda \to F^{\lambda^*}$ (a special case of *Frobenius reciprocity*) by

$$T \mapsto \widehat{T} \in F^{\lambda^*}, \quad \langle \widehat{T}, v\rangle = (Tv)(1) \quad \text{for } v \in F^\lambda .$$

This map obviously intertwines the action of K. It is injective, since $(Tv)(1) = 0$ for all $v \in F^\lambda$ implies

$$(Tv)(k) = (T\pi^\lambda(k)v)(1) = 0 \quad \text{for all } k \in K ,$$

and hence $T = 0$. The map is surjective, since $v^* \in F^{\lambda^*}$ defines $T \in E^\lambda$ by

$$(Tv)(k) = \langle v^*, \pi^\lambda(k)v\rangle .$$

Clearly $\widehat{T} = v^*$. Thus the decomposition (6), relative to the action of $K \times K$, is

$$\mathrm{Aff}(K) \cong \bigoplus_{\lambda \in \widehat{K}} F^{\lambda^*} \otimes F^\lambda \cong \bigoplus_{\lambda \in \widehat{K}} \mathrm{End}(F^\lambda) .$$

This shows that Aff(K) is *multiplicity free* as a representation of $K \times K$ and there is a duality correspondence $\lambda \longleftrightarrow \lambda^*$.

(4) (Harmonics on the zero-sphere) Let $G = \mathrm{O}(1) = \{\pm 1\}$ acting on $\mathbb{C}$, and take $L = \mathcal{P}(\mathbb{C})$. In this case

$$\widehat{G} = \{F^+, F^-\} \quad \text{(trivial, signum)} .$$

The G-isotypic decomposition of L is thus

$$L = L^+ \oplus L^- \quad \text{(even polynomials} \oplus \text{odd polynomials)}$$

and each component has infinite multiplicity. We apply the duality philosophy to explain the multiplicities by finding operators on L that commute with G. Let $\mathcal{PD}(\mathbb{C})$ be the polynomial coefficient differential operators on $\mathcal{P}(\mathbb{C})$. Then one has

(a) *The operators* $\Delta = (d/dx)^2$, *multiplication by* x^2, *and* $x(d/dx) + 1/2$ (*shifted Euler operator*) *commute with* G *and span a Lie algebra* $\mathfrak{g}' \cong \mathfrak{sl}(2, \mathbb{C})$ *in* $\mathcal{PD}(\mathbb{C})$.

(b) *The Lie algebra* $\mathfrak{g}'$ *generates the commutant* $\mathcal{PD}(\mathbb{C})^G$ *of* G.

The proof of (a) is an easy calculation. The proof of (b) follows by considering the *symbol* $f(x, \xi)$ of a differential operator and using the fact that the algebra of G-invariant polynomials in (x, ξ) is generated by the quadratic polynomials x^2, $x\xi$, and ξ^2.

We define the *G-harmonic polynomials*

$$\mathcal{H} = \mathrm{Ker}(\Delta) = (\mathbb{C} \cdot 1) \oplus (\mathbb{C} \cdot x)\,.$$

Since Δ commutes with G, we have $G \cdot \mathcal{H} = \mathcal{H}$. Also $\mathcal{H}$ is multiplicity-free as a G-module. Let

$$\mathcal{I} = \mathcal{P}(\mathbb{C})^G = \mathbb{C}[x^2]$$

(the G-invariant polynomials). Then we have the *Invariant-Harmonic Decomposition:*

$$\mathcal{P}(\mathbb{C}) = E^+ \oplus E^- \cong \mathcal{I} \otimes \mathcal{H}$$

where $E^+ = \mathbb{C}[x^2] \cdot 1$, $E^- = \mathbb{C}[x^2] \cdot x$. We view this decomposition from the perspective of duality as follows:

- E^+ is an irreducible $\mathfrak{g}'$ module generated by 1.
- E^- is an irreducible $\mathfrak{g}'$ module generated by x.
- $\mathcal{P}(\mathbb{C})$ is multiplicity-free as module for $\mathfrak{g}' \times G$:

$$\mathcal{P}(\mathbb{C}) = (E^+ \otimes F^+) \oplus (E^- \otimes F^-)$$

From the algebraic point of view, we now have a complete picture of $\mathcal{P}(\mathbb{C})$ as a module for G and $\mathfrak{g}'$. However, there is much more that can be seen on the analytical side. There is a pre-Hilbert space structure on $\mathcal{P}(\mathbb{C})$ given by the *Fischer inner product:*

$$\langle f \mid g \rangle = \partial(f) g^*(0) = \int_{\mathbb{C}} f(x)\overline{g(x)}\, d\mu(x)$$

(where $d\mu(x)$ is normalized Gaussian measure on $\mathbb{C}$). We define the Bargmann–Fock space $\mathbf{H}^2$ as the completion of $\mathcal{P}(\mathbb{C})$ in this norm. The elements of $\mathbf{H}^2$ are holomorphic functions on $\mathbb{C}$ that are square-integrable relative to Gaussian measure. Let

$$\mathfrak{g}'_0 = \{X \in \mathfrak{g}' : X \text{ is skew-Hermitian relative to } \langle \cdot \mid \cdot \rangle\}\,.$$

Then $\mathfrak{g}'_0$ is a real form of the Lie algebra $\mathfrak{g}'$ and is isomorphic to $\mathfrak{sl}(2,\mathbb{R})$. Let $G' = \mathrm{SL}(2,\mathbb{R})$ and let $\widetilde{G}'$ be the two-fold cover of G'. The analytic duality correspondence between G and $\widetilde{G}'$ is the following.

Theorem 2.5: *The representation of* $\mathfrak{g}'_0$ *on* $\mathcal{P}(\mathbb{C})$ *integrates to a unitary representation of* $\widetilde{G}'$ *on* $\mathbf{H}^2$ (*the* oscillator *or* metaplectic *representation*). *It decomposes under the action of* $\widetilde{G}' \times G$ *as a direct sum of irreducible Hilbert spaces*

$$(\mathbf{H}^2_+ \otimes F^+) \oplus (\mathbf{H}^2_- \otimes F^-) \quad \text{(multiplicity-free)}$$

This is a special case of *Howe duality* for unitary highest-weight representations. We will study it in full generality in later sections.

3. Schur–Weyl duality

3.1. *Commutant of* GL(n) *action on tensors*

Consider the action of $\mathrm{GL}(n,\mathbb{C})$ on $\bigotimes^k \mathbb{C}^n$ by the kth tensor power ρ_k of its defining representation:

$$\rho_k(g)(v_1 \otimes \cdots \otimes v_k) = gv_1 \otimes \cdots \otimes gv_k \quad \text{for } v_i \in \mathbb{C}^n\,.$$

The symmetric group $\mathfrak{S}_k$ acts on $\bigotimes^k \mathbb{C}^n$ by permuting the tensor positions:

$$\sigma_k(s)(v_1 \otimes \cdots \otimes v_k) = v_{s^{-1}(1)} \otimes \cdots \otimes v_{s^{-1}(k)}$$

(the vector in position i is moved to the vector in position $s(i)$). It is clear that $\sigma_k(s)\rho_k(g) = \rho_k(g)\sigma_k(s)$ for all $g \in \mathrm{GL}(n,\mathbb{C})$ and $s \in \mathfrak{S}_k$.

Proposition 3.1 (Schur): *Any linear transformation B on $\bigotimes^k \mathbb{C}^n$ that commutes with $\sigma_k(\mathfrak{S}_k)$ is a linear combination of the transformations $\rho_k(g)$, $g \in \mathrm{GL}(n,\mathbb{C})$.*

Proof: Let $\{e_i\}$ be the standard basis for $\mathbb{C}^n$. Then $\bigotimes^k \mathbb{C}^n$ has basis

$$e_I = e_{i_1} \otimes \cdots \otimes e_{i_k}\,, \quad \text{where } I = (i_1, \ldots, i_k) \text{ with } 1 \le i_j \le n\,.$$

For $s \in \mathfrak{S}$ we have $\sigma_k(s)e_I = e_{s\cdot I}$, where $s \cdot I = (i_{s^{-1}(1)}, \ldots, i_{s^{-1}(k)})$ (s moves the *positions* $1, \ldots, k$ of the indices; it does not permute their *values* $1, \ldots, n$). If we write $Be_J = \sum_I b^I_J e_I$, then the condition that B commute with $\mathfrak{S}_k$ is expressed by

$$b^I_J = b^{s\cdot I}_{s\cdot J} \quad \text{for all } s \in \mathfrak{S}_k \text{ and all indices } I, J\,. \tag{7}$$

Thus if we denote by $\Xi = \{(I,J)\}$ (all pairs of indices), then $\mathfrak{S}_k$ acts diagonally on Ξ and the matrix for B is an invariant function on Ξ.

Set $\mathcal{B} = \mathrm{End}_{\mathfrak{S}_k}(\bigotimes^k \mathbb{C}^n)$ and let $T \mapsto T^\natural$ be the projection from $\mathrm{End}(\bigotimes^k \mathbb{C}^n)$ onto $\mathcal{B}$. The bilinear form $(X, Y) = \mathrm{tr}(XY)$ on $\mathcal{B}$ is non-degenerate, since $(T, Y) = (T^\natural, Y)$ for $T \in \mathrm{End}(\bigotimes^k \mathbb{C}^n)$ and $Y \in \mathcal{B}$. Write

$$\mathcal{B}_0 = \mathrm{Span}\{\rho_k(g) \,:\, g \in \mathrm{GL}(n, \mathbb{C})\}\,.$$

Then to show $\mathcal{B}_0 = \mathcal{B}$, it suffices to show that if $B \in \mathcal{B}$ and $(B, \rho_k(g)) = 0$ for all $g \in \mathrm{GL}(n, \mathbb{C})$, then $B = 0$. Define $F(g) = (B, \rho_k(g))$. Since $g \mapsto F(g)$ is a polynomial function on the space $M_n(\mathbb{C})$ of $n \times n$ matrices and F vanishes on $\mathrm{GL}(n, \mathbb{C})$, it is identically zero. Hence $\{b^I_J\}$ satisfy the linear equations

$$\sum_{I,J} b^I_J x_{IJ} = 0 \quad \text{for all } x \in M_n(\mathbb{C}) \tag{8}$$

in addition to the invariance condition (7), where $x_{IJ} = x_{i_1 j_1} \cdots x_{i_k j_k}$.

It is easy to verify that $x_{IJ} = x_{I'J'}$ as functions on $M_n(\mathbb{C})$ if and only if $I = s \cdot I'$ and $J = s \cdot J'$ for some $s \in \mathfrak{S}_k$. Let Γ be a cross-section for the orbits of $\mathfrak{S}_k$ on Ξ. Then the set of monomials $\{x_\gamma \,:\, \gamma \in \Gamma\}$ are all distinct, and hence linearly independent functions on $M_n(\mathbb{C})$. Since b^I_J is constant on $\mathfrak{S}_k$ orbits by (7), equation (8) can be written as

$$\sum_{\gamma \in \Gamma} |\mathfrak{S}_k \cdot \gamma| b_\gamma x_\gamma = 0 \quad \text{for all } x \in M_n(\mathbb{C})\,.$$

Thus $b_\gamma = 0$ for all γ. □

Applying Proposition 3.1 and Theorem 1.4, we obtain a preliminary version of *Schur–Weyl duality:*

Corollary 3.2: *There are irreducible, mutually inequivalent $\mathfrak{S}_k$ modules E^λ and irreducible, mutually inequivalent $\mathrm{GL}(n, \mathbb{C})$ modules F^λ so that*

$$\bigotimes^k \mathbb{C}^n \cong \bigoplus_{\lambda \in \mathsf{Spec}(\rho_k)} E^\lambda \otimes F^\lambda$$

as a representation of $\mathfrak{S}_k \times \mathrm{GL}(n, \mathbb{C})$. The representation E^λ uniquely determines F^λ and conversely.

3.2. *Highest weight theory*

To make Schur–Weyl duality an effective tool, we will construct the irreducible regular representations of $\mathrm{GL}(n, \mathbb{C})$ by the *Theorem of the Highest Weight.* We give details for $\mathrm{GL}(n, \mathbb{C})$; analogous results hold for any complex reductive algebraic group (see [16, Chap. 5]). The starting point is the

Gauss decomposition. Let H be the subgroup of diagonal matrices, N the subgroup of upper-triangular unipotent matrices (all diagonal entries 1), and $\bar{N}$ the subgroup of lower triangular unipotent matrices. Then $\bar{N}HN$ is a Zariski-dense open subset of G, and a *generic* element $g \in G$ has a unique factorization $g = \bar{n}hn$.[b] Thus a regular representation of G is completely determined by its restriction to the subgroups $\bar{N}$, H, and N.

The subgroup H is a *maximal algebraic torus* in G. In particular, it is a reductive complex algebraic group. The irreducible representations of H are one-dimensional and given by

$$h = \mathrm{diag}[x_1, \ldots, x_n] \mapsto h^\mu = x_1^{m_1} \cdots x_n^{m_n}, \quad \text{where } \mu = [m_1, \ldots, m_n] \in \mathbb{Z}^n .$$

Thus we may identify $\widehat{H}$ with $\mathbb{Z}^n$. If (ρ, V) is a regular representation of G, then the restriction of ρ to H decomposes into *weight spaces:*

$$V = \bigoplus_{\mu \in \Phi(V)} V(\mu), \quad \text{where } V(\mu) \neq 0 \text{ and } \rho(h)v = h^\mu v \text{ for } v \in V(\mu).$$

We call $\Phi(V) \subset \widehat{H}$ the set of *weights* of V.

Let $\mathrm{Norm}_G(H)$ be the *normalizer* of H in G ($HgH = gH$), and $W = \mathrm{Norm}_G(H)/H$ the *Weyl group* of G. The elements of W permute the weight spaces and the weights of V. In this case, $W \cong \mathfrak{S}_n$ may be identified with the group of *permutation matrices* in G, and the action of W on H and $\widehat{H}$ is by the usual permutation of coordinates. Every W orbit in $\widehat{H}$ contains a unique *dominant weight*

$$\mu = [m_1, \ldots, m_n], \qquad m_1 \geq m_2 \geq \cdots \geq m_n .$$

We denote by $\mathbb{Z}^n_{++}$ the set of all such $\mu \in \mathbb{Z}^n$.

Examples

(1) Let $V = \mathbb{C}^n$ be the *defining representation* of G. Then

$$\Phi(V) = \{\varepsilon_1, \ldots, \varepsilon_n\}, \quad \text{where } \varepsilon(h) = x_i \text{ for } h = \mathrm{diag}[x_1, \ldots, x_n]$$

Here $\Phi(V) = W \cdot \varepsilon_1$ is a single W orbit with dominant weight ε_1.

(2) Let $V = \bigotimes^k \mathbb{C}^n$. The basis $\{e_I\}$ used in the proof of Proposition 3.1 diagonalizes $\rho_k(H)$. For an index $I = [i_1, \ldots, i_k]$, with $1 \leq i_j \leq n$, define

$$\mu_I = [\mu_1, \ldots, \mu_n], \qquad \text{where } \mu_p = \#\{j : i_j = p\}.$$

---

[b]The precise condition from linear algebra is that the principal minors $\Delta_i(g) \neq 0$ for $i = 1, 2, \ldots, n$.

Then $\rho_k(h)e_I = h^{\mu_I}e_I$ for $h \in H$. Hence for $\lambda \in \hat{H}$,

$$V(\lambda) = \mathrm{Span}\{e_I : \mu_I = \lambda\}.$$

In particular, $V(\lambda) \neq 0$ if and only if $\lambda_i \geq 0$ for $i = 1, \ldots, n$ and $|\lambda| = k$, where $|\lambda| = \lambda_1 + \cdots + \lambda_n$. Thus $\Phi(\bigotimes^k \mathbb{C}^n) = W \cdot \mathrm{Par}(k,n)$, where $\mathrm{Par}(k,n)$ is the set of all partitions of k with at most n parts. Each such partition defines a dominant weight μ of H such that $h \mapsto h^\mu$ is a *polynomial function* on H (no negative powers of the coordinates x_i).

(3) Let $\mathfrak{g} = \mathrm{Lie}(G) = M_n(\mathbb{C})$ be the Lie algebra of G, and let $\mathrm{Ad}(g)x = gxg^{-1}$ be the adjoint representation. The weights are 0 and $\{\varepsilon_i - \varepsilon_j : 1 \leq i \neq j \leq n\}$. We call the nonzero weights the *roots* of $\mathfrak{h}$ on $\mathfrak{g}$. The corresponding *root spaces* are

$$\mathfrak{g}_0 = \mathfrak{h} = \mathrm{Lie}(H), \quad \mathfrak{g}_{\varepsilon_i - \varepsilon_j} = \mathbb{C}E_{ij}$$

where E_{ij} is the usual elementary matrix with 1 in position (i,j) and zero elsewhere. If $\alpha = \varepsilon_i - \varepsilon_j$, then we say $\alpha > 0$ if $i < j$ (so E_{ij} is upper triangular) and $\alpha < 0$ if $i > j$. We denote the set of positive roots by Φ^+ and the set of negative roots by Φ^-. Thus

$$\mathfrak{n} = \mathrm{Lie}(N) = \bigoplus_{\alpha \in \Phi^+} \mathfrak{g}_\alpha, \qquad \bar{\mathfrak{n}} = \mathrm{Lie}(\bar{N}) = \bigoplus_{\alpha \in \Phi^-} \mathfrak{g}_\alpha.$$

The Lie algebra (additive) version of the Gauss decomposition is the *triangular decomposition* $\mathfrak{g} = \bar{\mathfrak{n}} \oplus \mathfrak{h} \oplus \mathfrak{n}$.

If (ρ, V) is any regular representation of G, then there is an associated Lie algebra representation $d\rho$ of $\mathfrak{g}$ defined by

$$d\rho(X)v = \frac{d}{dt}\rho(\exp tX)v\bigg|_{t=0}$$

Clearly $\rho(g)d\rho(X)v = d\rho(\mathrm{Ad}(g)X)\rho(g)v$. Hence if $h \in H$, $E_\alpha \in \mathfrak{g}_\alpha$ and $v \in V(\mu)$, then

$$\rho(h)d\rho(E_\alpha)v = d\rho(\mathrm{Ad}(h)E_\alpha)\rho(h)v = h^{\alpha+\mu}d\rho(E_\alpha)v.$$

This shows that $d\rho(\mathfrak{g}_\alpha)V(\mu) \subset V(\mu + \alpha)$. Thus

$$d\rho(\mathfrak{n})V(\mu) \subset \bigoplus_{\lambda \in \mu + \Phi^+} V(\lambda). \tag{9}$$

We call $\mu \in \Phi(V)$ an *N-extreme* weight if $\mu + \alpha \notin \Phi(V)$ for all $\alpha \in \Phi^+$.

Theorem 3.3: *Let (ρ, V) be an irreducible regular representation of G. Then there is a unique N-extreme weight $\mu_0 \in \Phi(V)$. This weight is dominant, the weight space $V(\mu_0) = V^N$ (the N-fixed vectors in V), and* $\dim V^N = 1$.

Proof: Define the *dominance order* on $\widehat{H}$ by $\nu \prec \mu$ if $\mu \neq \nu$ and $\mu - \nu$ is a nonnegative integer combination of positive roots. Since $\Phi(V)$ is a finite set, it contains an element μ_0 that is maximal relative to this partial order. Any maximal element must be dominant, since if μ_0 had a pair of coordinates $m_i < m_j$ for some $i < j$, then

$$\nu = \mu_0 + (m_j - m_i)(\varepsilon_i - \varepsilon_j) = s_{ij}\mu_0 \in \Phi(V),$$

where s_{ij} is the permutation $i \leftrightarrow j$. This would contradict the maximality of μ_0. Since no element of $\mu_0 + \Phi^+$ can be a weight of V, we see from (9) that $d\rho(\mathfrak{n})V(\mu_0) = 0$. But $N = \exp \mathfrak{n}$, so we conclude that $V(\mu_0) \subset V^N$.

Take any nonzero vector $v_0 \in V(\mu_0)$. Then by irreducibility and the Gauss Decomposition,

$$V = \mathrm{Span}\rho(G)v_0 = \mathrm{Span}\rho(\bar{N})v_0 = d\rho(U(\bar{\mathfrak{n}}))v_0\,,$$

where $U(\bar{\mathfrak{n}})$ is the universal enveloping algebra of $\bar{\mathfrak{n}}$. But the nonzero weights of H on $U(\bar{\mathfrak{n}})$ are positive integer combinations of negative roots, and the zero weight space is $\mathbb{C}1$. Hence

$$V = \mathbb{C}v_0 + \bigoplus_{\lambda \prec \mu_0} V(\lambda)\,.$$

It follows that μ_0 is the unique maximal weight and $V(\mu_0) = V^N$. □

We call μ_0 the *highest weight* of the representation (ρ, V). We next show that μ_0 determines (ρ, V) uniquely up to isomorphism. Let $s_0 \in W$ be the permutation $[1, 2, \ldots, n] \mapsto [n, \ldots, 2, 1]$. Then $\mathrm{Ad}(s_0)N = \bar{N}$ and $s_0 \cdot \Phi^+ = \Phi^-$ (this last property uniquely characterizes s_0). Since the weights and weight multiplicities of (ρ, V) are invariant under the Weyl group, we see from Theorem 3.3 that $\Phi(V)$ has a unique minimal element $-s_0\mu_0$ (the *lowest weight*), and $V(-s_0\mu_0) = V^{\bar{N}}$. The natural bilinear form on $V \times V^*$ is invariant under H, so its restriction to $V(\lambda) \times V^*(-\lambda)$ is nondegenerate. Thus $-\mu_0$ is the lowest weight of V^*.

Theorem 3.4: *Let (ρ, V) and (σ, U) be irreducible regular representations of G with the same highest weight λ. If $v_0 \in V$ and $u_0 \in U$ are highest weight vectors, then there exists a unique G-isomorphism $T : U \to V$ such that $Tu_0 = v_0$. Thus (ρ, V) is uniquely determined by its highest weight.*

Proof: We can take $v_0 \in V^N$ and $v_0^* \in V^{*\bar{N}}$ so that $\langle v_0, v_0^* \rangle = 1$. Set

$$\varphi(g) = \langle \rho(g)v_0, v_0^* \rangle \qquad \text{(generating function)}.$$

Then $\varphi \in \text{Aff}(G)$ and $\varphi(\bar{n}hn) = \langle \rho(h)v_0, \rho^*(\bar{n})v_0^* \rangle = h^\lambda$ for all $\bar{n}hn \in \bar{N}HN$. This equation uniquely determines φ in terms of the highest weight λ, and using U and u_0 would give the same generating function.

Define $T_0 : V \to \text{Aff}(G)$ by $T_0v(g) = \langle \rho(g)v, v_0^* \rangle$. Then T_0 intertwines the representation ρ on V with the right translation representation on $\text{Aff}(G)$. Since $T_0v_0 = \varphi$, the map T_0 is nonzero. Hence it is injective, by Schur's lemma, and $V \cong T_0(V)$. But $V = \text{Span}\{\rho(G)v\}$, so $T_0(V)$ is spanned by the right translates of the function φ. Let $T_1 : U \to \text{Aff}(G)$ be likewise defined. Then $T = T_0^{-1}T_1$ is the desired intertwining map. □

For applications to duality we will need the following sharpening of Theorem 3.3.

Theorem 3.5: *Let (ρ, L) be any regular representation of G. Suppose that $0 \neq v_0 \in L(\lambda)^N$. Then the subspace $V = \text{Span}\{\rho(G)v_0\}$ of L is an irreducible G-module with highest weight λ.*

Proof: From the proof of Theorem 3.3 we have a weight space decomposition

$$V = \mathbb{C}v_0 \oplus \bigoplus_{\mu \prec \lambda} V(\mu).$$

Let $U_0 \subset V$ be a proper G-invariant subspace. Then $v_0 \notin U_0$, so $U_0 = \bigoplus_{\mu \prec \lambda} U_0(\mu)$. Since G is reductive, there is a complementary G-invariant subspace U_1 so that $V = U_0 \oplus U_1$. Since U_1 is the direct sum of its weight spaces, we have $v_0 \in U_1$ and hence $U_1 = V$, proving irreducibility. □

Corollary 3.6: *For every dominant weight λ of H there exists an irreducible representation (π^λ, F^λ) with highest weight λ.*

Proof: Let $\Delta_k(g)$ be the kth principal minor of $g \in \text{GL}(n, \mathbb{C})$, for $k = 1, \ldots, n$. If $\lambda = [m_1, \ldots, m_n]$, set

$$f_\lambda = \Delta_1^{m_1 - m_2} \cdots \Delta_{n-1}^{m_{n-1} - m_n} \Delta_n^{m_n}.$$

Since λ is dominant, we have $m_1 \geq m_2 \cdots \geq m_n$. Also $\Delta_n(g) = \det(g) \neq 0$ for $g \in G$. Hence f_λ is a regular function on G. Also $f_\lambda(\bar{n}hn) = h^\lambda$ for $\bar{n}hn \in \bar{N}HN$.

Let $F^\lambda \subset \text{Aff}(G)$ be the span of the right translates of f_λ, and let π^λ be the restriction to F^λ of the right translation representation of G on

Aff(G). Then $\dim F^\lambda < \infty$ since f_λ is a regular function. Furthermore, π^λ is irreducible with highest weight λ by Theorem 3.5, since f_λ is N-fixed of weight λ. □

We shall refer to Theorems 3.3–3.5 and Corollary 3.6 collectively as the *Theorem of the Highest Weight.*

3.3. *Duality and N-fixed vectors*

Let (ρ, L) be any regular G-module. By the Theorem of the Highest Weight we can identify $\mathrm{Spec}(\rho)$ with the set of dominant weights λ such that $L(\lambda)^N \neq 0$. For $\lambda \in \mathrm{Spec}(\rho)$ set $E^\lambda = L(\lambda)^N$. This space is invariant under the commuting algebra $\mathrm{End}_G(L)$.

Theorem 3.7: *Under the joint action of* G *and* $\mathrm{End}_G(L)$ *the space* L *decomposes as*

$$L \cong \bigoplus_{\lambda\in\mathrm{Spec}(\rho)} E^\lambda \otimes F^\lambda\,. \tag{10}$$

Furthermore, E^λ *is an irreducible module for* $\mathrm{End}_G(L)$, *and distinct values of* λ *give inequivalent modules for* $\mathrm{End}_G(L)$.

Proof: Let $\lambda \in \mathrm{Spec}(\rho)$ and fix a highest weight vector $v_\lambda \in F^\lambda$. Define a linear map

$$\mathrm{Hom}_G(F^\lambda, L) \to L(\lambda)^N\,, \qquad T \mapsto Tv_\lambda\,.$$

This map is injective by the irreducibility of F^λ, surjective by Theorem 3.4, and it obviously intertwines the action of $\mathrm{End}_G(L)$. The theorem now follows from Theorem 1.4. □

We can now give a more precise version of Schur–Weyl duality (Corollary 3.2). A regular representation π of G is said to be *polynomial* if the matrix entries of π are polynomial functions on G (with no negative powers of $\det(g)$). When $\pi = \pi^\lambda$ is irreducible, it is a polynomial representation if and only if $\lambda_n \geq 0$. In this case λ corresponds to a partition of $|\lambda|$ with at most n parts.

Theorem 3.8 (Schur-Weyl Duality): *For* $\lambda \in \mathrm{Par}(k,n)$ *let* $(\sigma^\lambda, E^\lambda)$ *be the representation of* $\mathfrak{S}_k$ *on the space of* N*-fixed* k*-tensors of weight* λ, *where* N *is the upper triangular unipotent subgroup of* $\mathrm{GL}(n,\mathbb{C})$. *Let* $(\pi^\lambda, F^\lambda_{(n)})$ *be the irreducible representation of* $\mathrm{GL}(n,\mathbb{C})$ *with highest weight* λ. *Then* E^λ

is an irreducible $\mathfrak{S}_k$ module. Under the action of $\mathfrak{S}_k \times \mathrm{GL}(n, \mathbb{C})$, the space of k-tensors over $\mathbb{C}^n$ decomposes as

$$\bigotimes^k \mathbb{C}^n \cong \bigoplus_{\lambda \in \mathrm{Par}(k,n)} E^\lambda \otimes F^\lambda_{(n)} \,.$$

The representations E^λ are mutually inequivalent, and when $n \geq k$ they give all the irreducible representations of $\mathfrak{S}_k$. Furthermore, every irreducible polynomial representation of $\mathrm{GL}(n, \mathbb{C})$ occurs in $(\mathbb{C}^n)^{\otimes k}$ for some k.

Proof: From Example (2) of Section 3.2 the dominant weights of $(\mathbb{C}^n)^{\otimes k}$ are in $\mathrm{Par}(k, n)$. To verify that for every $\lambda \in \mathrm{Par}(k, n)$ there exists a nonzero N-fixed k-tensor of weight λ, we define

$$\varpi_i = \varepsilon_1 + \cdots + \varepsilon_i \,, \qquad u_i = e_1 \wedge \cdots \wedge e_i \qquad \text{for } i = 1, 2, \ldots, n \,.$$

Then ϖ_i is the highest weight of $\mathrm{GL}(n, \mathbb{C})$ on $\bigwedge^i \mathbb{C}^n$, and u_i is the corresponding highest weight vector. If we set $c_i = m_i - m_{i+1}$ (with $m_{n+1} = 0$), then $c_1, \ldots, c_n$ are nonnegative integers and $\lambda = c_1\varpi_1 + \cdots + c_n\varpi_n$. The tensor

$$u_\lambda = u_1^{\otimes c_1} \otimes u_2^{\otimes c_2} \otimes \cdots \otimes u_n^{\otimes c_n} \in \bigotimes^k \mathbb{C}^n$$

has weight λ and is N-fixed, since G acts as automorphisms of the tensor algebra over $\mathbb{C}^n$. It follows that $\mathrm{Spec}(\rho_k) = \mathrm{Par}(k, n)$ (we label elements of $\widehat{G}$ by their highest weights).

The assertions of the theorem now all follow from Corollary 3.2 and the Theorem of the Highest Weight, except for the fact that all irreducible representations of $\mathfrak{S}_k$ occur when $n \geq k$. To prove this, recall that the number of irreducible representations of a finite group is the same as the number of conjugacy classes (this common number is the dimension of the center of the group algebra). In the case of $\mathfrak{S}_k$, each conjugacy class corresponds to a cycle decomposition, and the lengths of the cycles determine a unique partition of k. Thus $\mathfrak{S}_k$ has $\#\mathrm{Par}(k)$ irreducible representations, where $\mathrm{Par}(k)$ denotes the set of all partitions of k. Since a partition of k has at most k parts, every partition occurs in $\mathrm{Par}(k, n)$ when $n \geq k$, and hence $(\mathbb{C}^n)^{\otimes k}$ contains all irreducible representations of $\mathfrak{S}_k$ in this case. □

Examples

(1) The group $\mathfrak{S}_k$ has two one-dimensional representations: the *trivial* representation and the *sign* representation. The corresponding subspaces of $\bigotimes^k \mathbb{C}^n$ are the *symmetric tensors* $S^k(\mathbb{C}^n)$ and (if $n \geq k$) the *skew symmetric tensors* $\bigwedge^k \mathbb{C}^n$. Hence these subspaces must be irreducible $\mathrm{GL}(n, \mathbb{C})$

modules, by Schur–Weyl duality (this is also easy to verify directly). The symmetric tensor $e_1^{\otimes k}$ is N-fixed with weight $k\varepsilon_1$, while the skew-symmetric tensor $e_1 \wedge \cdots \wedge e_k$ is N-fixed with weight $\varepsilon_1 + \cdots + \varepsilon_k$ (when $k \leq n$). Thus in the duality correspondence,

$$(\text{trivial}, \mathbb{C}) = (\sigma^{[k]}, E^{[k]}) \longleftrightarrow (\pi^{[k]}, S^k(\mathbb{C}^n))$$
$$(\text{sgn}, \mathbb{C}) = (\sigma^{[1^k]}, E^{[1^k]}) \longleftrightarrow \pi^{[1^k]}, \bigwedge^k \mathbb{C}^n) \quad \text{if } n \geq k\,.$$

When $k = 2$ and $n \geq 2$ this gives the complete decomposition of $\bigotimes^2 \mathbb{C}^n$: under $\mathfrak{S}_2 \times \mathrm{GL}(n, \mathbb{C})$:

$$\mathbb{C}^n \otimes \mathbb{C}^n \cong \left\{E^{[2,0]} \otimes S^2(\mathbb{C}^n)\right\} \oplus \left\{E^{[1,1]} \otimes \bigwedge^2 \mathbb{C}^n\right\}.$$

(2) Consider $\bigotimes^3 \mathbb{C}^n$ for $n \geq 3$. There are three partitions of 3, giving the decomposition

$$\bigotimes^3 \mathbb{C}^n \cong \left\{E^{[3,0]} \otimes S^3(\mathbb{C}^n)\right\} \oplus \left\{E^{[2,1]} \otimes F^{[2,1]}\right\} \oplus \left\{E^{[1,1,1]} \otimes \bigwedge{}^3\mathbb{C}^n\right\}$$

under $\mathfrak{S}_3 \times \mathrm{GL}(n, \mathbb{C})$. Here the representation $E^{[2,1]}$ of $\mathfrak{S}_3$ is the two-dimensional *standard representation* on $\mathbb{C}^3/\mathbb{C}[1, 1, 1]$.

We can view Schur–Weyl Duality as a method to construct representations of $G' = \mathfrak{S}_k$ from representations of $G = \mathrm{GL}(n, \mathbb{C})$ via the Theorem of the Highest Weight. Here we take the representations of G as the known objects, and the representations of G' as the unknown objects.[c] The relative size of n (the *rank* of G) and k then determines which representations of G' we get this way.

$n \geq k$: All partitions of k have at most k parts, so all representations of $\mathfrak{S}_k$ occur in $\bigotimes^k \mathbb{C}^n$ in this case.

$n \leq k$: Only those representations of $\mathfrak{S}_k$ occur in $\bigotimes^k \mathbb{C}^n$ that correspond to partitions of k with at most n parts.

To make this method effective, we will develop *character formulas* for the representations of $\mathfrak{S}_k$ in the next two sections, based on the celebrated *Weyl character formula* for $\mathrm{GL}(n, \mathbb{C})$.

[c]The representations of $\mathfrak{S}_k$ can be constructed directly by group-theoretic and combinatorial methods. Special elements of the group algebra $\mathbb{C}[\mathfrak{S}_k]$ (*Young symmetrizers*) project tensor space onto irreducible representations of $\mathrm{GL}(n, \mathbb{C})$ — the so-called *Weyl modules* — see [16, Sec. 9.3].

4. Commutant character formulas

4.1. *Characters*

Let G be a connected complex reductive algebraic group. Then G contains a maximal algebraic torus H and a maximal connected solvable subgroup HN (semidirect product), where N is the unipotent radical of HN. In fact, one can always embed G into $\mathrm{GL}(n, \mathbb{C})$ so that H consists of the diagonal matrices in G, and N the upper-triangular unipotent matrices in G, just as in the case of $\mathrm{GL}(n, \mathbb{C})$ treated in Section 3. Let $\mathfrak{h} = \mathrm{Lie}(H)$ and $\mathfrak{n} = \mathrm{Lie}(N)$.

The irreducible representations of the torus H are given by $h \mapsto h^\lambda$, where λ is in the *weight lattice* $P \subset \mathfrak{h}^*$ of H. By the Theorem of the Highest Weight (which is proved for G along the same lines as in Section 3 for $\mathrm{GL}(n, \mathbb{C})$), the irreducible regular representations of G are parameterized by the set P_{++} of *dominant weights* determined by the choice of N. For $\lambda \in P_{++}$ let (π^λ, F^λ) be the irreducible representation of G with highest weight λ.

Let (π, V) be a finite-dimensional rational representation of G. Set $\mathcal{B} = \mathrm{End}_G(V)$. From Theorem 3.7 V decomposes under the joint action of G and $\mathcal{B}$ into a multiplicity-free direct sum

$$V \cong \bigoplus_{\lambda \in \mathsf{Spec}(\pi)} E^\lambda \otimes F^\lambda . \tag{11}$$

Here $g \in G$ acts by $1 \otimes \pi^\lambda(g)$ and $b \in \mathcal{B}$ acts by $\sigma^\lambda(b) \otimes 1$ on the summands in (11). We may take $E^\lambda = V(\lambda)^N$ (the space of N-fixed vectors of weight λ in V) with $\sigma^\lambda(b)$ the natural action of $b \in \mathcal{B}$ on this space.

Finding the spaces $V(\lambda)^N$ explicitly is usually difficult. An easier problem is to calculate characters. For $\lambda \in P_{++}$ we write

$$\chi_\lambda(b) = \mathrm{tr}(\sigma^\lambda(b)) = \mathrm{tr}(b|_{V(\lambda)^N}), \quad \text{for } b \in \mathcal{B}.$$

4.2. *Frobenius and determinant character formulas*

We now obtain two formulas for the characters χ_λ that only involve the full H-weight spaces in V. Let Φ^+ be the weights of $\mathrm{Ad}(H)$ on $\mathfrak{n}$ and $\rho = \frac{1}{2}\sum_{\alpha\in\Phi^+} \alpha$. Let $W = \mathrm{Norm}_G(H)/H$ be the Weyl group of (G, H). Set

$$D(h) = \sum_{s \in W} \mathrm{sgn}(s)\, h^{s\cdot\rho} \qquad \text{for } h \in H$$

(the Weyl denominator). Here $s \mapsto \mathrm{sgn}(s) = \det(\mathrm{Ad}(s)|_{\mathfrak{h}})$ is the usual signum character on W.

Theorem 4.1 (Generalized Frobenius Formula): *For $\lambda \in P_{++}$ and $b \in \mathcal{B}$ one has*

$$\chi_\lambda(b) = \textit{coefficient of } h^{\lambda+\rho} \textit{ in } D(h)\,\mathrm{tr}_V(\pi(h)b) \tag{12}$$

(where $h \in H$).

Theorem 4.2 (Generalized Determinant Formula): *For $\lambda \in P_{++}$ and $b \in \mathcal{B}$ one has*

$$\chi_\lambda(b) = \sum_{s\in W} \mathrm{sgn}(s)\,\mathrm{tr}_{V(\lambda+\rho-s\cdot\rho)}(b)\,. \tag{13}$$

In particular,

$$\dim E^\lambda = \sum_{s\in W} \mathrm{sgn}(s)\,\dim V(\lambda+\rho-s\cdot\rho)\,. \tag{14}$$

4.3. *Proof of Frobenius character formula*

For $\lambda \in P_{++}$ we write

$$\chi^\lambda(g) = \mathrm{tr}(\pi^\lambda(g)) \quad \text{for } g \in G\,.$$

(the character of the representation with highest weight λ). We note from (11) that

$$\mathrm{tr}_V(\pi(g)b) = \sum_{\lambda\in\mathrm{Spec}(\pi)} \chi^\lambda(g)\,\chi_\lambda(b) \qquad \text{for } g \in G \text{ and } b \in \mathcal{B}\,. \tag{15}$$

By the Weyl character formula (WCF), we have

$$D(h)\,\chi^\lambda(h) = \sum_{s\in W} \mathrm{sgn}(s)\,h^{s\cdot(\lambda+\rho)} \qquad \text{for } h \in H\,.$$

Using the WCF in (15) we can write

$$D(h)\,\mathrm{tr}_V(\pi(h)b) = \sum_{\lambda\in\mathrm{Spec}(\pi)} \sum_{s\in W} \mathrm{sgn}(s)\,\chi_\lambda(b)\,h^{s\cdot(\lambda+\rho)}\,. \tag{16}$$

Due to the shift by ρ, the map $(s,\lambda) \mapsto s\cdot(\lambda+\rho)$ from $W \times P_{++} \to P$ is injective.[d] Hence the character $h \mapsto h^{\lambda+\rho}$ only occurs once in (16), and has coefficient $\chi_\lambda(b)$, as claimed. □

[d] In the case of $\mathrm{GL}(n,\mathbb{C})$ the partition $\lambda+\rho$ has all parts of *different* sizes.

4.4. *Proof of determinant character formula*

For the proof of Theorem 4.2, we need the following consequence of the WCF (which is, in fact, equivalent to the WCF).

Lemma 4.3: *Let $m_\lambda(\mu) = \dim F^\lambda(\mu)$ for $\mu \in P$ and $\lambda \in P_{++}$ (the multiplicity of the weight μ in F^λ). Then for $\lambda, \mu \in P_{++}$ one has*

$$\sum_{s \in W} \operatorname{sgn}(s)\, m_\mu(\lambda + \rho - s \cdot \rho) = \delta_{\lambda\mu} .$$

Proof: Write the Weyl denominator as an alternating sum over W of the characters $h \mapsto h^{s \cdot \rho}$. Multiplying this sum by χ^μ, we get

$$D(h)\chi^\mu(h) = \sum_{s \in W} \Big\{ \sum_{\nu \in P} \operatorname{sgn}(s)\, m_\mu(\nu)\, h^{\nu + s \cdot \rho} \Big\} \qquad \text{for } h \in H .$$

In the inner sum make the substitution $\nu \to \nu + \rho - s \cdot \rho$; then the sum on the right becomes

$$\sum_{\nu \in P} \Big\{ \sum_{s \in W} \operatorname{sgn}(s)\, m_\mu(\nu + \rho - s \cdot \rho) \Big\} h^{\nu + \rho} .$$

On the other hand, the WCF asserts that the coefficient of $h^{\lambda+\rho}$ in $D(h)\chi^\mu(h)$ is $\delta_{\lambda\mu}$ when $\lambda, \mu \in P_{++}$. □

Proposition 4.4 (Outer Multiplicity Formula)**:** *Let L be any regular G module. For $\lambda \in P_{++}$ let $\operatorname{mult}_L(\lambda)$ be the multiplicity of the representation F^λ. Then*

$$\operatorname{mult}_L(\lambda) = \sum_{s \in W} \operatorname{sgn}(s)\, \dim L(\lambda + \rho - s \cdot \rho) . \tag{17}$$

Proof: For $\nu \in P$ we have $\dim L(\nu) = \sum_{\mu \in P_{++}} \operatorname{mult}_L(\mu)\, m_\mu(\nu)$. Hence the right side of (17) is

$$\sum_{\mu \in P_{++}} \operatorname{mult}_L(\mu) \Big\{ \sum_{s \in W} \operatorname{sgn}(s) m_\mu(\lambda + \rho - s \cdot \rho) \Big\} .$$

But the inner sum is $\delta_{\lambda\mu}$ by Lemma 4.3, which proves (17). □

Proof of Theorem 4.2: Let $b \in \mathcal{B}$. Then b has a Jordan decomposition $b = b_s + b_n$, where b_s is semisimple, b_n is nilpotent, and b_s is a polynomial in b. Hence $b_s \in \mathcal{B}$ and $\chi_\lambda(b) = \chi_\lambda(b_s)$. So we may assume b is semisimple.

For $\zeta \in \mathbb{C}$ and $\lambda \in P_{++}$ define

$$V_\zeta = \{ v \in V : bv = \zeta v \} , \qquad E_\zeta^\lambda = \{ v \in E^\lambda : bv = \zeta v \} .$$

These spaces are G-invariant and

$$V_\zeta \cong \bigoplus_{\lambda \in P_{++}} E_\zeta^\lambda \otimes F^\lambda$$

as a G-module. In particular, the multiplicity of F^λ in V_ζ is $\dim E_\zeta^\lambda$. Now apply Proposition 4.4 to the G-module $L = V_\zeta$ to get

$$\begin{aligned}\chi_\lambda(b) &= \sum_{\zeta \in \mathbb{C}} \zeta \dim E_\zeta^\lambda \\ &= \sum_{s \in W} \operatorname{sgn}(s) \Big\{ \sum_{\zeta \in \mathbb{C}} \zeta \dim V_\zeta(\lambda + \rho - s \cdot \rho) \Big\} \\ &= \sum_{s \in W} \operatorname{sgn}(s) \operatorname{tr}_{V(\lambda+\rho-s\cdot\rho)}(b)\end{aligned}$$

which proves Theorem 4.2. □

5. Character formulas for Schur–Weyl duality

5.1. *Frobenius formula for $\mathfrak{S}_k$ characters*

We now apply the commutant character formulas to the Schur–Weyl duality between $G = \mathrm{GL}(n, \mathbb{C})$ and $\mathfrak{S}_k$, both acting on $V = \bigotimes^k \mathbb{C}^n$. Recall that the conjugacy classes in $\mathfrak{S}_k$ are described by cycle lengths. We denote by $C(1^{r_1} 2^{r_2} \cdots k^{r_k})$ the class of elements with r_j cycles of length j, where $r_1 + 2r_2 + 3r_3 + \cdots = k$. A permutation s in this class has r_1 fixed points, r_2 transpositions, and so on. To apply Theorem 4.1 in this context, we need to calculate the polynomial

$$h \mapsto \operatorname{tr}_V(\rho_k(h)\sigma_k(s)), \qquad \text{where } h = \operatorname{diag}[x_1, \ldots, x_n].$$

Recall that the tensors $\{e_I\}$ give a basis for V. The action of $h \in H$ and $s \in \mathfrak{S}_k$ is

$$\rho_k(h)e_I = x^{\mu(I)} e_I, \qquad \sigma_k(s)e_I = e_{s \cdot I}$$

where $\mu(I) = [\mu_1, \cdots, \mu_n]$ with $\mu_j = \#\{p : i_p = j\}$ and

$$s \cdot I = [i_{s^{-1}(1)}, \ldots, i_{s^{-1}(k)}].$$

Since $\sigma(s)$ permutes the basis $\{e_I\}$ and each e_I is a weight vector for H, it follows that

$$\operatorname{tr}_V(\rho_k(h)\sigma_k(s)) = \operatorname{tr}_{F_s}(\rho_k(h)) \qquad \text{for } h \in H, \tag{18}$$

where $F_s = \operatorname{Span}\{e_I : s \cdot I = I\}$. Let $V_j = \operatorname{Span}\{e_1^{\otimes j}, e_2^{\otimes j}, \cdots, e_n^{\otimes j}\} \subset \bigotimes^j \mathbb{C}^n$.

Lemma 5.1: *If* $s \in C(1^{r_1}2^{r_2}\cdots k^{r_k})$ *then*

$$F_s \cong V_1^{\otimes r_1} \otimes V_2^{\otimes r_2} \otimes \cdots \otimes V_k^{\otimes r_k}$$

as an H*-module.*

Proof: We may assume the cycle decomposition of s contains the integers $1, \ldots, k$ in increasing order, since replacing s by a conjugate element in $\mathfrak{S}_k$ does not change the H-module structure of F_s. In this case the condition $s \cdot I = I$ for an index I means that

$$I = [\underbrace{a_1, \ldots, a_{r_1}}_{r_1 \text{ singles}}, \underbrace{b_1, b_1, \ldots, b_{r_2}, b_{r_2}}_{r_2 \text{ pairs}}, \underbrace{c_1, c_1, c_1, \ldots, c_{r_3}, c_{r_3}, c_{r_3}}_{r_3 \text{ triples}}, \ldots]$$

where $a_i, b_i, c_i, \ldots$ range from 1 to n. Hence

$$e_I = e_{a_1} \otimes \cdots \otimes e_{a_{r_1}} \otimes e_{b_1}^{\otimes 2} \otimes \cdots \otimes e_{b_{r_2}}^{\otimes 2} \otimes e_{c_1}^{\otimes 3} \otimes \cdots \otimes e_{c_{r_3}}^{\otimes 3} \otimes \cdots$$

The lemma follows. □

For $h = \operatorname{diag}[x_1, \ldots, x_n]$ define

$$p_j(x) = \operatorname{tr}_{V_j}(\rho_k(h)) = x_1^j + \cdots + x_n^j \qquad (j\text{th power sum}).$$

Then Lemma 5.1 implies that

$$\operatorname{tr}_{F_s}(\rho_k(h)) = \prod_{j=1}^{k} \operatorname{tr}_{V_j}(\rho_k(h))^{r_j} = \prod_{j=1}^{k} p_j(x)^{r_j}.$$

Hence from Theorem 4.1 and (18) we obtain the *Frobenius character formula:*

Theorem 5.2: *Let* $s \in C(1^{r_1}2^{r_2}\cdots k^{r_k})$ *and* $\lambda \in \operatorname{Par}(k, n)$. *Then*

$$\chi_\lambda(s) = \textit{coefficient of } x^{\lambda+\rho_{[n]}} \textit{ in } D_n(x)\left\{\prod_{j=1}^{k} p_j(x)^{r_j}\right\}$$

where $\rho_{[n]} = [n-1, n-2, \ldots, 1, 0]$ *and* $D_n(x) = \prod_{1 \le i < j \le n}(x_i - x_j)$.

Examples

(1) Suppose $s = (1, 2, \ldots, m)(m+1)\cdots(k)$ is a single m-cycle with $k - m$ fixed points. Then

$$\chi_\lambda(s) = \text{coefficient of } x^{\lambda+\rho_{[n]}} \text{ in } (x_1 + \cdots + x_n)^{k-m}(x_1^m + \cdots + x_n^m) \times \prod_{1 \le i < j \le n}(x_i - x_j).$$

We call a monomial $x_1^{a_1} \cdots x_n^{a_n}$ *strictly dominant* if $a_1 > a_2 > \cdots > a_n$. For partitions λ with two parts and cycles of maximum length $m = k$, the strictly dominant terms in this formula are $x_1^{k+1} - x_1^k x_2$. Hence for $s = (1, 2, \ldots, k)$,

$$\chi_\lambda(s) = \begin{cases} -1 & \text{for } \lambda = [k-1, 1], \\ 0 & \text{for } \lambda = [k-j, j] \text{ with } j > 1. \end{cases}$$

(2) Consider the group $\mathfrak{S}_3$, which has three conjugacy classes: $C(1^3) = \{\text{identity}\}$, $C(1^1 2^1) = \{(12), (13), (23)\}$ and $C(3^1) = \{(123), (132)\}$. As we noted at the end of Section 3, the three representations of $\mathfrak{S}_3$ are $\sigma^{[3]}$ (the trivial representation), $\sigma^{[2,1]}$ (the two-dimensional standard representation), and $\sigma^{[1,1,1]}$ (the signum representation). To calculate the character of $\sigma^{[2,1]}$ by the Frobenius formula, we let $x = [x_1, x_2]$ and expand the polynomials

$$\begin{aligned} D_2(x) p_1(x)^3 &= x_1^4 + 2x_1^3 x_2 + \cdots \\ D_2(x) p_1(x) p_2(x) &= x_1^4 + \cdots \\ D_2(x) p_3(x) &= x_1^4 - x_1^3 x_2 + \cdots \end{aligned}$$

where $\cdots$ indicates non-dominant terms. By Theorem 4.1 the coefficients of the dominant terms in these formulas furnish the entries in the character table for $\mathfrak{S}_3$. We write χ_λ for the character of the representation σ^λ. For example, when $\lambda = [2, 1]$ we have $\lambda + \rho = [3, 1]$, so the coefficient of $x_1^3 x_2$ in $D_2(x) p_3(x)$ gives the value of $\chi_{[2,1]}$ on the conjugacy class $C(3^1)$. Table 1 gives all the characters, where the top row indicates the number of elements in each conjugacy class, and the rows in the table give the character values for each irreducible representation.

Table 1. Character table of $\mathfrak{S}_3$.

# elements:	1	3	2
conjugacy class:	$C(1^3)$	$C(1^1 2^1)$	$C(3^1)$
$\chi^{[3]}$	1	1	1
$\chi^{[2,1]}$	2	0	−1
$\chi^{[1,1,1]}$	1	−1	1

5.2. *Determinant formula for $\mathfrak{S}_k$ characters*

We next apply Theorem 4.2 to obtain an alternating sum formula for the characters of $\mathfrak{S}_k$. For this, we need to identify the weight spaces $V(\nu)$ as

$\mathfrak{S}_k$-modules. Here $\nu = [\nu_1, \ldots, \nu_n]$ with $\nu_i \geq 0$ and $\nu_1 + \cdots + \nu_n = k$. We have already seen that $V(\nu) = \mathrm{Span}\{e_I : \mu(I) = \nu\}$.

Lemma 5.3: *Let $\mathfrak{S}_\nu = \mathfrak{S}_{\nu_1} \times \cdots \times \mathfrak{S}_{\nu_n} \subset \mathfrak{S}_k$. Then $V(\nu) \cong \mathbb{C}[\mathfrak{S}_k/\mathfrak{S}_\nu]$ as a $\mathfrak{S}_k$-module.*

Proof: If I is a multi-index such that $\mu(I) = \nu$, then there is $s \in \mathfrak{S}_k$ so that

$$s \cdot I = [\underbrace{1, \ldots, 1}_{\nu_1}, \underbrace{2, \ldots, 2}_{\nu_2}, \underbrace{3, \ldots, 3}_{\nu_3}, \ldots].$$

Since $\sigma_k(s)e_I = e_{s \cdot I}$, the lemma follows. □

From the lemma we see that $\sigma_k(s)$ acts as a permutation matrix on $V(\nu)$, and hence

$$\mathrm{tr}_{V(\nu)}(\sigma_k(s)) = \#\{\text{fixed points of } s \text{ on } \mathfrak{S}_k/\mathfrak{S}_\nu\}.$$

The Weyl group for G is $\mathfrak{S}_n$ and acts on the weight lattice P as permutations of the coordinates of the weights. Applying Theorem 4.2 and using Lemma 5.3, we obtain the following character formula.

Theorem 5.4: *Let $\lambda \in \mathrm{Par}(k, n)$ and $s \in \mathfrak{S}_k$. Then*

$$\chi_\lambda(s) = \sum_t \mathrm{sgn}(t)\ \#\{\textit{fixed points of } s \textit{ on } \mathfrak{S}_k/\mathfrak{S}_{\lambda+\rho_{[n]}-t\cdot\rho_{[n]}}\}.$$

Here the sum is over all $t \in \mathfrak{S}_n$ such that all the coordinates of $\lambda + \rho_{[n]} - t\cdot\rho_{[n]}$ are nonnegative. In particular,

$$\dim E^\lambda = \sum_{t \in \mathfrak{S}_n} \mathrm{sgn}(t) \binom{k}{\lambda + \rho_{[n]} - t \cdot \rho_{[n]}}.$$

In Theorem 5.4 $\rho_{[n]} = [n-1, n-2, \ldots, 1, 0]$ and

$$\binom{k}{\nu} = \frac{k!}{\nu!} \qquad (\text{where } \nu! = \nu_1! \cdots \nu_n!)$$

is the multinomial coefficient (with the usual convention that it is zero if any entry in ν is negative). The dimension formula can be written as a determinant and then reduced to Vandermonde form. This gives the following *product formula* for the dimension of the representation E^λ that is analogous to the Weyl dimension formula for the representation F^λ.

Corollary 5.5: *Let $\lambda \in \mathrm{Par}(k, n)$. Then* $\dim E^\lambda = \frac{k!}{(\lambda+\rho_{[n]})!} D_n(\lambda + \rho_{[n]})$.

A partition $\lambda = [\lambda_1, \ldots, \lambda_n]$ can be represented in terms of its *Ferrers diagram*: a left-justified array of boxes, with λ_i boxes in the ith row (counting from the top down). Each box in the diagram has a *hook length:* the total number of boxes to the right and below the given box (including the box itself). We can then fill each box with its hook length. For example, $\lambda = [4,3,1] \in \mathrm{Par}(8,3)$ has Ferrers diagram and hook lengths

$$\begin{array}{|c|c|c|c|}\hline 6 & 4 & 3 & 1 \\ \hline 4 & 2 & 1 \\ \cline{1-3} 1 \\ \cline{1-1}\end{array}$$

From Corollary 5.5 one obtains (by induction on the number of columns of λ) the *Hook Length Formula*

$$\dim E^{\lambda} = \frac{k!}{\prod_{ij} h_{ij}(\lambda)} \tag{19}$$

where $h_{ij}(\lambda)$ is the hook length of the ij box in the Ferrers diagram of λ. By way of comparison, the *Weyl Dimension Formula* for $\mathrm{GL}(n, \mathbb{C})$ can be written as

$$\dim F^{\lambda}_{(n)} = \frac{(\lambda + \rho_{[n]})!}{\rho_{[n]}! \prod_{ij} h_{ij}(\lambda)} \tag{20}$$

(see [16, Sec. 9.1.4, Example #9]). For example, for $\mathfrak{S}_8 \times \mathrm{GL}(3, \mathbb{C})$ acting on $\bigotimes^8 \mathbb{C}^3$ we have $\rho_{[3]} = [2,1,0]$ and

$$\dim E^{[4,3,1]} = \frac{8!}{6 \cdot 4 \cdot 4 \cdot 3 \cdot 2} = 70, \quad \dim F^{[4,3,1]}_{(3)} = \frac{6! \cdot 4!}{2!(6 \cdot 4 \cdot 4 \cdot 3 \cdot 2)} = 15\,.$$

Thus $E^{[4,3,1]} \otimes F^{[4,3,1]}_{(3)}$ is an irreducible subspace of $\bigotimes^8 \mathbb{C}^3$ of dimension $(70) \cdot (15)$.

5.3. *Schur–Weyl duality and* GL(k)-GL(n) *duality*

There is another model for the irreducible representations of $\mathfrak{S}_k$ that comes from the identification of $\mathfrak{S}_k$ with the Weyl group of $\mathrm{GL}(k, \mathbb{C})$. Let $X = M_{k \times n}$ ($k \times n$ complex matrices) and let $\mathrm{GL}(k, \mathbb{C}) \times \mathrm{GL}(n, \mathbb{C})$ act on $\mathcal{P}(X)$ by

$$\rho(g_1, g_2) f(x) = f(g_1^t x g_2) \quad \text{for } g_1 \in \mathrm{GL}(k, \mathbb{C}) \text{ and } g_2 \in \mathrm{GL}(n, \mathbb{C})\,.$$

This representation is multiplicity free and decomposes as

$$\mathcal{P}(X) \cong \bigoplus_{\mu} F^{\mu}_{(k)} \otimes F^{\mu}_{(n)} \tag{21}$$

with the sum over all partitions μ with at most $\min\{k, n\}$ parts (see [16, Sec. 5.2.4] or the article by Benson–Ratcliff in this volume).

Let $H_k \subset \mathrm{GL}(k, \mathbb{C})$ be the maximal torus of diagonal matrices, and embed $\mathfrak{S}_k \subset \mathrm{GL}(k, \mathbb{C})$ as the permutation matrices. If we only consider the action of the subgroup $\mathrm{Norm}\,(H_k) \cong \mathfrak{S}_k \ltimes H_k$ of $\mathrm{GL}(k, \mathbb{C})$ on X together with the right action of $\mathrm{GL}(n, \mathbb{C})$, then

$$X \cong \underbrace{(\mathbb{C}^n)^* \oplus \cdots \oplus (\mathbb{C}^n)^*}_{k \text{ summands}} .$$

Hence

$$\mathcal{P}(X) \cong \underbrace{S(\mathbb{C}^n) \otimes \cdots \otimes S(\mathbb{C}^n)}_{k \text{ factors}}$$

as a representation of $\mathrm{Norm}\,(H_k) \times \mathrm{GL}(n, \mathbb{C})$. Here $\mathfrak{S}_k$ acts by permuting the tensor factors, while $h = \mathrm{diag}[x_1, \ldots, x_k] \in H_k$ acts by multiplication by x_j on the jth factor. The weight space decomposition of $\mathcal{P}(X)$ relative to the H_k action is thus

$$\mathcal{P}(X)(\mu) \cong S^{m_1}(\mathbb{C}^n) \otimes \cdots \otimes S^{m_k}(\mathbb{C}^n) \qquad \text{for } \mu = [m_1, \ldots, m_k] . \tag{22}$$

Here $\mathfrak{S}_k$ acts by permuting the factors in this decomposition while $\mathrm{GL}(n\, \mathbb{C})$ acts as usual on each copy of $\mathbb{C}^n$. In particular, the weight $\det_k = [1, 1, \ldots, 1]$ is fixed by $\mathfrak{S}_k$ and the corresponding weight space is

$$\mathcal{P}(X)(\det{}_k) \cong S^1(\mathbb{C}^n) \otimes \cdots \otimes S^1(\mathbb{C}^n) = (\mathbb{C}^n)^{\otimes k}$$

with the usual commuting actions of $\mathfrak{S}_k$ and $\mathrm{GL}(n, \mathbb{C})$. On the other hand, if we calculate this weight space using (21), we see that

$$(\mathbb{C}^n)^{\otimes k} \cong \bigoplus_{\lambda \in \mathrm{Par}(k,n)} F^{\lambda}_{(k)}(\det{}_k) \otimes F^{\lambda}_{(n)}$$

as a module for $\mathfrak{S}_k \times \mathrm{GL}(n, \mathbb{C})$. Invoking Theorem 3.8 we conclude: *For all $\lambda \in \mathrm{Par}(k, n)$,*

$$E^{\lambda} \cong F^{\lambda}_{(k)}(\det{}_k)$$

as a $\mathfrak{S}_k$ module, with the action of $\mathfrak{S}_k$ coming from its embedding into $\mathrm{GL}(k, \mathbb{C})$.

Examples

(1) Take $\lambda = [1, \ldots, 1] \in \mathrm{Par}(k)$. Then the representation $F^{\lambda}_{(k)}$ of $\mathrm{GL}(k, \mathbb{C})$ is $\bigwedge^k \mathbb{C}^k$, on which $\mathrm{GL}(k, \mathbb{C})$ acts by $g \mapsto \det(g)$. This shows once again that E^{λ} is the sgn representation of $\mathfrak{S}_k$.

(2) Now take $\lambda = [k]$. Then the representation $F^{\lambda}_{(k)}$ of $\mathrm{GL}(k, \mathbb{C})$ is $S^k(\mathbb{C}^k) \cong \mathcal{P}^k((\mathbb{C}^k)^*)$. The $\det_k$ weight space is one-dimensional and spanned by the monomial $x_1 \cdots x_k$, which is fixed by $\mathfrak{S}_k$. Again we see that $E^{[k]}$ is the trivial representation of $\mathfrak{S}_k$.

6. Polynomial invariants and FFT

6.1. *Invariant polynomials*

Let G be a reductive linear algebraic group. Recall from Section 1 that given a regular representation (π, V) of G, we have a locally regular representation ρ of G as automorphisms of the commutative algebra $\mathcal{P}(V)$ of complex-valued polynomial functions on V:

$$\rho(g)f(v) = f(g^{-1}v) \quad \text{for } f \in \mathcal{P}(V) \text{ and } g \in G$$

(here we write gv for $\pi(g)v$ when the action π is clear from the context). Since G acts by automorphisms of $\mathcal{P}(V)$, the space $\mathcal{J} = \mathcal{P}(V)^G$ of G-invariant polynomials is a subalgebra of $\mathcal{P}(V)$. Thus we can consider $\mathcal{P}(V)$ as a module for $\mathcal{J}$ under the action of pointwise multiplication, which commutes with the G action. Then in the isotypic decomposition

$$\mathcal{P}(V) = \bigoplus_{\lambda \in \widehat{G}} \mathcal{P}(V)_{(\lambda)}$$

each summand is invariant under $\mathcal{J}$. By Corollary 1.2 there is a projection $f \mapsto f^{\natural}$ from $\mathcal{P}(V)$ onto $\mathcal{J}$, with $\deg f^{\natural} \leq \deg f$. If $f \in \mathcal{P}(V)$ and $\varphi \in \mathcal{J}$ then

$$(\varphi f)^{\natural} = \varphi f^{\natural} \tag{23}$$

(Decompose $f = f^{\natural} + \cdots$ into isotypic components; then $\varphi f = \varphi f^{\natural} + \cdots$ is the isotypic decomposition of φf.)

Theorem 6.1 (Hilbert-Hurwitz): *$\mathcal{J}$ is finitely generated as an algebra over $\mathbb{C}$.*

Proof: Let $\mathcal{J}_+ = \{f \in \mathcal{J} : f(0) = 0\}$ and write $\mathcal{R} = \mathcal{P}(V)$. Since $\mathcal{R}$ is a polynomial ring in $\dim V$ generators, the Hilbert basis theorem implies that the ideal $\mathcal{R}\mathcal{J}_+$ is finitely generated as an $\mathcal{R}$ module: there exist $\varphi_j \in \mathcal{J}_+$ such that

$$\mathcal{R}\mathcal{J}_+ = \sum_{j=1}^{n} \mathcal{R}\varphi_j \,.$$

Furthermore, since $\mathcal{J}_+$ is invariant under the $\mathbb{C}^\times$ action on $\mathcal{R}$ ($f(v) \mapsto f(\zeta v)$ for $\zeta \in \mathbb{C}^\times$), we may take each φ_j to be homogeneous of some degree $d_j \geq 1$.

We claim that $\{\varphi_j\}$ generate $\mathcal{J}$ as an algebra over $\mathbb{C}$. Let $f \in \mathcal{J}$ be of degree d and assume inductively that all polynomials in $\mathcal{J}$ of degree less than d are polynomials in $\varphi_1, \ldots, \varphi_n$. We can find $f_j \in \mathcal{R}$ so that $f = \sum_j f_j \varphi_j$. Now project onto $\mathcal{J}$ and use (23):

$$f = f^\natural = \sum_j (f_j \varphi_j)^\natural = \sum_j f_j^\natural \varphi_j .$$

Since $\deg \varphi_j \geq 1$, we have $\deg f_j^\natural \leq \deg f_j < \deg f$. Hence by induction each $f_j^\natural$ is a polynomial in $\varphi_1, \ldots, \varphi_n$, so the same is true of f. □

We shall say that $\{\varphi_1, \ldots, \varphi_n\} \subset \mathcal{J}$ is a *basic set of G invariants* if

(i) $\{\varphi_1, \ldots, \varphi_n\}$ generates $\mathcal{J}$ as an algebra over $\mathbb{C}$
(ii) each φ_i is homogeneous (of some degree d_i)

and n is as small as possible, subject to (i) and (ii). By Theorem 6.1 there always exists a basic set of invariants (the polynomials φ_i are not unique but the set $\{d_1, \ldots, d_n\}$ of degrees is uniquely determined).

Example: Let $G = \mathfrak{S}_n$ and $V = \mathbb{C}^n$, with G acting as permutations of the coordinates. Then

$$\rho(s) f(x_1, \ldots, x_n) = f(x_{s(1)}, \ldots, x_{s(n)}) \quad \text{for } f \in \mathbb{C}[x_1, \ldots, x_n] \text{ and } s \in \mathfrak{S}_n$$

and $\mathcal{J}$ is the algebra of *symmetric functions* in n variables. By the *fundamental theorem of symmetric functions* one has $\mathcal{J} = \mathbb{C}[\sigma_1, \ldots, \sigma_n]$, where

$$\sigma_p(x_1, \ldots, x_n) = \sum_{1 \leq j_1 < \cdots < j_p \leq n} x_{j_1} \cdots x_{j_p} \qquad (p\text{th elementary symmetric function})$$

Furthermore, the functions $\{\sigma_1, \ldots, \sigma_n\}$ are algebraically independent, so they give a basic set of invariants with degrees $d_p = p$. (The function σ_p is the restriction to the diagonal matrices of the character of $\bigwedge^p \mathbb{C}^n$ as a representation of $\mathrm{GL}(n, \mathbb{C})$.)

6.2. *Invariants of vectors and covectors*

Take G as a classical group ($\mathrm{GL}(n, \mathbb{C})$, $\mathrm{O}(\mathbb{C}^n, B)$, or $\mathrm{Sp}(\mathbb{C}^n, \Omega)$ with n even) and $V = \mathbb{C}^n$ the *defining representation* of G. Let

$$V^m = \underbrace{V \oplus \cdots \oplus V}_{m \text{ column vectors}}, \qquad V^{*k} = \underbrace{V^* \oplus \cdots \oplus V^*}_{k \text{ row vectors}}$$

with the natural G action on each summand. Then $\mathcal{P}(V^{*k} \oplus V^m)^G$ is the algebra of G-invariant polynomial functions of k covectors and m vectors. The *First Fundamental Theorem* (FFT) of invariant theory for the group G gives an explicit description of sets of basic invariants (for all values of k and m).

There is an alternate picture that reveals the *hidden symmetries* in this situation and gives an obvious algebra of G-invariant polynomials together with a set of quadratic generators. Namely, we have the G-isomorphisms

$$V^{*k} \cong M_{k\times n} \quad \text{right } G \text{ action (matrix multiplication)}$$
$$V^{m} \cong M_{n\times m} \quad \text{left } G \text{ action (matrix multiplication)}$$

where $M_{k\times n}$ is the vector space of $k \times n$ complex matrices. In this picture we see that the reductive group $L = \mathrm{GL}(k,\mathbb{C}) \times \mathrm{GL}(m,\mathbb{C})$ acts on $M_{k\times n} \oplus M_{n\times m}$ by

$$(a,b)(x \oplus y) = ax \oplus yb^{-1} \quad \text{for } a \in \mathrm{GL}(k,\mathbb{C}) \text{ and } b \in \mathrm{GL}(m,\mathbb{C})\,.$$

This action obviously commutes with the G action. The induced action on functions makes $\mathcal{P}(M_{k\times n} \oplus M_{n\times m})^G$ into an L module. Note that the maximal torus of diagonal matrices in L acts in the original picture $V^{*k} \oplus V^m$ by scalar multiplication in each vector summand, while the Weyl group $\mathfrak{S}_k \times \mathfrak{S}_m$ of L acts by permutation of positions of the summands.

Define the *multiplication map*

$$\mu : M_{k\times n} \oplus M_{n\times m} \to M_{k\times m} \qquad x \oplus y \mapsto xy \quad \text{(matrix multiplication)}\,.$$

Obviously $\mu(xg \oplus g^{-1}y) = \mu(x \oplus y)$ for all $g \in \mathrm{GL}(n,\mathbb{C})$. Hence we have an algebra homomorphism

$$\mu^* : \mathcal{P}(M_{k\times m}) \to \mathcal{P}(V^{*k} \oplus V^m)^{\mathrm{GL}(n,\mathbb{C})}\,, \qquad \mu^*(f)(x \oplus y) = f(xy)\,.$$

In particular, if we take $f = x_{ij}$ (the (i,j) matrix entry function on $M_{k\times m}$), then

$$\mu^*(x_{ij})(v_1^*, \cdots, v_k^*, v_1, \cdots, v_m) = \langle v_i^*, v_j \rangle$$

(the *contraction* of the ith covector with the jth vector).

There is a natural action of L on $M_{k\times m}$ with $\mathrm{GL}(k,\mathbb{C})$ acting by left multiplication and $\mathrm{GL}(m,\mathbb{C})$ acting by right multiplication, Hence L acts on $\mathcal{P}(M_{k\times m})^G$. The map μ intertwines the two L actions.

6.3. *Polynomial FFT for* GL(n)

The FFT for $\mathrm{GL}(n,\mathbb{C})$ is the assertion that the method just indicated to construct invariants furnishes the complete algebra of polynomial invariants.

Theorem 6.2: *Let $G = \mathrm{GL}(n,\mathbb{C})$. Then the map μ^* is surjective. Hence the km quadratic polynomials $\phi_{ij} = \mu^*(x_{ij})$ with $1 \le i \le k$ and $1 \le j \le m$ give a set of basic invariants for $\mathcal{P}(M_{k\times n} \oplus M_{n\times m})^G$.*

After discussing tensor invariants in the next section we shall show there how this theorem is an immediate consequence of Proposition 3.1. At this point we observe that the image of μ consists of all $k \times m$ matrices x with $\mathrm{rank}(x) \le \min(k, m, n)$. This gives rise to the following dichotomy:

(1) If $n \ge \min(k, m)$, then μ is surjective. Hence μ^* is injective and

$$\mathcal{P}(M_{k\times n} \oplus M_{n\times m})^{\mathrm{GL}(n,\mathbb{C})} \cong \mathcal{P}(M_{k\times m})$$

is a polynomial algebra with km generators. One says that $M_{k\times m}$ is the *algebraic quotient* of $V^{*k} \oplus V^m$ by $\mathrm{GL}(n,\mathbb{C})$. The representation of L on $\mathcal{P}(M_{k\times m})$ is multiplicity-free (see [16, Theorem 5.2.7] or the article by Benson–Ratcliff in this volume).

(2) If $n < \min(k, m)$ then $\mathrm{Ker}(\mu^*) \ne 0$. The group L acts on $\mathrm{Ker}(\mu^*)$, and from the multiplicity-free decomposition of $\mathcal{P}(M_{k\times m})$ under L one finds that $\mathrm{Ker}(\mu^*)$ is a *determinantal ideal* generated by $(n+1)\times(n+1)$ minors. Thus $\mathcal{P}(M_{k\times n} \oplus M_{n\times m})^{\mathrm{GL}(n,\mathbb{C})}$ is the algebra of regular functions on the associated *determinantal variety.* This is the *Second Fundamental Theorem* (SFT) for $\mathrm{GL}(n,\mathbb{C})$ invariants (see [16, Theorem 5.2.15] for the complete statement).

6.4. *Polynomial FFT for the orthogonal group*

We next consider the full *orthogonal group* relative to the bilinear form $B(x,y) = x^t y$ on $V = \mathbb{C}^n$:

$$G = \mathrm{O}(n,\mathbb{C}) = \{g \in \mathrm{GL}(n,\mathbb{C}) \,:\, g^t g = I\}\,.$$

Since $V \cong V^*$ as a G-module, via the form B, it suffices to consider the invariants of k vector arguments $\mathcal{P}(V^k)^G = \mathcal{P}(M_{n\times k})^G$, where G acts on $M_{m\times k}$ by left multiplication. Define a map

$$\tau : M_{n\times k} \to SM_k \quad (k \times k \text{ symmetric matrices})\,, \quad \tau(x) = x^t x\,.$$

For $g \in G$ we have $\tau(gx) = x^t g^t g x = \tau(x)$. Hence

$$\tau^* : \mathcal{P}(SM_k) \to \mathcal{P}(V^k)^G$$

as in the case of $\mathrm{GL}(n, \mathbb{C})$. In particular, if we take $f = x_{ij}$ (the (i,j) matrix entry function on SM_k), then

$$\tau^*(x_{ij})(v_1, \cdots, v_k) = v_i^t v_j$$

(the inner product of the ith and jth vectors).

The map τ intertwines the right action of the *hidden symmetry group* $L = \mathrm{GL}(k, \mathbb{C})$ on $M_{n\times k}$. Here the action of L on SM_k is given by $x \mapsto bxb^t$ (for $b \in L$).

Theorem 6.3: *Let $G = \mathrm{O}(n, \mathbb{C})$. Then the map τ^* is surjective. Hence the $k(k+1)/2$ quadratic polynomials $\theta_{ij} = \tau^*(x_{ij})$ with $1 \le i \le j \le k$ give a set of basic invariants for $\mathcal{P}(M_{n\times k})^G$.*

Proof for the case $n \ge k$: There is a natural G-equivariant embedding $M_{n\times k} \subset M_{n\times n}$; just add $n-k$ columns of zeros on the right to make $x \in M_{n\times k}$ into an $n \times n$ matrix. Hence we may assume that $k = n$. Now see [16, Proposition 4.2.6] for the proof.[e] □

We shall complete the proof for the general case $n < k$ after discussing tensor invariants in the next section. Here we observe that the image of τ consists of all $k \times k$ symmetric matrices x with $\mathrm{rank}(x) \le \min(k, n)$. This gives rise to the following dichotomy:

(1) If $n \ge k$, then τ is surjective. Hence τ^* is injective and $\mathcal{P}(M_{n\times k})^G \cong \mathcal{P}(SM_k)$ is a polynomial algebra with $k(k+1)/2$ generators. One says that SM_k is the *algebraic quotient* of $M_{n\times k}$ by $\mathrm{O}(n, \mathbb{C})$. The representation of L on $\mathcal{P}(M_{n\times k})^G$ is multiplicity-free (see [16, Theorem 5.2.9] or the article by Benson–Ratcliff in this volume).

(2) If $n < k$ then $\mathrm{Ker}(\tau^*) \ne 0$. From the multiplicity-free decomposition of $\mathcal{P}(SM_k)$ under L one finds that $\mathrm{Ker}(\tau^*)$ is a *determinantal ideal* generated by $(n+1) \times (n+1)$ minors. Thus $\mathcal{P}(M_{n\times k})^G$ is the algebra of functions on the associated *symmetric determinantal variety.* This is the *Second Fundamental Theorem* (SFT) for $\mathrm{O}(n, \mathbb{C})$ invariants (see [16, Theorem 5.2.17] for the complete statement).

[e] The proof is by induction on n and can be viewed as an algebraic group version of the QR factorization for M_n and the *Cholesky Decomposition* for SM_n. This result is associated with a particular *partial compactification* of the symmetric space $\mathrm{GL}(n, \mathbb{C})/\mathrm{O}(n, \mathbb{C})$.

6.5. *Polynomial FFT for the symplectic group*

Now consider the *symplectic group* $G = \mathrm{Sp}(\mathbb{C}^n, \Omega)$, where $n = 2p$ is even and

$$\Omega(x, y) = x^t J y\,, \quad J = \begin{bmatrix} 0 & I_p \\ -I_p & 0 \end{bmatrix}.$$

Here I_p is the $p \times p$ identity matrix. Thus G is the subgroup of $\mathrm{GL}(n, \mathbb{C})$ defined by $g^t J g = J$. Since $(\mathbb{C}^n)^* \cong \mathbb{C}^n$ via the form Ω, it suffices to consider the invariants of k vector arguments $\mathcal{P}(V^k)^G = \mathcal{P}(M_{n\times k})$. Define a map

$$\gamma : M_{n\times k} \to AM_k \quad (k \times k \text{ skew-symmetric matrices})\,, \quad \gamma(x) = x^t J x\,.$$

For $g \in G$ we have $\gamma(gx) = x^t g^t J g x = \gamma(x)$. Hence

$$\gamma^* : \mathcal{P}(AM_k) \to \mathcal{P}(V^k)^G$$

as in the case of $\mathrm{O}(n, \mathbb{C})$. In particular, if we take $f = x_{ij}$ (the (i, j) matrix entry function on AM_k), then

$$\gamma^*(x_{ij})(v_1, \cdots, v_k) = \Omega(v_i, v_j)$$

(contraction of the ith and jth vectors by Ω).

The map τ intertwines the right action of the *hidden symmetry group* $L = \mathrm{GL}(k, \mathbb{C})$ on $M_{n\times k}$ with the action of L on AM_k given by $x \mapsto bxb^t$ (for $b \in L$).

Theorem 6.4: *Let $G = \mathrm{Sp}(\mathbb{C}^n, \Omega)$. Then the map γ^* is surjective. Hence the $k(k-1)/2$ quadratic polynomials $\omega_{ij} = \gamma^*(x_{ij})$ with $1 \le i < j \le k$ give a set of basic invariants for $\mathcal{P}(M_{n\times k})^G$.*

Proof for the case $n \ge k$: The same citation as for the orthogonal case. □

We shall complete the proof for the general case $n < k$ after discussing tensor invariants in the next section. Here we observe that the image of γ consists of all $k \times k$ skew-symmetric matrices x with $\mathrm{rank}(x) \le \min(k, n)$. This gives rise to the following dichotomy:

(1) If $n \ge k$, then γ is surjective. Hence γ^* is injective and $\mathcal{P}(M_{n\times k})^G \cong \mathcal{P}(AM_k)$ is a polynomial algebra with $k(k-1)/2$ generators. One says that AM_k is the *algebraic quotient* of $M_{n\times k}$ by $\mathrm{Sp}(\mathbb{C}^n, \Omega)$. The representation of L on $\mathcal{P}(M_{n\times k})^G$ is multiplicity-free (see [16, Theorem 5.2.11] or the article by Benson–Ratcliff in this volume).

(2) If $n < k$ then $\mathrm{Ker}(\gamma^*) \neq 0$. From the multiplicity-free decomposition of $\mathcal{P}(AM_k)$ under L one finds that $\mathrm{Ker}(\gamma^*)$ is generated by a set of *Pfaffian* polynomials of degree $n/2+1$. Thus $\mathcal{P}(M_{n\times k})^G$ is the algebra of functions on the associated *skew-symmetric Pfaffian variety*. This is the *Second Fundamental Theorem* (SFT) for $\mathrm{O}(n,\mathbb{C})$ invariants (see [16, Theorem 5.2.18] for the complete statement).

Summary: *For a classical group* G (general linear, orthogonal, symplectic), *the* G*-invariant polynomial functions of vectors and covectors are generated by all the possible* G*-invariant contractions of vectors and covectors.*

7. Tensor invariants and proof of FFT

7.1. *Tensor invariants for* GL(V)

We turn now from consideration of invariant polynomials to the general case of invariant tensors. Let $V = \mathbb{C}^n$ and consider the mixed tensor space $V^{\otimes m} \otimes V^{*\otimes k}$ as a $\mathrm{GL}(V)$ module. For $\zeta \in \mathbb{C}^\times$ the element ζI_n of $\mathrm{GL}(V)$ acts by ζ^{m-k} on this space. Hence there are no nonzero $\mathrm{GL}(V)$ invariant tensors if $m \neq k$ and we can assume $m = k$. In this case

$$V^{\otimes k} \otimes V^{*\otimes k} \cong \mathrm{End}(V^{\otimes k}) \tag{24}$$

as a $\mathrm{GL}(V)$ module, and hence

$$(V^{\otimes k} \otimes V^{*\otimes k})^{\mathrm{GL(V)}} \cong \mathrm{End}_{\mathrm{GL}(V)}(V^{\otimes k})\,.$$

By Schur duality (Corollary 1.6 and Proposition 3.1) we know that $\mathrm{End}_{\mathrm{GL}(V)}(V^{\otimes k})$ is spanned by the transformations $\sigma_k(s)$, $s \in \mathfrak{S}_k$.

Let $\{e_1, \ldots, e_n\}$ be the standard basis for $\mathbb{C}^n$ and let $\{e_1^*, \ldots, e_n^*\}$ be the dual basis. For an index $I = [i_1, \ldots, i_k]$ with $1 \leq i_p \leq n$ we set $|I| = k$ and

$$e_I = e_{i_1} \otimes \cdots \otimes e_{i_k} \in V^{\otimes k}\,, \quad e_I^* = e_{i_1}^* \otimes \cdots \otimes e_{i_k}^* \in V^{*\otimes k}\,.$$

Recall from Section 3 that the action of $s \in \mathfrak{S}_k$ on k-tensors is $\sigma_k(s)e_I = e_{s\cdot I}$. Define

$$C_s = \sum_{|I|=k} e_{s\cdot I} \otimes e_I^*\,.$$

Then C_s corresponds to $\sigma_k(s)$ under the isomorphism (24). Thus we obtain the *First Fundamental Theorem of Tensor Invariants* for $\mathrm{GL}(V)$:

Theorem 7.1: *For* $k \geq 1$ *one has* $(V^{\otimes k} \otimes V^{*\otimes k})^{\mathrm{GL}(V)} = \mathrm{Span}\{C_s : s \in \mathfrak{S}_k\}$.

The vector space $V^{\otimes k} \otimes V^{*\otimes k}$ is self-dual as a GL(V) module, and hence each of the mixed tensors C_s can also be viewed as a linear functional. This gives the alternate version of the Tensor FFT for GL(V) in terms of *total contractions* of vectors with covectors:

Corollary 7.2: *The space of* GL(V)*-invariant linear forms on* $V^{\otimes k} \otimes V^{*\otimes k}$ *is spanned by the contractions*

$$v_1 \otimes \cdots \otimes v_k \otimes v_1^* \otimes \cdots \otimes v_k^* \mapsto \prod_{j=1}^{k} \langle v_{s(j)}^*, v_j \rangle$$

for $s \in \mathfrak{S}_k$.

7.2. *Proof of polynomial FFT for* GL(V)

The polynomial form of the FFT for GL(V) is a consequence of Corollary 7.2. To prove this, we need to view GL(V)-invariant polynomials as tensors with additional symmetries, as we did in Section 5.3. Let $\mathbb{T}_{k,m} = (\mathbb{C}^\times)^k \times (\mathbb{C}^\times)^m$ and write $t \in \mathbb{T}_{k,m}$ as $t = [x_1, \ldots, x_k, y_1, \ldots, y_m]$. Denote the regular characters of $\mathbb{T}_{k,m}$ as

$$t \mapsto t^{[\mathbf{p},\mathbf{q}]} = \prod_{i=1}^{k} x_i^{p_i} \prod_{j=1}^{m} y_j^{q_j}$$

for $[\mathbf{p}, \mathbf{q}] \in \mathbb{Z}^k \times \mathbb{Z}^m$. Let $\mathbb{T}_{k,m}$ act on $z = [v_1, \ldots, v_k] \oplus [v_1^*, \ldots, v_m^*] \in V^k \oplus V^{*m}$ by

$$t \cdot z = [x_1 v_1, \ldots, x_k v_k] \oplus [y_1 v_1^*, \ldots, y_m v_m^*] .$$

This action commutes with the GL(V)-action on $V^k \oplus V^{*m}$, so GL(V) leaves invariant the weight spaces of $\mathbb{T}_{k,m}$ in $\mathcal{P}(V^k \oplus V^{*m})$. These weight spaces are described by the degrees of homogeneity of $f \in \mathcal{P}(V^k \oplus V^{*m})$ in v_i and v_j^* as follows. For $\mathbf{p} \in \mathbb{N}^k$ and $\mathbf{q} \in \mathbb{N}^m$ set

$$\mathcal{P}^{[\mathbf{p},\mathbf{q}]}(V^k \oplus V^{*m}) = \{ f \in \mathcal{P}(V^k \oplus V^{*m}) \,:\, f(t \cdot z) = t^{[\mathbf{p},\mathbf{q}]} f(z) \} .$$

Then

$$\mathcal{P}(V^k \oplus V^{*m}) = \bigoplus_{\mathbf{p} \in \mathbb{N}^k} \bigoplus_{\mathbf{q} \in \mathbb{N}^m} \mathcal{P}^{[\mathbf{p},\mathbf{q}]}(V^k \oplus V^{*m}) ,$$

and this decomposition is GL(V)-invariant. Thus

$$\mathcal{P}(V^k \oplus V^{*m})^G = \bigoplus_{\mathbf{p} \in \mathbb{N}^k} \bigoplus_{\mathbf{q} \in \mathbb{N}^m} \left[\mathcal{P}^{[\mathbf{p},\mathbf{q}]}(V^k \oplus V^{*m}) \right]^{\mathrm{GL}(V)} . \qquad (25)$$

We now give another realization of these weight spaces in terms of tensors. Given $\mathbf{p} \in \mathbb{N}^k$ and $\mathbf{q} \in \mathbb{N}^m$ we set

$$V^{*\otimes \mathbf{p}} \otimes V^{\otimes \mathbf{q}} = V^{*\otimes p_1} \otimes \cdots \otimes V^{*\otimes p_k} \otimes V^{\otimes q_1} \otimes \cdots \otimes V^{\otimes q_m}.$$

This space is isomorphic to $V^{*\otimes|\mathbf{p}|} \otimes V^{*\otimes|\mathbf{q}|}$ and is a $\mathrm{GL}(V)$ module with the usual action. Let $\mathfrak{S}_{\mathbf{p}} = \mathfrak{S}_{p_1} \times \cdots \times \mathfrak{S}_{p_k}$, with each factor acting as a group of permutations of the corresponding tensor factor in $V^{\otimes \mathbf{p}}$. This gives a representation of $\mathfrak{S}_{\mathbf{p}} \times \mathfrak{S}_{\mathbf{q}}$ on $V^{*\otimes \mathbf{p}} \otimes V^{\otimes \mathbf{q}}$ that commutes with the action of $\mathrm{GL}(V)$.

Lemma 7.3: *Let $\mathbf{p} \in \mathbb{N}^k$ and $\mathbf{q} \in \mathbb{N}^m$. There is a linear isomorphism*

$$\mathcal{P}^{[\mathbf{p},\mathbf{q}]}(V^k \oplus V^{*m})^{\mathrm{GL}(V)} \cong \left[\left(V^{*\otimes|\mathbf{p}|} \otimes V^{\otimes|\mathbf{q}|}\right)^{\mathrm{GL}(V)}\right]^{\mathfrak{S}_{\mathbf{p}} \times \mathfrak{S}_{\mathbf{q}}}. \tag{26}$$

Proof: We have the isomorphisms

$$\begin{aligned}\mathcal{P}(V^k \oplus V^{*m}) &\cong S((V^*)^k \oplus V^m) \\ &\cong \underbrace{S(V^*) \otimes \cdots \otimes S(V^*)}_{k \text{ factors}} \otimes \underbrace{S(V) \otimes \cdots \otimes S(V)}_{m \text{ factors}}\end{aligned}$$

as $\mathrm{GL}(V)$ modules. Hence

$$\mathcal{P}^{[\mathbf{p},\mathbf{q}]}(V^k \oplus V^{*m}) \cong S^{[\mathbf{p}]}(V^*) \otimes S^{[\mathbf{q}]}(V), \tag{27}$$

where $S^{[\mathbf{p}]}(V^*) = S^{p_1}(V^*) \otimes \cdots \otimes S^{p_k}(V^*)$ and $S^{[\mathbf{q}]}(V) = S^{q_1}(V) \otimes \cdots \otimes S^{q_m}(V)$. We also have a $\mathrm{GL}(V)$-module isomorphism

$$S^r(V) \cong \left[V^{\otimes r}\right]^{\mathfrak{S}_r} \subset V^{\otimes r}$$

with $\mathfrak{S}_r$ acting by permuting the tensor positions as usual. Combining this with (27) we obtain the linear isomorphisms

$$\begin{aligned}\mathcal{P}^{[\mathbf{p},\mathbf{q}]}(V^k \oplus V^{*m})^{\mathrm{GL}(V)} &\cong \left[S^{[\mathbf{p}]}(V^*) \otimes S^{[\mathbf{q}]}(V)\right]^{\mathrm{GL}(V)} \\ &\cong \left[\left(V^{*\otimes|\mathbf{p}|} \otimes V^{\otimes|\mathbf{q}|}\right)^{\mathfrak{S}_{\mathbf{p}} \times \mathfrak{S}_{\mathbf{q}}}\right]^{\mathrm{GL}(V)}.\end{aligned}$$

This implies (26) since the actions of $\mathrm{GL}(V)$ and $\mathfrak{S}_{\mathbf{p}} \times \mathfrak{S}_{\mathbf{q}}$ mutually commute. □

We now prove the polynomial version of the First Fundamental Theorem for $\mathrm{GL}(V)$. This theorem asserts that for each $\mathbf{p} \in \mathbb{N}^k$ and $\mathbf{q} \in \mathbb{N}^m$, the space

$$\mathcal{P}^{[\mathbf{p},\mathbf{q}]}(V^k \oplus V^{*m})^{\mathrm{GL}(V)}$$

is spanned by monomials of the form

$$\prod_{i=1}^{k}\prod_{j=1}^{m}\langle v_i, v_j^*\rangle^{r_{ij}} \tag{28}$$

for suitable choices of k, m and r_{ij}. The subgroup $\mathbb{T} = \{\zeta I_V : \zeta \in \mathbb{C}^\times\}$ of $\mathrm{GL}(V)$ acts on $\mathcal{P}^{[\mathbf{p},\mathbf{q}]}(V^k \oplus V^{*m})$ by the character $\zeta \mapsto \zeta^{|\mathbf{q}|-|\mathbf{p}|}$, so we may assume that $|\mathbf{p}| = |\mathbf{q}| = r$, say. By Lemma 7.3,

$$\mathcal{P}^{[\mathbf{p},\mathbf{q}]}(V^k \oplus V^{*m})^{\mathrm{GL}(V)} \cong \left[\left(V^{*\otimes r} \otimes V^{\otimes r}\right)^{\mathrm{GL}(V)}\right]^{\mathfrak{S}_{\mathbf{p}} \times \mathfrak{S}_{\mathbf{q}}}. \tag{29}$$

From Theorem 7.1 we know that the space $(V^{*\otimes r} \otimes V^{\otimes r})^{\mathrm{GL}(V)}$ is spanned by the complete contractions C_s for $s \in \mathfrak{S}_r$. Hence the right side of (29) is spanned by the tensors

$$\sum_{(g,h)\in\mathfrak{S}_{\mathbf{p}}\times\mathfrak{S}_{\mathbf{q}}} \sigma_r^*(g) \otimes \sigma_r(h) C_s$$

for $s \in \mathfrak{S}_r$. Under the isomorphism (29), the action of $\mathfrak{S}_{\mathbf{p}} \times \mathfrak{S}_{\mathbf{q}}$ disappears and these tensors correspond to the polynomials

$$\begin{aligned} F_s(v_1, \ldots, v_k, v_1^*, \ldots, v_m^*) &= C_s(v_1^{\otimes p_1} \otimes \cdots \otimes v_k^{\otimes p_k} \otimes v_1^{*\otimes q_1} \otimes \cdots \otimes v_m^{*\otimes q_m}) \\ &= \langle v_1^{\otimes p_1} \otimes \cdots \otimes v_k^{\otimes p_k}, w_1^* \otimes \cdots \otimes w_r^*\rangle \\ &= \prod_{i=1}^{n}\langle w_i, w_i^*\rangle, \end{aligned}$$

where each w_i is v_j for some j and each w_i^* is $v_{j'}^*$ for some j' (depending on s). Obviously F_s is of the form (28). □

7.3. *Tensor invariants for orthogonal and symplectic groups*

Consider now a nondegenerate bilinear form ω on V, which we assume to be either symmetric or skew-symmetric. Let G be the subgroup of $\mathrm{GL}(V)$ that preserves ω (so G is either the full orthogonal group or the symplectic group). Any mixed tensor that is invariant under $\mathrm{GL}(V)$ is also invariant under G, of course. To find additional tensor invariants, we can use the G-module isomorphism $V \cong V^*$ furnished by ω to restrict attention to $V^{*\otimes k}$. Furthermore, $(V^{*\otimes k})^G = 0$ if k is odd, since $-I \in G$. Hence we need only find a linear basis for $(V^{*\otimes 2k})^G$.

The given form $\omega \in (V^{*\otimes 2})^G$ by definition. Since G preserves tensor multiplication, it follows that

$$\theta_k = \omega^{\otimes k} \in (V^{*\otimes 2k})^G.$$

The representation σ_{2k} of $\mathfrak{S}_{2k}$ on $V^{*\otimes 2k}$ commutes with the action of G, of course, so the tensors $\sigma_{2k}(s)\theta_k$ are also G invariant, for every $s \in \mathfrak{S}_{2k}$.

Theorem 7.4: *For all integers $k \geq 1$ one has* $(V^{*\otimes 2k})^G = \mathrm{Span}\{\sigma_{2k}(s)\theta_k : s \in \mathfrak{S}_{2k}\}$.

Because of the symmetries of the tensor θ_k under the action of $\mathfrak{S}_{2k}$, there are redundancies in the spanning set of Theorem 7.4. A labeling that factors out these symmetries is the following, which we will also use in Section 13. Define a *two-partition* of the set $\{1, \ldots, 2k\}$ to be any set of k pairs $\xi = \{\{i_1, j_1\}, \ldots, \{i_k, j_k\}\}$ such that $\{i_1, j_1, \ldots, i_k, j_k\}$ consists of the integers $1, \ldots, 2k$. Denote the set of all 2-partitions of k by Ξ_k. For $\xi \in \Xi_k$ define the *complete contraction*

$$\lambda_\xi(v_1 \otimes \cdots \otimes v_{2k}) = \prod_{p=1}^{k} \omega(v_{i_p}, v_{j_p}) .$$

(We label the pairs in ξ so that $i_p < j_p$; then this definition has no sign ambiguity, even when ω is skew-symmetric.) The invariant tensors in Theorem 7.4 are just these contractions.

Corollary 7.5: *For all integers $k \geq 1$ one has* $(V^{*\otimes 2k})^G = \mathrm{Span}\{\lambda_\xi : \xi \in \Xi_k\}$.

Proof of Theorem 7.4: Following Attiyah–Bott–Patodi, we shift the action of G from $V^{*\otimes 2k}$ to $\mathrm{End}V$ by a tensor algebra version of the classical *polarization operators*.[f] This transforms the space of G-invariant tensors into a different space of $\mathrm{GL}(V)$-invariant tensors built from G-invariant polynomials on $\mathrm{End}(V)$. The results of Section 6 allow us to express these G-invariant polynomials as covariant tensors with no further G-invariance condition. By this means each G-invariant tensor gives rise to a unique mixed $\mathrm{GL}(V)$-invariant tensor. But we know that all such tensors are linear combinations of complete vector-covector contractions. Finally, specializing the polarization variables, we find that the original G-invariant tensor is in the span of the complete contractions relative to the form ω.

In more detail, given $\lambda \in V^{*\otimes m}$ we define $\Phi_\lambda \in \mathcal{P}^k(\mathrm{End}V) \otimes V^{*\otimes m}$ by

$$\Phi_\lambda(X, w) = \langle \lambda,\, X^{\otimes m} w \rangle \quad \text{for } X \in \mathrm{End}V \text{ and } w \in V^{\otimes m} .$$

Since $\langle \lambda, w \rangle = \Phi_\lambda(I, w)$, we see that the map $\lambda \mapsto \Phi_\lambda$ is injective.

[f] This can be viewed as *unseparation of variables*, and is another instance of a hidden symmetry.

Let $G \subset \mathrm{GL}(V)$ be any subgroup, for the moment. Let G act on $\mathcal{P}^m(\mathrm{End}V) \otimes V^{*\otimes m}$ by *left multiplication* on $\mathrm{End}(V)$ only (no action on $V^{*\otimes m}$). Let $\mathrm{GL}(V)$ act by *right multiplication* on the $\mathrm{End}V$ factor and in the usual way on $V^{*\otimes m}$. Since these actions of G and $\mathrm{GL}(V)$ mutually commute, we obtain a representation of the product group $G \times \mathrm{GL}(V)$. In particular,

$$(g,h)\cdot\Phi_\lambda(X,w) = \langle\lambda,\, (g^{-1}Xh)^{\otimes m}h^{-1}\cdot w\rangle = \Phi_{g\cdot\lambda}(X,w)$$

for $g \in G, h \in \mathrm{GL}(V)$. Hence Φ_λ is automatically invariant under $\mathrm{GL}(V)$ for any λ, while if λ is G-invariant, then so is Φ_λ. Conversely, if $\Phi \in \mathcal{P}^m(\mathrm{End}V) \otimes V^{*\otimes m}$ is invariant under $G \times \mathrm{GL}(V)$, then Φ is determined by the linear functional $\lambda : w \mapsto \Phi(I,w)$ since $\Phi(h,w) = \Phi(I, h\cdot w)$ for $h \in \mathrm{GL}(V)$ and $\mathrm{GL}(V)$ is dense in $\mathrm{End}V$. Furthermore, for $g \in G$ we have

$$\Phi(I,w) = \Phi(gI,w) = \Phi(Ig,w) = \Phi(I, g\cdot w)$$

(here we have used the inclusion $G \subset \mathrm{GL}(V)$ to pass from the left action of G on $\mathrm{End}V$ to the right action of $\mathrm{GL}(V)$ on $\mathrm{End}V$). Hence $\Phi = \Phi_\lambda$ with $\lambda \in (V^{*\otimes m})^G$. The map $\lambda \mapsto \Phi_\lambda$ thus gives a linear isomorphism

$$(V^{*\otimes m})^G \cong \left[\mathcal{P}^m(\mathrm{End}V)^{L(G)} \otimes V^{*\otimes m}\right]^{\mathrm{GL}(V)} \tag{30}$$

where $L(G)$ denotes the left-multiplication action of G on $\mathrm{End}(V)$.

Let $\lambda \in \left[V^{*\otimes 2k}\right]^G$. Then by (30) with $m = 2k$, we have

$$\Phi_\lambda \in \left[\mathcal{P}^{2k}(\mathrm{End}V)^G \otimes V^{*\otimes 2k}\right]^{\mathrm{GL}(V)}$$

and $\lambda = \Phi_\lambda(I)$. By Theorems 6.3 and 6.4 (which we have proved in the case $k = \dim V$), there is a polynomial F_λ on $SM_n \times V^{\otimes 2k}$ when $G = \mathrm{O}(V)$ or on $AM_n \times V^{\otimes 2k}$ when $G = \mathrm{Sp}(V)$, so that for all $X \in M_n$ and $w \in V^{\otimes 2k}$,

$$\Phi_\lambda(X,w) = \begin{cases} F_\lambda(X^tX,\, w) & \text{when } G = \mathrm{O}(V)\,, \\ F_\lambda(X^tJ_nX,\, w) & \text{when } G = \mathrm{Sp}(V)\,. \end{cases}$$

We view F_λ as an element of $\mathcal{P}^k(SM_n)\otimes V^{*\otimes 2k}$ (*resp.* of $\mathcal{P}^k(AM_n)\otimes V^{*\otimes 2k}$). Note that

$$\langle\lambda, w\rangle = \Phi_\lambda(I_n,w) = \begin{cases} F_\lambda(I_n,\, w) & \text{when } G = \mathrm{O}(V)\,, \\ F_\lambda(J_n,\, w) & \text{when } G = \mathrm{Sp}(V)\,. \end{cases}$$

The next step is to translate the $\mathrm{GL}(V)$-invariance of Φ_λ into an appropriate invariance property of F_λ. The map

$$\Theta : M_n \to V^* \otimes V^*\,, \quad \Theta(x) = \sum_{i,j} x_{ij}\, e_i^* \otimes e_j^*$$

furnishes GL(V)-module isomorphisms $AM_n \cong \bigwedge^2 V^*$ and $SM_n \cong S^2V^*$. Hence

$$\mathcal{P}^k(AM_n) \cong S^k(\textstyle\bigwedge^2 V), \qquad \mathcal{P}^k(SM_n) \cong S^k(S^2V).$$

Thus there is a GL(V)-invariant tensor $C \in V^{\otimes 2k} \otimes V^{*\otimes 2k}$ so that

$$F_\lambda(A, w) = \langle A^{\otimes k} \otimes w,\, C\rangle$$

for $w \in V^{\otimes 2k}$ and A in either S^2V^* or $\bigwedge^2 V^*$. By the tensor FFT for GL(V) (tensor form) we may assume that C is a *complete contraction*:

$$C = \sum_{|I|=2k} e^*_{s\cdot I} \otimes e_I$$

for some $s \in \mathfrak{S}_{2k}$. When $G = \mathrm{O}(V)$ we take $A = I_n$ to recover the original G-invariant tensor λ as

$$\begin{aligned}\langle \lambda, w\rangle &= F_\lambda(I_n, w) = \langle \Theta(I_n)^{\otimes k} \otimes w,\, C\rangle \\ &= \sum_{|I|=2k} \langle e_I,\, \Theta(I_n)^{\otimes k}\rangle \langle w,\, e^*_{s\cdot I}\rangle = \langle \sigma_{2k}(s)\Theta(I_n)^{\otimes k},\, w\rangle\,.\end{aligned}$$

When $G = \mathrm{Sp}(V)$, we likewise take $A = J_n$ to get $\langle \lambda, w\rangle = \langle \sigma_{2k}(s)\Theta(J_n)^{\otimes k}, w\rangle$. Since

$$\theta^*_k = \begin{cases} \Theta(I_n)^{\otimes k} & \text{when } G = \mathrm{O}(V) \\ \Theta(J_n)^{\otimes k} & \text{when } G = \mathrm{Sp}(V)\,, \end{cases}$$

we conclude that $\lambda = \sigma_{2k}(s)\theta^*_k$. □

7.4. *Proof of polynomial FFT for orthogonal and symplectic groups*

We finally complete the proof of the FFT for the action of $G = \mathrm{O}(V)$ or $G = \mathrm{Sp}(V)$ on $\mathcal{P}(V)$, using an argument similar to the case of GL(V) to deduce the polynomial version of the FFT from the tensor version. Let $\mathbf{p} \in \mathbb{N}^m$. Since $-I \in G$ and acts by $(-1)^{|\mathbf{p}|}$ on $\mathcal{P}^{[\mathbf{p}]}(V^m)$, we may assume that $|\mathbf{p}| = 2k$. We now show that the space $\mathcal{P}^{[\mathbf{p}]}(V^m)^G$ is spanned by monomials

$$\varphi(v_1, \ldots, v_m) = \prod_{i,j=1}^{m} \omega(v_i, v_j)^{r_{ij}} \tag{31}$$

of weight $\mathbf{p}$. This will prove the FFT (polynomial version) for G.

By Lemma 7.3,

$$\mathcal{P}^{[\mathbf{p}]}(V^m)^G \cong \left[\left(V^{*\otimes 2k}\right)^G\right]^{\mathfrak{S}_{\mathbf{p}}}. \tag{32}$$

The space $(V^{*\otimes 2k})^G$ is spanned by the tensors $\sigma_{2k}^*(s)\theta_k^*$ for $s \in \mathfrak{S}_{2k}$ (Theorem 7.4). Hence the right side of (32) is spanned by the tensors

$$\sum_{t \in \mathfrak{S}_{\mathbf{p}}} \sigma_{2k}^*(ts)\theta_k^*$$

for $s \in \mathfrak{S}_{2k}$. Under the isomorphism (32), the action of $\mathfrak{S}_{\mathbf{p}}$ disappears and these tensors correspond to the polynomials

$$F_s(v_1, \ldots, v_m) = \sigma_{2k}^*(s)\theta_k^*(v_1^{\otimes p_1} \otimes \cdots \otimes v_m^{\otimes p_m}) = \prod_{i=1}^{k} \omega(u_i, u_{k+i}),$$

where each u_i is v_j for some j (depending on s). Thus F_s is of the form (31). □

8. Weyl algebra and Howe duality

8.1. *Duality in the Weyl algebra*

We shall now apply the general duality theorem from Section 1 to the following situation. Let V be an n-dimensional vector space over $\mathbb{C}$ and let $x_1, \ldots, x_n$ be coordinates on V relative to a basis $\{e_1, \ldots, e_n\}$. Let $\xi_1, \ldots, \xi_n$ be the coordinates for V^* relative to the dual basis $\{e_1^*, \ldots, e_n^*\}$. We denote by $\mathcal{PD}(V)$ the algebra of *polynomial coefficient differential operators* on V. This is the subalgebra of $\mathrm{End}(\mathcal{P}(V))$ generated (as an associative algebra) by the operators

$$D_i = \frac{\partial}{\partial x_i}, \qquad M_i = \text{multiplication by } x_i \quad (i = 1, \ldots, n).$$

Since $(\partial/\partial x_i)(x_j f) = (\partial x_j/\partial x_i)f + x_j(\partial f/\partial x_i)$ for $f \in \mathcal{P}(V)$, these operators satisfy the *Heisenberg commutation relations*

$$[D_i, M_j] = \delta_{ij} I \quad \text{for } i, j = 1, \ldots, n \tag{33}$$

(the algebra $\mathcal{PD}(V)$ is often called the *Weyl algebra*).

Define $\mathcal{PD}_0(V) = \mathbb{C}I$ and for $k \geq 1$ let $\mathcal{PD}_k(V)$ be the linear span of all products of k or fewer operators from the generating set $\{D_1, \ldots, D_n, M_1, \ldots, M_n\}$. This defines an increasing filtration of the algebra $\mathcal{PD}(V)$:

$$\mathcal{PD}_0(V) \subset \cdots \subset \mathcal{PD}_k(V) \subset \mathcal{PD}_{k+1}(V) \subset \cdots \quad \text{with } \bigcup_{k \geq 0} \mathcal{PD}_k(V) = \mathcal{PD}(V)$$

$$\mathcal{PD}_k(V) \cdot \mathcal{PD}_m(V) \subset \mathcal{PD}_{k+m}(V).$$

Let $\mathrm{Gr}(\mathcal{PD}(V)) = \bigoplus_{k\geq 0} \mathrm{Gr}^k(\mathcal{PD}(V))$ be the associated graded algebra. If $T \in \mathcal{PD}(V)$ then we say T has *filtration degree* k if $T \in \mathcal{PD}_k(V)$ but $T \notin \mathcal{PD}_{k-1}(V)$, and we write $\deg(T) = k$. We write

$$\mathrm{Gr}(T) = T + \mathcal{PD}_{k-1}(V) \in \mathrm{Gr}^k(\mathcal{PD}(V))$$

when $\deg(T) = k$. The map $T \mapsto \mathrm{Gr}(T)$ is a linear isomorphism (but not an algebra homomorphism) from $\mathcal{PD}(V)$ to $\mathrm{Gr}(\mathcal{PD}(V))$. From (33) it is easily verified that

$$\deg(M^\alpha D^\beta) = |\alpha| + |\beta| \quad \text{for } \alpha, \beta \in \mathbb{N}^n$$

and the set of operators $\{M^\alpha D^\beta : \alpha, \beta \in \mathbb{N}^n\}$ is a (vector-space) basis for $\mathcal{PD}(V)$, where we write

$$M^\alpha = M_1^{\alpha_1} \cdots M_n^{\alpha_n}\,, \qquad D^\beta = D_1^{\beta_1} \cdots D_n^{\beta_n}\,.$$

Let ρ be the representation of $\mathrm{GL}(V)$ on $\mathcal{P}(V)$ with

$$\rho(g)f(x) = f(g^{-1}x) \quad \text{for } f \in \mathcal{P}(V)\,.$$

We view $\mathcal{PD}(V)$ as a $\mathrm{GL}(V)$-module relative to the action

$$g \cdot T = \rho(g)T\rho(g^{-1}) \quad \text{for } T \in \mathcal{PD}(V),\ g \in \mathrm{GL}(V)\,.$$

For $g \in \mathrm{GL}(V)$ with matrix $[g_{ij}]$ relative to the basis $\{e_1, \ldots, e_n\}$, we calculate that

$$\rho(g)D_j\rho(g^{-1}) = \sum_{i=1}^{n} g_{ij}D_i\,, \qquad \rho(g)M_i\rho(g^{-1}) = \sum_{j=1}^{n} g_{ij}M_j\,. \tag{34}$$

The set $\{\mathrm{Gr}(M^\alpha D^\beta) : |\alpha| + |\beta| = k\}$ is a basis for $\mathrm{Gr}^k(\mathcal{PD}(V))$. Thus the nonzero operators of filtration degree k are those of the form

$$T = \sum_{|\alpha|+|\beta|\leq k} c_{\alpha\beta} M^\alpha D^\beta \tag{35}$$

with $c_{\alpha\beta} \neq 0$ for some pair α, β with $|\alpha| + |\beta| = k$ (note that the filtration degree of T is generally larger than the order of T as a differential operator). If T in (35) has filtration degree k then we define the *symbol* of T to be the polynomial $\sigma(T) \in \mathcal{P}^k(V \oplus V^*)$ given by

$$\sigma(T) = \sum_{|\alpha|+|\beta|=k} c_{\alpha\beta} x^\alpha \xi^\beta\,.$$

Lemma 8.1: *The symbol map gives a linear isomorphism* $\mathcal{PD}(V) \cong \mathcal{P}(V \oplus V^*)$ *as* $\mathrm{GL}(V)$*-modules.*

Proof: Using (33), one shows by induction on k that any monomial of degree k in the operators $D_1, \ldots, D_n$, $M_1, \ldots, M_n$ is congruent (modulo $\mathcal{PD}_{k-1}(V)$) to a unique ordered monomial $M^\alpha D^\beta$ with $|\alpha| + |\beta| = k$. Hence $\sigma(T) = \sigma(S)$ if $\mathrm{Gr}(T) = \mathrm{Gr}(S)$. Thus σ gives a linear isomorphism

$$\mathrm{Gr}^k(\mathcal{PD}(V)) \xrightarrow{\cong} \mathcal{P}^k(V \oplus V^*)\,. \tag{36}$$

Since $\rho(g)\mathcal{PD}_k(V)\rho(g^{-1}) = \mathcal{PD}_k(V)$, there is a representation of $\mathrm{GL}(V)$ on $\mathrm{Gr}^k(\mathcal{PD}(V))$ for each k. From (34) we see that D_i transforms as the vector e_i under conjugation by $\mathrm{GL}(V)$, while M_i transforms as the dual vector e_i^*. Since $\mathrm{GL}(V)$ acts by algebra automorphisms on $\mathrm{Gr}(\mathcal{PD}(V))$ and on $\mathcal{P}(V \oplus V^*)$, this implies that (36) is an isomorphism of $\mathrm{GL}(V)$ modules. Now compose these maps with the canonical quotient maps $\mathcal{PD}_k(V) \to \mathrm{Gr}^k(\mathcal{PD}_k(V))$. □

We can now obtain the general *Weyl algebra duality theorem*:

Theorem 8.2: *Let G be a reductive algebraic group acting regularly on V. Then there is a multiplicity-free decomposition*

$$\mathcal{P}(V) \cong \bigoplus_{\lambda \in \Sigma(V)} E^\lambda \otimes F^\lambda\,, \tag{37}$$

as a module under the joint actions of $\mathcal{PD}(V)^G$ and $\mathbb{C}[G]$. Here $\Sigma(V) \subset \widehat{G}$, F^λ is an irreducible regular G-module of type λ, and E^λ is an irreducible module for $\mathcal{PD}(V)^G$ that uniquely determines λ.

Proof: We apply Theorem 1.4, with $L = \mathcal{P}(V)$ and $\mathcal{R} = \mathcal{PD}(V)$. Note that L is the direct sum of the finite-dimensional G-invariant subspaces $L_k = \mathcal{P}^k(V)$ of homogeneous polynomials of degree k. The action of G on each L_k is regular, so L is a locally regular G module.

We shall show that $\mathcal{R}$ satisfies conditions (i) and (ii) of Theorem 1.4. Let $0 \neq f \in \mathcal{P}(V)$ be of degree d. Then there is some $\alpha \in \mathbb{N}^n$ with $|\alpha| = d$ such that $0 \neq D^\alpha f \in \mathbb{C}$. Given any $g \in \mathcal{P}(V)$, let $M_g \in \mathcal{PD}(V)$ be the operator of multiplication by g. Then $g \in \mathbb{C}M_g D^\alpha f$. This proves that $\mathcal{R}$ acts irreducibly on $\mathcal{P}(V)$ (condition (i)). The algebra $\mathcal{R}$ is the union of the finite-dimensional G-invariant subalgebras $\mathcal{PD}_k(V)$, and the action of G on $\mathcal{PD}_k(V)$ is regular by Lemma 8.1. Hence $\mathcal{R}$ also satisfies condition (ii). □

To use Theorem 8.2 effectively for a particular G-module V we need a more explicit description of the algebra $\mathcal{PD}(V)^G$. The following result is a first step in that direction.

Theorem 8.3: *Let $\{\psi_1, \ldots, \psi_r\}$ generate the algebra $\mathcal{P}(V \oplus V^*)^G$. Suppose $T_j \in \mathcal{PD}(V)^G$ are such that $\sigma(T_j) = \psi_j$ for $j = 1, \ldots, r$. Then $\{T_1, \ldots, T_r\}$ generates the algebra $\mathcal{PD}(V)^G$.*

Proof: Let $\mathcal{J} \subset \mathcal{PD}(V)^G$ be the subalgebra generated by $T_1, \ldots, T_r$. Then $\mathcal{PD}_0(V)^G = \mathbb{C}I \subset \mathcal{J}$. Let $S \in \mathcal{PD}_k(V)^G$ have filtration degree k. We may assume by induction on k that $\mathcal{PD}_{k-1}(V)^G \subset \mathcal{J}$. Since $\sigma(S) \in \mathcal{P}^k(V \oplus V^*)^G$ by Lemma 8.1, we can write

$$\sigma(S) = \sum_{j_1, \ldots, j_r} c_{j_1 \cdots j_r} \, \psi_1^{j_1} \cdots \psi_r^{j_r}$$

where $c_{j_1 \cdots j_r} \in \mathbb{C}$. Set

$$R = \sum_{j_1, \ldots, j_r} c_{j_1 \cdots j_r} \, T_1^{j_1} \cdots T_r^{j_r} \, .$$

Although R is not unique (it depends on the enumeration of the T_j), we have $\sigma(R) = \sigma(S)$ since σ is an algebra homomorphism. Hence $R - S \in \mathcal{PD}_{k-1}(V)$ by Lemma 8.1. By the induction hypothesis, $R - S \in \mathcal{J}$, so we have $S \in \mathcal{J}$. □

Corollary 8.4: (Notation as in Theorem 8.3) *Suppose $T_1, \ldots, T_r$ can be chosen so that*

$$\mathfrak{g}' = \mathrm{Span}\{T_1, \ldots, T_r\}$$

is a Lie subalgebra of $\mathcal{PD}(V)^G$. Then in the canonical decomposition (37) *the spaces E^λ are irreducible modules for the Lie algebra $\mathfrak{g}'$, and λ is uniquely determined by the equivalence class of E^λ as a $\mathfrak{g}'$-module. Hence there is a bijection (duality correspondence)*

$$\Sigma(V) \leftrightarrow \Lambda(V)\,,$$

where $\Lambda(V)$ is the set of irreducible representations of $\mathfrak{g}'$ that occur in $\mathcal{P}(V)$.

Proof: The representations of the Lie algebra $\mathfrak{g}'$ are the same as the representations of the universal enveloping algebra $U(\mathfrak{g}')$. Let $\rho' : U(\mathfrak{g}') \to \mathrm{End}(\mathcal{P}(V))$ be the representation associated to the action of $\mathfrak{g}'$ as differential operators. The assumption on $T_1, \ldots, T_r$ implies that $\rho'(U(\mathfrak{g}')) = \mathcal{PD}(V)^G$. Hence the irreducible $\mathcal{PD}(V)^G$-modules are the same as irreducible $\mathfrak{g}'$-modules. □

Remark: Theorem 8.2 is also valid when V is any smooth connected affine G-variety. Here we take $\mathcal{R} = \mathbb{D}(V)$ to be the ring of *algebraic differen-*

tial operators on V and use Theorem 1.4.[g] The algebra $\mathbb{D}(V)^G$ in this case (G connected, reductive), has been studied by Knop [25]. He proves that its center $\mathfrak{Z}_G(V)$ is a polynomial ring in $\mathrm{rank}_G(V)$ generators, where $\mathrm{rank}_G(V) = \dim Bx - \dim Nx$ for a generic point $x \in V$ (here B is a Borel subgroup of G with nilradical N). Furthermore, $\mathbb{D}(V)^G$ is a free module over $\mathfrak{Z}_G(V)$ (this is a generalization of results of Kostant [26] for the case $V = G$, with G acting by left multiplication). The representation theory of $\mathbb{D}(V)^G$ seems to be unknown, in general, although special cases have been studied by I. Agricola, F. Knop, T. Levasseur, G. Schwarz, J. Stafford, and others.

8.2. *Howe duality for orthogonal/symplectic groups*

We now determine the structure of $\mathcal{PD}(V)^G$ when G is an orthogonal or symplectic group and V is the sum of n copies of the fundamental representation of G. Using the First Fundamental Theorem of classical invariant theory, we will show that the assumptions of Corollary 8.4 are satisfied. This will give the *Howe duality* between the (finite-dimensional) regular representations of G occurring in $\mathcal{P}(V)$ and a set of irreducible representations of the *dual* Lie algebra $\mathfrak{g}'$.

Let ω be a nondegenerate bilinear form on $\mathbb{C}^k$ that is either symmetric or skew symmetric, and let $G \subset \mathrm{GL}(k, \mathbb{C})$ be the isometry group of ω. Thus G is the (complex) orthogonal group when ω is symmetric, and G is the (complex) symplectic group when ω is skew (and k even). Let $V = (\mathbb{C}^k)^n$. Then

$$\mathcal{P}(V \oplus V^*) = \mathcal{P}(\underbrace{\mathbb{C}^k \oplus \cdots \oplus \mathbb{C}^k}_{n \text{ copies}} \oplus \underbrace{(\mathbb{C}^k)^* \oplus \cdots \oplus (\mathbb{C}^k)^*}_{n \text{ copies}}).$$

Hence if $T \in \mathcal{PD}(V)$ then the symbol of T is a polynomial function

$$f(\mathbf{x}_1, \ldots, \mathbf{x}_n, \xi_1, \ldots, \xi_n), \quad \text{where } \mathbf{x}_i \in \mathbb{C}^k, \quad \xi_j \in (\mathbb{C}^k)^*.$$

From Section 6 we know that the algebra of G-invariant polynomials $P(V \oplus V^*)^G$ is generated by three types of quadratic polynomials:

evaluation of ω on two vectors:	$r_{ij}(\mathbf{x}_1, \ldots, \mathbf{x}_n, \xi_1, \ldots, \xi_n) = \omega(\mathbf{x}_i, \mathbf{x}_j)$
evaluation of ω^* on two covectors:	$\rho_{ij}(\mathbf{x}_1, \ldots, \mathbf{x}_n, \xi_1, \ldots, \xi_n) = \omega^*(\xi_i, \xi_j)$
contraction of vector-covector:	$c_{ij}(\mathbf{x}_1, \ldots, \mathbf{x}_n, \xi_1, \ldots, \xi_n) = \langle \mathbf{x}_i, \xi_j \rangle$

[g]See [1]; the smoothness assumption on V is essential here, since the action of $\mathbb{D}(V)$ on $\mathrm{Aff}(V)$ can fail to be irreducible when V is not smooth.

where $1 \le i, j \le n$ and ω^* is the form on $(\mathbb{C}^k)^*$ dual to ω. There is a canonical GL(V)-module isomorphism ∂ from $\mathcal{P}(V^*) \cong S(V)$ to the algebra of constant-coefficient differential operators on V. The linear span of the quadratic invariant polynomials above furnish symbols for the following Lie algebras of G-invariant differential operators:

$$\begin{aligned} \mathfrak{p}_- &= \mathrm{Span}\{\text{multiplication by } r_{ij} \,:\, 1 \le i, j \le n\} \\ \mathfrak{p}_+ &= \mathrm{Span}\{\text{differentiation by } \Delta_{ij} = \partial(\rho_{ij}) \,:\, 1 \le i, j \le n\} \\ \mathfrak{k} &= \mathrm{Span}\{E_{ij} + \tfrac{k}{2}\delta_{ij} \,:\, 1 \le i, j \le n\} \end{aligned}$$

Here it is convenient to identify V with $M_{n\times k}$, with G acting by right multiplication. If $\mathbf{x}_i$ denotes the ith row of $x \in M_{n\times k}$, then

$$E_{ij} = \mathbf{x}_i \cdot \nabla_{\mathbf{x}_j} = \sum_{r=1}^{k} x_{ir} \frac{\partial}{\partial x_{jr}} .$$

The operators E_{ij}, which correspond to vector-covector contractions, commute with the right action of all of GL$(k, \mathbb{C})$ (E_{ij} is the classical *polarization operator*). Obviously $[\mathfrak{p}_-, \mathfrak{p}_-] = 0$ and $[\mathfrak{p}_+, \mathfrak{p}_+] = 0$. An easy calculation shows that

$$[\mathfrak{k}, \mathfrak{p}_\pm] = \mathfrak{p}_\pm \,, \quad [\mathfrak{p}_-, \mathfrak{p}_+] = \mathfrak{k} \,.$$

The choice of shift $\frac{n}{2}\delta_{ij}$ for the operators in $\mathfrak{k}$ arises from the last commutation relation.

Theorem 8.5: *Set $\mathfrak{g}' = \mathfrak{p}_- + \mathfrak{k} + \mathfrak{p}_+$. Then $\mathfrak{g}'$ is a Lie algebra and it generates the associative algebra $\mathcal{PD}(M_{n\times k})^G$. Furthermore,*

$$\mathfrak{g}' \cong \begin{cases} \mathfrak{sp}(n, \mathbb{C}) & \text{when } \omega \text{ is symmetric} \\ \mathfrak{so}(2n, \mathbb{C}) & \text{when } \omega \text{ is skew} \end{cases}$$

The subalgebra $\mathfrak{k} \cong \mathfrak{gl}(n, \mathbb{C})$ acts on $\mathcal{P}(M_{n\times k})$ by the differential of the representation

$$\rho(g) f(x) = (\det g)^{-k/2} f(g^{-1}x)$$

of $K = \mathrm{GL}(n, \mathbb{C})$ (replace K by its two-fold cover if k is odd).

Proof: The first statement follows from Corollary 8.4. The other parts are easy calculations (see [16, Sec. 4.5] for details). □

We call $\mathfrak{g}'$ the *Howe dual* of $\mathfrak{g} = \mathrm{Lie}(G)$ associated to the representation of $\mathfrak{g}$ on V. Notice that the correspondence between $\mathfrak{g}$ and $\mathfrak{g}'$ interchanges orthogonal and symplectic Lie algebras. There is an asymmetry between $\mathfrak{g}$ and $\mathfrak{g}'$, however. The action of $\mathfrak{g}$ on $\mathcal{P}(V)$ is by vector fields (corresponding to the representation of G on V), whereas the subalgebras $\mathfrak{p}_\pm$ of $\mathfrak{g}'$ act by second-order differential operators and multiplication by quadratic polynomials, which do not come from a geometric action on V. We will show in Section 11 how to *exponentiate* the action of a real form of $\mathfrak{g}'$ on $\mathcal{P}(V)$ to a unitary representation of an associated real Lie group on a Hilbert-space completion of $\mathcal{P}(V)$.

8.3. *Howe duality for* GL(k)

Now consider $G = \mathrm{GL}(k,\mathbb{C})$ acting on

$$V = \underbrace{\mathbb{C}^k \oplus \cdots \oplus \mathbb{C}^k}_{p \text{ copies}} \oplus \underbrace{(\mathbb{C}^k)^* \oplus \cdots \oplus (\mathbb{C}^k)^*}_{q \text{ copies}}.$$

In this case the symbol of $T \in \mathcal{PD}(V)$ is a polynomial function

$$f(\mathbf{x}_1,\ldots,\mathbf{x}_p,\eta_1,\ldots,\eta_q,\xi_1,\ldots,\xi_p,\mathbf{y}_1,\ldots,\mathbf{y}_q),$$

where $[\mathbf{x}_1,\ldots,\mathbf{x}_p,\eta_1,\ldots,\eta_q] \in V$ and $[\xi_1,\ldots,\xi_p,\mathbf{y}_1,\ldots,\mathbf{y}_q] \in V^*$ ($\mathbf{x}_i$, $\mathbf{y}_j$ are vectors in $\mathbb{C}^k$ and ξ_i, η_j are covectors in $(\mathbb{C}^k)^*$). Theorem 6.2 asserts that the algebra of G-invariant polynomials on $V \oplus V^*$ is generated by contractions of a vector with a covector. Now there are four possibilities for contractions:

(1)	vector and covector in V:	$\langle \mathbf{x}_i, \eta_j\rangle$	for $1 \le i \le p$ and $1 \le j \le q$
(2)	vector and covector in V^*:	$\langle \mathbf{y}_j, \xi_i\rangle$	for $1 \le i \le p$ and $1 \le j \le q$
(3)	vector from V, covector from V^*:	$\langle \mathbf{x}_i, \xi_j\rangle$	for $1 \le i, j \le p$
(4)	covector from V, vector from V^*:	$\langle \mathbf{y}_i, \eta_j\rangle$	for $1 \le i, j \le q$

We can identify V with $M_{(p+q)\times k}$ if we make $g \in G$ act on the right by

$$\begin{bmatrix} x \\ \eta \end{bmatrix} \cdot g = \begin{bmatrix} x(g^t)^{-1} \\ \eta g \end{bmatrix} \quad \text{for } x \in M_{p\times k}, \quad \eta \in M_{q \times k}.$$

Here $\mathbf{x}_i$ is the ith row of x and η_j the jth row of η. Contractions of type (1) and (2) furnish symbols for the G-invariant operators

$$\begin{aligned} \mathfrak{p}_- &= \mathrm{Span}\{\text{ multiplication by } r_{ij} = \langle \mathbf{x}_i, \eta_j\rangle\} \\ \mathfrak{p}_+ &= \mathrm{Span}\{\text{ differentiation by } \Delta_{ij}\}, \end{aligned}$$

where

$$\Delta_{ij} = \nabla_{\mathbf{x}_i} \cdot \nabla_{\eta_j} = \sum_{r=1}^{k} \frac{\partial}{\partial x_{ir}} \frac{\partial}{\partial \eta_{jr}} \qquad \text{for } 1 \le i \le p \text{ and } 1 \le j \le q.$$

The linear span of contractions of type (3) and (4) furnishes symbols for the G-invariant operators

$$\mathfrak{k} = \mathrm{Span}\{E_{ij}^{(x)} + \tfrac{k}{2}\delta_{ij} \, : \, 1 \le i,j \le p\} \oplus \mathrm{Span}\{E_{ij}^{(\eta)} + \tfrac{k}{2}\delta_{ij} \, : \, 1 \le i,j \le q\}\,,$$

where $E_{ij}^{(x)}$ is the polarization operator for the x variables and $E_{ij}^{(\eta)}$ for the η variables. By the same argument as in Theorem 8.5 we conclude that $\mathcal{PD}(V)^G$ is generated by

$$\mathfrak{g}' = \mathfrak{p}_- + \mathfrak{k} + \mathfrak{p}_+\,.$$

These subalgebras have the commutation relations

$$[\mathfrak{k}, \mathfrak{p}_\pm] = \mathfrak{p}_\pm\,, \quad [\mathfrak{p}_-, \mathfrak{p}_+] \subset \mathfrak{k}\,.$$

In this case $\mathfrak{g}'$ is isomorphic to $\mathfrak{gl}(p+q,\mathbb{C})$, with $\mathfrak{k} \cong \mathfrak{gl}(p,\mathbb{C}) \oplus \mathfrak{gl}(q,\mathbb{C})$. The action of $\mathfrak{k}$ on $\mathcal{P}(M_{p\times k} \oplus M_{q\times k})$ is the differential of the representation

$$\rho(g,h)f(x,\eta) = (\det g \det h)^{-k/2} f(g^{-1}x, h^{-1}\eta)$$

for $(g,h) \in K = \mathrm{GL}(p,\mathbb{C}) \times \mathrm{GL}(q,\mathbb{C})$. (We must replace K by the two-fold covers of each factor when n is odd).

9. Harmonic duality

9.1. *Harmonic polynomials*

Let G be $\mathrm{O}(\mathbb{C}^k,\omega)$, $\mathrm{Sp}(\mathbb{C}^k,\omega)$, or $\mathrm{GL}(k,\mathbb{C})$ acting on $V = M_{n\times k}$ on the right. In the case of $\mathrm{GL}(\mathbb{C}^k)$ the first p rows of $x \in M_{n\times k}$ transform as vectors, whereas the remaining q rows transform as covectors. From Section 8 the *Howe dual* to G is, respectively,

$$\mathfrak{g}' \cong \mathfrak{sp}(n,\mathbb{C}),\ \mathfrak{so}(2n,\mathbb{C}),\ \text{or } \mathfrak{gl}(p+q,\mathbb{C}) \quad \text{with } p+q=n\,.$$

We will assume that $p > 0$ and $q > 0$ in the third case.[h] With G fixed, the spectrum $\Sigma(V)$ of G on $\mathcal{P}(V)$ only depends on n (or the pair p,q in the third case); we can thus denote it by $\Sigma(n)$ (or $\Sigma(p,q)$). From the dual point of view if we fix $\mathfrak{g}'$, then the set $\Lambda(V)$ of irreducible representations of $\mathfrak{g}'$ that occur in $\mathcal{P}(V)$ only depends on k; we can thus denote it as $\Lambda(k)$. The general duality theorem gives a bijection $\Sigma(n) \leftrightarrow \Lambda(k)$. We now show how to express this bijection in terms of *harmonic duality.*

[h] If $p = 0$ or $q = 0$ then $\mathfrak{k} = \mathfrak{g}'$ and the modules $E^\lambda \otimes F^\lambda$ that occur in the decomposition of $\mathcal{P}(V)$ are finite-dimensional. This is the well-known $\mathrm{GL}(n)$-$\mathrm{GL}(k)$ duality (see the lectures of Benson–Ratcliff in this volume).

In all cases there is a triangular decomposition

$$\mathfrak{g}' = \mathfrak{p}_- \oplus \mathfrak{k} \oplus \mathfrak{p}_+ .$$

Here $\mathfrak{k}$ is the Lie algebra of the reductive group K (a two-fold cover of $\mathrm{GL}(n, \mathbb{C})$ or $\mathrm{GL}(p, \mathbb{C}) \times \mathrm{GL}(q, \mathbb{C})$ in general). The representation of K on $\mathcal{P}(V)$ is the natural representation associated with the left multiplication action of $\mathrm{GL}(n, \mathbb{C})$ on V tensored with the one-dimensional representation

$$g \mapsto (\det g)^{-k/2} \quad \text{or} \quad (g, h) \mapsto (\det g \det h)^{-k/2} .$$

Let δ denote this character, viewed as a weight of the maximal torus of K. The subalgebra $\mathfrak{p}_-$ acts by multiplication by G-invariant quadratic polynomials, whereas $\mathfrak{p}_+$ acts by G-invariant constant-coefficient Laplace operators $\{\Delta_{ij}\}$.

We define the *G-harmonic polynomials* to be

$$\mathcal{H} = \mathcal{P}(V)^{\mathfrak{p}_+} = \bigcap\nolimits_{i,j} \mathrm{Ker}(\Delta_{ij})$$

Since $\mathrm{Ad}(K)\mathfrak{p}_+ = \mathfrak{p}_+$, the space $\mathcal{H}$ is invariant under the reductive group $K \times G$. In this section we will show that $\mathcal{H}$ gives a multiplicity-free duality pairing between irreducible representations of K and G; furthermore, the decomposition of $\mathcal{H}$ generates the decomposition of $\mathcal{P}(V)$ under $\mathfrak{g}'$ and G.

Let $U(k) \subset \mathrm{GL}(k, \mathbb{C})$ denote the unitary group. Then $K_0 = K \cap U(n)$ is a compact real form of K. We assume that the bilinear form ω is chosen so that $G_0 = G \cap U(k)$ is a compact real form of G and ω is real on $\mathbb{R}^k$. Define an inner product on $M_{n\times k}$ by

$$(x \mid y) = \mathrm{tr}(y^* x) \quad \text{for } x, y \in M_{n\times k} \quad (y^* = \bar{y}^t) .$$

This inner product is invariant under $U(n) \times U(k)$, acting by left and right multiplication, hence it is invariant under $K_0 \times U_0$. We set $||x||^2 = (x \mid x)$.

Let $f(x) = \sum_\alpha c_\alpha x^\alpha$ be in $\mathcal{P}(V)$, where α is a multi-index and $x^\alpha = \prod_\alpha (x_{ij})^{\alpha_{ij}}$ as usual (x_{ij} are the matrix entry functions on $M_{n\times k}$). Define the constant-coefficient differential operator

$$\partial(f) = \sum_\alpha c_\alpha \left(\frac{\partial}{\partial x}\right)^\alpha .$$

Then $f \mapsto \partial(f)$ is an algebra isomorphism from $\mathcal{P}(V)$ to the constant-coefficient differential operators on V that is equivariant relative to the action of $U(k) \times U(n)$. Set $g^*(x) = \overline{g(\bar{x})}$ for $g \in \mathcal{P}(V)$. If $g(x) = \sum_\alpha d_\alpha x^\alpha$, then

$$(\partial(f)g^*)(0) = \sum_\alpha \alpha!\, c_\alpha \overline{d_\alpha} . \tag{38}$$

We define $\langle f \mid g \rangle = (\partial(f)g^*)(0)$. From (38) we see that this is a positive definite Hermitian inner product on $\mathcal{P}(V)$, called the *Fischer inner product.* We note that

$$\langle fg \mid h \rangle = \langle f \mid \partial(g^*)h \rangle \tag{39}$$

for all f, g, $h \in \mathcal{P}(V)$.

The Fischer inner product has the following analytic definition. Denote Lebesgue measure on V by $d\lambda(z)$, where we identify V with $\mathbb{R}^{2nk}$ via the real and imaginary parts of the matrix coordinates.

Lemma 9.1: *For $f, g \in \mathcal{P}(V)$ one has*

$$\langle f \mid g \rangle = \frac{1}{\pi^d} \int_V f(z)\overline{g(z)}\, e^{-\|z\|^2}\, d\lambda(z)$$

($d = \dim_{\mathbb{C}} V = nk$).

Proof: See the lectures of Benson–Ratcliff in this volume. □

9.2. *Main theorem*

We now apply the Weyl algebra theorem from Section 8 to obtain multiplicity-free decompositions of the harmonic polynomials and the entire space $\mathcal{P}(V)$.

Theorem 9.2 (Harmonic Duality): (1) *The space $\mathcal{H}$ of G-harmonic polynomials on V decomposes under $K \times G$ into mutually orthogonal subspaces (relative to the Fischer inner product) as*

$$\mathcal{H} = \bigoplus_{\sigma \in \Sigma(V)} \mathcal{E}^{\tau(\sigma)+\delta} \otimes \mathcal{F}^{\sigma}\,.$$

Here $\Sigma(V) \subset \widehat{G}$ is the spectrum of $\mathcal{P}(V)$ as a G module, $\mathcal{F}^\sigma \subset \mathcal{H}$ is an irreducible G-module of type σ, and $\mathcal{E}^{\tau(\sigma)+\delta} \subset \mathcal{H}$ is an irreducible finite-dimensional K-module with highest weight $\tau(\sigma) + \delta$. In particular, every irreducible representation of G in $\mathcal{P}(V)$ is realized in the harmonic polynomials.

(2) *Set $E^{\tau(\sigma)+\delta} = \mathcal{P}(V)^G \cdot \mathcal{E}^{\tau(\sigma)+\delta}$. Then $E^{\tau(\sigma)+\delta}$ is an irreducible $\mathfrak{g}'$ module and*

$$\mathcal{P}(V) = \bigoplus_{\sigma \in \Sigma(V)} E^{\tau(\sigma)+\delta} \otimes \mathcal{F}^{\sigma}$$

is an orthogonal decomposition of $\mathcal{P}(V)$ (relative to the Fischer inner product) under the mutually commuting actions of $\mathfrak{g}'$ and G.

(3) *The map $\sigma \mapsto \tau(\sigma)$ from $\Sigma(V) \to \widehat{K}$ is injective. Thus $\mathcal{H}$ is multiplicity-free as a $K \times G$ module.*

Proof: Since $\mathcal{H}$ is an invariant subspace for the reductive group $K \times G$, there is a subset $\Gamma \subset \widehat{K} \times \widehat{G}$ and multiplicity function $m : \Gamma \to \{1, 2, \ldots\}$ such that

$$\mathcal{H} \cong \bigoplus_{(\mu,\sigma)\in\Gamma} \mathbb{C}^{m(\mu,\sigma)} \otimes \mathcal{E}^{\mu+\delta} \otimes \mathcal{F}^{\sigma} \tag{40}$$

with $K \times G$ acting trivially on the multiplicity spaces $\mathbb{C}^{m(\mu,\sigma)}$. Indeed, we first consider $\mathcal{H}$ as a locally-finite $K \times G$-module relative to the natural left-right action on $V = M_{n\times k}$ (omitting the determinant twist from the K representation) and use Proposition 1.1 and Example 1 in Section 2.3. Then tensor with the character $\det^{-k/2}$ of K to shift the highest weights from μ to $\mu + \delta$. To prove that $\Gamma = \{(\tau(\sigma), \sigma) : \sigma \in \Sigma(V)\}$ and $m(\tau(\sigma), \sigma) = 1$, we need to examine the action of $\mathfrak{g}'$ on $\mathcal{P}(V)$ in more detail.

Let $\mathcal{J} = \mathcal{P}(V)^G$ be the G-invariant polynomials, and let $\mathcal{J}_j$ be the homogeneous polynomials of degree j in $\mathcal{J}$. Then $\mathcal{J}_j = 0$ for j odd, and $(\mathfrak{p}_-)^j$ acts by multiplication by $\mathcal{J}_{2j}$ on $\mathcal{P}(V)$. Since the bilinear form ω is real on $\mathbb{R}^k$, we have $\mathcal{J}^* = \mathcal{J}$ and $\mathcal{H}^* = \mathcal{H}$. Let $\mathcal{J}_+ = \{f \in \mathcal{J} : f(0) = 0\}$. We claim that

$$\mathcal{H}^{\perp} = \mathcal{J}_+ \cdot \mathcal{P}(V) \tag{41}$$

(orthogonal complement relative to the Fischer inner product). Indeed, if $f \in \mathcal{J}_+ \cdot \mathcal{P}(V)$ and $h \in \mathcal{H}$ then $\partial(f)h = 0$ by definition of $\mathcal{H}$, and thus $f \perp h$. Conversely, if $h \perp \mathcal{J}_+ \cdot \mathcal{P}(V)$ then for all $f \in \mathcal{P}(V)$ and $g \in \mathcal{J}_+$,

$$0 = \langle fg \mid h\rangle = \langle f \mid \partial(g^*)h\rangle \,.$$

Hence $\partial(g)h = 0$ for all $g \in \mathcal{J}_+$, so we have $h \in \mathcal{H}$.

We can now determine the general structure of the irreducible $\mathfrak{g}'$-modules in $\mathcal{P}(V)$. The commutation relations in $\mathfrak{g}'$ can be expressed as

$$\mathfrak{p}_+\mathfrak{p}_- \subset \mathfrak{p}_-\mathfrak{p}_+ + \mathfrak{k}\,, \qquad \mathfrak{k}\,\mathfrak{p}_- \subset \mathfrak{p}_-(\mathfrak{k} + 1)$$

in the universal enveloping algebra $U(\mathfrak{g}')$. Hence by induction, one has

$$\mathfrak{p}_+(\mathfrak{p}_-)^m \subset (\mathfrak{p}_-)^m\mathfrak{p}_+ + (\mathfrak{p}_-)^{m-1}(\mathfrak{k}+1)\,, \qquad \mathfrak{k}\,(\mathfrak{p}_-)^m \subset (\mathfrak{p}_-)^m(\mathfrak{k}+1) \tag{42}$$

for all integers $m \geq 1$. Thus if $Z \subset \mathcal{H}$ is any $\mathfrak{k}$-invariant linear subspace, then (42) implies that

$$\mathfrak{p}_+(\mathfrak{p}_-)^m \cdot Z \subset (\mathfrak{p}_-)^{m-1} \cdot Z \quad \text{and} \quad \mathfrak{k}\,(\mathfrak{p}_-)^m \cdot Z \subset (\mathfrak{p}_-)^m \cdot Z \tag{43}$$

for all $m \geq 1$.

(a) *Let $\mathcal{E} \subset \mathcal{H}$ be any $\mathfrak{k}$-irreducible subspace. Set $E = \mathcal{J} \cdot \mathcal{E}$. Then E is an irreducible $\mathfrak{g}'$-module and $\mathcal{E} = E \cap \mathcal{H}$.*

Indeed, (43) implies that E is invariant under $\mathfrak{g}'$. Also every $f \in E$ is of the form

$$f = \sum_{j=0}^{m} g_j h_j \quad \text{where } 0 \neq g_j \in \mathcal{J}_{2j} \text{ and } h_j \in \mathcal{E}\,. \tag{44}$$

Suppose $F \subset E$ is a nonzero $\mathfrak{g}'$-invariant subspace. Take $f \in F$ so that the integer m in (44) is minimal. Then (43) implies that $\mathfrak{p}_+ f = 0$. Hence $f \in \mathcal{H}$. Thus

$$\sum_{j=1}^{m} g_j h_j = f - g_0 h_0 \in \mathcal{H}\,.$$

Since the left side is in $\mathcal{J}_+ \cdot \mathcal{P}(V)$, it must be zero by (41). Hence we conclude that $f \in \mathcal{E}$. But $\mathfrak{k}$ acts irreducibly on $\mathcal{E}$, so $U(\mathfrak{k})f = \mathcal{E}$ and thus $F = E$. The same argument shows that $E \cap \mathcal{H} = \mathcal{E}$, completing the proof of (a).

(b) *Let $E \subset \mathcal{P}(V)$ be an irreducible $\mathfrak{g}'$-module. Set $\mathcal{E} = E \cap \mathcal{H}$. Then $\mathcal{E}$ is an irreducible $\mathfrak{k}$-module and $E = \mathcal{J} \cdot \mathcal{E}$.*

Note that the action of $\mathfrak{p}_+$ on $\mathcal{P}(V)$ lowers the degree of polynomials, so $\mathcal{E} \neq 0$. If $0 \neq \mathcal{F} \subset \mathcal{E}$ were a proper $\mathfrak{k}$-submodule, then $\mathcal{J} \cdot \mathcal{F} \subset E$ would be a proper irreducible $\mathfrak{g}'$-submodule by (a). Hence $\mathcal{E}$ must be irreducible as a $\mathfrak{k}$-module and $E = \mathcal{J} \cdot \mathcal{E}$, proving (b).

(c) *Let $\mathcal{E}$ and $\mathcal{F}$ be $\mathfrak{k}$-invariant subspaces of $\mathcal{H}$. Assume that $\mathcal{E} \perp \mathcal{F}$ (relative to the Fischer inner product). Set $E = \mathcal{J} \cdot \mathcal{E}$ and $F = \mathcal{J} \cdot \mathcal{F}$. Then $E \perp F$.*

By (41) we have the orthogonal decompositions

$$E = \mathcal{E} \oplus \mathcal{J}_+ \cdot \mathcal{E} \qquad F = \mathcal{F} \oplus \mathcal{J}_+ \cdot \mathcal{F}\,.$$

Thus $E \perp \mathcal{F}$ and $F \perp \mathcal{E}$, so we only need to verify that $\mathcal{J}_+ \cdot \mathcal{E} \perp \mathcal{J}_+ \cdot \mathcal{F}$. Now

$$\langle \mathcal{J}_+ \cdot \mathcal{E} \mid \mathcal{J}_+ \cdot \mathcal{F} \rangle = \langle \mathcal{E} \mid \partial(\mathcal{J}_+)\mathcal{J}_+ \cdot \mathcal{F} \rangle\,.$$

But $\partial(\mathcal{J}_+)\mathcal{J}_+ \cdot \mathcal{F} \subset F$ since $\mathcal{F}$ is $\mathfrak{k}$-invariant. Hence $\mathcal{E} \perp \partial(\mathcal{J}_+)\mathcal{J}_+ \cdot \mathcal{F}$, proving (c).

We now complete the proof of the theorem. It is clear from the integral formula for the Fischer inner product (Lemma 9.1) that G_0 and K_0 act by unitary operators on $\mathcal{P}(V)$, hence the decomposition (40) of $\mathcal{H}$ is orthogonal relative to the Fischer inner product because G_0 and K_0 have the same finite-dimensional invariant subspaces in $\mathcal{P}(V)$ as G and K, respectively (see [16, Sec. 2.4.4]). Also, since K is connected, a finite-dimensional subspace of $\mathcal{P}(V)$ is invariant under K if and only if it is invariant under $\mathfrak{k}$.

Let $(\mu, \sigma) \in \Gamma$ occur in (40). By (a), (b) and Theorem 3.4 we know that the irreducible $\mathfrak{g}'$-module $E^{\mu+\delta} = \mathcal{J} \cdot \mathcal{E}^{\mu+\delta}$ uniquely determines μ. On the other hand, Theorem 8.2 asserts that $\mathcal{P}(V)$ is semisimple as a $\mathfrak{g}'$-module, with the $\mathfrak{g}'$ multiplicity spaces being irreducible regular G-modules corresponding bijectively to the associated $\mathfrak{g}'$-modules. Hence Γ is determined by its projection onto $\widehat{G}$. If we call this projection Σ and write the elements of Γ as $(\tau(\sigma), \sigma)$, then the map $\sigma \mapsto \tau(\sigma)$ is injective. The multiplicities $m(\tau(\sigma), \sigma) = 1$ for all $\sigma \in \Sigma$, since otherwise (a) and (c) would imply that $\mathcal{F}^\sigma$ is paired with more than one copy of an irreducible $\mathfrak{g}'$ module, contradicting Theorem 8.2. Finally, (b) implies that $\Sigma = \Sigma(V)$, since $\mathcal{P}(V)$ is semisimple as a $\mathfrak{g}'$-module □

Remarks: (1) Theorem 9.2 was obtained by Howe in his influential paper [21] (which circulated as a preprint for more than a decade); his proof used an argument based on a filtration by finite-dimensional subspaces and the classical double commutant theorem, instead of Theorem 8.2. Knowing that the decomposition of the harmonics is multiplicity free simplifies the task of finding harmonic highest weight vectors, as we will see in Section 10.

(2) The shift by δ in the highest weights for K in the harmonic decomposition would appear to be a minor nuisance. In fact, it plays an important analytic role. For $T \in \mathcal{PD}(V)$ let T^* denote the adjoint of T relative to the Fischer inner product:

$$\langle Tf \mid g \rangle = \langle f \mid T^* g \rangle .$$

If we write T in polarized form as $T = \sum_j f_j \partial(g_j)$ with f_j, $g_j \in \mathcal{P}(V)$, then we see from (39) that $T^* = \sum_j g_j^* \partial(f_j^*)$. Hence $(\mathfrak{p}_\pm)^* = \mathfrak{p}_\mp$ and $\mathfrak{k}^* = \mathfrak{k}$. It follows that

$$\mathfrak{g}'_0 := \{T \in \mathfrak{g}' \, : \, T^* = -T\}$$

is a real form of $\mathfrak{g}'$. We will show in Section 11 that the irreducible representation of $\mathfrak{g}'_0$ on the space E^λ, with $\lambda = \tau(\sigma) + \delta$, can be integrated to a unitary representation π^λ of a (noncompact) real group G'_0 with Lie algebra $\mathfrak{g}'_0$. The shift by δ controls the rate of decay at infinity on G'_0 of the matrix entries of π^λ. We will show in Section 12 that for k large enough (relative to n), the representations π^λ are square-integrable (recall that $\delta = k\delta_0$, where δ_0 is a fixed weight of $\mathfrak{g}'$).

(3) The injective map $\sigma \mapsto \pi^{\tau(\sigma)+\delta}$ is called the *theta-correspondence* (more precisely, the *local* theta-correspondence over $\mathbb{R}$) because of the connection between the oscillator representation and theta-functions (see [5]). There

are many recent papers devoted to the problem of understanding the theta-correspondence from a geometric orbit perspective (see [19, Chap. 12] for a survey).

10. Decomposition of harmonic polynomials

We now turn to the explicit determination of the harmonic duality from Section 9 when G is the orthogonal group and $\mathfrak{g}'$ the symplectic Lie algebra (for the other two cases, when G is the symplectic or general linear group, see [22] and [6]). It is convenient to take G as the orthogonal group $\mathrm{O}(\mathbb{C}^k, \omega)$ for the symmetric form $\omega(x, y) = x^t C_k y$ on $\mathbb{C}^k$, where

$$C_k = \begin{bmatrix} 0 & I_l \\ I_l & 0 \end{bmatrix} \quad \text{when } k = 2l, \quad C_k = \begin{bmatrix} 0 & I_l & 0 \\ I_l & 0 & 0 \\ 0 & 0 & 1 \end{bmatrix} \quad \text{when } k = 2l + 1 .$$

Here I_l denotes the $l \times l$ identity matrix. This choice of ω ensures that the diagonal matrices in G give a maximal torus. Also G is a *self-adjoint* matrix group (invariant under $g \mapsto g^*$), so the subgroup $G_0 = G \cap U(k)$ is a compact real form of G, and ω is real on the real matrices, as we assumed in Section 8.

In accordance with the block decomposition of C_k, we write elements $z \in M_{n\times k}$ as

$$z = [\, x \quad y \,] \quad \text{when } k = 2l, \qquad z = [\, x \quad y \quad t \,] \quad \text{when } k = 2l + 1, \tag{45}$$

where $x, y \in M_{n \times l}$ and $t \in \mathbb{C}^n$. Define the map

$$\beta : M_{n\times k} \to SM_n , \qquad \beta(z) = zC_k z^t .$$

From Theorem 6.3 we know that the algebra of G-invariant polynomials on $M_{n\times k}$ (relative to right G-multiplication) is generated by the matrix entries of β:

$$\beta(z)_{pq} = \begin{cases} \sum_{s=1}^{l} (y_{ps}\, x_{qs} + x_{ps}\, y_{qs}) & \text{when } k = 2l , \\ \sum_{s=1}^{l} (y_{ps}\, x_{qs} + x_{ps}\, y_{qs}) + t_p t_q & \text{when } k = 2l + 1 . \end{cases} \tag{46}$$

We denote by $\Delta_{pq} = \partial(\beta_{pq})$ the corresponding constant-coefficient differential operators, as in Section 9.1.

The space of *G-harmonic polynomials* is

$$\mathcal{H} = \{ f \in \mathcal{P}(M_{n\times k}) : \Delta_{pq} f = 0 \quad \text{for } 1 \le p, q \le n \} .$$

Denote by $\mathcal{H}^{(j)}$ the G-harmonic polynomials that are homogeneous of degree j. The space $\mathcal{H}$ is invariant under $\mathrm{GL}(n,\mathbb{C}) \times G$ with the action

$$\pi(h,g)f(z) = f(h^{-1}zg) \quad \text{for } h \in \mathrm{GL}(n,\mathbb{C}) \text{ and } g \in G\,.$$

(Note that we have omitted the factor $(\det h)^{-k/2}$ that occurs in Theorem 8.5, so π is single-valued on $\mathrm{GL}(n,\mathbb{C})$ even when k is odd). From Theorem 9.2 we know that $\mathcal{H}$ decomposes under the representation π as a multiplicity-free direct sum[i]

$$\mathcal{H} = \bigoplus_{\sigma\in\Sigma} \mathcal{E}^{\tau(\sigma)} \otimes \mathcal{F}^{\sigma}\,.$$

We will now determine Σ and the duality correspondence $\sigma \mapsto \tau(\sigma)$. The key point is to find generators for the algebra $\mathcal{H}^{N_n\times N}$ of harmonic highest weight vectors, relative to a Borel subgroup $B_n \times B \subset \mathrm{GL}(n,\mathbb{C}) \times G$. Here $B_n = D_nN_n$ is the upper-triangular subgroup of $\mathrm{GL}(n,\mathbb{C})$ (D_n the diagonal matrices, N_n the unipotent upper-triangular matrices), and $B = HN$ is a Borel subgroup of G. The fact that $\mathcal{H}$ is multiplicity-free under $\mathrm{GL}(n,\mathbb{C})\times G$ will play a crucial role.

Notation: We denote by ε_j the character $\mathrm{diag}[a_1,\dots,a_n] \mapsto a_j$ of D_n. We write $\mathbb{N}^p_{++}$ for the integer p-tuples $\lambda = [m_1,\dots,m_p]$ with $m_1 \ge m_2 \ge \cdots \ge m_p \ge 0$. Set $|\lambda| = m_1+\cdots+m_p$ and define the *depth* of λ to be the smallest integer i such that $m_i > 0$ (if $\lambda = 0$, set $\mathrm{depth}(0) = 0$).

10.1. O(k) *Harmonics* (*k odd*)

Assume that $k = 2l+1$ is odd. Then $G = G^\circ \times \{\pm I\}$, where $G^\circ = \mathrm{SO}(\mathbb{C}^k,\omega)$ is the identity component of G. We fix the Borel subgroup $B = HN \subset G^\circ$ as follows. The maximal torus H consists of the diagonal matrices

$$h = \mathrm{diag}[x_1,\dots,x_l,x_1^{-1},\dots,x_l^{-1},1]\,, \quad x_i \in \mathbb{C}^\times\,.$$

The unipotent radical N has Lie algebra $\mathfrak{n}$ consisting of the matrices with block decomposition

$$\begin{bmatrix} a & b & c \\ 0 & -a^t & 0 \\ 0 & -c^t & 0 \end{bmatrix}, \quad \begin{array}{l} a \in M_{l\times l} \text{ strictly upper-triangular,} \\ b = -b^t \in M_{l\times l},\ c \in \mathbb{C}^l\,. \end{array} \tag{47}$$

---

[i]The irreducible $\mathrm{GL}(n,\mathbb{C})$-module $\mathcal{E}^{\tau(\sigma)} \cong F^{\tau(\sigma)}_{(n)}$, in the notation of Section 3.

The weights of H are parameterized by $\mathbb{Z}^l$. For $h \in H$ and $\lambda = [m_1, \ldots, m_l] \in \mathbb{Z}^l$ we set $h^\lambda = x_1^{m_1} \cdots x_l^{m_l}$ for the corresponding character of H.

The irreducible representations of G remain irreducible on restriction to G° and $\widehat{G}$ is parameterized as $\{\pi^{\lambda,\epsilon}\}$, where $\lambda \in \mathbb{N}^l_{++}$ is the highest weight for G°, $\epsilon = \pm 1$, and $\pi^{\lambda,\epsilon}(-I) = \epsilon(-I)^{|\lambda|}$. Thus $\widehat{G} = \widehat{G}_1 \cup \widehat{G}_{-1}$, where

$$\widehat{G}_1 = \{(\lambda, 1) : \lambda \in \mathbb{N}^l_{++}\}, \qquad \widehat{G}_{-1} = \{(\lambda, -1) : \lambda \in \mathbb{N}^l_{++}\},$$

(see [16, Sec. 5.2.2]).

Theorem 10.1: ($G = \mathrm{O}(\mathbb{C}^k, \omega)$, $k = 2l + 1$) *Let Σ be the spectrum of G on the G-harmonic polynomials $\mathcal{H} \subset \mathcal{P}(M_{n\times k})$.*

(a) *Assume $k \le n$. Then $\Sigma = \widehat{G}$ and hence Σ does not depend on n (G-stable range).*

(b) *Assume $l < n < k$. Then $\widehat{G}_1 \subset \Sigma$ and*

$$\Sigma \cap \widehat{G}_{-1} = \{(\lambda, -1) : k - n \le \operatorname{depth}(\lambda) \le l\}$$

(unstable range: Σ depends on k and n).

(c) *Assume $n \le l$. Then $\Sigma \cap \widehat{G}_{-1} = \emptyset$ and*

$$\Sigma \cap \widehat{G}_1 = \{(\lambda, 1) : \operatorname{depth}(\lambda) \le n\}.$$

Thus Σ does not depend on k ($\mathrm{GL}(n)$-stable range).

The duality correspondence is given as follows: Let $\lambda = [m_1, \ldots, m_d, 0, \ldots, 0] \in \mathbb{N}^l_{++}$ have depth d with $0 \le d \le \min\{l, n\}$. Then

$$\tau(\sigma) = \begin{cases} [\underbrace{0, \ldots, 0}_{n-d}, -m_d, \ldots, -m_1] & \text{for } \sigma = (\lambda, 1) \in \Sigma \cap \widehat{G}_1, \\ [\underbrace{0, \ldots, 0}_{n-k+d}, \underbrace{-1, \ldots, -1}_{k-2d}, -m_d, \ldots, -m_1] & \text{for } \sigma = (\lambda, -1) \in \Sigma \cap \widehat{G}_{-1}. \end{cases}$$

Remark: The parameter ϵ for the representation $\pi^{\lambda,\epsilon}$ is determined by the corresponding $\mathrm{GL}(n)$ highest weight $\tau(\lambda, \epsilon)$, since left multiplication by $-I_n$ on $M_{n\times k}$ is the same as right multiplication by $-I_k$. Hence $\mathcal{H}$ is also multiplicity-free as a module for $\mathrm{GL}(n, \mathbb{C}) \times G^\circ$ (this property will be used in the proof).

The first step in the proof of Theorem 10.1 is to find a set of generators for the joint eigenfunctions of $B_n \times B$ in $\mathcal{H}$. Just as in the case of $\mathrm{GL}(n) \times$

GL(k) duality (see the article by Benson–Ratcliff), the general strategy is to take appropriate minor determinants. By (46) the operators Δ_{pq} are given in coordinates as

$$\Delta_{pq} = \sum_{s=1}^{l} \left(\frac{\partial}{\partial y_{ps}} \frac{\partial}{\partial x_{qs}} + \frac{\partial}{\partial x_{ps}} \frac{\partial}{\partial y_{qs}} \right) + \frac{\partial^2}{\partial t_p t_q} . \tag{48}$$

The minors of $z = [x \quad y \quad t]$ are linear functions of each column of the matrix components x, y, t. If the minors are chosen to depend only on x or to be linear in t, then they will obviously be harmonic. If they depend on both x and y, then interchanging an x column for a y column will change the sign of the minor but not change the action of the operators $\Delta_{pq} = \Delta_{qp}$, so once again the minor will be harmonic.

We now proceed to carry out this program. Let $p \leq n$ and $q \leq l$. For $u \in M_{n \times l}$ define $p \times q$ submatrices $L_{p,q}(u)$ and $R_{p,q}(u)$ of u by

$$u = \begin{bmatrix} * & * \\ L_{p,q}(u) & * \end{bmatrix} = \begin{bmatrix} * & * \\ * & R_{p,q}(u) \end{bmatrix} .$$

For $t \in \mathbb{C}^n$ and $j \leq n$ define

$$t_{(j)} = \begin{bmatrix} t_{n-j+1} \\ \vdots \\ t_n \end{bmatrix} \in \mathbb{C}^{n-j}$$

(the bottom j entries of t).

Let $z = [x \quad y \quad t] \in M_{n \times k}$ as in (45). Define

$$f_j(z) = \det L_{j,j}(x) \quad \text{for } 1 \leq j \leq \min\{l, n\} .$$

If $n \geq l+1$ then we also define

$$g_j(z) = \begin{cases} \det \begin{bmatrix} L_{j,l}(x) & t_{(j)} \end{bmatrix} & \text{for } j = l+1 , \\ \det \begin{bmatrix} L_{j,l}(x) & R_{j,\,j-l-1}(y) & t_{(j)} \end{bmatrix} & \text{for } l+2 \leq j \leq \min\{n, k\} . \end{cases}$$

Lemma 10.2: (a) *Let $1 \leq j \leq \min\{l, n\}$. Then $f_j \in \mathcal{H}^{(j)}$ and f_j is a $B_n \times B$ eigenfunction of weight (μ, ν), where*

$$\mu = -\varepsilon_{n-j+1} - \cdots - \varepsilon_n \quad \textit{and} \quad \nu = \varepsilon_1 + \cdots + \varepsilon_j .$$

(b) *Assume $n > l$ and let $l+1 \leq j \leq \min\{n, k\}$. Then $g_j \in \mathcal{H}^{(j)}$ and g_j is a $B_n \times B$ eigenfunction of weight (μ, γ), where*

$$\mu = -\varepsilon_{n-j+1} - \cdots - \varepsilon_n \quad \textit{and} \quad \gamma = \varepsilon_1 + \cdots + \varepsilon_{k-j}$$

(here $\gamma = 0$ if $j = k$).

Proof: Assume $1 \le j \le \min\{l, n\}$. It is clear from (48) that $\Delta_{pq} f_j(z) = 0$ for $1 \le p, q \le n$, since $f_j(z)$ only depends on x. The diagonal matrices in B_n and B act on $f \in \mathcal{P}(M_{n\times k})$ by

$$f(x, y, t) \mapsto f(a^{-1}xb,\, a^{-1}yb^{-1},\, a^{-1}t), \quad \text{with } a \in D_n,\, b \in D_l\,. \tag{49}$$

Since f_j involves columns $1, \dots, j$ and rows $n - j + 1, \dots, n$ of x, it has weight (μ, ν) as stated (the columns of x transform as vectors under H, whereas the rows of x transform as covectors under D_n). To verify that f_j is fixed under the left action of N_n, note that $u \in N_n$ acts by $x \mapsto ux$. Since u is unipotent upper triangular, this action transforms the ith row of x by adding multiples of rows below the ith row, so it fixes f_j.

To verify that f_j is fixed under the right action of the unipotent radical $N = \exp \mathfrak{n}$ of B, we observe from (47) that N is generated by the subgroups N_A, N_B, N_C consisting of the matrices

$$\begin{bmatrix} a & 0 & 0 \\ 0 & (a^t)^{-1} & 0 \\ 0 & 0 & 1 \end{bmatrix}, \quad \begin{bmatrix} I & b & 0 \\ 0 & I & 0 \\ 0 & 0 & 1 \end{bmatrix}, \quad \begin{bmatrix} I & -\frac{1}{2}cc^t & c \\ 0 & I & 0 \\ 0 & -c^t & 1 \end{bmatrix}, \tag{50}$$

respectively, where a is upper-triangular unipotent, $b^t = -b$, and $c \in \mathbb{C}^l$. The elements of the subgroup N_A act by $[x \;\; y \;\; t] \mapsto [xa \;\; y \;\; t]$, while the subgroups N_B and N_C fix x. Hence the action of N transforms the ith column of x by adding multiples of columns to the left of the ith column, so f_j is invariant under N. This proves part (a) of the lemma.

Now assume $n > l$ and $l + 1 \le j \le \min\{k, n\}$. Then $g_j(z)$ is a linear function of t and does not depend on the variables y_{rs} for $s \le k - j$. Hence by (48) we have $\Delta_{pq} g_j(z) = 0$ if $\min\{p, q\} \le k - j$. If $\min\{p, q\} \ge k - j + 1$, then

$$\Delta_{pq}\, g_j(z) = \sum_{s=k-j+1}^{l} \left(\frac{\partial}{\partial y_{ps}} \frac{\partial}{\partial x_{qs}} + \frac{\partial}{\partial x_{ps}} \frac{\partial}{\partial y_{qs}} \right) g_j(z)\,.$$

Fix s with $k - j + 1 \le s \le l$. If the column #s of x and column #s of y are interchanged in the determinant defining $g_j(z)$, then $g_j(z)$ changes sign. Hence the function

$$h(z) = \left(\frac{\partial}{\partial y_{ps}} \frac{\partial}{\partial x_{qs}} + \frac{\partial}{\partial x_{ps}} \frac{\partial}{\partial y_{qs}} \right) g_j(z)$$

likewise changes sign since the differential operator is symmetric in the variables x and y. But $h(z)$ is of degree zero in the variables x_{ps}, x_{qs}, y_{ps},

y_{qs} since $g_j(z)$ depends linearly on each variable. Hence $h(z) = 0$. This proves that $g_j(z)$ is G-harmonic.

Since g_j involves columns $1, \ldots, l$ of x and columns $k-j+1, \ldots, l$ of y, we see from (49) that g_j transforms under H by the weight

$$\gamma = (\varepsilon_1 + \cdots + \varepsilon_l) - (\varepsilon_{k-j+1} + \cdots + \varepsilon_l) = \varepsilon_1 + \cdots + \varepsilon_{k-j} .$$

Since g_j involves rows $n-j+1, \ldots, n$ of z, it transforms under D_n by the same weight μ as does f_j.

It is clear that g_j is fixed under the left action of N_n. To verify that g_j is fixed under the right action of N, it suffices to check the action of the matrices in (50). These give the transformations

$$z \mapsto [xa \quad y \quad t], \quad z \mapsto [x \quad y+xb \quad t], \quad z \mapsto \left[x \quad y - \frac{1}{2}xcc^t - tc^t \quad t+xc\right],$$

respectively. The determinant defining g_j involves all the columns of x and t. Since the columns of xcc^t, xc, and tc^t are linear combinations of the columns of x and t, it is clear that these transformations fix g_j. □

Corollary 10.3: *Let* $\mathbf{m} = [m_1, \ldots, m_r] \in \mathbb{N}^r_{++}$, *where* $r = \min\{l, n\}$. *Assume that* $\mathbf{m}$ *has depth* d *and set* $\lambda = [\mathbf{m}, 0, \ldots, 0] \in \mathbb{N}^l_{++}$. *Define* $\varphi_{\mathbf{m}} = f_1^{m_1-m_2} \cdots f_{d-1}^{m_{d-1}-m_d} f_d^{m_d}$ *(when* $\mathbf{m} = 0$ *set* $\varphi_0(z) = 1$*).*

(a) $\varphi_{\mathbf{m}}$ *is a* G*-harmonic polynomial, homogeneous of degree* $|m|$. *Thus* $\varphi_{\mathbf{m}}(-z) = (-1)^{|\mathbf{m}|}\varphi_{\mathbf{m}}(z)$ *for* $z \in M_{n\times k}$. *Furthermore,* $\varphi_{\mathbf{m}}$ *is a* $B_n \times B$ *eigenfunction of weight* (α, λ), *where*

$$\alpha = [\underbrace{0, \ldots, 0}_{n-d}, -m_d, \ldots, -m_1]$$

(when $\mathbf{m} = 0$ *take* $\alpha = 0$*).*

(b) *Suppose* $n > l$ *and* $n - k + d \geq 0$. *For* $\mathbf{m} \neq 0$, *define* $\psi_{\mathbf{m}} = \varphi_{\mathbf{m}}\, g_{k-d}/f_d$ *(when* $\mathbf{m} = 0$ *set* $\psi_0 = g_k$*). Then* $\psi_{\mathbf{m}}$ *is a* G*-harmonic polynomial, homogeneous of degree* $|m| + k - 2d$. *Thus* $\psi_{\mathbf{m}}(-z) = -(-1)^{|\mathbf{m}|}\psi_{\mathbf{m}}(z)$ *for* $z \in M_{n\times k}$. *Furthermore,* $\psi_{\mathbf{m}}$ *is a* $B_n \times B$ *eigenfunction of weight* (β, λ), *where*

$$\beta = [\underbrace{0, \ldots, 0}_{n-k+d}, \underbrace{-1, \ldots, -1}_{k-2d}, -m_d, \ldots, -m_1] .$$

(When $\mathbf{m} = 0$ *take* $\beta = [\underbrace{0, \ldots, 0}_{n-k}, \underbrace{-1, \ldots, -1}_{k}]$*.)*

Proof: Since $\varphi_{\mathbf{m}}(z)$ is a function of x alone (where $z = [x \quad y \quad t]$ as above), it is clear that $\varphi_{\mathbf{m}}$ is G-harmonic. For the same reason,

$$\Delta_{pq}\,\psi_{\mathbf{m}} = (\varphi_{\mathbf{m}}/f_d)\Delta_{pq}\,g_{k-d} = 0\,.$$

Thus we see that $\psi_{\mathbf{m}}$ is also G-harmonic. The other assertions are immediate consequences of Lemma 10.2 □

Now we turn to the proof of Theorem 10.1. A $B_n \times B$ joint eigenfunction generates an irreducible subspace under the action of $\mathrm{GL}(n) \times G^\circ$ by Theorem 3.5. Since the space of harmonic polynomials on $M_{n\times k}$ is a multiplicity-free $\mathrm{GL}(k,\mathbb{C}) \times G^\circ$ module, it follows that a $B_k \times B$ eigenfunction is uniquely determined (up to a scalar multiple) by its weight and parity. If $\mathbf{m} \in \mathbb{N}^l_{++}$ has depth $d \le n$, then from Corollary 10.3 we see that the right translates of $\varphi_{\mathbf{m}}$ under G span an irreducible space of type $(\mathbf{m}, 1)$, while the right translates of $\psi_{\mathbf{m}}$ under G span an irreducible space of type $(\mathbf{m}, -1)$. When $k \le n$, then the conditions $n > l$ and $n - k + d \ge 0$ in part (b) of Corollary 10.3 are automatic. Thus every irreducible representation of G occurs in $\mathcal{H}$ in this case, as asserted in part (a) of the theorem.

To prove parts (b) and (c) of the theorem, assume that $k > n$. Let $f \in \mathcal{H}$ be a $B_n \times B$ eigenfunction. Define a polynomial $\widetilde{f}$ on $M_{k\times k}$ by

$$\widetilde{f}\left(\begin{bmatrix} z' \\ z'' \end{bmatrix}\right) = f(z'') \quad \text{for } z' \in M_{(k-n)\times k} \text{ and } z'' \in M_{n\times k}\,.$$

We claim that $\widetilde{f}$ is G-harmonic. Indeed, if $\min\{p,q\} \le k-n$ then $\Delta_{pq}\widetilde{f} = 0$ since $\widetilde{f}$ does not depend on the variables z_{pq} for $p \le k-n$. On the other hand, if $\min\{p,q\} > k-n$ then

$$\Delta_{pq}\widetilde{f}(z) = \Delta_{p'q'}\,f(z'') = 0$$

(where $p' = p - k + n$ and $q' = j - k + n$), since f is G-harmonic. To see that $\widetilde{f}$ is a $B_k \times B$ eigenfunction, write $b \in B_k$ as

$$b = \begin{bmatrix} \alpha & \beta \\ 0 & \delta \end{bmatrix}, \quad \text{where } \alpha \in B_{k-n},\ \beta \in M_{(k-n)\times n},\ \delta \in B_n\,.$$

Then $\widetilde{f}(b^{-1}zb') = f(\delta^{-1}z''b')$ for $b' \in B$. Since f is $B_n \times B$ eigenfunction, it follows that $\widetilde{f}$ is a $B_k \times B$-eigenfunction. Furthermore this shows that B_k weight μ of $\widetilde{f}$ is of the form

$$\mu = [\underbrace{0,\ldots,0}_{k-n}, a_n, \ldots, a_1] \quad \text{with } a_n \ge \cdots \ge a_1\,.$$

Thus by part (a) we know that $\widetilde{f}$ is a multiple of either $\varphi_{\mathbf{m}}$ or $\psi_{\mathbf{m}}$ for some $\mathbf{m} \in \mathbb{N}^l_{++}$ of depth $d \le n$, since f is homogeneous.

If $l < n < k$, then $\varphi_{\mathbf{m}}$ is defined for all $\mathbf{m} \in \mathbb{N}^l_{++}$, but $\psi_{\mathbf{m}}$ is only defined when the depth d of $\mathbf{m}$ satisfies $k - n \le d \le l$. This implies part (b) of the theorem. If $n \le l$, then $\varphi_{\mathbf{m}}$ is defined for all $\mathbf{m}$ of depth $d \le n$, but in this case $\psi_{\mathbf{m}}$ is never defined. This implies part (c) of the theorem. The formula for the map τ follows from the formulas for α and β in Corollary 10.3. □

10.2. O(k) *Harmonics* (k *even*)

We now assume that $k = 2l$ is even. We take the Borel subgroup $B \subset G$ whose Lie algebra consists of the matrices with block decomposition (block sizes $l \times l$)

$$\begin{bmatrix} a & b \\ 0 & -a^t \end{bmatrix} \qquad (a \text{ upper-triangular, } b^t = -b)$$

(If $k = 2$ then $b = 0$ and $B \cong \mathbb{C}^\times$). Let $N \subset B$ be the unipotent radical (the matrices as above with a upper-triangular unipotent). Recall that

$$\mathrm{O}(\mathbb{C}^k, \omega) = G^\circ \rtimes \{I, s\}$$

where $G^\circ = \mathrm{SO}(\mathbb{C}^k, \omega)$ is the identity component and $s \in G$ is the reflection interchanging the basis vectors e_l and e_{2l} and fixing all other basis vectors e_i. Since s normalizes B it acts on the characters of B. Let $\lambda = [m_1, \ldots, m_l] \in \mathbb{N}^l_{++}$. If $m_l \neq 0$, then $s \cdot \lambda \neq \lambda$ (since s changes m_l to $-m_l$). In this case there is a unique irreducible G representation $\pi^{\lambda,0}$ such that

$$\pi^{\lambda,0}|_{G^\circ} = \pi^\lambda \oplus \pi^{s\cdot\lambda}$$

(where π^μ denotes the irreducible G° representation with highest weight μ). If $m_l = 0$, then there are two irreducible representations $\pi^{\lambda,\epsilon}$ ($\epsilon = \pm 1$) of G whose restriction to G° is π^λ. They are related by

$$\pi^{\lambda,\epsilon} = \det \otimes \pi^{\lambda,-\epsilon}$$

and labeled so that $\pi^{\lambda,\epsilon}(s)$ acts by ϵ on the G° highest weight vector (see [16, Sec. 5.2.2]). Thus $\widehat{G}$ can be written as a disjoint union $\widehat{G} = \widehat{G}_{-1} \cup \widehat{G}_0 \cup \widehat{G}_1$, where

$$\widehat{G}_{\pm 1} = \{\pi^{\lambda,\epsilon} : \mathrm{depth}(\lambda) < l, \epsilon = \pm 1\}, \quad \widehat{G}_0 = \{\pi^{\lambda,\epsilon} : \mathrm{depth}(\lambda) = l, \epsilon = 0\}.$$

Theorem 10.4: $(G = \mathrm{O}(\mathbb{C}^k, \omega),\ k = 2l)$ *Let Σ be the spectrum of G on the G-harmonic polynomials $\mathcal{H} \subset \mathcal{P}(M_{n\times k})$.*

(a) *Assume $k \le n$. Then $\Sigma = \widehat{G}$ and thus Σ does not depend on n (G-stable range).*

(b) *Assume $l < n < k$. Then $\widehat{G}_1 \cup \widehat{G}_0 \subset \Sigma$, whereas*

$$\Sigma \cap \widehat{G}_{-1} = \{(\lambda, -1) : k - n \le \mathrm{depth}(\lambda) < l\}$$

(unstable range: Σ depends on k and n).

(c) *Assume $n = l$. Then $\Sigma = \widehat{G}_1 \cup \widehat{G}_0$.*

(d) *Assume $n < l$. Then $\Sigma = \{(\lambda, 1) : \mathrm{depth}(\lambda) \le n\} \subset \widehat{G}_1$.*

Thus Σ does not depend on k when $n \le l$ ($\mathrm{GL}(n)$-stable range). The duality correspondence is given as follows: Let $\lambda = [m_1, \ldots, m_d, 0, \ldots, 0] \in \mathbb{N}^l_{++}$ have depth d with $1 \le d \le \min\{l, n\}$. Then

$$\tau(\sigma) = \begin{cases} [\underbrace{0, \ldots, 0}_{n-d}, -m_d, \ldots, -m_1] & \text{for } \sigma = (\lambda, \epsilon) \in \Sigma \cap (\widehat{G}_1 \cup \widehat{G}_0)\,, \\ [\underbrace{0, \ldots, 0}_{n-k+d}, \underbrace{-1, \ldots, -1}_{k-2d}, -m_d, \ldots, -m_1] & \text{for } \sigma = (\lambda, -1) \in \Sigma \cap \widehat{G}_{-1}\,. \end{cases}$$

To prove the theorem, we will find a set of generators for the joint eigenfunctions of $B_n \times B$ in $\mathcal{H}$. For $u \in M_{n\times l}$, $p \le n$, and $q \le l$, define matrices $L_{p,q}(u)$ and $R_{p,q}(u)$ as in the proof of Theorem 10.1. In this case we write $z = [\,x \quad y\,]$ as in (45) and we define

$$f_j(z) = \det L_{j,j}(x) \quad \text{for } 1 \le j \le \min\{l, n\}\,.$$

If $n \ge l + 1$ then we also define

$$g_j(z) = \det \left[\, L_{j,l}(x) \quad R_{j,j-l}(y) \,\right] \quad \text{for } l + 1 \le j \le \min\{k, n\}\,.$$

Lemma 10.5: (a) *Let $1 \le j \le \min\{l, n\}$. Then $f_j \in \mathcal{H}^{(j)}$ and f_j is a $B_n \times B$ eigenfunction of weight (μ, ν), where*

$$\mu = -\varepsilon_{n-j+1} - \cdots - \varepsilon_n \quad \text{and} \quad \nu = \varepsilon_1 + \cdots + \varepsilon_j\,.$$

Furthermore $f_j(zs) = f_j(z)$ if $j < l$, where $s \in G$ is the reflection $e_l \leftrightarrow e_{2l}$.

(b) *Suppose $n \ge l + 1$ and take $l + 1 \le j \le \min\{n, k\}$. Then $g_j \in \mathcal{H}^{(j)}$ and g_j is a $B_n \times B$ eigenfunction of weight (μ, γ), where*

$$\mu = -\varepsilon_{n-j+1} - \cdots - \varepsilon_n \quad \text{and} \quad \gamma = \varepsilon_1 + \cdots + \varepsilon_{k-j}$$

(here $\gamma = 0$ if $j = k$). Furthermore $g_j(zs) = -g_j(z)$, with $s \in G$ as in (a).

Proof: Essentially the same as the proof of Lemma 10.2. Note that in this case the unipotent radical N of B is generated by the transformations

$$z \mapsto [xa \quad y]\,, \quad z \mapsto [x \quad y + xb]\,,$$

with a upper-triangular unipotent and $b^t = -b$. □

Corollary 10.6: *Let* $\mathbf{m} = [m_1, \ldots, m_r] \in \mathbb{N}^r_{++}$, *where* $r = \min\{l, n\}$. *Assume* $\mathbf{m}$ *has depth* d *and set* $\lambda = [\mathbf{m}, 0, \ldots, 0] \in \mathbb{N}^l_{++}$. *Define* $\varphi_{\mathbf{m}} = f_1^{m_1 - m_2} \cdots f_{d-1}^{m_{d-1} - m_d} f_d^{m_d}$ *(when* $\mathbf{m} = 0$ *set* $\varphi_0 = 1$*).*

(a) $\varphi_{\mathbf{m}}$ *is a G-harmonic polynomial, homogeneous of degree* $|\mathbf{m}|$. *Furthermore,* $\varphi_{\mathbf{m}}$ *is a* $B_n \times B$ *eigenfunction of weight* (α, λ), *where*

$$\alpha = [\underbrace{0, \ldots, 0}_{n-d}, -m_d, \ldots, -m_1]\,,$$

(when $\mathbf{m} = 0$ *take* $\alpha = 0$*). Set* $\varphi^s_{\mathbf{m}}(z) = \varphi_{\mathbf{m}}(zs)$. *Then* $\varphi^s_{\mathbf{m}} = \varphi_{\mathbf{m}}$ *when* $d < l$.

(b) *Let* $n > l$. *If* $d < l$ *and* $n - k + d \geq 0$ *define* $\psi_{\mathbf{m}} = \varphi_{\mathbf{m}}\, g_{k-d}/f_d$ *(when* $\mathbf{m} = 0$ *set* $\psi_0 = g_k$*). Then* $\psi_{\mathbf{m}}$ *is a G-harmonic polynomial, homogeneous of degree* $|m| + k - 2d$. *Furthermore* $\psi_{\mathbf{m}}$ *is a* $B_n \times B$ *eigenfunction of weight* (β, λ), *where*

$$\beta = [\underbrace{0, \ldots, 0}_{n-k+d}, \underbrace{-1, \ldots, -1}_{k-2d}, -m_d, \ldots, -m_1]$$

(when $\mathbf{m} = 0$ *take* $\beta = [\underbrace{0, \ldots, 0}_{n-k}, \underbrace{-1, \ldots, -1}_{k}]$*). Set* $\psi^s_{\mathbf{m}}(z) = \psi_{\mathbf{m}}(zs)$. *Then* $\psi^s_{\mathbf{m}} = -\psi_{\mathbf{m}}$.

Proof: This follows from Lemma 10.5 by the same arguments as in the proof of Corollary 10.3. □

To prove Theorem 10.4, assume first that $n \geq k$. By Corollary 10.6 the functions $\varphi_{\mathbf{m}}$ are defined for all $\mathbf{m} \in \mathbb{N}^l_{++}$. If $\mathbf{m}$ has depth l then the right G-translates of $\varphi_{\mathbf{m}}$ span an irreducible subspace of type $(\mathbf{m}, 0)$. If $\mathbf{m}$ has depth less than l then $\psi_{\mathbf{m}}$ is also defined. In this case the right G-translates of $\varphi_{\mathbf{m}}$ span a G-irreducible subspace of type $(\mathbf{m}, 1)$, whereas the right G-translates of $\psi_{\mathbf{m}}$ span an irreducible subspace of type $(\mathbf{m}, -1)$. Thus we get all irreducible representations of G in $\mathcal{H}$, as asserted in part (a) of the theorem.

The argument when $n < k$ proceeds as in the proof of Theorem 10.1 by lifting harmonic $B_n \times B$ eigenfunctions from $M_{n \times k}$ to $M_{k \times k}$. Note that

$\varphi_{\mathbf{m}}$ is defined for all $\mathbf{m}$ of depth $d \leq \min\{n, l\}$, whereas $\psi_{\mathbf{m}}$ is only defined when $n > l$ and $k - n \leq d < l$. We omit the details. □

10.3. *Examples of harmonic decompositions*

(1) Assume $n \leq l$ and $k = 2l + 1$ or $2l$, so that we are in case (c) of Theorem 10.1 or cases (c) and (d) of Theorem 10.4. The restrictions to $\mathrm{SO}(k)$ of the representations in Σ are the *class* n representations of $\mathrm{SO}(k)$ — those that have a vector fixed under the subgroup $\mathrm{SO}(k-n)$. This follows from the branching law (see [16, Sec. 8.1]). In this case the harmonic polynomials on $M_{n\times k}$ decompose under $\mathrm{GL}(n) \times \mathrm{SO}(k)$ as

$$\mathcal{H} \cong \bigoplus_{\lambda \in \mathbb{N}^n_{++}} \mathcal{E}^{\lambda^{\check{}}} \otimes \mathcal{V}^{\lambda}. \tag{51}$$

Here $\lambda^{\check{}} = [-m_n, \ldots, -m_1]$ and $\mathcal{V}^\lambda$ is the irreducible $\mathrm{SO}(k)$ module with highest weight λ (when $n = l$, $k = 2l$ is even and $m_l \neq 0$, then $\mathcal{V}^\lambda$ is the sum of the irreducible representations with highest weights λ and $s\cdot\lambda$). For $n = 1$, (51) is the classical spherical harmonic decomposition and gives the decomposition of polynomials restricted to the sphere $\mathrm{SO}(k)/\mathrm{SO}(k-1)$. For $n > 1$ (51) gives the decomposition of polynomials restricted to the *Stiefel manifold* $\mathrm{SO}(k)/\mathrm{SO}(k-n)$. This decomposition was obtained by Gelbart [12] and Ton-That [32] before Kashiwara and Vergne [22] worked out the general case that we have presented here.

(2) Now assume $l < n < k$, so that we are in case (b) of Theorems 10.1 and 10.4. The decomposition of the harmonics in this case was obtained by Strichartz [29]. For example, let $n = 2$ and $k = 3$. Then we have the decomposition

$$\mathcal{H} = \Big\{ \bigoplus_{m\geq 0} \mathcal{E}^{[0,-m]} \otimes \mathcal{V}^{[m]} \Big\} \oplus \Big\{ \bigoplus_{m\geq 1} \mathcal{E}^{[-1,-m]} \otimes \mathcal{V}^{[m]} \Big\}$$

of the harmonic polynomials on $M_{2\times 3}$. Here $\mathcal{V}^{[m]}$ denotes the irreducible $\mathrm{SO}(3)$ representation with highest weight $m\varepsilon_1$. The $B_2 \times B$ harmonic eigenfunction $\varphi_{(m)}(z) = x_2^m$ generates the summand $\mathcal{E}^{[0,-m]} \otimes \mathcal{V}^{[m]}$. The $B_2 \times B$ harmonic eigenfunction $\psi_{(m)}(z) = x_2^{m-1}(x_1 t_2 - x_2 t_1)$ generates the summand $\mathcal{E}^{[-1,-m]} \otimes \mathcal{V}^{[m]}$. Here we write

$$z = \begin{bmatrix} x_1 & y_1 & t_1 \\ x_2 & y_2 & t_2 \end{bmatrix}.$$

(3) Let $n = 3$ and $k = 3$ so that we are in case (a) of Theorem 10.1. Then we have the decomposition

$$\mathcal{H} = \left\{ \bigoplus_{m \geq 0} \mathcal{E}^{[0,0,-m]} \otimes \mathcal{V}^{[m]} \right\} \oplus \left\{ \mathcal{E}^{[-1,-1,-1]} \otimes \mathcal{V}^{[0]} \right\}$$
$$\oplus \left\{ \bigoplus_{m \geq 1} \mathcal{E}^{[0,-1,-m]} \otimes \mathcal{V}^{[m]} \right\}$$

of the harmonic polynomials on $M_{3\times 3}$ as a module for $\mathrm{GL}(3) \times \mathrm{SO}(3)$. The $B_3 \times B$ harmonic eigenfunction $\varphi_{(m)}(z) = x_3^m$ generates the summand $\mathcal{E}^{[0,0,-m]} \otimes \mathcal{V}^{[m]}$. The $B_3 \times B$ harmonic eigenfunction $\psi_{(m)}(z) = x_3^{m-1}(x_2 t_3 - x_3 t_2)$ generates the summand $\mathcal{E}^{[0,-1,-m]} \otimes \mathcal{V}^{[m]}$. Here we write

$$z = \begin{bmatrix} x_1 & y_1 & t_1 \\ x_2 & y_2 & t_2 \\ x_3 & y_3 & t_3 \end{bmatrix}.$$

For $m = 0$ the function $\psi_{(0)}(z) = \det z$ generates the one-dimensional summand $\mathcal{E}^{[-1,-1,-1]} \otimes \mathcal{V}^{[0]}$.

Let $\mathbb{C}^{(m)}$ denote the one-dimension representation $g \mapsto (\det g)^m$ of $\mathrm{GL}(3)$. Let $\varpi_1 = [1,0,0]$ and $\varpi_2 = [1,1,0]$. Then the $\mathrm{GL}(3)$ representations occurring in $\mathcal{H}$ are $\mathbb{C}^{(-1)}$,

$$\mathcal{E}^{[0,0,-m]} \cong \mathbb{C}^{(-m)} \otimes \mathcal{E}^{m\varpi_2}$$

for all $m \geq 0$, and

$$\mathcal{E}^{[0,-1,-m]} \cong \mathbb{C}^{(-m)} \otimes \mathcal{E}^{\varpi_1 + (m-1)\varpi_2}$$

for all $m \geq 1$.

11. Symplectic group and oscillator representation

We now turn to the functional-analytic aspects of the harmonic duality decomposition in Theorem 9.2 (recall Example 4 in Section 2.3). If we replace the complex group G by its compact real form $G_0 = G \cap \mathrm{U}(V)$ then the finite-dimensional representations $\mathcal{F}^\sigma$ remain irreducible under G_0 and the action of G_0 is unitary relative to the Fischer inner product.

We would like to have a similar picture for the dual representations E^λ (where $\lambda = \tau(\sigma) + \delta$). At the Lie algebra level it is clear that to obtain a unitary representation, we should take the real form $\mathfrak{g}_0'$ of $\mathfrak{g}'$ that acts by skew-hermitian operators relative to the Fischer inner product. The analytic problem is to construct a unitary representation of an associated real Lie group G'_0 on the completion of $\mathcal{P}(V)$, and to describe its action on the Hilbert space completions of the infinite-dimensional spaces E^λ.

We will construct G'_0 as a subgroup of the *metaplectic group* $\mathrm{Mp}(nk, \mathbb{R})$ (the two-sheeted cover of the real symplectic group $\mathrm{Sp}(nk, \mathbb{R})$). The associated unitary representation will be the restriction to G'_0 of the *oscillator representation* of the metaplectic group. This representation already appears in the harmonic decomposition as a Lie algebra representation by elements of degree 2 in the Weyl algebra. However, when we try to exponentiate it to a unitary group representation, we encounter the conflict between the particle and the wave description of quantum mechanics; the representation has a simple description (the holomorphic model) relative to the maximal compact subgroup $K_0 \cong \mathrm{U}(n)$ of $\mathrm{Sp}(n, \mathbb{R})$, and another simple description (the real-wave model) relative to the maximal parabolic subgroup $P \cong \mathrm{GL}(n, \mathbb{R}) \ltimes SM_n(\mathbb{R})$ of $\mathrm{Sp}(n, \mathbb{R})$. In both descriptions $K_0 \cap P \cong \mathrm{O}(n)$ acts geometrically, but some of the remaining group elements act in a more subtle way. Thus it will be necessary to consider two matrix forms of the real symplectic group and the intertwining operator (the Bargmann–Segal transform) that relates the two versions of the oscillator representation.

11.1. *Real symplectic group*

Let $\mathrm{Sp}(n, \mathbb{C})$ be the subgroup of $\mathrm{GL}(2n, \mathbb{C})$ that preserves the skew-form

$$\Omega(x, y) = \sum_{i=1}^{n} (x_i y_{n+i} - x_{n+i} y_i)$$

on $\mathbb{C}^{2n}$. Thus $g \in \mathrm{Sp}(n, \mathbb{C})$ if and only if $g^t J_n g = J_n$, where g^t denotes matrix transpose and J_n is the matrix

$$J_n = \begin{bmatrix} 0 & I_n \\ -I_n & 0 \end{bmatrix}.$$

We can also describe $\mathrm{Sp}(n, \mathbb{C})$ as the fixed-point group of the involution $\tau : g \mapsto J_n (g^t)^{-1} J_n^{-1}$ on $\mathrm{GL}(2n, \mathbb{C})$.

The Lie algebra $\mathfrak{sp}(n, \mathbb{C})$ of $\mathrm{Sp}(n, \mathbb{C})$ consists of all $X \in M_{2n}$ such that $J_n X + X^t J_n = 0$. These matrices have block form

$$X = \begin{bmatrix} A & B \\ C & -A^t \end{bmatrix} \quad \text{with } A \in M_n \text{ and } B, C \in SM_n .$$

Here we use the notation M_n for the $n \times n$ complex matrices and SM_n for the $n \times n$ symmetric complex matrices.

The *real symplectic group* $\mathrm{Sp}(n, \mathbb{R}) = \mathrm{Sp}(n, \mathbb{C}) \cap \mathrm{GL}(2n, \mathbb{R})$. Its Lie algebra $\mathfrak{sp}(n, \mathbb{R})$ consists of all the real matrices in $\mathfrak{sp}(n, \mathbb{C})$.

Maximal Compact Subgroup. A fundamental technique for studying a unitary representation of a real reductive group such as $\mathrm{Sp}(n, \mathbb{R})$ is to restrict the representation to a maximal compact subgroup, under which the representation space decomposes as the (Hilbert-space) direct sum of multiples of irreducible (finite-dimensional) subspaces. The *real orthogonal group* $\mathrm{O}(k) \subset \mathrm{U}(k)$ is the subgroup of real unitary matrices. Since $\mathrm{Sp}(n, \mathbb{C})$ and $\mathrm{Sp}(n, \mathbb{R})$ are invariant under the map $g \mapsto g^*$, the groups

$$\mathrm{Sp}(n) = \mathrm{Sp}(n, \mathbb{C}) \cap \mathrm{U}(2n) \quad \text{and} \quad \mathrm{Sp}(n, \mathbb{R}) \cap \mathrm{U}(2n) = \mathrm{Sp}(n, \mathbb{R}) \cap \mathrm{O}(2n)$$

are maximal compact subgroups of $\mathrm{Sp}(n, \mathbb{C})$ and $\mathrm{Sp}(n, \mathbb{R})$, respectively (see [24, Proposition 1.2]). The subgroup of diagonal matrices in $\mathrm{Sp}(n)$ is a maximal torus in $\mathrm{Sp}(n)$. However, the subgroup of diagonal matrices in $\mathrm{Sp}(n, \mathbb{R}) \cap \mathrm{O}(2n)$ is finite and is *not* a maximal torus in $\mathrm{Sp}(n, \mathbb{R})$. Hence it is convenient to replace $\mathrm{Sp}(n, \mathbb{R})$ by an isomorphic real form G_0 so that the diagonal matrices in $G_0 \cap \mathrm{U}(2n)$ comprise a maximal (compact) torus in G_0.

Define

$$I_{n,n} = \begin{bmatrix} I_n & 0 \\ 0 & -I_n \end{bmatrix}$$

and let σ be the *conjugation* (conjugate-holomorphic involution) $\sigma(g) = I_{n,n}(g^*)^{-1} I_{n,n}$ on $\mathrm{GL}(2n, \mathbb{C})$. The fixed-point set of σ is the real form $\mathrm{U}(n,n)$ of $\mathrm{GL}(2n, \mathbb{C})$. Set

$$K_n = I_{n,n} J_n = \begin{bmatrix} 0 & I_n \\ I_n & 0 \end{bmatrix}.$$

Then $J_n^{-1} I_{n,n} = K_n$, so it follows that $\sigma\tau = \tau\sigma$. Hence σ leaves $\mathrm{Sp}(n, \mathbb{C})$ invariant and its restriction to $\mathrm{Sp}(n, \mathbb{C})$ defines a conjugation of $\mathrm{Sp}(n, \mathbb{C})$ which we continue to denote as σ. If $g \in \mathrm{Sp}(n, \mathbb{C})$ then $\sigma(g) = \sigma\tau(g) = K_n \bar{g} K_n$. In terms of the $n \times n$ block decomposition, σ acts by

$$\sigma \begin{bmatrix} A & B \\ C & D \end{bmatrix} = \begin{bmatrix} \bar{D} & \bar{C} \\ \bar{B} & \bar{A} \end{bmatrix}.$$

Define $G_0 = \{g \in \mathrm{Sp}(n, \mathbb{C}) : \sigma(g) = g\}$. Then G_0 is a real form of $\mathrm{Sp}(n, \mathbb{C})$. Its Lie algebra $\mathfrak{g}_0 = \mathrm{Lie}(G_0)$ consists of all matrices $X \in \mathfrak{sp}(n, \mathbb{C})$ such that $\sigma(X) = X$. In terms of the block decomposition, $\mathfrak{g}_0$ consists of the matrices

$$X = \begin{bmatrix} A & B \\ \bar{B} & \bar{A} \end{bmatrix}, \quad A^* = -A,\ B = B^t. \tag{52}$$

Lemma 11.1: *The subgroup $K_0 = G_0 \cap \mathrm{U}(2n)$ is a maximal compact subgroup of G_0 and consists of all matrices*

$$\begin{bmatrix} A & 0 \\ 0 & \bar{A} \end{bmatrix}, \quad \textit{with } A \in \mathrm{U}(n).$$

Hence $K_0 \cong \mathrm{U}(n)$ and the subgroup of diagonal matrices in K_0 is a maximal compact torus of G_0.

Proof: Since $\sigma(g^*) = \sigma(g)^*$ for $g \in \mathrm{GL}(2n, \mathbb{C})$, the group G_0 is invariant under $g \mapsto g^*$. Hence K_0 is a maximal compact subgroup of G_0. Write $g \in \mathrm{GL}(2n, \mathbb{C})$ as

$$g = \begin{bmatrix} A & B \\ C & D \end{bmatrix} \quad \text{with } A, B, C, D \in M_n.$$

Then $g \in \mathrm{U}(n,n)$ if and only if $g^* I_{n,n} g = I_{n,n}$. This condition can be written as

$$A^*A - C^*C = I_n, \quad B^*B - D^*D = -I_n, \quad A^*B - C^*D = 0. \tag{53}$$

On the other hand, $g \in \mathrm{U}(2n)$ if and only if $g^*g = I_{2n}$. This condition can be written as

$$A^*A + C^*C = I_n, \quad B^*B + D^*D = I_n, \quad A^*B + C^*D = 0. \tag{54}$$

If $g \in \mathrm{U}(2n) \cap \mathrm{U}(n,n)$ then from (53) and (54) we have $C^*C = 0$ and $B^*B = 0$. Hence $B = 0$ and $C = 0$, so $A^*A = I_n$ and $D^*D = I_n$. Thus $\mathrm{U}(2n) \cap \mathrm{U}(n,n) = \mathrm{U}(n) \times \mathrm{U}(n)$. But if $g \in \mathrm{Sp}(n, \mathbb{C})$ is in block-diagonal form, then

$$g = \begin{bmatrix} A & 0 \\ 0 & (A^t)^{-1} \end{bmatrix}, \quad \text{with } A \in \mathrm{GL}(n, \mathbb{C}).$$

Hence $g \in K_0$ if and only if $A \in \mathrm{U}(n)$. □

Define an involution θ on $\mathrm{Sp}(n, \mathbb{C})$ by

$$\theta(g) = I_{n,n} g I_{n,n}$$

(note that θ is an inner automorphism of $\mathrm{Sp}(n, \mathbb{C})$). If $g \in G_0$ then $(g^t)^{-1} = J_n g J_n^{-1}$ and $\bar{g} = K_n g K_n$. Hence $(g^*)^{-1} = J_n \bar{g} J_n^{-1} = J_n K_n g K_n J_n^{-1}$. Since $J_n K_n = I_{n,n}$, it follows that

$$\theta(g) = (g^*)^{-1} \quad \text{for } g \in G_0.$$

Thus the maximal compact subgroup K_0 is the fixed-point set of θ in G_0. Its complexification is

$$K = \{g \in \mathrm{Sp}(n, \mathbb{C}) \, : \, \theta(g) = g\}.$$

Note that if $g \in \mathrm{GL}(2n, \mathbb{C})$, then $\theta(g) = g$ if and only if

$$g = \begin{bmatrix} a & 0 \\ 0 & d \end{bmatrix}, \quad a, d \in \mathrm{GL}(n, \mathbb{C}).$$

If $g \in K$, then in this block decomposition $d = (a^t)^{-1}$. Hence $K \cong \mathrm{GL}(n, \mathbb{C})$ via the homomorphism

$$a \mapsto \begin{bmatrix} a & 0 \\ 0 & (a^t)^{-1} \end{bmatrix}.$$

The complexification of the Lie algebra $\mathfrak{k}_0$ of K_0 is the Lie algebra $\mathfrak{k} \cong \mathfrak{gl}(n, \mathbb{C})$ of K.

The involution θ gives a decomposition of $\mathfrak{sp}(n, \mathbb{C})$. The $+1$ eigenspace of θ on $\mathfrak{sp}(n, \mathbb{C})$ is $\mathfrak{k}$, whereas the -1 eigenspace is

$$\mathfrak{p} = \left\{ \begin{bmatrix} 0 & B \\ C & 0 \end{bmatrix} : B, C \in SM_n \right\}.$$

We have $\mathfrak{sp}(n, \mathbb{C}) = \mathfrak{k} \oplus \mathfrak{p}$ with commutation relations

$$[\mathfrak{k}, \mathfrak{k}] \subset \mathfrak{k}, \quad [\mathfrak{k}, \mathfrak{p}] \subset \mathfrak{p}, \quad [\mathfrak{p}, \mathfrak{p}] \subset \mathfrak{k}.$$

The center of $\mathfrak{k}$ is spanned by $I_{n,n}$ and $\mathfrak{k} = \mathbb{C}I_{n,n} \oplus [\mathfrak{k}, \mathfrak{k}]$, with the derived algebra $[\mathfrak{k}, \mathfrak{k}] \cong \mathfrak{sl}(n, \mathbb{C})$. The ± 1 eigenspaces of $\mathrm{ad} I_{n,n}$ on $\mathfrak{p}$ are

$$\mathfrak{p}_+ = \left\{ \begin{bmatrix} 0 & B \\ 0 & 0 \end{bmatrix} : B \in SM_n \right\}, \quad \mathfrak{p}_- = \left\{ \begin{bmatrix} 0 & 0 \\ C & 0 \end{bmatrix} : C \in SM_n \right\}.$$

These subspaces are invariant under $\mathfrak{k}$ and have the commutation relations

$$[\mathfrak{p}_+, \mathfrak{p}_+] = 0, \quad [\mathfrak{p}_-, \mathfrak{p}_-] = 0, \quad [\mathfrak{p}_+, \mathfrak{p}_-] \subset \mathfrak{k}.$$

Thus there is a triangular decomposition

$$\mathfrak{sp}(n, \mathbb{C}) = \mathfrak{p}_- \oplus \mathfrak{k} \oplus \mathfrak{p}_+$$

(as we already noted in Section 8). The conjugation σ interchanges $\mathfrak{p}_+$ and $\mathfrak{p}_-$, since

$$\sigma\left(\begin{bmatrix} 0 & B \\ 0 & 0 \end{bmatrix} \right) = \begin{bmatrix} 0 & 0 \\ \bar{B} & 0 \end{bmatrix}$$

for $B \in SM_n$.

We can describe these decompositions in terms of root spaces as follows (see [16, Sec. 2.3.1]). The complexification T of T_0 is a maximal (algebraic) torus in $\mathrm{Sp}(n, \mathbb{C})$ and has Lie algebra

$$\mathfrak{t} = \{X = \mathrm{diag}[x_1, \ldots, x_n, -x_1, \ldots, -x_n] : x_j \in \mathbb{C}\}.$$

The set of roots $\Phi = \Phi(\mathfrak{g}, \mathfrak{t})$ of $\mathfrak{t}$ on $\mathfrak{g}$ is $\pm\varepsilon_i \pm \varepsilon_j$ for $1 \le i, j \le n$, where $\varepsilon_i(X) = x_i$ for $X \in \mathfrak{t}$ as above. We have $\Phi = \Phi_c \cup \Phi_n$, where

$$\Phi_c = \{\pm(\varepsilon_i - \varepsilon_j) \, : \, 1 \le i < j \le n\}$$

is the set of *compact* roots (the roots of $\mathfrak{t}$ on $\mathfrak{k}$) and

$$\Phi_n = \{\pm(\varepsilon_i + \varepsilon_j) \, : \, 1 \le i \le j \le n\}$$

is the set of *noncompact* roots (the roots of $\mathfrak{t}$ on $\mathfrak{p}$).

Take the set of positive roots Φ^+ to be $\varepsilon_i \pm \varepsilon_j$ for $1 \le i \le j \le n$, and let Φ_c^+ (*respectively* Φ_n^+) be the positive compact (*respectively* noncompact) roots. Then

$$\mathfrak{k} = \mathfrak{t} + \sum_{\alpha \in \Phi_c} \mathfrak{g}_\alpha \,, \qquad \mathfrak{p}_\pm = \sum_{\beta \in \Phi_n^+} \mathfrak{g}_{\pm\beta} \,.$$

The simple roots in Φ^+ are $\alpha_1, \ldots, \alpha_n$, where

$$\alpha_i = \varepsilon_i - \varepsilon_{i+1} \quad \text{for } i = 1, \ldots, n-1 \text{ and } \alpha_n = 2\varepsilon_n \,.$$

The unique simple noncompact root is the long root α_n, and the highest root is $\gamma = 2\varepsilon_1$ (it is noncompact). Let ρ be one-half the sum of the positive roots. Then

$$\rho = n\varepsilon_1 + (n-1)\varepsilon_2 + \cdots + \varepsilon_n \,. \tag{55}$$

Cayley Transform. We now show that the group $G_0 = \mathrm{Sp}(n, \mathbb{C}) \cap \mathrm{U}(n, n)$ is conjugate to $\mathrm{Sp}(n, \mathbb{R})$ within $\mathrm{Sp}(n, \mathbb{C})$. To understand this in terms of the adjoint representation of $\mathfrak{sp}(n, \mathbb{C})$, consider first the case $n = 1$ (recall that $\mathrm{Sp}(1, \mathbb{C}) = \mathrm{SL}(2, \mathbb{C})$). Set

$$\mathbf{k} = \begin{bmatrix} 0 & i \\ -i & 0 \end{bmatrix}, \quad \mathbf{x} = \frac{1}{2}\begin{bmatrix} 1 & -i \\ -i & -1 \end{bmatrix}, \quad \mathbf{y} = \frac{1}{2}\begin{bmatrix} 1 & i \\ i & -1 \end{bmatrix},$$

where i is a fixed choice of $\sqrt{-1}$. Then $[\mathbf{k}, \mathbf{x}] = 2\mathbf{x}$, $[\mathbf{k}, \mathbf{y}] = -2\mathbf{y}$, $[\mathbf{x}, \mathbf{y}] = \mathbf{k}$, so $\{\mathbf{x}, \mathbf{y}, \mathbf{k}\}$ is a TDS (three-dimensional simple) triple. Furthermore, the one-parameter subgroup

$$t \mapsto \exp(it\mathbf{k}) = \begin{bmatrix} \cos t & -\sin t \\ \sin t & \cos t \end{bmatrix}, \quad t \in \mathbb{R},$$

is a maximal compact torus in $\mathrm{SL}(2, \mathbb{R})$. Let

$$\mathbf{h} = \begin{bmatrix} 1 & 0 \\ 0 & -1 \end{bmatrix}, \quad \mathbf{e} = \begin{bmatrix} 0 & 1 \\ 0 & 0 \end{bmatrix}, \quad \mathbf{f} = \begin{bmatrix} 0 & 0 \\ 1 & 0 \end{bmatrix}$$

be the standard TDS in $\mathfrak{sl}(2,\mathbb{C})$. We can conjugate $\{\mathbf{x},\mathbf{y},\mathbf{k}\}$ to $\{\mathbf{e},\mathbf{f},\mathbf{h}\}$ as follows: Since $\mathbf{x}+\mathbf{y}=\mathbf{h}$, we have

$$(\mathrm{ad}(\mathbf{y}-\mathbf{x}))(\mathbf{k}) = 2\mathbf{h}, \quad (\mathrm{ad}(\mathbf{y}-\mathbf{x}))(\mathbf{h}) = -2\mathbf{k}\,,$$

and so

$$e^{t\,\mathrm{ad}(\mathbf{y}-\mathbf{x})}\mathbf{k} = (\cos 2t)\mathbf{h} + (\sin 2t)\mathbf{k}\,, \quad \text{for } t\in\mathbb{C}\,.$$

Setting $t=\pi/4$, we obtain

$$e^{(\pi/4)\mathrm{ad}(\mathbf{y}-\mathbf{x})}\mathbf{k} = \mathbf{h}\,.$$

Since $(\mathbf{y}-\mathbf{x})^2 = -I$, we have

$$\exp[t(\mathbf{y}-\mathbf{x})] = (\cos t)I + (\sin t)(\mathbf{y}-\mathbf{x})\,, \quad \text{for } t\in\mathbb{C}\,.$$

Define

$$\mathbf{c} = \exp[\frac{\pi}{4}(\mathbf{y}-\mathbf{x})] = \frac{1}{\sqrt{2}}\begin{bmatrix}1 & i\\ i & 1\end{bmatrix}.$$

Then $\mathbf{c}(i\mathbf{k})\mathbf{c}^{-1} = i\mathbf{h}$. Thus $\mathbf{c}$ conjugates the compact torus in $\mathrm{SL}(2,\mathbb{R})$ generated by $i\mathbf{k}$ to the compact torus in G_0 generated by $i\mathbf{h}$:

$$\mathbf{c}\begin{bmatrix}\cos t & -\sin t\\ \sin t & \cos t\end{bmatrix}\mathbf{c}^{-1} = \begin{bmatrix}e^{it} & 0\\ 0 & e^{-it}\end{bmatrix}, \quad t\in\mathbb{R}\,.$$

The automorphism $g\mapsto \mathbf{c}g\mathbf{c}^{-1}$ is called the *Cayley transform.*

A similar construction works in $\mathrm{Sp}(n,\mathbb{C})$ (and for other real semisimple Lie groups). Set

$$\mathbf{c} = \frac{1}{\sqrt{2}}\begin{bmatrix}I_n & iI_n\\ iI_n & I_n\end{bmatrix}.$$

Then $\mathbf{c}\in\mathrm{Sp}(n)$ and $\mathbf{c}^{-1} = \bar{\mathbf{c}}$.

Lemma 11.2: *Let* $G_0 = \mathrm{U}(n,n)\cap\mathrm{Sp}(n,\mathbb{C})$. *Then* $\mathbf{c}^{-1}G_0\mathbf{c} = \mathrm{Sp}(n,\mathbb{R})$.

Proof: Since G_0 is closed under $g\mapsto g^*$ it has a polar decomposition $G_0 = K_0\exp(\mathfrak{p}_0)$. We already have shown that $K_0\cong\mathrm{U}(n)$, hence K_0 is connected. Thus G_0 is also connected. So it will suffice to show that

$$\mathrm{Ad}(\mathbf{c})^{-1}\mathfrak{g}_0 = \mathfrak{sp}(n,\mathbb{R})\,.$$

Let $X\in\mathfrak{g}_0$ be given by (52). Then one calculates that

$$\mathbf{c}^{-1}X\mathbf{c} = \frac{1}{2}\begin{bmatrix}(A+\bar{A})+i(B-\bar{B}) & i(A-\bar{A})+(B+\bar{B})\\ i(\bar{A}-A)+(B+\bar{B}) & (A+\bar{A})-i(B-\bar{B})\end{bmatrix}.$$

Set $u = \frac{1}{2}(A + \bar{A}) + \frac{i}{2}(B - \bar{B})$, $v = \frac{i}{2}(A - \bar{A}) + \frac{1}{2}(B + \bar{B})$, and $w = \frac{i}{2}(\bar{A} - A) + \frac{1}{2}(B + \bar{B})$. Since $\bar{A} = -A^t$ and $B = B^t$, we have

$$u = \bar{u}\,, \quad v = \bar{v} = v^t\,, \quad w = \bar{w} = w^t\,.$$

Also $\frac{1}{2}(A + \bar{A}) - \frac{i}{2}(B - \bar{B}) = -u^t$. Hence

$$\mathbf{c}^{-1}X\mathbf{c} = \begin{bmatrix} u & v \\ w & -u^t \end{bmatrix} \in \mathfrak{sp}(n, \mathbb{R})\,.$$

Since $\mathfrak{g}_0$ and $\mathfrak{sp}(n, \mathbb{R})$ are real forms of $\mathfrak{sp}(n, \mathbb{C})$, they have the same real dimension. Hence the map $X \mapsto \mathbf{c}^{-1}X\mathbf{c}$ is a real Lie algebra isomorphism from $\mathfrak{g}_0$ to $\mathfrak{sp}(n, \mathbb{R})$. □

Maximal Parabolic Subgroup. Let P be the subgroup of $\mathrm{Sp}(n, \mathbb{R})$ consisting of the matrices

$$\begin{bmatrix} A & B \\ 0 & (A^t)^{-1} \end{bmatrix}, \quad A \in \mathrm{GL}(n, \mathbb{R}),\ B \in SM_n(\mathbb{R})\,,$$

where $SM_n(\mathbb{R})$ denotes the real $n \times n$ symmetric matrices. The group P is a *maximal parabolic subgroup* of $\mathrm{Sp}(n, \mathbb{R})$. It has the structure of a semidirect product MN, where M consists of the block-diagonal matrices

$$\begin{bmatrix} A & 0 \\ 0 & (A^t)^{-1} \end{bmatrix}, \quad A \in \mathrm{GL}(n, \mathbb{R})$$

and N consists of the matrices

$$\begin{bmatrix} I_n & B \\ 0 & I_n \end{bmatrix}, \quad B \in SM_n(\mathbb{R})\,.$$

Thus as Lie groups $M \cong \mathrm{GL}(n, \mathbb{R})$ and $N \cong SM_n(\mathbb{R})$ (an abelian group). The group $K_0 \cap P$ consists of all matrices

$$\begin{bmatrix} A & 0 \\ 0 & A \end{bmatrix}, \quad A \in \mathrm{O}(n)$$

where $\mathrm{O}(n) = \{g \in \mathrm{GL}(n, \mathbb{R}) \ :\ g^t g = I_n\}$ is the usual real orthogonal group. Define $N^- = N^t$. Thus N^- is the group of matrices

$$\begin{bmatrix} I_n & 0 \\ C & I_n \end{bmatrix}, \quad C \in SM_n(\mathbb{R})\,.$$

Then $P^- = MN^-$ is the *opposite parabolic* subgroup to P and $P \cap P^- = M$. Note that

$$J_n \begin{bmatrix} I_n & B \\ 0 & I_n \end{bmatrix} J_n^{-1} = \begin{bmatrix} I_n & 0 \\ -B & I_n \end{bmatrix}$$

and

$$J_n \begin{bmatrix} A & 0 \\ 0 & (A^t)^{-1} \end{bmatrix} J_n^{-1} = \begin{bmatrix} (A^t)^{-1} & 0 \\ 0 & A \end{bmatrix}.$$

Thus $P^- = J_n P J_n^{-1}$ is conjugate to P in $\mathrm{Sp}(n, \mathbb{R})$.

Let

$$g = \begin{bmatrix} A & B \\ C & D \end{bmatrix}$$

be in $\mathrm{Sp}(n, \mathbb{R})$. If $\det A \neq 0$, then $D = (A^t)^{-1} + CB$ (this follows from $g^t J_n g = J_n$). Hence we can factor g as

$$g = \begin{bmatrix} I_n & 0 \\ C & I_n \end{bmatrix} \begin{bmatrix} A & B \\ 0 & (A^t)^{-1} \end{bmatrix} \in N^- M N.$$

Thus the subset $N^- M N$ is open and dense in $\mathrm{Sp}(n, \mathbb{R})$. This shows that a continuous representation π of $\mathrm{Sp}(n, \mathbb{R})$ is uniquely determined by its restriction to the subgroup P together with the single operator $\pi(J_n)$.

11.2. *Holomorphic (coherent-state) model for oscillator representation*

We write $\mathbb{H}^2(\mathbb{C}^n) = \mathbb{H}^2(\mathbb{C}^n, e^{-\|z\|^2} d\lambda(z))$ for the Hilbert-space completion of $\mathcal{P}(\mathbb{C}^n)$ relative to the Fischer inner product introduced in Section 9. The elements of this space are naturally identified with holomorphic functions f on $\mathbb{C}^n$ such that

$$\int_{\mathbb{C}^n} |f(z)|^2 e^{-\|z\|^2} d\lambda(z) < \infty.$$

For each $w \in \mathbb{C}^n$ the function $K_w(z) = e^{\langle z | w \rangle}$ is in $\mathbb{H}^2(\mathbb{C}^n)$ and plays the role of *reproducing kernel* for this space:

$$f(w) = \langle f \mid K_w \rangle \quad \text{for } f \in \mathbb{H}^2(\mathbb{C}^n)$$

(see [7, Proposition XI.1.1]).

Define the *annihilation operators* A_j and *creation operators* $A_j^\dagger$ by

$$A_j f(z) = \frac{\partial}{\partial z_j} f(z), \quad A_j^\dagger f(z) = z_j f(z) \quad \text{for } f \in \mathcal{P}(\mathbb{C}^n).$$

These operators satisfy the commutation relations $[A_i, A_j^\dagger] = \delta_{ij} I$. They are mutually adjoint relative to the Fischer inner product:

$$\langle A_j \varphi \mid \psi \rangle = \langle \varphi \mid A_j^\dagger \psi \rangle \tag{56}$$

for $\varphi, \psi \in \mathcal{P}(\mathbb{C}^n)$. The operators $\{A_1, \ldots, A_n, A_1^\dagger, \ldots, A_n^\dagger\}$ generate the Weyl algebra $\mathcal{PD}(\mathbb{C}^n)$.

Define a representation ϖ of $\mathfrak{sp}(n, \mathbb{C})$ on $\mathcal{P}(\mathbb{C}^n)$ as follows. For $b \in SM_n$ let $Q_b(z) = z^t b z$ be the quadratic form on $\mathbb{C}^n$ defined by b. Let

$$X = \begin{bmatrix} 0 & b \\ 0 & 0 \end{bmatrix} \in \mathfrak{p}_+ , \quad Y = \begin{bmatrix} 0 & 0 \\ c & 0 \end{bmatrix} \in \mathfrak{p}_- ,$$

where $b, c \in SM_n$. Define operators $\varpi(X)$ and $\varpi(Y)$ on $\mathcal{P}(\mathbb{C}^n)$ by

$$\varpi(X) f(z) = \frac{1}{2i} \partial(Q_b) f(z) , \quad \varpi(Y) f(z) = \frac{1}{2i} Q_c(z) f(z)$$

where $i = \sqrt{-1}$ (recall the map $Q \mapsto \partial(Q)$ from $\mathcal{P}(\mathbb{C}^n)$ to constant-coefficient differential operators on $\mathcal{P}(\mathbb{C}^n)$ that was introduced in the proof of Theorem 9.2). We calculate that

$$[\varpi(X), \varpi(Y)] = -\sum_{j,k} (bc)_{kj}\, z_j \frac{\partial}{\partial z_k} - \frac{1}{2} \mathrm{tr}(bc) .$$

Let $H = \begin{bmatrix} h & 0 \\ 0 & -h^t \end{bmatrix} \in \mathfrak{k}$, where $h \in M_n$. Define the operator $\varpi(H)$ by

$$\varpi(H) = -\sum_{j,k} h_{kj} z_j \frac{\partial}{\partial z_k} - \frac{1}{2} \mathrm{tr}(h) .$$

Since $[X, Y] = \begin{bmatrix} bc & 0 \\ 0 & -cb \end{bmatrix}$, we see that $[\varpi(X), \varpi(Y)] = \varpi([X, Y])$. One calculates that

$$[\varpi(H), \varpi(X)] = \varpi([H, X]) , \quad [\varpi(H), \varpi(Y)] = \varpi([H, Y])$$

(note that $H \mapsto \varpi(H)$ is the standard representation of $\mathfrak{k} \cong \mathfrak{gl}(n, \mathbb{C})$ on $\mathcal{P}(\mathbb{C}^n)$ tensored with the one-dimensional representation $H \mapsto -\frac{1}{2}\mathrm{tr}(h)$). Thus ϖ is a representation of $\mathfrak{sp}(n, \mathbb{C})$ on $\mathcal{P}(\mathbb{C}^n)$ (this is the representation of $\mathfrak{g}' = \mathfrak{sp}(n, \mathbb{C})$ in Theorem 8.5 for the case $G = \mathrm{O}(1, \mathbb{C})$).

We can write the representation ϖ in terms of the annihilation and creation operators as

$$\varpi(X) = \frac{1}{2i}\sum_{j,k} b_{jk}A_jA_k\,, \quad \varpi(Y) = \frac{1}{2i}\sum_{j,k} c_{jk}A_j^\dagger A_k^\dagger$$

$$\varpi(H) = -\frac{1}{2}\sum_{j,k} h_{jk}(A_jA_k^\dagger + A_k^\dagger A_j)$$

for X, Y, H as above. Now take $Y = \sigma(X)$, where σ is the conjugation defining the real form $\mathfrak{g}_0$ of $\mathfrak{sp}(n,\mathbb{C})$. Then $c_{jk} = \bar{b}_{jk}$, so from (56) we see that

$$\langle \varpi(X)\varphi \mid \psi\rangle = -\langle \varphi \mid \varpi(\sigma(X))\psi\rangle \tag{57}$$

for all $X \in \mathfrak{p}_+$ and $\varphi, \psi \in \mathcal{P}(\mathbb{C}^n)$. Since $\sigma(\mathfrak{p}_\pm) = \mathfrak{p}_\mp$ and $\mathfrak{p}_+, \mathfrak{p}_-$ generate $\mathfrak{sp}(n,\mathbb{C})$, it follows that relation (57) holds for all $X \in \mathfrak{sp}(n,\mathbb{C})$. Since $\mathfrak{g}_0$ is the fixed-point set of σ, we conclude that

$$\langle \varpi(Z)\varphi \mid \psi\rangle = -\langle \varphi \mid \varpi(Z)\psi\rangle \quad \text{for } Z \in \mathfrak{g}_0\,.$$

One says that ϖ is a *unitarizable* representation of $\mathfrak{g}_0$.

By the Cartan decomposition $G_0 = K_0\exp(\mathfrak{p}_0)$ the group K_0 is a topological retract of G_0. Since $\mathfrak{k}_0$ has a one-dimensional center, it follows that G_0 has an m-sheeted covering group for every integer m. Let $\gamma : \mathrm{Mp}(n,\mathbb{R}) \to G_0$ be the two-sheeted covering and let $\widetilde{K}_0 = \gamma^{-1}(K_0)$. Then $\widetilde{K}_0$ is a two-sheeted covering of $\mathrm{U}(n)$. If $\widetilde{k} \in \widetilde{K}_0$ and $z \in \mathbb{C}^n$ we set $\widetilde{k}\cdot z = uz$, where

$$\gamma(\widetilde{k}) = \begin{bmatrix} u & 0 \\ 0 & \bar{u}\end{bmatrix} \quad \text{with } u \in \mathrm{U}(n)\,.$$

The function $\chi : \widetilde{k} \mapsto \det(u)^{-1/2}$ is a (single-valued) character of $\widetilde{K}_0$.

Theorem 11.3: *There is a unitary representation ϖ of $\mathrm{Mp}(n,\mathbb{R})$ on $\mathbb{H}^2(\mathbb{C}^n)$ whose differential is the representation ϖ of $\mathfrak{g}_0$. If $\widetilde{k} \in \widetilde{K}_0$ then*

$$\varpi(\widetilde{k})f(z) = \chi(\widetilde{k})f(\widetilde{k}^{-1}\cdot z) \quad \textit{for } f \in \mathbb{H}^2(\mathbb{C}^n)\,.$$

Proof: Let

$$H_0 = \begin{bmatrix} iI_n & 0 \\ 0 & -iI_n\end{bmatrix} \in \mathfrak{k}_0\,.$$

Take the basis for $\varpi(\mathfrak{sp}(n,\mathbb{C}))$ to be the operators A_iA_j and $A_i^\dagger A_j^\dagger$ for $1 \le i \le j \le n$ and $\frac{1}{2}(A_i^\dagger A_j + A_jA_i^\dagger)$ for $1 \le i, j \le n$. Define the seminorms ρ_k on $\mathcal{P}(\mathbb{C}^n)$ in terms of this basis and the norm $||\varphi||$ as in Section 11.5.

Lemma 11.4: *The inequality*

$$\rho_1(\varphi) \leq ||\varphi|| + ||\varpi(H_0)\varphi|| \tag{58}$$

holds for all $\varphi \in \mathcal{P}(\mathbb{C}^n)$.

Proof of Lemma 11.4: Let $\varphi = \sum_\alpha c_\alpha z^\alpha$. Since $\varpi(iH_0)$ acts on z^α by $|\alpha| + \frac{n}{2}$, we have

$$||\varpi(H_0)\varphi||^2 = \sum_\alpha \left(|\alpha| + \frac{n}{2}\right)^2 \alpha!\,|c_\alpha|^2 .$$

We calculate that

$$||A_iA_j\varphi||^2 = \begin{cases} \sum_\alpha (\alpha_i\alpha_j)\,\alpha!\,|c_\alpha|^2 & \text{if } i \neq j, \\ \sum_\alpha \alpha_i(\alpha_i - 1)\,\alpha!\,|c_\alpha|^2 & \text{if } i = j. \end{cases}$$

Likewise,

$$||A_i^\dagger A_j^\dagger\varphi||^2 = \begin{cases} \sum_\alpha (\alpha_i + 1)(\alpha_j + 1)\,\alpha!\,|c_\alpha|^2 & \text{if } i \neq j, \\ \sum_\alpha (\alpha_i + 1)(\alpha_i + 2)\,\alpha!\,|c_\alpha|^2 & \text{if } i = j. \end{cases}$$

Finally,

$$||\frac{1}{2}(A_i^\dagger A_j + A_jA_i^\dagger)\varphi||^2 = \begin{cases} \sum_\alpha (\alpha_i + 1)\alpha_j\,\alpha!\,|c_\alpha|^2 & \text{if } i \neq j, \\ \sum_\alpha \left(\alpha_i + \frac{1}{2}\right)^2\,\alpha!\,|c_\alpha|^2 & \text{if } i = j. \end{cases}$$

The estimate (58) now follows from these formulas and the inequality $2ab \leq (a + b)^2$. □

Proof of Theorem 11.3: From Lemma 11.4 and Corollary 11.9 we obtain a unitary representation ϖ of the universal covering group of $\mathrm{Sp}(n, \mathbb{R})$ whose differential is the representation ϖ. Now H_0 spans the center of $\mathfrak{k}_0$ and $\exp(2\pi H_0) = I$ in $\mathrm{Sp}(n, \mathbb{C})$. Since the one-parameter unitary group $\varpi(\exp(tH_0))$ acts by $e^{it(k+n/2)}$ on $\mathcal{P}^k(\mathbb{C}^n)$, we have $\varpi(\exp(4\pi H_0)) = I$. Hence the representation ϖ descends to a single-valued representation of $\mathrm{Mp}(n, \mathbb{R})$. □

We will call ϖ the *oscillator representation* of $\mathrm{Sp}(n, \mathbb{R})$. We write $\varpi^{(n)}$ if the dependence on n is not evident from the context. An explicit formula for $\varpi(g)$ as an integral operator was obtained by Bargmann (see [8, Theorem 4.37]).

11.3. *Bargmann–Segal transform*

To obtain a realization of the oscillator representation in which the action of the maximal parabolic subgroup P of $\mathrm{Sp}(n,\mathbb{R})$ is easily described, we use another representation of the creation and annihilation operators. Let $\mathcal{S}(\mathbb{R}^n)$ denote the Schwartz space of rapidly-decreasing smooth functions on $\mathbb{R}^n$. Define *creation* and *annihilation* operators

$$a_j^\dagger = \frac{1}{\sqrt{2}}\left(-\frac{\partial}{\partial x_j} + x_j\right), \qquad a_j = \frac{1}{\sqrt{2}}\left(\frac{\partial}{\partial x_j} + x_j\right)$$

on $\mathcal{S}(\mathbb{R}^n)$ for $1 \leq j \leq n$. These operators satisfy the commutation relations $[a_i,\, a_j^\dagger] = \delta_{ij} I$ and they are mutually adjoint relative to the $L^2(\mathbb{R}^n,\, d\lambda(x))$ inner product. They leave invariant the space $\mathcal{P}(\mathbb{R}^n)$ of (complex-valued) polynomial functions on $\mathbb{R}^n$ and act irreducibly on this space.

Following [2] and [28], we construct a unitary operator

$$B : L^2(\mathbb{R}^n,\, d\lambda(x)) \to \mathbb{H}^2(\mathbb{C}^n,\, e^{-\|z\|^2} d\lambda(z))$$

that intertwines a_j with A_j and $a_j^\dagger$ with $A_j^\dagger$. Since the space $L^2(\mathbb{R}^n)$ (*respectively* $\mathbb{H}^2(\mathbb{C}^n)$) is the n-fold Hilbert-space tensor product of the space $L^2(\mathbb{R})$ (*respectively* $\mathbb{H}^2(\mathbb{C})$), it suffices to do the calculation for the case $n = 1$.

Because $\mathbb{H}^2$ has a reproducing kernel, any such operator B will be given as

$$Bf(z) = \int_{-\infty}^{\infty} B(z,x) f(x)\, dx\,.$$

To intertwine the two pairs of creation-annihilation operators, the kernel $B(z,x)$ must be a holomorphic function of z and smooth function of x that satisfies

$$\frac{\partial}{\partial z} B(z,x) = \frac{1}{\sqrt{2}}\left(-\frac{\partial}{\partial x} + x\right) B(z,x)$$

$$zB(z,x) = \frac{1}{\sqrt{2}}\left(\frac{\partial}{\partial x} + x\right) B(z,x)\,.$$

These equations imply that

$$\frac{\partial B}{\partial z} = (\sqrt{2}x - z)B \quad \text{and} \quad \frac{\partial B}{\partial x} = (\sqrt{2}z - x)B\,.$$

The solution is easily found to be

$$B(x,z) = C \exp\left\{\sqrt{2}xz - \frac{1}{2}(x^2 + z^2)\right\}$$

with C a constant. It remains to verify that the operator defined by this kernel is unitary (with appropriate choice of C). For this, take $p \in \mathcal{P}(\mathbb{R})$ and the normalized Gaussian (ground state)

$$\varphi_0(x) = \frac{1}{\sqrt{\pi}} \exp\left(-\frac{1}{2}x^2\right)$$

(note that $a_1\varphi_0 = 0$). Then

$$B(p\varphi_0)(z) = \frac{C}{\sqrt{\pi}} \int_{-\infty}^{\infty} p(x) \exp\left\{-x^2 + \sqrt{2}xz - \frac{1}{2}z^2\right\} dx\,.$$

Completing the square in the exponential and using the translation-invariance of the measure dx, we obtain

$$B(p\varphi_0)(z) = \frac{C}{\sqrt{\pi}} \int_{-\infty}^{\infty} p\left(x + \frac{z}{\sqrt{2}}\right) e^{-x^2}\, dx\,.$$

The right side of this equation is obviously a polynomial of the same degree as p, so it follows that B is a bijection from $\mathcal{P}(\mathbb{R})\varphi_0$ onto $\mathcal{P}(\mathbb{C})$. Furthermore $B\varphi_0(z) = C$. Since φ_0 has L^2-norm 1 and the constant function 1 has $\mathbb{H}^2$-norm 1, we conclude that $C = 1$.

To complete the proof that B is a unitary operator, we observe that $\mathcal{P}(\mathbb{R})\varphi_0$ is the cyclic space generated by φ_0 under the action of the operators a_1 and $a_1^\dagger$. Likewise $\mathcal{P}(\mathbb{C})$ is the cyclic space generated by the constant function 1 under the action of the operators A_1 and $A_1^\dagger$. The creation-annihilation operators act irreducibly on these spaces. Since B^* intertwines the pair $\{a_1, a_1^\dagger\}$ with $\{A_1, A_1^\dagger\}$, it follows that B^*B commutes with a_1 and $a_1^\dagger$ on $\mathcal{P}(\mathbb{R})\varphi_0$, while BB^* commutes with A_1 and $A_1^\dagger$ on $\mathcal{P}(\mathbb{C})$. Thus B^*B and BB^* are both multiples of the identity by Lemma 2.1. Since we have normalized the kernel $B(x, z)$ so that B carries the unit vector φ_0 to the unit vector 1, it follows that B is unitary in the case $n = 1$. This implies that B is unitary for any n as remarked above.

For any integer $n \geq 1$ and polynomial $p \in \mathcal{P}(\mathbb{R}^n)$, define

$$B(\varphi_0 p)(z) = \pi^{-n/2} \int_{\mathbb{R}^n} p\left(x + \frac{z}{\sqrt{2}}\right) e^{-x^t x}\, d\lambda(x) \tag{59}$$

for $z \in \mathbb{C}^n$, where $\varphi_0(x) = \pi^{-n/2} \exp\left(-\frac{1}{2}x^t x\right)$ for $x \in \mathbb{R}^n$. Notice that if $g \in \mathrm{O}(n)$ then $\varphi_0(gx) = \varphi_0(x)$. Thus if we set $f_g(x) = f(g^{-1}x)$, then $B(f_g) = B(f)_g$ for $f = \varphi_0 p$ and $g \in \mathrm{O}(n)$. We have proved the following.

Theorem 11.5 (Bargmann–Segal Transform): *The operator B maps the space $\mathcal{P}(\mathbb{R}^n)\varphi_0$ onto $\mathcal{P}(\mathbb{C}^n)$ bijectively. It extends to a unitary operator from $L^2(\mathbb{R}^n; dx)$ onto $\mathbb{H}^2(\mathbb{C}^n)$. If $f \in \mathcal{S}(\mathbb{R}^n)$ then*

$$Bf(z) = \int_{\mathbb{R}^n} f(x) \exp\left\{\sqrt{2}x^t z - \frac{1}{2}(x^t x + z^t z)\right\} d\lambda(x) .$$

Furthermore, B intertwines the representations $f \mapsto f_g$ of $\mathrm{O}(n)$ on $L^2(\mathbb{R}^n; dx)$ and on $\mathbb{H}^2(\mathbb{C}^n)$. Also $Ba_j B^{-1} = A_j$ and $Ba_j^\dagger B^{-1} = A_j^\dagger$ on $\mathcal{P}(\mathbb{C}^n)$.

Remark: Since B is unitary, $B^{-1} = B^*$ is an integral operator on $\mathbb{H}^2(\mathbb{C}^n)$ with kernel $\overline{B(z,x)} = B(\bar{z}, x)$.

11.4. *Real (oscillatory-wave) model for oscillator representation*

We now use the Bargmann–Segal transform to obtain the real-wave (Schrödinger) model of the oscillator representation.

Let $\gamma : \mathrm{Mp}(n,\mathbb{R}) \to \mathrm{Sp}(n,\mathbb{R})$ be the covering homomorphism and let $\mathbf{c} \in \mathrm{Sp}(n)$ be the Cayley transform. We define a unitary representation π of $\mathrm{Mp}(n,\mathbb{R})$ on $L^2(\mathbb{R}^n)$ by

$$\pi(g) = B^{-1}\varpi(\mathbf{c}g\mathbf{c}^{-1})B \tag{60}$$

(here $\mathbf{c}g\mathbf{c}^{-1} \in \mathrm{Mp}(n,\mathbb{R})$ is the element such that $\gamma(\mathbf{c}g\mathbf{c}^{-1}) = \mathbf{c}\gamma(g)\mathbf{c}^{-1}$). For $f \in \mathcal{S}(\mathbb{R}^n)$ let $\mathcal{F}f$ be the Fourier transform

$$\mathcal{F}(f)(x) = \left(\frac{1}{\pi}\right)^{n/2} \int_{\mathbb{R}^n} e^{ix^t y} f(y)\, dy .$$

Theorem 11.6: *The action of $\pi(g)$ on $f \in \mathcal{S}(\mathbb{R}^n)$ is as follows:*

(1) *If* $\gamma(g) = \begin{bmatrix} A & 0 \\ 0 & (A^t)^{-1} \end{bmatrix}$ *with* $A \in \mathrm{GL}(n,\mathbb{R})$ *then* $\pi(g)f(x) = (\det A)^{-1/2} f(A^{-1}x)$.

(2) *If* $\gamma(g) = \begin{bmatrix} 1 & 0 \\ b & 1 \end{bmatrix}$ *with* $b \in SM_n(\mathbb{R})$ *then* $\pi(g)f(x) = e^{-(i/2)x^t bx} f(x)$.

(3) *If* $\gamma(g) = \begin{bmatrix} 1 & b \\ 0 & 1 \end{bmatrix}$ *with* $b \in SM_n(\mathbb{R})$ *then* $\mathcal{F}(\pi(g)f)(x) = e^{(i/2)x^t bx} \mathcal{F}f(x)$.

These formulas uniquely determine π.

Proof: For $X \in \mathfrak{sp}(n, \mathbb{R})$ write $d\pi(X) = B\varpi(\mathrm{Ad}(\mathbf{c})X)B^{-1}$ for the Lie algebra representation corresponding to π. Here the operator $d\pi(X)$ acts on $\mathcal{S}(\mathbb{R}^n)$.

(1) Let $X = \begin{bmatrix} h & 0 \\ 0 & -h \end{bmatrix}$ with $h = \mathrm{diag}[h_1, \ldots, h_n]$ and $h_j \in \mathbb{R}$. Then $\mathrm{Ad}(\mathbf{c})X = \begin{bmatrix} 0 & -ih \\ ih & 0 \end{bmatrix}$. Hence the formulas of Section 11.2 give

$$\begin{aligned} \varpi(\mathrm{Ad}(\mathbf{c})X) &= -\frac{1}{2}\sum_{j=1}^{n} h_j \left\{(A_j)^2 - (A_j^\dagger)^2\right\} \\ &= -\frac{1}{2}\sum_{j=1}^{n} h_j (A_j + A_j^\dagger)(A_j - A_j^\dagger) - \frac{1}{2}\mathrm{tr}(h)\,. \end{aligned}$$

Now $B^{-1}A_jB = a_j$ and $B^{-1}A_j^\dagger B = a_j^\dagger$. Also we have

$$(a_j + a_j^\dagger)f(x) = \sqrt{2}x_j f(x)\,, \qquad (a_j - a_j^\dagger)f(x) = \sqrt{2}\frac{\partial f}{\partial x_j}(x) \tag{61}$$

for $f \in \mathcal{S}(\mathbb{R}^n)$. Thus we obtain

$$d\pi(X) = -\sum_{j=1}^{n} h_j x_j \frac{\partial}{\partial x_j} - \frac{1}{2}\mathrm{tr}(h)\,.$$

This shows that (1) is true for $A = \exp h$ (in this case $\det A > 0$ and there is no need to pass to the metaplectic group). If $A \in \mathrm{O}(n)$ then (1) holds for g, since B intertwines the action of $\mathrm{O}(n)$ on $L^2(\mathbb{R}^n)$ and $\mathbb{H}^2(\mathbb{C}^n)$ (note that $\gamma(g)$ is in the maximal compact subgroup K_0 of G_0, so $\varpi(g)$ is described in Theorem 11.3). By the polar decomposition,

$$\mathrm{GL}(n, \mathbb{R}) = \mathrm{O}(n)(\exp \mathfrak{a})\mathrm{O}(n)\,,$$

where $\mathfrak{a}$ is the subspace of real diagonal matrices. Hence (1) holds for all $A \in \mathrm{GL}(n, \mathbb{R})$.

(2) Let $X = \begin{bmatrix} 0 & 0 \\ b & 0 \end{bmatrix}$ with $b \in SM_n(\mathbb{R})$. Then $\mathrm{Ad}(\mathbf{c})X = \frac{1}{2}\begin{bmatrix} ib & b \\ b & -ib \end{bmatrix}$. Hence the formulas of Section 11.2 give

$$\begin{aligned} \varpi(\mathrm{Ad}(\mathbf{c})X) &= \frac{1}{4i}\sum_{j,k=1}^{n} b_{jk}\left\{A_jA_k + A_j^\dagger A_k^\dagger + A_jA_k^\dagger + A_k^\dagger A_j\right\} \\ &= \frac{1}{4i}\sum_{j,k=1}^{n} b_{jk}(A_j + A_j^\dagger)(A_k + A_k^\dagger)\,. \end{aligned}$$

Now applying the Bargmann–Segal transform and using (61), we obtain

$$d\pi(X)f(x) = \frac{1}{2i}\Big\{\sum_{j,k=1}^{n} b_{jk}x_j x_k\Big\}f(x)\,.$$

This proves (2).

(3) Let $X = \begin{bmatrix} 0 & b \\ 0 & 0 \end{bmatrix}$ with $b \in SM_n(\mathbb{R})$. Then $\mathrm{Ad}(\mathbf{c})X = \frac{1}{2}\begin{bmatrix} -ib & b \\ b & ib \end{bmatrix}$. Thus

$$\begin{aligned} \varpi(\mathrm{Ad}(\mathbf{c})X) &= \frac{1}{4i}\sum_{j,k=1}^{n} b_{jk}\left\{A_jA_k + A_j^{\dagger}A_k^{\dagger} - A_jA_k^{\dagger} - A_k^{\dagger}A_j\right\} \\ &= \frac{1}{4i}\sum_{j,k=1}^{n} b_{jk}(A_j - A_j^{\dagger})(A_k - A_k^{\dagger})\,. \end{aligned}$$

Applying the Bargmann–Segal transform and using (61) again, we obtain

$$d\pi(X)f(x) = \frac{1}{2i}\left\{\sum_{j,k=1}^{n} b_{jk}\,\frac{\partial}{\partial x_j}\frac{\partial}{\partial x_k}\right\}f(x)$$

for $f \in \mathcal{S}(\mathbb{R}^n)$. Since $\mathcal{F}(\partial^2/\partial x_j\partial x_k)\mathcal{F}^{-1}$ is the operator of multiplication by $-x_jx_k$, this proves (3).

Formulas (1), (2), and (3) uniquely determine π since N^-MN is dense in $\mathrm{Sp}(n,\mathbb{R})$. □

11.5. *Analytic vectors*

Here we present refinements of some results of Nelson [27] concerning exponentiation of Lie algebra representations, following the approach in [14]. Suppose $\mathfrak{g}_0$ is a real finite-dimensional Lie algebra, represented as skew-Hermitian (unbounded) operators on a complex inner product space $\mathcal{V}$ (not assumed complete). Let $\mathfrak{g}$ be the complexification of $\mathfrak{g}_0$. Then $X \mapsto X^*$ (the Hermitian adjoint of X relative to the inner product on $\mathcal{V}$) is a conjugate-linear anti-automorphism of $\mathfrak{g}$ such that $X^* = -X$ for $X \in \mathfrak{g}_0$.

Fix a basis $\{X_1, \ldots, X_d\}$ for $\mathfrak{g}$. Define seminorms ρ_n on $\mathcal{V}$ by setting $\rho_0(v) = ||v||$ and

$$\rho_n(v) = \max_{i_1,\ldots,i_n} ||X_{i_1}\cdots X_{i_n}v||$$

for $n = 1, 2, \ldots$. Here $i_1, \ldots, i_n$ run over $1, 2, \ldots, d$ and $||u||$ denotes the norm of $u \in \mathcal{V}$. Let V be the Hilbert-space completion of $\mathcal{V}$ relative to the norm $||v||$ and let V^∞ be the completion of $\mathcal{V}$ relative to the family of seminorms $\{\rho_n\}$. Then the representation of $\mathfrak{g}$ on $\mathcal{V}$ extends continuously to a representation on V^∞. The seminorm $\rho_n(v)$ is also defined for $v \in V^\infty$.

One says that $v \in V^\infty$ is an *analytic vector* for $\mathfrak{g}$ if there is an $r > 0$ such that

$$\sum_{n=0}^{\infty} \frac{r^n}{n!} \rho_n(v) < \infty \,. \tag{62}$$

Let $V_r^\omega \subset V^\infty$ be the subspace for which (62) holds. The space $V^\omega = \bigcup_{r>0} V_r^\omega$ of analytic vectors for $\mathfrak{g}$ is invariant under $\mathfrak{g}$.

Theorem 11.7 (Nelson): *Suppose there exists an $r > 0$ so that V_r^ω is dense in V. Then the representation of $\mathfrak{g}_0$ integrates to a strongly continuous unitary representation on V of the simply-connected Lie group $\widetilde{G}_0$ with Lie algebra $\mathfrak{g}_0$.*

Proof: (Sketch) Define a norm on $\mathfrak{g}$ by $||\sum_{i=1}^d c_i X_i|| = \sum_{i=1}^d |c_i|$. If $X \in \mathfrak{g}$ then

$$||X^n v|| \le ||X||^n \rho_n(v) \,.$$

Hence the operator e^X is defined from V_r^ω to V by the exponential series provided $||X|| < r$. The map $\exp X \mapsto e^X$ defines a local representation of the complex Lie group germ corresponding to $\mathfrak{g}$ (the rearrangement of the exponential series needed for the Campbell–Hausdorff formula is justified by convergence of (62); see [3] or [18]). The operator e^X is unitary for $X \in \mathfrak{g}_0$, since X is skew-Hermitian, and the local representation extends to a strongly continuous unitary representation of $\widetilde{G}_0$ on W. □

Suppose there is an element $H_0 \in \mathfrak{g}$ such that

$$\rho_1(v) \le ||v|| + ||H_0 v|| \quad \text{for all } v \in \mathcal{V} \,. \tag{63}$$

Let $A = \max_{1 \le i \le d} ||[H_0, X_i]||$ (the norm of $\mathrm{ad} H_0$ on $\mathfrak{g}$). Note that $A = 0$ if and only if H_0 is in the center of $\mathfrak{g}$, and this case is of no interest here. So we assume $A > 0$.

Theorem 11.8: *Every analytic vector for H_0 is an analytic vector for $\mathfrak{g}$. More precisely, if*

$$\sum_{n=0}^{\infty} \frac{s^n}{n!} ||H_0^n v|| < \infty \,, \quad \textit{for some } s > 0 \,, \tag{64}$$

then $v \in V_r^\omega$ for all $r < \min\{A^{-1}, A^{-1}(1 - e^{-As})\}$.

Remark: If s can be arbitrarily large in (64) (one says that v is an *entire vector* for H_0 in this case), then $v \in V_r^\omega$ for all $r < A^{-1}$. Note that this upper bound for r is controlled by the non-commutativity of $\mathfrak{g}$ and it is finite if $A \neq 0$. In general v is *not* an entire vector for $\mathfrak{g}$ (see [15] for more precise results along this line).

Proof: Let Y_j be any of the basis elements X_i. Then the *a priori* estimate (63) implies that

$$||Y_{m+1}Y_m \cdots Y_1 v|| \leq ||Y_m \cdots Y_1 v|| + ||H_0 Y_m \cdots Y_1 v||$$

for all $v \in V^\infty$. Now

$$H_0 Y_m \cdots Y_1 = Y_m \cdots Y_1 H_0 + \sum_{k=1}^{m} Y_m \cdots Y_{k-1}[H_0, Y_k]Y_{k+1} \cdots Y_1 \,,$$

and by definition of ρ_m and the constant A we have $||Y_m \cdots [H_0, Y_k] \cdots Y_1 v|| \leq A\rho_m(v)$. Hence

$$\begin{aligned} ||Y_{m+1} \cdots Y_1 v|| &\leq ||Y_m \cdots Y_1 H_0 v|| + (1 + mA)\rho_m(v) \\ &\leq \rho_m(H_0 v) + (1 + mA)\rho_m(v) \,. \end{aligned}$$

Since this holds for any choice of $Y_1, \ldots, Y_{m+1}$, it implies

$$\rho_{m+1}(v) \leq \rho_m(H_0 v) + (1 + mA)\rho_m(v) \quad \text{for all } v \in \mathcal{V} \,. \tag{65}$$

Now fix $v \in V^\infty$ and set $a_{m,n} = \rho_m(H_0^n v)$. Replacing v by $H_0^n v$ in (65), we see that the sequence $\{a_{m,n}\}$ satisfies the recursive inequalities

$$a_{m+1,n} \leq a_{m,n+1} + (1 + mA)a_{m,n} \,.$$

To estimate the rate of growth of $a_{m,n}$, we introduce the *majorant sequence* $b_{m,n}$ defined by $b_{0,n} = a_{0,n}$ for all n and

$$b_{m+1,n} = b_{m,n+1} + (1 + mA)b_{m,n} \quad \text{for all } m \geq 0, n \geq 0 \,.$$

Clearly $a_{m,n} \leq b_{m,n}$ for all m, n.

Consider the generating function

$$\varphi(x, y) = \sum_{m=0}^{\infty} \sum_{n=0}^{\infty} \frac{x^m}{m!} \frac{y^n}{n!} b_{m,n} \,.$$

The recursion for $b_{m,n}$ implies that (as a formal series)

$$(1 - Ax)\frac{\partial}{\partial x}\varphi(x, y) = \frac{\partial}{\partial y}\varphi(x, y) + \varphi(x, y) \,.$$

We assume that the series $f(y) = \varphi(0, y)$ converges for $y = s$. The Cauchy problem for this analytic first-order p.d.e. is easily solved by the method of characteristics, and one obtains

$$\varphi(x, y) = (1 - Ax)^{-1/A} f(y - A^{-1}\log(1 - Ax)) .$$

(the analytic solution must agree with the formal solution since the line $x = 0$ is non-characteristic). Setting $y = 0$, we see that the series for $\varphi(x, 0)$ converges absolutely for $|x| \leq r$ provided

$$r < \min\{A^{-1}, A^{-1}(1 - e^{-As})\} .$$

Since $a_{m,0} \leq b_{m,0}$ this proves the theorem. □

Corollary 11.9: *Suppose $H_0 \in \mathfrak{g}$ and $\mathcal{V}$ has an (algebraic) basis consisting of eigenvectors for H_0. Then $\mathcal{V} \subset V_r^\omega$ for all $r < A^{-1}$ and hence V_r^ω is dense in V for all $r < A^{-1}$. Thus the representation of $\mathfrak{g}_0$ integrates to a strongly continuous unitary representation of $\widetilde{G}_0$ on V.*

Proof: If $H_0 v = \lambda v$, then the left side of (64) is $e^{s|\lambda|}$, and hence is finite for all $s > 0$. By the remark after Theorem 11.8 this implies that (62) holds for all $r < A^{-1}$. Now apply Theorem 11.7. □

12. Dual pair Sp(*n*, R)–O(*k*)

The oscillator representation has many applications to analysis and physics (see [8] and [20], for example). Here we apply it in the context of unitary representation theory and highest weight representations (see [6] for more on this point). To determine which of the representations that occur in the decomposition of the oscillator representation are square-integrable, we apply Harish–Chandra's criterion to the explicit formula for the θ-correspondence that we calculated in Theorems 10.1 and 10.4. In particular, we show that all the square-integrable highest-weight representations of $\mathrm{Sp}(n, \mathbb{R})$ occur in the duality correspondence with $\mathrm{O}(2n)$ (this was first proved by Gelbart [11]).

12.1. *Decomposition of* $\mathrm{H}^2(M_{n\times k})$ *under* $\mathrm{Mp}(n, \mathbb{R}) \times \mathrm{O}(k)$

Let $G = \mathrm{O}(k, \mathbb{C}) = \{g \in \mathrm{GL}(k, \mathbb{C}) : gg^t = I\}$ and let $G' = \mathrm{Sp}(n, \mathbb{C}) \subset \mathrm{GL}(2n, \mathbb{C})$ be the symplectic group relative to the skew-form with matrix J_n as in Section 11.1. Define a skew form Ω on $M_{2n\times k}$ by

$$\Omega(w, z) = \mathrm{tr}(w^t J_n z) \quad \text{for } w, z \in M_{2n\times k} .$$

Then Ω is nondegenerate. We embed $G' \times G$ into $\mathrm{Sp}(M_{2n\times k}, \Omega)$ as follows. Let $g \in G$ and $h \in G'$. Then

$$\Omega(hwg, hzg) = \mathrm{tr}(w^t(h^t J_n h)zgg^t) = \Omega(w, z)$$

since $gg^t = I$ and $h^t J_n h = J_n$. Hence we have an injective regular homomorphism $L \times R : G' \times G \to \mathrm{Sp}(M_{2n\times k}, \Omega)$ given by

$$R(g)z = zg^{-1}, \quad L(h)z = hz \quad \text{for } g \in G,\ h \in G',\ z \in M_{2n\times k}\,.$$

We identify $M_{2n\times k}$ with $\mathbb{C}^{2nk}$ by the map

$$z = [\mathbf{z}_1, \ldots, \mathbf{z}_k] \mapsto \tilde{z} = \begin{bmatrix} \mathbf{z}_1 \\ \vdots \\ \mathbf{z}_k \end{bmatrix} \in \mathbb{C}^{2nk}$$

where $\mathbf{z}_j \in \mathbb{C}^{2n}$ is the jth column of z. It is easy to check that

$$\Omega(z, w) = \tilde{z}^t J_{nk} \tilde{w}\,,$$

so $\mathrm{Sp}(M_{2n\times k}, \Omega)$ becomes $\mathrm{Sp}(nk, \mathbb{C})$ under this identification. Thus we will view z either as a $2n \times k$ matrix or a vector in $\mathbb{C}^{2nk}$, whichever is more convenient for the calculation at hand.

Define a hermitian form on $M_{2n\times k}$ by

$$(z, w) = \mathrm{tr}(w^* I_{n,n} z)\,,$$

where $I_{n,n}$ is the matrix in Section 11.1. We have $(z, w) = \tilde{w}^* I_{nk,nk} \tilde{z}$, so when $z \in M_{2n\times k}$ is identified with $\tilde{z} \in \mathbb{C}^{2nk}$, the form (z, w) becomes the one used in Section 11 to define the group $U(nk, nk)$. Thus we will denote the isometry group of this form as $\mathrm{U}(nk, nk)$. If $g \in \mathrm{U}(n, n)$ then

$$(gz, gw) = \mathrm{tr}(w^* g^* I_{n,n} g z) = (z, w)$$

since $g^* I_{n,n} g = I_{n,n}$. Thus the left multiplication homomorphism $L : \mathrm{GL}(2n, \mathbb{C}) \to \mathrm{GL}(M_{2n\times k})$ carries $\mathrm{U}(n, n)$ into $\mathrm{U}(nk, nk)$. If $h \in \mathrm{U}(k)$ then

$$(zh, wh) = \mathrm{tr}(w^* I_{n,n} z h h^*) = (z, w)\,.$$

Furthermore,

$$\begin{bmatrix} x \\ y \end{bmatrix} h = \begin{bmatrix} xh \\ yh \end{bmatrix}.$$

Hence the right multiplication homomorphism $R : \mathrm{GL}(k, \mathbb{C}) \to \mathrm{GL}(M_{2n\times k})$ carries $\mathrm{U}(k)$ into the maximal compact subgroup $\mathrm{U}(nk) \times \mathrm{U}(nk)$ of $\mathrm{U}(nk, nk)$.

Let $G_0 = G \cap \mathrm{U}(k) = \mathrm{O}(k)$ be the compact real form of G, and let $G_0' = G' \cap U(n,n) \cong \mathrm{Sp}(n,\mathbb{R})$ be the real form of $\mathrm{Sp}(n,\mathbb{C})$ as in Section 11.1. Let $K_0 = G_0' \cap \mathrm{U}(2n) \cong \mathrm{U}(n)$ be the maximal compact subgroup of G_0'. Then the embedding $L \times R : \mathrm{Sp}(n,\mathbb{C}) \times \mathrm{O}(k,\mathbb{C}) \to \mathrm{Sp}(nk,\mathbb{C})$ gives an embedding of the real forms

$$G_0' \times G_0 \longrightarrow \mathrm{Sp}(nk,\mathbb{C}) \cap \mathrm{U}(nk,nk) \cong \mathrm{Sp}(nk,\mathbb{R})$$

and carries the maximal compact subgroup $K_0 \times G_0$ into the maximal compact subgroup $\mathrm{Sp}(nk,\mathbb{C}) \cap (\mathrm{U}(nk) \times \mathrm{U}(nk))$ of $\mathrm{Sp}(nk,\mathbb{C}) \cap \mathrm{U}(nk,nk)$. If $u \in \mathrm{U}(n)$ and $k_0 = \begin{bmatrix} u & 0 \\ 0 & \bar{u} \end{bmatrix}$ is the corresponding element of K_0, then the pair $(k_0, g) \in K_0 \times G_0$ acts on $M_{2n\times k}$ by

$$L(k_0)R(g) \begin{bmatrix} x \\ y \end{bmatrix} = \begin{bmatrix} uxg^t \\ \bar{u}yg^t \end{bmatrix} . \tag{66}$$

We now calculate the restriction of the oscillator representation $\varpi^{(nk)}$ to $L(K_0) \times R(G_0)$ in the holomorphic model on $\mathcal{P}(V)$, where $V = M_{n\times k}$. Let $(k_0, g) \in K_0 \times G_0$. From (66) and Theorem 11.3 we see that

$$\varpi^{(nk)}(L(k_0)R(g))f(x) = (\det u)^{-k/2}(\det g)^{n/2} f(u^{-1}xg) \qquad \text{for } f \in \mathcal{P}(V) \tag{67}$$

(note that the determinant of the map $x \mapsto u^{-1}xg$ is $(\det u)^{-k}(\det g)^n$). If k and n are both even, formula (67) defines a representation of $K_0 \times G_0$. For the general case, let $\widetilde{K}_0 \subset \mathrm{Mp}(n,\mathbb{R})$ be the two-sheeted cover of K_0 and let $\widetilde{G}_0$ be the lift of $R(G_0)$ to $\mathrm{Mp}(nk,\mathbb{R})$. Then (67) gives a single-valued unitary representation of $\widetilde{K}_0 \times \widetilde{G}_0$. In the following we shall simply drop the factor $(\det g)^{n/2}$ from (67) to make the representation single-valued on G_0. However, the factor $(\det u)^{-k/2}$ is essential for extending the representation from $\widetilde{K}_0$ to $\mathrm{Mp}(n,\mathbb{R})$. Let δ denote the differential of this character of $\widetilde{K_0}$.

Let $\mathfrak{g}_0'$ be the Lie algebra of G_0', The complexification of $\mathfrak{g}_0'$ is $\mathfrak{g}' = \mathfrak{sp}(n,\mathbb{C})$. We now calculate the action of $\varpi^{(nk)}(L(\mathfrak{g}'))$. Let $X = \begin{bmatrix} 0 & b \\ 0 & 0 \end{bmatrix}$ and $Y = \begin{bmatrix} 0 & 0 \\ c & 0 \end{bmatrix}$ with $b, c \in SM_n$. Then

$$L(X) = \begin{bmatrix} 0 & L(b) \\ 0 & 0 \end{bmatrix}, \quad L(Y) = \begin{bmatrix} 0 & 0 \\ L(c) & 0 \end{bmatrix},$$

(where $L(b)x = bx$ for $x \in M_{n\times k}$). The quadratic form on V associated with $L(b)$ is

$$Q_{L(b)}(x) = x^t bx = \sum_{i,j} b_{ij} \left\{ \sum_{p=1}^{n} x_{ip}x_{jp} \right\} .$$

Hence

$$\varpi^{(nk)}(L(X))f(x) = \frac{1}{2}\sum_{i,j} b_{ij}\left\{ \sum_{p=1}^{n} \frac{\partial^2}{\partial x_{ip}\partial x_{jp}} \right\},$$

and $\varpi^{(nk)}(L(Y))$ is the operator of multiplication by $-\frac{1}{2}Q_{L(c)}$. Since $\mathfrak{p}_\pm$ generate $\mathfrak{g}'$, these operators determine $\varpi^{(nk)}(\mathfrak{g}')$. From Theorem 8.5 we conclude that the algebra $\mathcal{PD}(V)^G$ is generated by $\varpi^{(nk)}(\mathfrak{g}')$.

We recall some notation that was introduced earlier. Let $\mathcal{H}$ denote the space of G-harmonic polynomials on $M_{n\times k}$ and let $\Sigma \subset \widehat{G}$ be the spectrum of G on $\mathcal{H}$. Let the map

$$\tau : \Sigma \to \Lambda \subset \mathbb{Z}^n_{++}$$

be as in Theorems 10.1 and 10.4. Let $\mathcal{F}^\sigma \subset \mathcal{H}$ be an irreducible G-module in the class σ. Let $\mathcal{E}^{\tau(\sigma)+\delta} \subset \mathcal{H}$ be the irreducible finite-dimensional $\widetilde{K}$-module with highest weight $\tau(\sigma)+\delta$, as in Theorem 9.2.

Let $V = M_{n\times k}$. Consider the unitary representation of $\mathrm{Mp}(n,\mathbb{R}) \times \mathrm{O}(k)$ on $\mathbb{H}^2(V)$, where $\mathrm{Mp}(n,\mathbb{R})$ acts by the restriction of the oscillator representation $\varpi^{(nk)}$ and $\mathrm{O}(k)$ acts geometrically by right multiplication on V. For $\sigma \in \Sigma$ let $\mathbb{E}^{\tau(\sigma)+\delta}$ be the closure in $\mathbb{H}^2(V)$ of the $\mathfrak{g}'$-irreducible subspace $\mathcal{P}(V)^G \cdot \mathcal{E}^{\tau(\sigma)+\delta}$.

Theorem 12.1: *The spaces* $\mathbb{E}^{\tau(\sigma)+\delta}$, *for* $\sigma \in \Sigma$, *are irreducible and mutually inequivalent unitary representations of* $\mathrm{Mp}(n,\mathbb{R})$. *Furthermore,* $\mathbb{H}^2(V)$ *decomposes as a multiplicity-free Hilbert space orthogonal sum*

$$\mathbb{H}^2(V) = \bigoplus_{\sigma\in\Sigma} \mathbb{E}^{\tau(\sigma)+\delta} \otimes \mathcal{F}^\sigma \tag{68}$$

under the action of $\mathrm{Mp}(n,\mathbb{R}) \times \mathrm{O}(k)$.

Remark: When k is even the character $u \mapsto (\det u)^{-k/2}$ occuring in the oscillator representation is well-defined on $U(n)$ and $\mathbb{E}^{\tau(\sigma)+\delta}$ gives an irreducible unitary representation of $\mathrm{Sp}(n,\mathbb{R})$.

Proof: The key point is the following density result:

(∗) *Suppose* $E \subset \mathbb{H}^2(V)$ *is a closed subspace that is invariant under* $\mathrm{Mp}(n,\mathbb{R})$. *Set* $E_0 = E \cap \mathcal{P}(V)$. *Then* E_0 *is dense in* E *and is invariant under* $\mathfrak{g}'$.

To prove this, let $f(x) = \sum_\alpha c_\alpha x^\alpha$ be in $\mathbb{H}^2(V)$ and set

$$f_q(x) = \sum_{|\alpha| \le q} c_\alpha x^\alpha \quad \text{for } q = 0, 1, 2, \ldots .$$

Then $f_q \in \mathcal{P}(V)$ and it is clear from (38) that $||f - f_q|| \to 0$ as $q \to \infty$. Now take

$$H_0 = \begin{bmatrix} iI_n & 0 \\ 0 & -iI_n \end{bmatrix} \in \mathfrak{k}_0 ,$$

as in the proof of Theorem 11.3. Since

$$\exp(tH_0) = \begin{bmatrix} e^{it}I_n & 0 \\ 0 & e^{-it}I_n \end{bmatrix} \in K_0 ,$$

we see from (67) that $\varpi^{(nk)}(H_0)$ acts by $-i(j + \frac{k}{2})$ on $\mathcal{P}^j(V)$. Hence if we define

$$P_q = \sum_{j=0}^{q} \frac{1}{4\pi} \int_0^{4\pi} e^{it(j+k/2)} \varpi^{(nk)}(\exp tH_0)\, dt , \tag{69}$$

then P_q is a bounded operator on $\mathbb{H}^2(V)$ and $P_q f = f_q$ for $q = 0, 1, 2, \ldots$. Since E is closed and invariant under $\mathrm{Mp}(n,\mathbb{R})$, we know that E is invariant under the one-parameter unitary group $t \mapsto \varpi^{(nk)}(\exp tH_0)$. Hence $P_q E \subset E$ for $q = 0, 1, 2, \ldots$. This shows that $E_0 = \mathcal{P}(V) \cap E$ is dense in E.

To prove that E_0 is invariant under $\mathfrak{g}'$, take $\varphi \in E_0$ and $\psi \in E^\perp$. If $X \in \mathfrak{g}_0'$, then

$$\langle \varpi^{(nk)}(\exp tX)\varphi \mid \psi \rangle = 0 \quad \text{for all } t \in \mathbb{R} , \tag{70}$$

since E is invariant under $\mathrm{Mp}(n,\mathbb{R})$. But since $\varphi \in \mathcal{P}(V)$, the left side of (70) is an analytic function of t for $|t|$ near zero by Corollary 11.9. Taking the derivative in t and setting $t = 0$, we conclude that

$$\langle \varpi^{(nk)}(X)\varphi \mid \psi \rangle = 0 \quad \text{for all } \psi \in E^\perp .$$

Hence $\varpi^{(nk)}(X)\varphi \in E$, completing the proof of (∗).

To prove the theorem, first observe that if $E \subset \mathbb{E}^{\tau(\sigma)+\delta}$ is a proper closed subspace that is invariant under $\mathrm{Mp}(n,\mathbb{R})$, then by (∗) the space E_0 is invariant under $\mathfrak{g}'$. Hence $E_0 = 0$ by the irreducibility of $\mathcal{E}^{\tau(\sigma)+\delta}$. But E_0 is dense in E, so $E = 0$, showing that $\mathbb{E}^{\tau(\sigma)+\delta}$ is irreducible.

If $\sigma, \sigma' \in \Sigma$ and T is a bounded $\mathrm{Mp}(n, \mathbb{R})$ intertwining operator from $\mathbb{E}^{\tau(\sigma)+\delta}$ to $\mathbb{E}^{\tau(\sigma')+\delta}$, then T commutes with the projection operators P_q in (69). Hence T maps $\mathcal{E}^{\tau(\sigma)+\delta}$ to $\mathcal{E}^{\tau(\sigma')+\delta}$. Since the functions in $\mathcal{P}(V)$ are analytic vectors for $\mathrm{Mp}(n, \mathbb{R})$, we conclude (as in the proof of (70)) that T intertwines the $\mathfrak{g}'$ actions. Hence $\sigma = \sigma'$ by Theorem 9.2. The orthogonality of the decomposition also follows from Theorem 9.2. □

12.2. *Square-integrable representations of* $\mathrm{Sp}(n, \mathbb{R})$

The irreducible unitary representations of $\widetilde{G}_0 = \mathrm{Mp}(n, \mathbb{R})$ that occur in Theorem 12.1 are called *highest-weight* representations. Some of them also appear as discrete summands in the decomposition of the left regular representation of $\widetilde{G}_0$ on $L^2(\widetilde{G}_0)$ (these representations are called *square-integrable*). We now apply Harish–Chandra's criterion [17] to determine which of the representations $\mathbb{E}^{\tau(\sigma)+\delta}$ are square-integrable. It is convenient to give separate statements of the result depending on the parity of k.

Theorem 12.2: (notation of Theorem 10.1) *Let* $k = 2l + 1$ *be odd. Let* $\sigma \in \Sigma$.

(a) *If* $n > l + 1$ *then* $\mathbb{E}^{\tau(\sigma)+\delta}$ *is never square-integrable.*

(b) *If* $n = l+1$ *then* $\mathbb{E}^{\tau(\sigma)+\delta}$ *is square-integrable if and only if* $\sigma = (\lambda, -1) \in \widehat{G}_{-1}$ *and* $\mathrm{depth}(\lambda) = l$.

(c) *If* $n \leq l$ *then* $\mathbb{E}^{\tau(\sigma)+\delta}$ *is square-integrable for all* $\sigma \in \Sigma$.

Proof: The general condition on the highest weight λ for square-integrability is

$$\langle \lambda + \rho, \gamma^{\check{}} \rangle < 0\,, \tag{71}$$

where ρ is one-half the sum of the positive roots and $\gamma^{\check{}}$ is the co-root to the highest noncompact root γ. For $\mathfrak{sp}(n, \mathbb{R})$ we have $\rho = [n, n-1, \ldots, 2, 1]$ and $\gamma = 2\varepsilon_1$, so $\gamma^{\check{}} = \varepsilon_1$ (see Section 11.1). We must check this condition when $\lambda = \tau(\sigma) + \delta$, with $\delta = [-k/2, \ldots, -k/2]$.

Let $\tau(\sigma)_1$ denote the first coordinate of $\tau(\sigma)$. Then $\langle \lambda + \rho, \gamma^{\check{}} \rangle = \tau(\sigma)_1 - k/2 + n$. Since $k = 2l + 1$, the Harish–Chandra condition (71) is

$$\tau(\sigma)_1 < l + 1 - n\,. \tag{72}$$

Case (a): $n > l+1$. The formulas for $\tau(\sigma)$ in Theorem 10.1 show that $\tau(\sigma)_1$ is either 0 or -1 in this case. But $l - n + 1 \leq -1$, so (72) is never satisfied.

Case (b): $n = l + 1$. Now the right side of (72) is zero. The formulas for $\tau(\sigma)$ show that $\tau(\sigma)_1 < 0$ if and only if $\sigma = (\lambda, -1) \in \widehat{G}_{-1}$ with $d = l$.

Case (c): $n \leq l$. Now the right side of (72) is positive. The formulas for $\tau(\sigma)$ show that $\tau(\sigma)_1 \leq 0$ for all $\sigma \in \Sigma$, so (72) is always satisfied. □

Theorem 12.3: (notation of Theorem 10.4) *Let* $k = 2l$ *be even. Let* $\sigma \in \Sigma$.

(a) *If* $n > l$ *then* $\mathbb{E}^{\tau(\sigma)+\delta}$ *is never square-integrable.*

(b) *If* $n = l$ *then* $\mathbb{E}^{\tau(\sigma)+\delta}$ *is square-integrable if and only if* $\sigma = (\lambda, 0) \in \widehat{G}_0$ *and* $\mathrm{depth}(\lambda) = l$.

(c) *If* $n < l$ *then* $\mathbb{E}^{\tau(\sigma)+\delta}$ *is square-integrable for all* $\sigma \in \Sigma$.

Proof: When $k = 2l$ the Harish–Chandra condition (71) becomes

$$\tau(\sigma)_1 < l - n\,. \tag{73}$$

Case (a): $n > l$. The formulas for $\tau(\sigma)$ in Theorem 10.4 show that $\tau(\sigma)_1$ is either 0 or -1 in this case. But $l - n \leq -1$, so (73) is never satisfied.

Case (b): $n = l$. Now the right side of (73) is zero. The formulas for $\tau(\sigma)$ show that $\tau(\sigma)_1 < 0$ if and only if $\sigma = (\lambda, 0) \in \widehat{G}_0$ and $\mathrm{depth}(\lambda) = l$.

Case (c): $n < l$. Now the right side of (73) is positive. The formulas for $\tau(\sigma)$ show that $\tau(\sigma)_1 \leq 0$ for all $\sigma \in \Sigma$, so (73) is always satisfied. □

Examples

(1) Assume k is even. Then the oscillator representation $\varpi^{(nk)}$ is single-valued on $\mathrm{Sp}(n, \mathbb{R})$. If $k \geq 2n$ then we see from the formula for the θ-correspondence that every $\mathrm{GL}(n, \mathbb{C})$-highest weight λ that satisfies the Harish–Chandra inequality is of the form $\tau(\sigma) + \delta$, for some $\sigma \in \Sigma$. Thus every highest-weight discrete-series representation of $\mathrm{Sp}(n, \mathbb{R})$ occurs in the reduction of $\varpi^{(nk)}$ in this case.

(2) Let σ be the trivial representation of $\mathrm{O}(k)$ (denoted by $\pi^{(0,1)}$ in Sections 10.1 and 10.2). Then $\sigma \in \Sigma$ for all $n \geq 1$ and $\tau(\sigma) = 0$, by Theorems 10.1 and 10.4. The representation, call it π^+, of $\mathrm{Mp}(n, \mathbb{R})$ that corresponds to σ is square integrable if and only if $2n < k$. It occurs with multiplicity one in $\mathbb{H}^2(M_{n\times k})$, and has highest weight $\delta = [-k/2, \ldots, -k/2]$. This weight parameterizes the one-dimensional representation $g \mapsto \det(g)^{-k/2}$ of the maximal compact subgroup of $\mathrm{Mp}(n, \mathbb{R})$. Since $\mathcal{F}^\sigma$ consists of the constant functions, the space $\mathbb{H}^2_+ := \mathbb{E}^\delta$ of π^+ is the completion (in the Fischer norm) of the space $\mathcal{P}(M_{n\times k})^G$, where $G = \mathrm{O}(k, \mathbb{C})$.

(3) Let σ be the representation $g \mapsto \det(g)$ of $\mathrm{O}(k)$ (denoted by $\pi^{(0,-1)}$ in Sections 10.1 and 10.2). Then $\sigma \in \Sigma$ if and only if $n \geq k$. In this case

$$\tau(\sigma) = [\underbrace{0,\ldots,0}_{n-k}, \underbrace{-1,\ldots,-1}_{k}]$$

by Theorems 10.1 and 10.4. The representation, call it π^-, of $\mathrm{Mp}(n,\mathbb{R})$ that corresponds to σ is never square integrable. It occurs with multiplicity one in $\mathbf{H}^2(M_{n\times k})$ when $n \geq k$, and it has highest weight

$$\lambda = [\underbrace{-k/2,\ldots,-k/2}_{n}] + [\underbrace{0,\ldots,0}_{n-k}, \underbrace{-1,\ldots,-1}_{k}].$$

This is the highest weight of the representation $(\det)^{-k/2} \otimes \bigwedge^k(\mathbb{C}^n)^*$ of the maximal compact subgroup of $\mathrm{Mp}(n,\mathbb{R})$. In this case $\mathcal{F}^\sigma = \mathbb{C}g_k$, where $g_k(z)$ is the determinant of the bottom $k \times k$ block of $z \in M_{n\times k}$ when we take the orthogonal group $G = \mathrm{O}(\mathbb{C}^k, \omega)$ as in Section 10. Thus the space $\mathbb{H}^2_- := \mathbb{E}^\lambda$ of π^- is the completion (in the Fischer norm) of the space $(\mathcal{P}(M_{n\times k})^G\, g_k) \otimes \bigwedge^k(\mathbb{C}^n)^*$. Note that for fixed n, one obtains a distinguished set of n irreducible unitary highest-weight representations of $\mathrm{Mp}(n,\mathbb{R})$ this way by taking $k = 1,\ldots,n$.

(4) By the harmonic duality theorem, $(\mathbb{H}^2_\pm, \pi^\pm)$ are the only irreducible $\mathrm{Mp}(n,\mathbb{R})$ modules that occur with multiplicity one in $\mathbb{H}^2(M_{n\times k})$. More details and other models for the representations $\mathbb{E}^{\tau(\sigma)+\delta}$ can be found in [6].

Final Remarks: In Schur–Weyl duality we took tensor powers of the representation of $\mathrm{GL}(n,\mathbb{C})$ on $\mathbb{C}^n$ (the representation of smallest dimension) to obtain all the irreducible finite-dimensional polynomial representations of $\mathrm{GL}(n,\mathbb{C})$. The two irreducible components $\pi^\pm$ of the oscillator representation on $\mathbb{H}^2(\mathbb{C}^n)$ are the smallest unitary highest-weight representations of $\mathrm{Mp}(n,\mathbb{R})$ in the sense of *Gelfand-Kirillov dimension* (see [31]). As we already noted in Section 11.3 the representation on $\mathbb{H}^2(M_{n\times k})$ is the k-fold tensor product of this representation:

$$\mathbb{H}^2(M_{n\times k}) = \bigotimes^k \mathbb{H}^2(\mathbb{C}^n) \quad \text{(Hilbert-space tensor product)}. \tag{74}$$

Thus the action of the group $\mathrm{O}(k)$ on the right-side of (74) is another instance of a *hidden symmetry*.[j]

[j] The unitary representations that occur in the decomposition of this tensor product are the mathematical analog of the elementary particles, some familiar and some exotic, that physicists create by high-energy collisions of the basic particles.

13. Brauer algebra and tensor harmonics

In this final section we use duality to decompose the space of k-tensors under the action of the orthogonal or symplectic group G. This was first done by Brauer [4], who determined the generators and relations of the G-centralizer algebra. The complication here is that this algebra is not a group algebra (as was the case when $G = \mathrm{GL}(n, \mathbb{C})$). However, just as in the case of Howe duality, there is an analog of the *harmonic duality* of Section 9 in this situation. The centralizer algebra contains $\mathbb{C}[\mathfrak{S}_k]$ as a subalgebra, and there is a subspace of *harmonic tensors* (in Weyl's terminology *completely traceless*) which decomposes in a multiplicity-free way under the jointly commuting actions of G and $\mathfrak{S}_k$. The full space of k-tensors then decomposes as the sum of spaces of *partially harmonic* tensors (see [16, Sec. 10.3], [9] and [10] for details).

13.1. *Centralizer algebra and Brauer diagrams*

Let G be the full isometry group of a nondegenerate bilinear form ω on a finite-dimensional complex vector space V. We assume ω to be either symmetric or skew-symmetric. For $f \in V^*$ define $f^\flat \in V$ by

$$\omega(f^\flat, v) = \langle f, v \rangle \quad \text{for all } v \in V\,.$$

The map $f \mapsto f^\flat$ is then a G-isomorphism between V^* and V. Define a G-module isomorphism $T : V^{*\otimes 2k} \to \mathrm{End}(V^{\otimes k})$ by

$$T(f_1 \otimes \cdots \otimes f_{2k})u = \omega(f_2^\flat \otimes f_4^\flat \otimes \cdots \otimes f_{2k}^\flat,\, u) f_1^\flat \otimes f_3^\flat \otimes \cdots \otimes f_{2k-1}^\flat \quad (75)$$

for $f_i \in V^*$ and $u \in V^{\otimes k}$. Here we have extended ω to a bilinear form on $V^{\otimes k}$ by

$$\omega(u_1 \otimes \cdots \otimes u_k,\, v_1 \otimes \cdots \otimes v_k) = \prod_{i=1}^{k} \omega(u_i, v_i) \quad \text{for } u_i, v_i \in V\,.$$

Theorem 13.1: *Let Ξ_k be the set of two-partitions of $\{1, \ldots, 2k\}$. For $\xi \in \Xi_k$ let $\lambda_\xi \in (V^{*\otimes 2k})^G$ be the corresponding complete contraction. Then*

$$\mathrm{End}_G(V^{\otimes k}) = \mathrm{Span}\{T(\lambda_\xi) : \xi \in \Xi_k\}\,.$$

Proof: Since T is a G-module isomorphism this is a immediate consequence of Corollary 7.5. □

Theorem 13.1 only gives a spanning set for the centralizer algebra $\mathrm{End}_G(V^{\otimes k})$ as a vector space. To describe the multiplicative structure of this algebra it is convenient to introduce a graphic presentation of the set of two-partitions. We display the set $\{1, 2, \ldots, 2k\}$ as an array of two rows of k labeled dots, with the dots in the top row labeled $1, 3, \ldots, 2k-1$ from left to right, and the dots in the bottom row labeled $2, 4, \ldots, 2k$. Consider the set X_k of all (unoriented) graphs whose vertices are the two rows of dots, and such that each dot is connected with exactly one other dot by an edge. (A dot in the top row can be connected either with another dot in the top row or with a dot in the bottom row.) An example with $k = 5$ is shown in Figure 1. We call an element of X_k a *Brauer diagram.*[k] Thus we can identify the set Ξ_k of two-partitions with X_k; if $\xi \in \Xi_k$ corresponds to the Brauer diagram $x \in X_k$, we shall write λ_x for the complete contraction λ_ξ.

$x_1 =$ 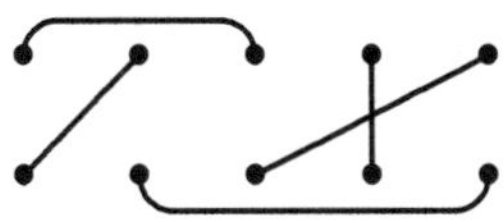

Fig. 1. A Brauer diagram.

The group $\mathfrak{S}_{2k}$ acts transitively on X_k by permuting the dots according to their labels. If $x \in X_k$ and $s \in \mathfrak{S}_{2k}$ then $s \cdot x$ is the graph obtained by permuting the dots by s and maintaining the edge connections (dot $s(i)$ is connected to dot $s(j)$ in $s \cdot x$ if and only if dot i is connected to dot j in x). Clearly

$$\sigma_{2k}^*(s)\lambda_x = \lambda_{s \cdot x} \qquad \text{for } s \in \mathfrak{S}_{2k} \text{ and } x \in X_k\,. \tag{76}$$

Here σ_k denotes the representation of $\mathfrak{S}_k$ on $V^{\otimes k}$, as in Section 3, and σ_{2k}^* is the contragredient representation on $V^{*\otimes 2k}$. Let x_0 be the graph with each dot in the top row connected with the dot below it (see Figure 2 for

$x_0 =$ 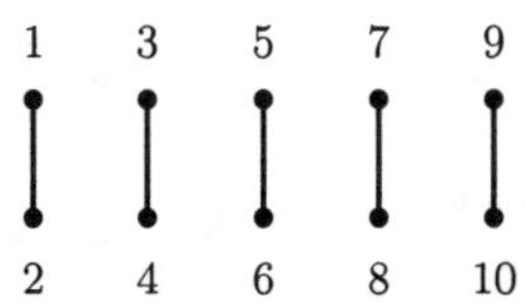

Fig. 2. Basic diagram with labeled dots.

[k]Kerov [23] uses the term *chip* because of the analogy with an integrated circuit chip, where the top row of dots are the input ports and the bottom row of dots the output ports. For a development of Kerov's approach, see [9] and [10].

the case $k = 5$). Then the Brauer diagram x_1 in Figure 1 is $s \cdot x_0$ where $s \in \mathfrak{S}_{10}$ is the cyclic permutation (2594).

Let $\tau : \mathfrak{S}_k \to \mathfrak{S}_{2k}$ be defined by $\tau(s)(2j-1) = 2s(j)-1$ and $\tau(s)(2j) = 2j$ for $j = 1, \ldots, k$. If $s \in \mathfrak{S}_k$, then $\tau(s)$ acts on a Brauer diagram by permuting the top row of dots according to s while leaving each dot in the bottom row fixed. Clearly τ is an injective homomorphism, and from (75) and (76) we see that

$$\sigma_k(s)T(\lambda_x) = T(\lambda_{\tau(s)\cdot x}) \quad \text{for } s \in \mathfrak{S}_k \text{ and } x \in X_k\,. \tag{77}$$

For the basic diagram x_0 we have $\tau(s) \cdot x_0 = x_0$ if and only if s is the identity, so the permutations in $\mathfrak{S}_k$ correspond to the diagrams in the orbit $\tau(\mathfrak{S}_k)\cdot x_0$ (these diagrams are just the *two-line* notation for a permutation). Hence by (77) the operators in $\mathrm{End}_G(V^{\otimes k})$ associated with the orbit of x_0 come from the natural action of $\mathfrak{S}_k$ on $V^{\otimes k}$. In particular, the basic diagram x_0 corresponding to the G-invariant tensor $\omega^{\otimes k}$ gives the identity operator on $V^{\otimes k}$.

The complete set of $\tau(\mathfrak{S}_k)$ orbits on X_k can be described as follows. For $x \in X_k$ let r be the number of edges in the diagram of x that connect a dot in the top row with another dot in the top row (call such an edge a *top bar*). The bottom row of x also must have r such edges (call them *bottom bars*), and we call x an *r-bar diagram.* All diagrams in the $\tau(\mathfrak{S}_k)$-orbit of x also have r top bars, and there is a unique z in this orbit with all its edges either horizontal or vertical (that is, if z is considered as a two-partition of $2k$, then every odd-even pair $\{2i-1, 2j\}$ that occurs in z has $i = j$). We will call such a Brauer diagram (or two-partition) *normalized.* The normalized diagrams give a set of representatives for the $\tau(\mathfrak{S}_k)$ orbits on X_k. For example, when $k = 3$ and $r = 1$ then there are three orbits of 1-bar diagrams, with normalized representatives indicated in Figure 3.

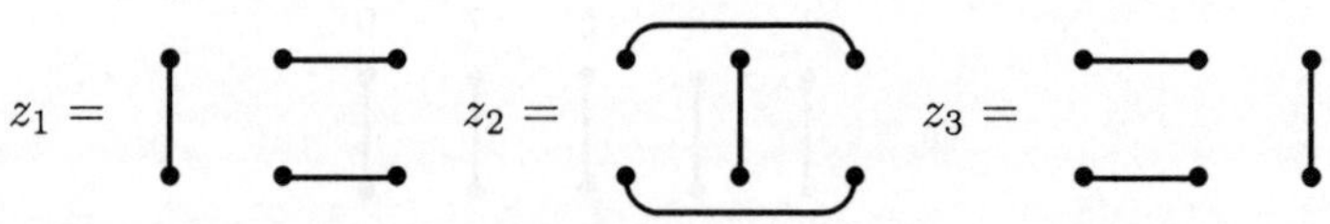

Fig. 3. Normalized 1-bar Brauer diagrams ($k = 3$).

These orbits correspond to the two-partitions

$$z_1 = \{\{1,2\},\{3,5\},\{4,6\}\}\,, \qquad z_2 = \{\{1,5\},\{2,6\},\{3,4\}\}\,,$$
$$z_3 = \{\{1,3\},\{2,4\},\{5,6\}\}\,.$$

If z is a normalized Brauer diagram, then for every top bar in z joining the dots numbered $2i-1$ and $2j-1$ there is a corresponding bottom bar joining the dots numbered $2i$ and $2j$. We will say that z contains an (i,j)-*bar* in this case (with the convention that $i<j$). For example, the normalized diagram in the orbit $\tau(\mathfrak{S}_5)x_1$ (with x_1 from Figure 1) is shown in Figure 4; it contains a $(2,5)$-bar.

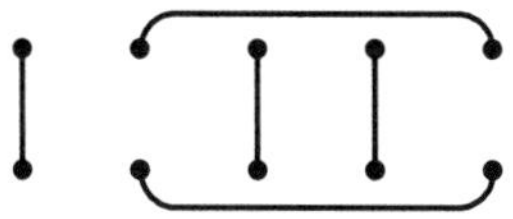

Fig. 4. Normalized Brauer diagram with $(2,5)$-bar.

13.2. *Generators for the centralizer algebra*

A normalized Brauer diagram determines an element of the algebra $\mathrm{End}_G(V^{\otimes k})$ by Theorem 13.1. For example, the diagram shown in Figure 5

Fig. 5. Brauer diagram for $\tau_{12} = D_{12}C_{12}$.

contains a single $(1,2)$-bar corresponding to the tensor $\sigma^*_{2k}(23)\omega^{\otimes k}$, where (23) is the transposition $2 \leftrightarrow 3$. Since $\sigma^*_{2k}(23)\omega^{\otimes k} = (\sigma^*_4(23)\omega^{\otimes 2})\otimes\omega^{\otimes(k-2)}$, we have

$$T(\sigma^*_{2k}(23)\omega^{\otimes k})\, v_1 \otimes v_2 \otimes u = \left\{ \sum_{p_2} \omega(v_1, f_{p_2})\omega(v_2, f^{p_2}) \right\} \sum_{p_1} f_{p_1} \otimes f^{p_1} \otimes u$$
$$= \omega(v_1, v_2)\theta \otimes u$$

for $v_1, v_2 \in V$ and $u \in V^{\otimes(k-2)}$. Here $\{f_p\}$ and $\{f^p\}$ are bases for V with $\omega(f_p, f^q) = \delta_{pq}$, and

$$\theta = \sum_p f_p \otimes f^p \in (V\otimes V)^G$$

is the tensor dual to ω. Thus this 1-bar diagram gives an operator $\tau_{12} = T(\sigma_{2k}(23)\theta_k)$ which is the composition

$$V^{\otimes k} \xrightarrow{C_{12}} V^{\otimes(k-2)} \xrightarrow{D_{12}} V^{\otimes k},$$

with $C_{12}(v_1 \otimes v_2 \otimes u) = \omega(v_1, v_2)u$ a *contraction operator* (contract the first and second tensor positions by ω) and $D_{12}(u) = \theta \otimes u$ an *expansion operator* (multiply on the left by θ). These operators obviously intertwine the actions of G on $V^{\otimes k}$ and $V^{\otimes(k-2)}$.

In general, for any pair $1 \leq i < j \leq k$ we define the *ij-contraction operator* $C_{ij} : V^{\otimes k} \to V^{\otimes(k-2)}$ by

$$C_{ij}(v_1 \otimes \cdots \otimes v_k) = \omega(v_i, v_j)\, v_1 \otimes \cdots \otimes \widehat{v_i} \otimes \cdots \otimes \widehat{v_j} \otimes \cdots \otimes v_k$$

(omit v_i and v_j in tensor product) and the *ij-expansion operator* D_{ij} : $V^{\otimes(k-2)} \to V^{\otimes k}$ by

$$D_{ij}(v_1 \otimes \cdots \otimes v_{k-2}) = \sum_{p=1}^{n} v_1 \otimes \cdots \otimes \underbrace{f_p}_{i\text{th}} \otimes \cdots \otimes \underbrace{f^p}_{j\text{th}} \otimes \cdots \otimes v_{k-2}\,.$$

These operators intertwine the action of G and are mutually adjoint, relative to the invariant form ω on $V^{\otimes k}$:

$$\omega(C_{ij}u,\, w) = \omega(u,\, D_{ij}w) \qquad \text{for } u \in V^{\otimes k},\ w \in V^{\otimes(k-2)}\,. \tag{78}$$

Set $\tau_{ij} = D_{ij}C_{ij} \in \mathrm{End}_G(V^{\otimes k})$. If $u = v_1 \otimes \cdots \otimes v_k$ with $v_i \in V$, then

$$\tau_{ij}(u) = \omega(v_i, v_j) \sum_{p=1}^{n} v_1 \otimes \cdots \otimes \underbrace{f_p}_{i\text{th}} \otimes \cdots \otimes \underbrace{f^p}_{j\text{th}} \otimes \cdots \otimes v_k\,. \tag{79}$$

The contraction and expansion operators satisfy the symmetry properties

$$C_{ij} = \varepsilon C_{ji}\,, \quad D_{ij} = \varepsilon D_{ji}\,, \tag{80}$$

since $\sum_p f_p \otimes f^p = \varepsilon \sum_p f^p \otimes f_p$. Hence $\tau_{ij} = \tau_{ji}$, so the operator τ_{ij} only depends on the set $\{i, j\}$.

Let $Z_{k,r} \subset X_k$ be the set of normalized r-bar Brauer diagrams, and set

$$Z_k = \bigcup_{r=0}^{[k/2]} Z_{k,r}\,.$$

Lemma 13.2: *Suppose that $z \in Z_{k,r}$ is a normalized r-bar Brauer diagram with bars $\{i_1, j_1\}, \ldots, \{i_r, j_r\}$. Then*

$$\tau_{i_p j_p}\tau_{i_q j_q} = \tau_{i_q j_q}\tau_{i_p j_p} \quad \textit{for } p \neq q\,. \tag{81}$$

Thus the operator $\tau_z = \tau_{i_1 j_1} \cdots \tau_{i_r j_r}$ only depends on z and not on the enumeration of the bars in z. Furthermore, $\tau_z = T(\lambda_z)$.

Proof: The commutativity relation (81) is clear since τ_{ij} only operates on the ith and jth tensor positions. □

Proposition 13.3: *The algebra* $\mathrm{End}_G(V^{\otimes k})$ *is spanned by the set of operators* $\sigma_k(s)\tau_z$ *with* $s \in \mathfrak{S}_k$ *and* $z \in Z_k$. *Furthermore, if* $\dim V \geq 2k$ *then this set of operators is linearly independent.*

Proof: Since Z_k gives a cross-section for the $\tau(\mathfrak{S}_k)$ orbits on X_k, the first statement follows from Theorem 13.1, Lemma 13.2, and the intertwining relation (77). The proof of linear independence when $\dim V \geq 2k$ is straightforward (see [16, Corollary 10.1.4]). □

13.3. *Relations in the centralizer algebra*

We next determine the algebraic relations among the operators in Proposition 13.3.

Lemma 13.4: *Let* $n = \dim V$ *and set* $\varepsilon = 1$ *if* ω *is symmetric and* $\varepsilon = -1$ *if* ω *is skew. The operators* τ_{ij} *(where* $1 \leq i, j \leq k$ *and* $i \neq j$*) satisfy the following relations, where* (il) *denotes the transposition of* i *and* l*:*

(1) $\tau_{ij} = \tau_{ji}$ *and* $\tau_{ij}^2 = n\,\tau_{ij}$

(2) $\tau_{ij}\tau_{lm} = \tau_{lm}\tau_{ij}$ *for distinct* i, j, l, m

(3) $\tau_{ij}\tau_{jl} = \sigma_k(il)\tau_{jl}$ *for distinct* i, j, l

(4) $\sigma_k(s)\tau_{ij}\sigma_k(s)^{-1} = \tau_{s(i),s(j)}$ *for all* $s \in \mathfrak{S}_k$

(5) $\sigma_k(ij)\tau_{ij} = \varepsilon\tau_{ij}$

Proof: The contraction and expansion operators satisfy

$$C_{ij}D_{ij} = nI\,, \tag{82}$$

which follows from $\sum_{p=1}^{n} \omega(f_p, f^p) = n$. This implies property (1). Property (2) was already checked in Lemma 13.2. To verify (3), note that

$$\tau_{ij}\tau_{jl}(v_1 \otimes \cdots \otimes v_k) = \omega(v_j, v_l) \sum_{p,q} \omega(v_i, f_p) u_{pq}\,,$$

where $u_{pq} = v_1 \otimes \cdots \otimes \underbrace{f_q}_{i\text{th}} \otimes \cdots \otimes \underbrace{f^q}_{j\text{th}} \otimes \cdots \otimes \underbrace{f^p}_{l\text{th}} \otimes \cdots \otimes v_k$. But

$$\sum_p \omega(v_i, f_p) u_{pq} = \varepsilon v_1 \otimes \cdots \otimes \underbrace{f_q}_{i\text{th}} \otimes \cdots \otimes \underbrace{f^q}_{j\text{th}} \otimes \cdots \otimes \underbrace{v_i}_{l\text{th}} \otimes \cdots \otimes v_k\,,$$

which gives (3). Relations (4) and (5) are simple calculations from the definition of τ_{ij}. □

Define the *Brauer Algebra* $\mathcal{B}_k(\varepsilon, n)$ with parameters k, ε, n to be the associative algebra generated by $\mathfrak{S}_k$ and elements $\{\tau_{ij} : 1 \leq i < j \leq k\}$

subject to the relations (1)–(5) in Lemma 13.4; here n can be any complex number and $\varepsilon = \pm 1$. From these relations it is clear that $\mathcal{B}_k(\varepsilon, n)$ is finite-dimensional. If $\mathcal{T}$ is the subalgebra generated by $\{\tau_{ij}\}$, then $\mathcal{T}$ is an ideal and we have the decomposition

$$\mathcal{B}_k(\varepsilon, n) = \mathbb{C}[\mathfrak{S}_k] \oplus \mathcal{T}\,. \tag{83}$$

From Proposition 13.3 and Lemma 13.4 we see that there is a surjective algebra homomorphism

$$\mathcal{B}_k(\varepsilon, n) \longrightarrow \mathrm{End}_G(V^{\otimes k}) \qquad (n = \dim V)$$

with $\varepsilon = \pm 1$ determined as in Lemma 13.4 (5). The two algebras are isomorphic if $n \geq 2k$. In any case, the centralizer algebra $\mathrm{End}_G(V^{\otimes k})$ is the quotient of the associated Brauer algebra by a two-sided ideal, so the representations of the centralizer algebra can be viewed as representations of the Brauer algebra.[1]

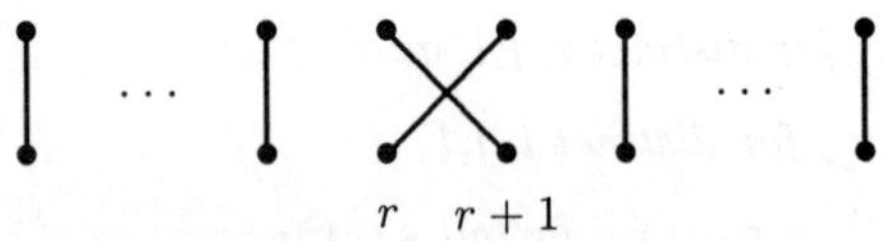

Fig. 6. Brauer diagram for s_r.

We can describe the multiplication in $\mathcal{B}_k(\varepsilon, n)$ and the relations in Lemma 13.4 in terms of concatenation of Brauer diagrams. Let $s_r \in \mathfrak{S}_k$ be the transposition $r \leftrightarrow r+1$. It corresponds to the Brauer diagram shown in Figure 6. Let $z_r = \tau_{r,r+1}$ be the operator corresponding to the normalized Brauer diagram with a single $(r, r+1)$ bar, as in Figure 7. Since $\mathfrak{S}_k$ is generated by $s_1, \ldots, s_{k-1}$, we see from Proposition 13.3 and property (3) in Lemma 13.4 that the algebra $\mathcal{B}_k(\varepsilon n)$ is generated by the operators $s_1, \ldots, s_{k-1}$ and $z_1, \ldots, z_{k-1}$. If x, y are Brauer diagrams, then their product xy in the Brauer algebra is obtained by placing the x above y and joining the lower row of dots in x to the upper row of dots in y. When x, y correspond to elements of $\mathfrak{S}_k$ (no bars) this procedure obviously gives the multiplication in $\mathfrak{S}_k$. When x or y have bars, we remove the closed loops from the concatenated graph using relation (1) in Lemma 13.4.

[1]See [9] and [10] for recent work on the representation theory of the Brauer algebra and citations of earlier work.

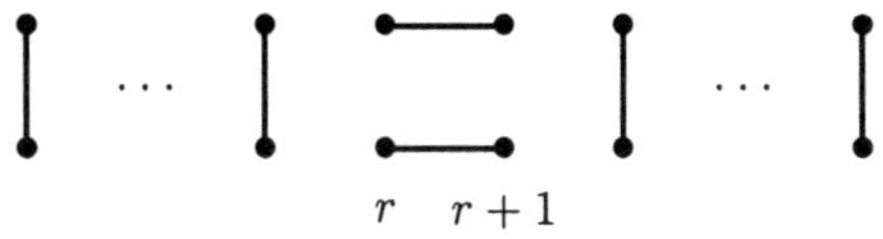

Fig. 7. Brauer diagram for z_r.

The general recipe for transforming the concatenated Brauer diagrams of x and y into a scalar multiple of the Brauer diagram for xy is as follows:

(1) Delete each closed loop in the concatenated diagram and multiply by a scalar factor of n^r if there are r such loops.
(2) Multiply by a factor of ε for every path in the concatenated diagram that begins and ends on the top row of x or on the bottom row of y.

For example, if $x = \sigma(236)\tau_{35}\tau_{46}$ and $y = \sigma(46)\tau_{12}\tau_{34}\tau_{56}$, then xy is obtained as shown in Figure 8 (see [16, Sec. 10.1.2 and Exercises 10.1.3] for further details and examples).

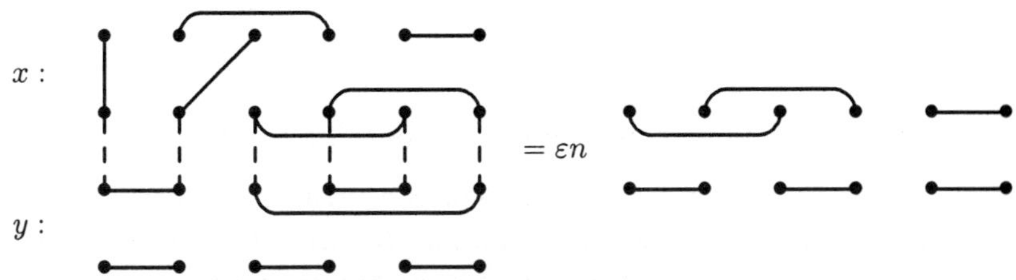

Fig. 8. The relation $(\sigma(236)\tau_{35}\tau_{46}) \cdot (\sigma(46)\tau_{12}\tau_{34}\tau_{56}) = \varepsilon n\, \sigma(23)\, \tau_{12}\tau_{34}\tau_{56}$.

13.4. *Harmonic tensors*

Let $k \geq 2$. A tensor $u \in V^{\otimes k}$ is called *ω-harmonic*[m] if it is annihilated by all the contraction operators C_{ij}. Denote by

$$\mathcal{H}(V^{\otimes k}, \omega) = \bigcap_{1 \leq i < j \leq k} \operatorname{Ker}(C_{ij})$$

the space of all ω-harmonic k-tensors. We will simply call these tensors *harmonic* and write $\mathcal{H}(V^{\otimes k}, \omega) = \mathcal{H}(V^{\otimes k})$ when ω is clear from the context.

Example: Assume ω is symmetric and let $v \in V$. Then $C_{ij}v^{\otimes k} = \omega(v,v)v^{\otimes(k-2)}$. Thus the symmetric tensor $v^{\otimes k}$ is harmonic if and only

[m] Weyl uses the term *traceless*; we prefer the term harmonic because when ω is symmetric the contraction operators C_{ij} act as Laplacians on the symmetric tensors.

if v is an *isotropic vector* for ω. This is the same as the polynomial function $\xi \mapsto \langle \xi, x \rangle^k$ on V^* being harmonic relative to the Laplace operator defined by ω. On the other hand, every skew-symmetric tensor is harmonic when ω is symmetric.

Theorem 13.5 (Harmonic tensor duality): *The space $\mathcal{H}(V^{\otimes k})$ is invariant under $\mathfrak{S}_k \times G$ and decomposes as*

$$\mathcal{H}(V^{\otimes k}) \cong \bigoplus_{\lambda \in \Lambda} E^\lambda \otimes U^\lambda . \tag{84}$$

Here $\Lambda \subset \mathrm{Par}(k)$, E^λ is the irreducible $\mathfrak{S}_k$-module corresponding to the partition λ by Schur–Weyl duality, and U^λ is an irreducible G-module. Furthermore, the modules U^λ are all distinct.

Proof: Since $C_{ij}\tau_{ij} = C_{ij}D_{ij}C_{ij} = nC_{ij}$, we have $\mathrm{Ker}(C_{ij}) = \mathrm{Ker}(\tau_{ij})$. Hence u is harmonic if and only if $\tau_{ij}u = 0$ for $1 \le i < j \le k$. Since τ_{ij} commutes with $\rho_k(G)$, we see that $\mathcal{H}(V^{\otimes k})$ is invariant under G. Proposition 13.3 and Lemma 13.4 imply that $\mathcal{H}(V^{\otimes k})$ is invariant under $\mathcal{B}_k(\varepsilon, n)$ and the action of $\mathcal{B}_k(\varepsilon, n)$ on $\mathcal{H}(V^{\otimes k})$ reduces to the action of $\mathfrak{S}_k$. Now apply Theorem 1.4. □

13.5. *Decomposition of harmonic tensors for* Sp(V)

We now determine the set Λ of partitions of k occurring in Theorem 13.5 and the corresponding irreducible representations U^λ when G is the symplectic group.[n] We take $V = \mathbb{C}^n$ with $n = 2l$ and the bilinear form

$$\omega(x, y) = \sum_{i=1}^{l} x_i y_{n+1-i} - y_i x_{n+1-i}$$

(so the standard basis vectors e_i are ω-isotropic and e_1 is paired with e_n, e_2 is paired with e_{n-1}, and so forth). Let $G = \mathrm{Sp}(V, \omega)$ and let H be the diagonal matrices in G. Then H is a maximal torus whose elements are of the form

$$h = \mathrm{diag}[x_1, \ldots, x_l, x_l^{-1}, \ldots, x_1^{-1}], \quad x_i \in \mathbb{C}^\times . \tag{85}$$

Following the notation in Section 10, we let D_n be the diagonal matrices, B_n the upper-triangular matrices, and N_n the upper-triangular unipotent matrices in $\mathrm{GL}(V)$. With our choice of ω the group $B = G \cap B_n$ is a Borel

[n] Similar methods work for the orthogonal groups but the details are considerably more intricate — see [16, Sec. 10.2] for a full treatment.

subgroup of G and $N = G \cap N_n$ is its unipotent radical. The weight lattice of H is identified with $\mathbb{Z}^l$, where $\lambda = [m_1, \ldots, m_l]$ gives the character

$$h \mapsto x_1^{m_1} \cdots x_l^{m_l}$$

when h is given by (85). The Weyl group W of G acts by all permutations and sign changes of the coordinations of λ. The set of B-dominant weights is thus identified with $\mathbb{N}^l_{++}$ (see [16, Sec. 2.5]).

For $\lambda \in \mathbb{N}^l_{++}$ let (π^λ, U^λ) be the irreducible representation of G with highest weight λ. If $|\lambda| = k$ then we view λ as a partition of k with at most l parts. Let $E^\lambda = (V^{\otimes k})^{N_n}(\lambda)$ be the corresponding irreducible representation of $\mathfrak{S}_k$ on the space of $\mathrm{GL}(n, \mathbb{C})$ highest weight vectors of weight λ, as in Theorem 3.8.

Theorem 13.6: *Let $\lambda \in \mathrm{Par}(k, n)$. Then $E^\lambda \subset \mathcal{H}(V^{\otimes k})$ if and only if λ has at most l parts. Furthermore, the space of ω-harmonic k-tensors has isotypic decomposition*

$$\mathcal{H}(V^{\otimes k}) \cong \bigoplus_{\lambda \in \mathrm{Par}(k,l)} E^\lambda \otimes U^\lambda \tag{86}$$

under $\mathfrak{S}_k \times \mathrm{Sp}(V, \omega)$. Thus all the irreducible representations of $\mathrm{Sp}(V, \omega)$ occur in the decomposition of the harmonic k-tensors, for $k = 1, 2, \ldots$.

The general form of decomposition (86) follows from Theorem 13.5. To determine the spectrum Λ of $\mathrm{Sp}(V, \omega)$ on the harmonic tensors, we will compare the spaces of B eigenvectors in $V^{\otimes k}$ with the spaces E^λ. If $\mu = [m_1, \ldots, m_n] \in \mathbb{Z}^n$ is a weight of D_n, then we denote by $\bar{\mu}$ the restriction of μ to H. From (85)

$$\bar{\mu} = [m_1 - m_n, m_2 - m_{n-1}, \ldots, m_l - m_{l+1}]\,. \tag{87}$$

Hence if $\mu \in \mathbb{N}^n_{++}$ is a B_n-dominant weight, then $\bar{\mu}$ is a B-dominant weight. We introduce the notation

$$W^k(\lambda) = (V^{\otimes k})^N(\lambda)\,, \quad \text{for } \lambda \in \mathbb{N}^l_{++}\,.$$

Since $N \subset N_n$, we have $E^\mu \subset W^k(\bar{\mu})$.

Proposition 13.7: *There are the following dichotomies:*

(1) *Assume $\lambda \in \mathbb{N}^l_{++}$. Then either $W^k(\lambda) \cap \mathcal{H}(V^{\otimes k}) = 0$ or else $W^k(\lambda) \subset \mathcal{H}(V^{\otimes k})$.*

(2) *Assume $\mu \in \mathrm{Par}(k, n)$. Then either $E^\mu \cap \mathcal{H}(V^{\otimes k}) = 0$ or else $E^\mu = W^k(\bar{\mu})$.*

Proof: (1) By Theorems 3.7 and 13.1 we know that $W^k(\lambda)$ is an irreducible module for $\mathcal{B}_k(\varepsilon, n)$. Since $W^k(\lambda) \cap \mathcal{H}(V^{\otimes k})$ is a $\mathcal{B}_k(\varepsilon, n)$-invariant subspace of $W^k(\lambda)$, it must be 0 or $W^k(\lambda)$.

(2) Assume $E^\mu \cap \mathcal{H}(V^{\otimes k}) \neq 0$. Since $E^\mu \subset W^k(\bar{\mu})$, it follows by (1) that $W^k(\bar{\mu}) \subset \mathcal{H}(V^{\otimes k})$. Furthermore, $W^k(\bar{\mu})$ is irreducible under $\mathfrak{S}_k$. Indeed, it is irreducible under $\mathcal{B}_k(\varepsilon, n)$ by (1), and on the harmonic tensors the action of $\mathcal{B}_k(\varepsilon, n)$ is the same as the action of $\mathfrak{S}_k$. By Theorem 3.8 it follows that $W^k(\bar{\mu}) = E^\mu$. □

Proposition 13.8: *Let $\mu \in \operatorname{Par}(k, n)$. Then $E^\mu \subset \mathcal{H}(V^{\otimes k})$ if and only if μ has at most l parts.*

Proof: Let $\mu = [m_1, \ldots, m_n]$. Define $u_\mu = u_1^{\otimes c_1} \otimes \cdots \otimes u_n^{\otimes c_n}$ as in Theorem 3.8, where $u_p = e_1 \wedge \cdots \wedge e_p$ and $c_j = m_j - m_{j+1}$ (with $m_{n+1} = 0$). Then $u_\mu \in E^\mu$ and the depth of μ is the largest integer d such that $c_d \neq 0$.

We first verify that u_p is harmonic if and only if $p \leq l$. To see this, take any pair i, j with $1 \leq i < j \leq p$. If $p \leq l$, then $C_{ij} u_p = 0$ since $\omega(e_i, e_j) = 0$. Conversely, if $p > l$ then

$$C_{l,l+1} u_p = e_1 \wedge \cdots \wedge e_{l-1} \wedge e_{l+2} \wedge \cdots \wedge e_p \neq 0\,,$$

since $\omega(e_l, e_{l+1}) = 1$. So u_p is not harmonic in this case.

Thus to finish the proof of the proposition, we may assume that μ has at most l parts. Let $1 \leq p \leq q \leq l$. We now show that

$$C_{ij}(u_p \otimes u_q) = 0 \quad \text{for all } 1 \leq i \leq p < j \leq p + q\,.$$

Set $v = u_p \otimes u_q$. In terms of the basis $\{e_I\}$, v is obtained by a double alternation:

$$v = \frac{1}{p!q!} \sum_{s \in \mathfrak{S}_p} \sum_{t \in \mathfrak{S}_q} \operatorname{sgn}(s) \operatorname{sgn}(t)\, e_{s(1)} \otimes \cdots \otimes e_{s(p)} \otimes e_{t(1)} \otimes \cdots \otimes e_{t(q)}\,.$$

The contraction operator C_{ij} removes $e_{s(i)}$ and $e_{t(j-p)}$ from each term of the sum and multiplies the resulting $(p+q-2)$-tensor by $\omega(e_{s(i)}, e_{t(j-p)})$. But $s(i) + t(j-p) \leq p + q \leq n$, while $\omega(e_a, e_b) = 0$ unless $a + b = n + 1$. Hence $C_{i,j}(v) = 0$. □

We now complete the proof of Theorem 13.6. By Propositions 13.7 and 13.8, it will suffice to prove the following:

(*) *If $\lambda \in \mathbb{N}^l_{++}$ is such that $0 \neq W^k(\lambda) \subset \mathcal{H}(V^{\otimes k})$, then $|\lambda| = k$.*

To establish $(*)$, take a nonzero tensor $u \in W^k(\lambda)$ and decompose u under the action of D_n as

$$u = \sum_{\mu} u_\mu\,, \quad \text{where } \mu \in \mathrm{Par}(k,n) \text{ and } u_\mu \in V^{\otimes k}(\mu)\,.$$

Fix some $\mu = [m_1, \ldots, m_n]$ such that $u_\mu \neq 0$. Then $\bar{\mu} = \lambda$ and from (87) we see that

$$|\lambda| = \sum_{i=1}^{l} (m_i - m_{n+1-i}) = k - 2r\,, \quad \text{where } r = \sum_{i=l+1}^{n} m_i\,.$$

Thus $\lambda \in \mathrm{Par}(k-2r, l)$ so from Theorem 3.8 we know that $0 \neq E^\lambda \subset V^{\otimes(k-2r)}$. Suppose for the sake of contradiction that $r > 0$. Since the expansion operator D_{12} is injective and intertwines the action of G on tensors, we have

$$0 \neq (D_{12})^r E^\lambda \subset W^k(\lambda)\,.$$

Since $C_{12}D_{12} = nI$, this implies that

$$C_{12}(D_{12})^r E^\lambda = (D_{12})^{r-1} E^\lambda \neq 0\,.$$

But we have assumed that $W^k(\lambda)$ in contained in the harmonic tensors, a contradiction. Thus $r = 0$ and $(*)$ is proved. □

Examples

(1) Assume $k \leq l$ and take $\lambda = [1, \ldots, 1] \in \mathrm{Par}(k)$. Then E^λ is the sgn representation of $\mathfrak{S}_k$ and U^λ is the kth *fundamental representation* of $\mathrm{Sp}(V, \omega)$. From Theorem 13.6 we know that U^λ is the sgn-isotypic component for $\mathfrak{S}_k$ in $\mathcal{H}(V^{\otimes k})$. Hence

$$U^\lambda = \mathcal{H}(V^{\otimes k}) \cap \textstyle\bigwedge^k V = \mathcal{H}_{\mathrm{skew}}(V^{\otimes k})$$

is the space of *harmonic skew k-tensors.*

(2) Let $k = 2$ and assume $l \geq 2$. Then the two partitions of 2 give the trivial and sgn representations of $\mathfrak{S}_k$, respectively. Because the form ω is skew-symmetric, every symmetric tensor is harmonic. Hence by Theorem 13.6 we have

$$\mathcal{H}(V^{\otimes 2}) = S^2(V) \oplus \mathcal{H}_{\mathrm{skew}}(V)\,.$$

The summands are the irreducible representations of G with highest weights $2\varepsilon_1$ and $\varepsilon_1 + \varepsilon_2$.

References

1. I. Agricola, Invariante Differentialoperatoren und die Frobenius-Zerlegung einer G-Varietät, *J. Lie Th.* **11** (2001) 81–109.
2. V. Bargmann, On a Hilbert space of analytic functions and an associated integral transform, Part I, *Comm. Pure Appl. Math.* **14** (1961) 187–214.
3. N. Bourbaki, *Éléments de mathématique, Fascicule XXXVII, Groupes et algèbres de Lie*, Ch. 2 & 3, Hermann, Paris 1972.
4. R. Brauer, On algebras which are connected with semisimple continuous groups, *Ann. of Math.* **38** (1937) 857–872.
5. P. Cartier, Quantum mechanical commutation relations and theta functions, in *Algebraic Groups and Discontinuous Subgroups* (Proc. Symp. in Pure Mathematics IX), Amer. Math. Soc., 1966, 361–386.
6. M. Davidson, T. Enright and R. Stanke, *Differential Operators and Highest Weight Representations* (Memoirs of the AMS # 455), Amer. Math. Soc., 1991.
7. J. Faraut and A. Korányi, *Analysis on Symmetric Cones*, Oxford University Press, 1994.
8. G. B. Folland, *Harmonic Analysis in Phase Space* (Annals of Math. Study # 122), Princeton University Press, 1989.
9. F. Gavarini and P. Papi, Representations of the Brauer algebra and Littlewood's restriction rules, *J. Algebra* **194** (1997) 275–298.
10. F. Gavarini, A Brauer algebra theoretic proof of Littlewood's restriction rules, *J. Algebra* **212** (1999) 240–271.
11. S. Gelbart, Holomorphic discrete series for the real symplectic group, *Invent. Math.* **19** (1973) 49–58.
12. S. Gelbart, A theory of Stiefel harmonics, *Trans. Amer. Math. Soc.* **192** (1974) 29–50.
13. S. Gelbart, Examples of dual reductive pairs, in *Automorphic Forms, Representations and L-Functions* (Proc. of Symp. in Pure Math. XXXIII, vol. 1), Amer. Math. Soc., 1979, 287–296.
14. R. Goodman, Analytic domination by fractional powers of a positive operator, *J. Funct. Anal.* **3** (1969) 246–264.
15. R. Goodman, Analytic and entire vectors for representations of Lie groups, *Trans. Amer. Math. Soc.* **143** (1969) 55–76.
16. R. Goodman and N.R. Wallach, *Representations and Invariants of the Classical Groups*, Cambridge University Press, 1998.
17. Harish-Chandra, Representations of semisimple Lie groups V, *Amer. J. Math.* **78** (1956) 1–41.
18. G. Hochschild, *The Structure of Lie Groups*, Holden-Day, 1965.
19. J.-S. Huang, *Lectures on Representation Theory*, World Scientific, 1999.
20. R. Howe, Dual pairs in physics: harmonic oscillators, photons, electrons, and singletons, in *Applications of Groups Theory in Physics and Mathematical Physics*, Lectures in Appl. Math. **21**, Amer. Math. Soc., 1985, 179–206.
21. R. Howe, Remarks on classical invariant theory, *Trans. Amer. Math. Soc.* **313** (1989) 539–570.

22. M. Kashiwara and M. Vergne, On the Segal–Shale–Weil representations and harmonic polynomials, *Invent. Math.* **44** (1978) 1–47.
23. S. V. Kerov, Realizations of representations of the Brauer semigroup, *J. Soviet Math.* **47** (1989) 2503–2507.
24. A. W. Knapp, *Representation Theory of Semisimple Groups*, Princeton University Press, 1986.
25. F. Knop, A Harish–Chandra homomorphism for reductive group actions, *Ann. of Math.* **140** (1994) 253–288.
26. B. Kostant, Lie group representations on polynomial rings, *Amer. J. Math.* **85** (1963) 327–404.
27. E. Nelson, Analytic Vectors, *Ann. of Math.* **70** (1959) 572–615.
28. I. E. Segal, Tensor algebras over Hilbert spaces I, *Trans. Amer. Math. Soc.* **81** (1956) 106–134.
29. R. Strichartz, The explicit Fourier decomposition of $L^2(\mathrm{SO}(n)/\mathrm{SO}(n-m))$, *Canad. J. Math.* **27** (1975) 294–310.
30. S. Sternberg, *Group Theory and Physics*, Cambridge University Press, 1994.
31. E.-C. Tan and C.-B. Zhu, Poincaré series of holomorphic representations, *Indag. Mathem.*, N.S., **7** (1996) 111–126.
32. T. Ton-That, Lie group representations and harmonic polynomials of a matrix variable, *Trans. Amer. Math. Soc.* **216** (1976) 1–46.